Quantum Entanglement and Interferometry

An Introduction to Multi-Photon Superposition

Quantum
Entanglement
and
Interferometry

An Introduction to
Multi-Photon Superposition

Yanhua Shih
University of Maryland, USA

World Scientific

NEW JERSEY · LONDON · SINGAPORE · BEIJING · SHANGHAI · HONG KONG · TAIPEI · CHENNAI · TOKYO

Published by

World Scientific Publishing Co. Pte. Ltd.
5 Toh Tuck Link, Singapore 596224
USA office: 27 Warren Street, Suite 401-402, Hackensack, NJ 07601
UK office: 57 Shelton Street, Covent Garden, London WC2H 9HE

British Library Cataloguing-in-Publication Data
A catalogue record for this book is available from the British Library.

QUANTUM ENTANGLEMENT AND INTERFEROMETRY
An Introduction to Multi-Photon Superposition

ISBN 978-981-98-0041-4 (hardcover)
ISBN 978-981-98-0042-1 (ebook for institutions)
ISBN 978-981-98-0043-8 (ebook for individuals)

For any available supplementary material, please visit
https://www.worldscientific.com/worldscibooks/10.1142/14030#t=suppl

Typeset by Stallion Press
Email: enquiries@stallionpress.com

Preface

Thanks to World Scientific Publishing for encouraging me to publish this book. This book is a selection of lectures I gave at the University of Maryland on quantum entanglement and interferometry.

What is quantum entanglement? Why is it so nonclassical? The concept of quantum entanglement started from Einstein, Podolsky, and Rosen (EPR). In 1935, EPR suggested a *gedankenexperiment* and introduced an entangled two-particle system based on the superposition of two-particle wavefunctions. Indeed, it is very different, far beyond not only our everyday common sense, but also that of ordinary quantum theory in Einstein's time. The two-particle wavefunction, or the two-particle state, does not specify either momentum or position of the subsystems, however, if the momentum or position of particle 1 is observed with a certain value, the momentum or position of particle 2 must take a corresponding unique value, even if the two particles are far apart in distance without any interaction. Does nature have such a peculiar two-particle system? Yes, it does. In light of new technology, we can now even artificially create such entangled two-particle state. For instance, a laser beam is able to produce a pair of entangled photons from a nonlinear optical process, technically known as spontaneous parametric down-conversion (SPDC). The entangled photon pair, namely the signal and idler, is characterized by a nonfactorizable pure state of quantum mechanics:

$$|\Psi\rangle = \Psi_0 \sum_{s,i} \delta\left(\omega_s + \omega_i - \omega_p\right) \delta\left(\mathbf{k}_s + \mathbf{k}_i - \mathbf{k}_p\right) a_s^\dagger(\mathbf{k}_s)\, a_i^\dagger(\mathbf{k}_i)\, |\,0\rangle$$

where ω_j, $\mathbf{k}_j$, $j = s, i, p$, are the frequency (energy) and wavevector (momentum) of the signal (s), idler (i), and pump (p), $a_s^\dagger$ and $a_i^\dagger$ are creation operators for the signal and the idler photon, respectively, and

Ψ_0 is a normalization constant. We have assumed a CW monochromatic laser pump, i.e., ω_p and $\mathbf{k}_p$ are considered as constants. What is so special (nonclassical) about this state? The following two points are typical statements. Point-1: Neither the energy–momentum of the signal photon nor the energy–momentum of the idler photon is specified in the state; however, if the energy–momentum of the signal (idler) photon is observed with a certain value, the energy–momentum of the idler (signal) photon must take a corresponding unique value. Point-2: This state is a coherent superposition of a large number of sub-states of the signal photon and the idler photon. Although each sub-state is the product of two individual states of the signal photon and the idler photon, as a result of the coherent superposition, the entangled two-photon state cannot be factorized into the product state of the signal photon and the idler photon. Quantum theory does not prevent such states.

Historically, for some reason, the most common concerns, especially in philosophical matters, about entanglement has been focused on Point-1. In fact, a realistic classical model, such as the hidden variable theory, is able to simulate an ensemble of mixed twin-particle system that satisfies Point-1 without any "surprises": nature may produce a pair of twin-particle at a time, and each particle has defined values of energy–momentum that satisfy the laws of energy conservation and momentum conservation. In this realistic model, while nature knows the energy–momentum of each particle pair during its preparation, we do not. Until we measure the energy–momentum of a particle, there is an equal chance of observing any value; however, if a measurement of one particle gives a specific value, the other particle must be found with a defined unique value. The true "surprise" is in Point-2. The two-particle state is not an statistical mixture but rather a pure state of coherent superposition of two-particle states, and this coherent superposition may occur at distant space-time coordinates. A pure state characterizes each and all particle pair, but not a mixed ensemble, indicating that each particle pair must have all possible energies and momentums "simultaneously". This means that not only do we not know what the energy–momentum of the particle pair is, but neither does also nature itself. Compare to the case of single-particle states, this is equivalent to claiming that a cat, namely, Schrödinger's cat, must be "simultaneously" alive and dead in a pure state of coherent superposition of $|\text{alive}\rangle$ and $|\text{dead}\rangle$.

In quantum mechanics, we use two-photon state, either entangled state or mixed state, to calculate the probability of observing a joint-photodetection event from a pair of photons. The entangled two-photon state described above is the result of a "coherent superposition" of all possible different yet indistinguishable alternatives in which a pair of photons may produce a joint-photodetection event at distant space-time coordinates. In each different yet indistinguishable alternatives, the signal photon and the idler photon take defined values of energy–momentum satisfying the conservation laws. The individual probability for the signal–idler pair to be in a certain defined values of energy–momentum is specified by the state. However, the "total" probability is not the sum of these individual probabilities; quantum mechanics says, the "total" probability of observing a joint detection event is the normal square of the sum of the individual "probability amplitudes" as the result of "coherent superposition". The probability amplitude is a complex number that contains an important physical parameter, the "phase", which indicates the nature of wave and is a key parameter when dealing with interference phenomena. Indeed, this coherent superposition represents a two-photon interference: A pair of photons interferes with the pair itself. The superposition of quantum amplitudes is a common phenomenon in the quantum world. It occurs between either single-particle amplitudes or multi-particle amplitudes. It is the multi-particle interference that produces the "surprises" of quantum entanglement. For instance, the superposition of two-photon amplitudes of a signal–idler photon pair is able to produce a point-to-point correlation between two distant optical planes, i.e., the object plane and the image plane, with the help of an imaging lens. This point-to-point correlation plays the role of image-form function producing a "ghost" image in the joint-photodetection of two distant photodetectors. "Ghost" images were demonstrated experimentally in 1995 as an example of nonlocal two-photon interference of a pair of entangled photons.

Does two-photon interference only occur with entangled photon pair? We have asked ourselves this question ever since we started experimental study of entangled two-photon states. We realized that the historical Hanbury Brown and Twiss (HBT) effect is a two-photon interference phenomenon. The HBT correlation is the result of a randomly created and randomly paired photons interferes with the pair itself. Traditionally, we have overemphasized the statistical nature of the phenomenon and, on this basis, have viewed the HBT effect as a statistical correlation of intensity fluctuations of thermal field. In this book, we provided sufficient theoretical and experimental evidences to emphasize the coherence nature of this phenomenon. In Chapter 5, we pointed out that the HBT spatial correlation, a point-to-point correlation between two distance planes equidistant from the thermal light source, is a nonlocal two-photon interference phenomenon: a randomly created and randomly paired photons interferes with the pair itself at distance. Similar to the ghost imaging of entangled states, this nonlocal two-photon interference induced point-to-point correspondence may also serve the role of image-forming function producing a "ghost" image in the photon number correlation, or intensity correlation, measurement of two distant photodetectors. We named it "lensless ghost imaging". "Ghost" image of thermal light was demonstrated experimentally in 2005, 10 years after the "ghost" imaging of entangled photons. Two-photon interferometer were soon demonstrated after the lensless ghost imaging. We found a few interesting properties of thermal light two-photon interferometer: (1) two-photon interference is observable from first-order incoherent light; (2) two-photon interference can be turbulence-free, i.e., any rapid phase variations along the optical path due to random changes in composition, density, length, index of refraction, or medium vibration, which completely blurs first-order interference, do not affect the visibility and contrast of the two-photon interference.

Can a coherent CW laser beam produce intensity fluctuation correlation like thermal light? This book gives a firm affirmative answer. In Chapter 8, we analyzed a recent experimental demonstration of a ghost frequency comb (GFC). The experiment is very simple: a multi-longitudinal-mode CW laser beam is split into two, and directed to two point-like photodetectors through kilometers long optical fibers. The two photodetectors are used to measure the local intensities, respectively, and the nonlocal intensity correlation of the CW laser beam, jointly, at kilometers distance. In their local measurements, both detectors observe constant intensities as expected from a CW laser beam. Surprisingly, a 50% contrast periodic pulse train of comb-like, ultra-narrow peaks are observed from their joint nonlocal correlation measurement. Examining the correlation function, it appears that the intensity fluctuations of the laser beam are correlated only within these periodic, precise, and narrow time windows. The intensity fluctuations of the laser beam become uncorrelated as long as the relative delay of the two photodetections falls into the region between the periodic sharp correlation peaks. Does any CW laser beam have such peculiar statistical behavior with respect to intensity fluctuations? Of course not. In fact, in a CW laser beam, the intensity fluctuations are much smaller than the intensity, $|\Delta n| \ll \bar{n}$, and can be ignored. Questions naturally arise: Can the observed GFC be considered as correlation of intensity fluctuations of CW laser beams? If the GFC is not caused by the statistical intensity fluctuations of the laser beam, then what is the cause of the observed GFC? From the perspective of quantum coherence, the observed GFC is the result of a nonlocal interference: a pair of randomly created and randomly paired indistinguishable photon groups, or cavity modes, interferes with the pair itself. The observation of GFC is strong evidence of nonlocal quantum interference.

This book aims to introduce the basic theory and concepts of multi-photon entanglement and interferometry. Differing from most traditional textbooks on this subject, it places greater emphasis on experiments and observations. All fundamental concepts are introduced in the process of analyzing typical experimental measurements and observations. The basic methods of classical and quantum mechanical treatments for optical measurements are naturally and gradually explored in the analysis of typical experiments. This attempt is aimed at (1) helping students and young scientists analyze, summarize, and resolve quantum optical problems and (2) encouraging students and young researchers to be more open minded in looking for the truth and improving their ability in making new discoveries in the field of physics.

In this regard, this book attempts to provide a number of nontraditional treatments and interpretations in certain historical and recent experimental discoveries in the field of quantum optics. The reader may find the following differences between this book and other traditional books on this subject: (1) This book introduces Einstein's granularity picture of light in terms of quantized bundle of rays or subfields and relates atomic transition to sub-sources. This attempt is aimed at preparing a general physical picture and background for introducing the concept of photon and the quantum mechanical theory of light. (2) It connects the interference phenomenon among a large number of sub-radiations with the concept of statistical ensemble average in the classical treatments of optical measurements and optical coherence. What is the physical cause of intensity fluctuation correlation of thermal field? This book gives a nontraditional but perhaps more reasonable answer. (3) It distinguishes quantum mechanical multi-photon interference from classical statistical correlation of intensities. From the point of view of classical theory, the joint detection by two or more pho-todetectors measures the statistical correlation of intensities. Any nontrivial second-order or higher-order correlation of light is caused by the nontrivial correlation of intensity fluctuations. From the point of view of quantum theory, the joint detection by two or more photodetectors measures the probability of jointly having two or more photons contribute to a joint-photodetection event. If more than one multi-photon amplitude contributes to the event, the superposition between these quantum amplitudes results in a multi-photon constructive-destructive interference effect, which may not be considered or may not be explainable in the classical statistical theory of intensity fluctuation correlation. It does not seem difficult to distinguish multi-photon interference from statistical correlation in the measurement of entangled photon pairs. However, it is definitely not easy to appreciate the quantum interference picture in the measurement of classical thermal light, even if the measurement is at the single photon level. This book gives much experimental evidence and theoretical analysis in supporting the viewpoint of quantum mechanics. I hope that this effort would help readers see a general quantum interference picture of light, which perhaps reflects the physical truth behind all of the optical observations.

In addition, this book introduced a new type of entangled state, namely the entangled coherent states, or entangled laser beams. The entangled coherent state can be generated from a CW monochromatic laser beam pumped multi-longitudinal-mode optical parametric oscillator (OPO):

$$|\Psi\rangle = \prod_{s,i} \delta\left(\omega_s + \omega_i - \omega_p\right) \delta\left(\mathbf{k}_s + \mathbf{k}_i - \mathbf{k}_p\right) |\alpha_s(\mathbf{k}_s)\rangle |\alpha_i(\mathbf{k}_i)\rangle,$$

where $|\alpha_s(\mathbf{k}_s)\rangle$ and $|\alpha_i(\mathbf{k}_i)\rangle$, $|\alpha| \gg 1$, are the state vectors of the signal–idler modes of the OPO cavity in the coherent state representation; ω_j and $\mathbf{k}_j$, for $j = s, i, p$, are the frequency and wavevector of the signal (s) modes, idler (i) modes, and the pump (p) laser beam. Considering each cavity mode a group of indistinguishable photons, this state indicates: (1) the cavity modes are created in pairs, each group of identical photons, i.e., a signal mode represented by $|\alpha_s(\mathbf{k}_s)\rangle$, is entangled with another unique group of identical photons, i.e., an idler mode represented by $|\alpha_i(\mathbf{k}_i)\rangle$, by means of energy conservation $\hbar\omega_s + \hbar\omega_i = \hbar\omega_p$ and momentum conservation $\hbar\mathbf{k}_s + \hbar\mathbf{k}_i = \hbar\mathbf{k}_p$; and (2) a large number of such signal–idler cavity-mode pairs are created and superposed coherently to form a pair of entangled signal–idler laser beams. The entangled coherent state is no longer at single photon's level, it consists of a large number of groups of indistinguishable photons in coherent states. A GFC of 100% contrast is observable from the joint photodetection of the signal–idler laser beams, although each single detector measures constant intensities as expected from a CW laser beam. Examining the comb-like second-order correlation function $G^{(2)}(\tau)$, we find that the entangled signal–idler laser beams are periodically correlated and anti-correlated. The signal and idler beams are correlated within each precise and narrow time windows. However, as long as the relative delay of $\tau = \tau_1 - \tau_2$ falls into the region between the periodic sharp correlation peaks, the signal and idler become anti-correlated. The periodic correlation-anti-correlation function of GFC poses more challenges to conventional thinking. We know that most of the observed correlation functions of entangled photon pairs, either created from atomic cascade decay or generated from SPDC, are Gaussian-like functions with a single peak at zero relative temporal delay. A widely accepted idea is that this is due to the simultaneous generation of entangled photon pairs at the source. Since the entangled photon pair is produced simultaneously in the source, they must be jointly detected by the detectors at zero relative temporal delays. Following this line of thought, can we assume that the entangled signal–idler identical photon groups are generated simultaneously and periodically by the CW OPO? Not likely! CW OPOs do not work that way. Furthermore, the contrast of the GFC is 100%, which means that no joint detection event occurs occur in these regions between the periodic sharp correlation peaks. From the perspective of classical thinking, it is absolutely impossible for zero-coincidence to occur in the measurement of CW laser beams. Even if somehow there are chances to observe zero-coincidences, why are they periodic in such a peculiar manner?

Introducing the basic concepts, tools, and the exciting developments of quantum optics to the reader, this book starts from Maxwell's equations, which explain the electromagnetic wave nature of light. We then review the historic blockbody radiation problem and photoelectric effect to address a critical question of why the quantization of light is necessary. Einstein's granularity picture of light is followed immediately as a semi-quantum theory. Although Einstein's concept of quantized subfield is still in the framework of electromagnetic theory, the physical behavior of his subfields, especially the self-interference of a single subfield or a group of subfields, is consistent with that of quantum mechanical concept of photon or a group of photons, except for a serious problem of locality. Having such a picture in mind, it would be natural to introduce the concept of field quantization and the concepts and tools of quantum optics based on the principles and rules of quantum mechanics. In this book, we introduce the concept of effective wavefunction of a photon or a group of photons and compare it with Einstein's subfield or subfields. There is no surprise that the effective wavefunction of a photon or a group of identical photons in thermal state is mathematically the same as that of Einstein's subfield. Based on these basic understandings, we introduce the concepts of coherent and incoherent radiations in terms of coherently and incoherently radiated subfields and Fourier modes. After the introduction of typical quantum states of quantized field or photon, we study the first-order coherence and the second-order coherence or correlation of light in terms of the self-interference of a single photon and the self-interference of a pair of photons in different quantum states. We then generalize the concept to Nth-order coherence or correlation in terms of N-photon interference: a group of N photons interfering with the group itself. Standing solidly on these fundamental concepts and theories, we then concentrate on the recent research topics of quantum optics, including quantum entanglement, multi-photon interferometry, and optical tests of foundations of quantum theory. I hope to not only introduce these exciting developments to the reader but also give the reader an opportunity to practice the concepts and tools learned from this book through the analysis of these fascinating observations.

I hope that this book, which has been written to the best of my ability and knowledge, can be helpful to students, researchers, and all readers in general, in their efforts to understand, develop, and advance the field of optical science, which is currently undergoing tremendous change.

Acknowledgments

First of all, I would like to thank my outstanding teachers Carroll O. Alley, David N. Klyshko, Morton H. Rubin, and John A. Wheeler, who not only taught me physics, but also passed on to me the spirit of physicists: seek the truth, and only the truth. My thanks also go to my students, postdoctors, and coworkers for their impressive experimental and theoretical research work, which have provided great support to this book. At the same time, my apologizes to them for not having included all of their research findings due to limited space and time; only those experiments and theories that are directly relevant to the contents and discussion points have been selected. For the same reason, I would like to apologize to all other researchers in the field of quantum optics for not being able to include their research. I would also like to thank Ling-An Wu, Jason Simon, Tom Smith, and Liang Shih for helping me with language editing issues.

About the Author

Yanhua Shih, Professor of Physics, received his Ph.D. in 1987 from the Department of Physics and Astronomy, University of Maryland College Park. He started the Quantum Optics Laboratory at the University of Maryland Baltimore County (UMBC) in the fall of 1989. His group has been recognized as one of the leading groups in the field of quantum optics that attempts to probe the foundations of quantum theory. His pioneering research on multi-photon entanglement, multi-photon interferometry, quantum imaging, and optical tests of foundations of quantum theory has attracted great attention from the physics and engineering community. In thirty years of teaching and research, he pubished hundreds of experimental and theoretical works in leading refereed journals on the subject of quantum optics and has given hundreds of invited lectures and presentations at national and international professional conferences and workshops. Yanhua Shih received the Willis Lamb Medal in 2002 for his pioneer contributions to quantum optics, especially the study of coherence effects of multi-photon entangled states.

Contents

Chapter 1

Quantum Theory of Light: Light Quantum and Field Quantization

In the beginning of 20th century, the successful introduction of the concept of light quantum, or photon, inspired a new foundation of physics, namely, the quantum theory. Today, quantum theory has turned out to be the overarching principle of modern physics. It would be difficult to find a single subject among the physical sciences which is not affected in its foundations or in its applications by quantum theory.

After 100 years of studies, how much do we know about the photon? The photon is a wave: It has no mass, it travels at the highest speed in the universe, and it interferes with itself. The photon is a particle: It has well-defined values of momentum and energy, and it even "spins" like a particle. The photon is neither a wave nor a particle because whichever we think it is, we would be tripped into difficulties in explaining the other part of its behavior. The photon is a wave-like-particle and/or a particle-like-wave: A photon can never be divided into parts, but interference of a single photon can be easily observed in a modern laboratory. It seems that a photon passes both paths of an interferometer when interference patterns are observed; however, if the interferometer is set in such a way that its two paths are "distinguishable", the photon "knows" which path to follow and never passes through both paths. Apparently, a photon has to make a choice in its behavior when facing an interferometer: a choice of "both-path" like a wave or "which-path" like a particle. Surprisingly, the choice is not necessary before passing through the interferometer. It has been experimentally demonstrated that the choice of "which-path" and/or "both-path" can be delayed until after the photon has passed through the interferometer. More surprisingly, the which-path information can even be "erased" after the annihilation of the photon itself. The behavior of a photon

1

apparently does not follow any of the basic criterion, reality, causality, and locality, of our everyday life. Of course, the peculiarity of wave–particle duality is not only the property of photons, it belongs to all quanta in the quantum world. Perhaps, it is easy to accept the particle picture of an electron with mass, m_e, and charge, e; it is definitely not easy to accept the particle nature of a photon. On the other hand, perhaps, it is easy to accept the wave picture of a photon with frequency, ω, and wavevector, $\mathbf{k}$; it is definitely not easy to accept the wave nature of an electron.

Although questions regarding the fundamental issues about the concept of a photon still exist, the quantum theory of light has contributed perhaps the most influential and successful yet controversial part to quantum mechanics.

In this chapter, we constrain ourselves to the following basic questions about the quantum theory of light: (1) Why is quantization of light radiation necessary, (2) how to quantize the radiation field, (3) how to describe the state of the quantized field, and (4) how to physically model and mathematically formulate a photodetection event or a joint-photodetection event?

1.1   The Experimental Foundation — I: Blackbody Radiation

Experimental physicists observed an unexpected phenomenon around 1900: The experimentally observed spectral distribution of blackbody radiation does not match the theoretical predictions of all existing physical theories at that time.

A "blackbody" is a perfect absorber that absorbs all incident radiation. The best approximation to a blackbody is a tiny pinhole in the wall of a hollow enclosure, or cavity. The intensity, $I(\nu)$, of radiation per unit solid angle, coming from the pinhole, in the frequency range between ν and $\nu + d\nu$ can be accurately measured. It was found experimentally, under the condition of thermodynamic equilibrium, that any blackbody has the same characteristic emission function $I(\nu)$. Typical curves of $I(\nu)$ are shown in Fig. 1.1.1. $I(\nu)$ depends only on the temperature of the walls of the enclosure, neither on the material of which the enclosure is made nor on the shape of the cavity. At a particular temperature $I(\nu) \propto \nu^2$ for low frequencies, while at high frequencies $I(\nu)$ drops off exponentially. Another interesting feature of the spectral distribution is that the maximum $I(\nu)$ shifts to higher frequencies while the temperature of the blackbody

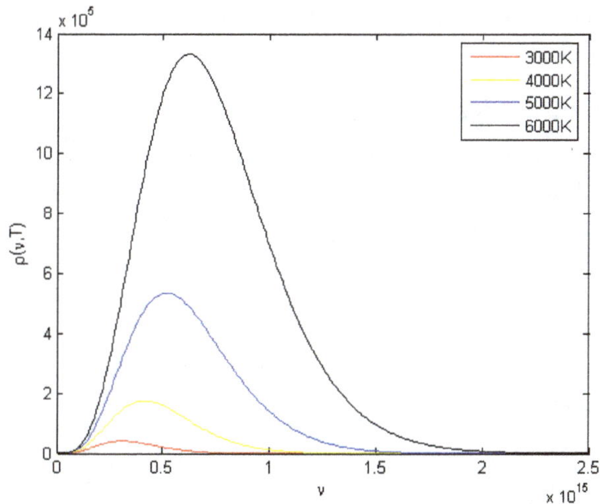

Fig. 1.1.1   Blackbody radiation curves. $I(\nu)$ depends only on the temperature of the walls of the enclosure. At a particular temperature, $I(\nu) \propto \nu^2$ for low frequencies, while at high frequencies $I(\nu)$ drops off exponentially. Another interesting feature of the spectrum is that the maximum $I(\nu)$ is shifted toward higher frequencies while the temperature of the blackbody is raised.

increases. The four curves in Fig. 1.1.1 clearly indicate these characteristics of the blackbody radiation.

It was indeed a surprise at that time that the theory of classical mechanics, electrodynamics, and thermodynamics all together failed to explain this simple phenomenon.

In the frame work of classical electrodynamics, the intensity of the blackbody radiation per unit solid angle is related to the energy density of the radiation in the cavity, or the enclosure:

$$I(\nu) = \frac{c}{4\pi} \, u(\nu), \tag{1.1.1}$$

where c is the speed of light and $u(\nu)$ is the energy density of frequency ν. Therefore, the measured intensity of blackbody radiation $I(\nu)$ is determined by $u(\nu)$ of the cavity. The source of the radiation energy in the cavity is obviously the walls of the enclosure, which continually emit waves of every possible frequency and wave vector, or say all possible modes. In thermodynamic equilibrium, the amount of energy $u(\nu)d\nu$, in the frequency range between ν and $\nu + d\nu$ is easily calculated. What we need is to (1) estimate the number of permissible modes under the boundary condition of the cavity and (2) calculate the mean energy of the permissible modes.

(1) Number of modes

Suppose we have a rectangular cavity of dimension L_x, L_y, and L_z. Applying the boundary condition required by electromagnetic theory, the allowed wavevector $\mathbf{k}$ is thus

$$k_x = \frac{2\pi l}{L_x} \quad k_y = \frac{2\pi m}{L_y} \quad k_z = \frac{2\pi n}{L_z}, \tag{1.1.2}$$

where l, m, and n are integers taking values from $-\infty$ to ∞. The number of allowed modes is therefore

$$\Delta N = \Delta l \, \Delta m \, \Delta n = \frac{V}{(2\pi)^3} \, dk_x \, dk_y \, dk_z, \tag{1.1.3}$$

where $V = L_x L_y L_z$. To estimate the number of permissible modes in the frequency range between ν and $\nu + d\nu$, it is more convenient to adopt polar coordinates in $\mathbf{k}$ space by considering the volume element $dk_x \, dk_y \, dk_z = k^2 \, dk \, d\Omega$, where $d\Omega$ is the element of solid angle. The number of permissible modes in the frequency range between ν and $\nu + d\nu$ is thus

$$\Delta N = \frac{4\pi V}{(2\pi)^3} \, k^2 \, dk = \frac{4\pi V}{c^3} \, \nu^2 \, d\nu,$$

where we have integrated over $d\Omega$, since we are not interested in the direction of the wavevector $\mathbf{k}$. Taking into consideration the polarization for each mode, the number of permissible modes in the frequency range between ν and $\nu + d\nu$ will be doubled:

$$\Delta N = \frac{8\pi V}{c^3} \, \nu^2 \, d\nu. \tag{1.1.4}$$

(2) Mean energy of each mode

To calculate the mean energy of each mode, we adopt the results of classical statistical mechanics. We consider each mode of radiation to be in thermodynamic equilibrium with the walls of the enclosure which is treated as a heat reservoir. A heat reservoir is defined as a very large system with constant temperature. Physically, this means that the temperature of the enclosure remains unaffected by whatever small amount of energy it gives to the radiation mode. Under the condition of equilibrium, the probability of finding the radiation mode between E and $E + dE$ follows the canonical distribution:

$$P(E)\,dE = \frac{e^{-E/kT}\,dE}{\int_0^\infty e^{-E/kT}\,dE}. \qquad (1.1.5)$$

The mean value of the energy $\bar{E}$ is calculated by weighting each possible energy according to its probability:

$$\bar{E} = \frac{\int_0^\infty E\,e^{-E/kT}\,dE}{\int_0^\infty e^{-E/kT}\,dE} = kT\frac{\int_0^\infty \epsilon\,e^{-\epsilon}\,d\epsilon}{\int_0^\infty e^{-\epsilon}\,d\epsilon} = kT, \qquad (1.1.6)$$

where $\epsilon = E/kT$. In Eq. (1.1.6), we have used the solution of the Γ-function, $\Gamma(n+1) = n!$ and $\Gamma(1) = 1$.

The mean energy of each radiation mode is kT. This is a good example of the theorem of equipartition of energy. The amount of radiation energy in the frequency range between ν and $\nu + d\nu$ is thus

$$u(\nu)\,d\nu = \bar{E}\Delta N = \frac{8\pi V}{c^3}\,kT\,\nu^2\,d\nu \qquad (1.1.7)$$

which is called the Rayleigh–Jean's law.

Comparing with the blackbody radiation curves shown in Fig. 1.1.1, Rayleigh–Jean's law only agrees with the experimental observations at low frequencies; it gives too much power of radiation for high frequencies. In addition, if we integrate over all frequencies to calculate the total energy, the result diverges, meaning an infinite amount of energy contained in the cavity.

Wien attempted a different classical approach. Based on classical thermodynamic arguments, Wien showed that the blackbody radiation distribution must be of the form $u(\nu) = \nu^3\,f(\nu/T)$. The function $f(\nu/T)$, however, cannot be determined from thermodynamics alone. Wien obtained a distribution function that was later named as Wien's law:

$$u(\nu)\,d\nu \sim \nu^3\,e^{-h\nu/kT}\,d\nu, \qquad (1.1.8)$$

where h is a constant determined experimentally by data fitting. This constant later turned out to be the symbolic constant of quantum theory and was named Planck constant. Wien's distribution improved the high-frequency spectrum fitting, but got worse at low frequencies.

In history, all classical attempts, whether the electrodynamic approaches or the thermodynamic treatments, failed to give an accurate distribution function to fit the observation curves of blackbody radiation, within experimental error. We thus conclude that the concepts we have used to derive these laws, or distributions, may not be adequate to describe the behavior of blackbody radiation.

In the year 1900, Planck decided to abandon the classical tradition and in doing so he succeed in fitting the experimentally measured blackbody radiation spectrum. Planck's hypothesis was very simple. He assumed that a radiation mode can only take energy values of an integer multiple of a basic unit of energy, $E = nh\nu$, where n is an integer running from 0 to ∞. This basic energy unit $h\nu$ is not the same for all modes but rather is proportional to the frequency of the mode. With this assumption, phenomenologically, Planck explained the blackbody radiation by accurately fitting the experimentally measured distribution curves, within experimental error.

Planck's assumption is truly inconsistent with classical concepts. According to classical mechanics and classical electrodynamics, there are no restrictions on the energy of a radiation mode. The only "restriction" regarding the energy of a radiation mode is the mean value, kT, which is independent of the frequency of the radiation. Planck's assumption also seems inconsistent with many of our everyday experiences. For instance, the output power of an AM or FM radio oscillator may have any value. There is no experimental evidence that the energy of a radio oscillator must be quantized to $E = nh\nu$. Does it mean that Planck's theory is inadequate to describe the behavior of radio waves? The answer is NO. The apparent inconsistency arises from the fact that h is a very small quantity, $h \sim 6.6 \times 10^{-34}$ J s. In the *AM* and *FM* radio frequencies, for example, $\nu \sim 10^6$ Hz and $\nu \sim 10^8$ Hz, the basic unit of energy $h\nu$ is on the order of 6.6×10^{-28} J and 6.6×10^{-26} J which are not detectable by any available sensitive detection apparatus. With light waves, however, the values of $h\nu$ increase significantly, for $\nu \sim 10^{15}$ Hz, $h\nu \sim 10^{-19}$ J. This value is measurable by modern measurement devices. Therefore, as we go to higher frequencies, Planck's quantization hypothesis is easier to verify.

We now derive the spectral distribution function for blackbody radiation by following Planck's energy quantization. Similar to what we did in the early classical analysis, we recalculate the mean energy per mode by applying the canonical distribution. The probability for a mode to be in a given energy $E_n = nh\nu$ is then

$$P(n) \propto e^{-E_n/kT} = e^{-nh\nu/kT}. \tag{1.1.9}$$

Normalizing Eq. (1.1.9), we obtain

$$P(n) = \frac{e^{-nh\nu/kT}}{\sum_{n=0}^{\infty} e^{-nh\nu/kT}} = \frac{e^{-nh\nu/kT}}{(1 - e^{-h\nu/kT})^{-1}}. \qquad (1.1.10)$$

The mean energy per mode is then

$$\bar{E} = \sum_{n=0}^{\infty} E_n \, P(n) = (1 - e^{-h\nu/kT}) \sum_{n=0}^{\infty} nh\nu \, e^{-nh\nu/kT}$$

$$= h\nu \, (1 - e^{-h\nu/kT}) \sum_{n=0}^{\infty} n \, e^{-nh\nu/kT}. \qquad (1.1.11)$$

To evaluate Eq. (1.1.11), we can write

$$\sum_{n=0}^{\infty} n \, e^{-nh\nu/kT} = -\frac{d}{d\alpha} \sum_{n=0}^{\infty} e^{-n\alpha} = -\frac{d}{d\alpha}\left(\frac{1}{1 - e^{-\alpha}}\right) = \frac{e^{-\alpha}}{(1 - e^{-\alpha})^2},$$

where $\alpha = h\nu/kT$. The mean energy per mode, $\bar{E}$, is then

$$\bar{E} = \frac{h\nu \, e^{-h\nu/kT}}{1 - e^{-h\nu/kT}}. \qquad (1.1.12)$$

The amount of radiation energy in the frequency range between ν and $\nu + d\nu$ is found to be

$$u(\nu) \, d\nu = \bar{E}\Delta N = \frac{8\pi V}{c^3} \, h\nu^3 \, \frac{e^{-h\nu/kT}}{1 - e^{-h\nu/kT}} \, d\nu.$$

We thus obtain the Planck distribution

$$u(\nu) = \frac{8\pi V}{c^3} \, \frac{h\nu^3 \, e^{-h\nu/kT}}{1 - e^{-h\nu/kT}} \qquad (1.1.13)$$

which is an exact fit, within experimental error, to the distribution of blackbody radiation.

1.2 The Experimental Foundation — II: Photoelectric Effect

The study of blackbody radiation concluded indirectly that electromagnetic waves may increase or decrease energy only in the units of $h\nu$. The discovery of the photoelectric effect confirms this surprising conclusion in a more direct way. In fact, the photoelectric effect was first reported by Hertz in 1887, more than 10 years before Planck's work. The quantum explanation of the effect was given later by Einstein in 1905 as a result

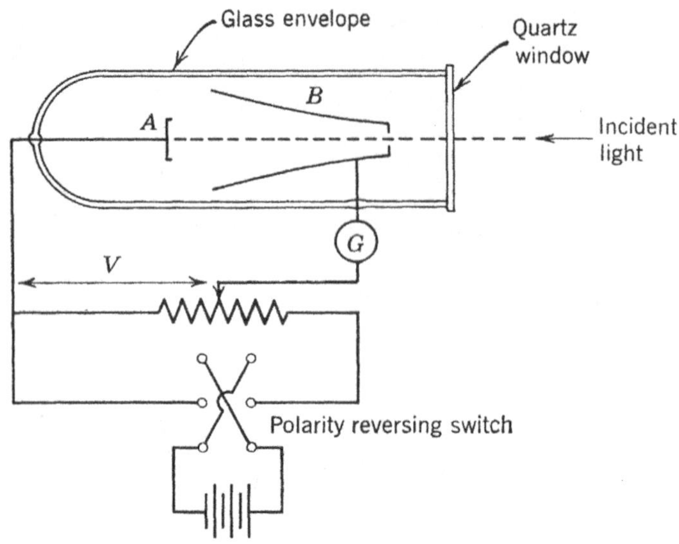

Fig. 1.2.1   Typical schematic experimental setup for observing the photoelectric effect.

of five years of thinking about Planck's hypothesis. Figure 1.2.1 shows a typical experimental setup for observing the photoelectric effect. A simple vacuum tube, containing a metal plate, A, and an anode, B, is used for the experimental observation. Monochromatic light is incident through the quartz window of the vacuum tube on the metal A. The photoelectrons liberated from the surface of the metal A are collected by the anode B. An adjustable potential difference V is applied between the metal plate A and the anode B. The output photocurrent of the anode is monitored by a sensitive ammeter G.

Figure 1.2.2 is a typical observation of the photoelectric current, i, as a function of the applied potential difference, V. When V is positive and takes large enough values, the photocurrent i reaches a saturated value, which means all liberated photoelectrons are collected by the anode A. The saturated value of the photoelectric current i is proportional to the intensity of the incident light. This result is reasonable because a large intensity should indeed eject more photoelectrons. The surprise, however, comes when V is reversed, making it negative, and adjusted to reach the stopping potential V_0 where the photoelectric current drops to zero, $i \sim 0$. It was found that V_0 is independent of the intensity of the incident light, as shown in Fig. 1.2.2. When a negative potential, V, is applied, the

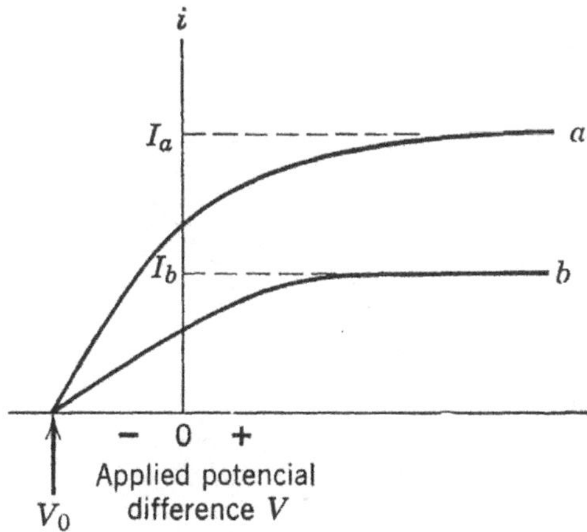

Fig. 1.2.2   Typical measured photoelectric current, i, as a function of the applied voltage, V, between the metal plate, A, and the anode, B. Note that the voltage can be switched to negative or positive. The two curves, a and b, correspond to two different incident light intensities: $I_a = 2I_b$. It is a surprise to find that V_0 is independent of the intensity of the incident light.

photoelectric current does not immediately drop to zero. This suggests the electron escapes from the surface of the metal with a certain kinetic energy. Some of the escaping electrons can still reach the anode, B, if their kinetic energies are large enough, $K_{\max} > eV_0$, to overcome the applied electric potential against their motion. If the negative potential V is made large enough to be equal or greater than the maximum kinetic energy of the escaping electrons, $eV_0 \geq K_{\max}$, no electrons can reach the anode, B, and consequently the photoelectric current i drops to zero. It is a surprise from the classical point of view that the stopping potential V_0 and consequently the kinetic energy of a liberated electron do not dependent on the intensity of the incident light. In classical electromagnetic wave theory, $I \propto |E|^2$. Since the force applied to an electron is eE, the kinetic energy of the photoelectron should increase as the intensity of the incident light increases.

More surprises came later from Millikan's work. Millikan's experiment showed that the stopping potential, V_0, is linearly dependent on the frequency of the incident light and there exists, for each different metal plate, a characteristic cutoff frequency ν_0. For any frequency lower than ν_0, photoelectric effect stops occurring, no matter how intense the incident light is. Figure 1.2.3 shows a typical measurement of V_0 as a linear function of the incident light frequency. The slope of the experimental curve is on

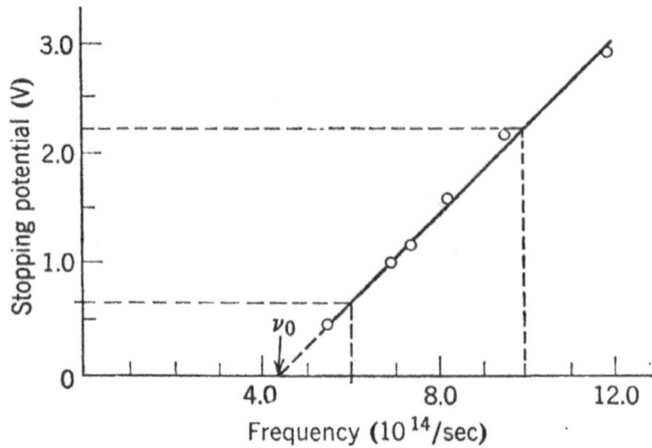

Fig. 1.2.3   Typical measurement of V_0 as a linear function of the frequency of the incident light. The slope of the experimental curve is on the order of 3.9×10^{-15} V s. There exists, for each different metal plate, a characteristic cutoff frequency ν_0. For any frequency less than ν_0, the photoelectric effect stops occurring, no matter how intense the incident light is.

the order of 3.9×10^{-15} V s. The existence of the cutoff frequency, ν_0, is inconsistent with classical concepts of the electromagnetic wave theory. According to classical electrodynamics, the photoelectric effect should be observable for any frequency, provided the incident light is intense enough to give the necessary amount of kinetic energy to the photoelectron.

In 1905, Einstein proposed a theory which successfully explained the photoelectric effect. In his theory, Einstein quantized the radiation energy into localized "bundles", $E = h\nu$, which were later named "photons" (1926). Einstein assumed that one photon, individually, is completely absorbed by one excited electron in the process of a photoelectron ejection. When the electron is ejected from the surface of the metal, its kinetic energy is given by

$$K = h\nu - W, \qquad (1.2.1)$$

where W is the the work required to overcome the attractive forces of the atoms that bind the electron to the metal. W is called the work function.

Einstein was the first physicist to relate the photoelectric effect with Plank's hypothesis. Einstein's Eq. (1.2.1) can be rewritten as

$$V_0 = \frac{h}{e}\nu - \frac{W}{e}, \qquad (1.2.2)$$

where we have substituted eV_0 for $K_{\max}$, and V_0 is the applied stopping potential at which the photoelectric current drops to zero. Equation (1.2.2) indicates a linear relationship between the stopping potential V_0 and the frequency of the incident light, in agreement with Millikan's experimental results, see Fig. 1.2.3. The measured slop of the experimental curve in Millikan's experiment is

$$\frac{h}{e} \sim \frac{2.20V - 0.65V}{(10.0 - 6.0) \times 10^{14}/\text{s}} \simeq 3.9 \times 10^{-15} \text{ V s.}$$

Multiplying the measured slope by the electronic charge, e, yields $h \sim 6.2 \times 10^{-34}$ J s which is close to the value $h \sim 6.6 \times 10^{-34}$ J s, appearing in Plank's distribution function of 1900. Later, more accurate photoelectric experiments measured $h \sim 6.6262 \times 10^{-34}$ J s. The agreement between the two constants, h, appearing in the photoelectric experiments and the black-body radiation observation is a strong confirmation that h is a universal constant and the radiation field can exchange energy only in units of $h\nu$.

1.3 Einstein's Granularity Picture of Light

After five years of thinking about the blackbody radiation hypothesis of Planck and the experimental studies of Hertz on the photoelectric effect, in 1905, Einstein introduced a granularity to radiation, abandoning the continuum interpretation of Maxwell. This led to a microscopic picture of radiation and a statistical view of light. Although Einstein did not name his "bundle of ray", which is an English translation from German "strahlenbündel", a photon, at the heart of Einstein's theory, is the particle picture of radiation. Einstein assumed that the radiation energy is quantized into localized bundles as light quanta. The energy of the bundle, or light quantum, is initially localized in a small volume of space and remains localized as it moves away from the radiation source with velocity c. The energy of the bundle, or light quantum, is related to its frequency by multiplication of a universal constant h. In the photoelectric process, one bundle of energy $E = h\nu$, or one photon, is completely absorbed by one electron originally bound with the metal. How can one bounded electron, which is a particle localized within a very small volume, completely absorb a photon to become an photoelectron? The simplest physical picture is that the particle-like photon transferred all its energy and momentum to the electron during a *collision*. Is the photon a localized object? Yes and no. When facing its particle-like behavior, it must be localized. When facing its wave-like behavior, it cannot be localized. The biggest confusion, perhaps, happens when the photon is treated as a particle in the interpretation but is treated as a wave in the calculation.

Standing on the shoulders of giants, now we have a better picture on Einstein's concept of granularity. The following picture may not be exactly the same as the original idea of Einstein, but we have enough reason to name it Einstein's picture of light.

In Einstein's picture, a natural light source, such as the sun or a distant star, consists of many point-like sub-sources ("atomic transitions" in modern language), each of which emits its own subfields (originally labeled in German by "strahlenbündel," translated to "bundle of ray" in English, and now named as "photon" in modern language) that carry energy $E = h\nu$ in a random manner: The mth subfield (photon) that is emitted from the mth point-like sub-source (an atomic transition in which the atom changes its state from higher energy level E_2 to lower energy level E_1 and releases a photon with energy $h\nu = E_2 - E_1$) may propagate in all possible directions with all possible random phases. Einstein's concept of "bundle of ray" or "subfield" refers a quantized microscopic realistic substance of electromagnetic field, corresponding to the quantum mechanical concept of photon; precisely, it has the same mathematical expression as the effective wavefunction of photon in thermal state. Later in this book, we introduce a quantum mechanical concept of "effective wavefunction" from the quantum theory of optical coherence. It is interesting to find that the effective wavefunction of a photon plays the same role as Einstein's subfield, and the effective wavefunction of a photon in the thermal state is mathematically the same function of space–time as that of Einstein's subfield of natural light. It should be emphasized that the physical means of the two are fundamentally different: Einstein's subfield $E_m(\mathbf{r}, t)$ is a quantized microscopic realistic entity of electromagnetic wave propagating in space–time. The quantum mechanical effective wavefunction is the probability amplitude for a photon to produce a photodetection event at space–time coordinate $(\mathbf{r}, t)$.

In Einstein's picture, the radiation measured at coordinate $(\mathbf{r}, t)$ is the result of a superposition among a large number of subfields:

$$\mathbf{E}(\mathbf{r}, t) = \sum_{m=1}^{M} \mathbf{E}_m(\mathbf{r}, t), \qquad (1.3.1)$$

where $\mathbf{E}_m(\mathbf{r}, t)$ is the observed mth subfield at space–time $(\mathbf{r}, t)$, and the subscript index m, $m = 1, 2, \ldots, M$, labels the mth subfield that is created from the mth point-like sub-source (atomic transition) at space–time coordinate $(\mathbf{r}_{0m}, t_{0m})$ with energy $\hbar\omega = E_2 - E_1$. M is the total number of quantized subfields contributed to the measurement. In the following discussions, we simplify the expression of sum from $\sum_{m=1}^{M}$ to $\sum_m$. For natural light, the incoherent superposition of quantized subfields is treated as a discontinuous sum, and the coherent superposition of Fourier modes within an atomic transition, however, is treated as a continues integral:

$$\mathbf{E}_m(\mathbf{r}, t) = \int d\mathbf{k}\, \hat{\mathbf{e}}_{m,\mathbf{k}}\, E_m(\mathbf{k})\, g_m(\mathbf{k}; \mathbf{r}, t)$$

$$\simeq \int d\mathbf{k}\, \hat{\mathbf{e}}_{m,\mathbf{k}}\, E_m(\mathbf{k})\, e^{-i[\omega(t-t_{0m})-\mathbf{k}\cdot(\mathbf{r}-\mathbf{r}_{0m})]}, \qquad (1.3.2)$$

where $E_m(\mathbf{k}) = a_m(\mathbf{k})e^{\varphi_m(\mathbf{k})}$ is the complex amplitude of the $\mathbf{k}$-mode of the mth subfield; $g_m(\mathbf{k}; \mathbf{r}, t)$ is Green's function that propagates the $\mathbf{k}$-mode of the mth subfield from the mth sub-source to space–time coordinate $(\mathbf{r}, t)$ of the observation. The approximation in the last line of Eq. (1.3.2) is made by assuming a point-to-point propagation. In Eq. (1.3.2), we have considered the energy–time uncertainty of the atomic transition $\Delta E \Delta t \geq h$: When the life time of $\Delta t \neq \infty$, $\Delta E_2 \neq 0$. The mth quantized subfield, or photon, emitted from the mth atomic transition may carry any value of energy between $\hbar\omega = (E_2 - \Delta E_2/2) - E_1$ and $\hbar\omega = (E_2 + \Delta E_2/2) - E_1$, while E_1 is considered the ground level. It is helpful to define a central frequency of the subfield by means of $\hbar\omega_0 = (\bar{E}_2 - E_1)/2$ with $\bar{E}_2$ the means value of E_2. In Eq. (1.3.2), we have also considered the momentum–position uncertainty: When the mth subfield is emitted from a point-like sub-source, it may propagate to any direction within $\Omega = 4\pi$ solid angle. The uncertainty in energy and momentum restrict the integrals of $d\omega$ (frequency) and $d\Omega$ (direction) and thus $d\mathbf{k}$.

In terms of the stochastically created subfields $E_m(\mathbf{r}_{0m}, t_{0m})$ at their creation space–time points $(\mathbf{r}_{0m}, t_{0m})$, we may relate the observed field $E(\mathbf{r}, t)$ at their observation space–time point $(\mathbf{r}, t)$ with $E_m(\mathbf{r}_{0m}, t_{0m})$ by the following propagator:

$$E(\mathbf{r}, t) = \sum_m \int d\mathbf{k}\, \hat{\mathbf{e}}_{m,\mathbf{k}}\, \frac{E_m(\mathbf{k}; \mathbf{r}_{0m}, t_{0m})}{|\mathbf{r} - \mathbf{r}_{0m}|}\, e^{-i[\omega(t-t_{0m})-\mathbf{k}\cdot(\mathbf{r}-\mathbf{r}_{0m})]}$$

$$= \sum_m \int d\mathbf{k}\, \hat{\mathbf{e}}_{m,\mathbf{k}}\, E_m(\mathbf{k}; \mathbf{r}_{0m}, t_{0m})\, G_m(\mathbf{k}; \mathbf{r}, t), \qquad (1.3.3)$$

where

$$G_m(\mathbf{k}; \mathbf{r}, t) = \frac{e^{-i[\omega(t-t_{0m})-\mathbf{k}\cdot(\mathbf{r}-\mathbf{r}_{0m})]}}{|\mathbf{r}-\mathbf{r}_{0m}|}$$

is Green's function of free propagation that propagates the $\mathbf{k}$-mode of the mth subfield $E_m(\mathbf{k}; \mathbf{r}_m, t_m)$ from $(\mathbf{r}_{0m}, t_{0m})$ to $(\mathbf{r}, t)$. Formally, Eq. (1.3.3) can be written as

$$E(\mathbf{r}, t) = \sum_m E_m(\mathbf{r}_m, t_m)\, G_m(\mathbf{r}, t), \qquad (1.3.4)$$

with a common statement: Green's function $G_m(\mathbf{r}, t)$ propagates the mth subfield from $(\mathbf{r}_m, t_m)$ to $(\mathbf{r}, t)$.

In most optical measurements, it is not the real-positive amplitude $a_m(\mathbf{k})$ but the phase $\varphi_m(\mathbf{k})$ playing the significant role determining the coherent or incoherent nature of the radiation. To simplify the notation, we write Eq. (1.3.3) as

$$E(\mathbf{r}, t) = \sum_m \int d\mathbf{k}\, \hat{\mathbf{e}}_{m,\mathbf{k}}\, E_m(\mathbf{k})\, g_m(\mathbf{k}; \mathbf{r}, t)$$

$$= \sum_m \int d\mathbf{k}\, \hat{\mathbf{e}}_{m,\mathbf{k}}\, a_m(\mathbf{k})\, e^{i\varphi_m(\mathbf{k})}\, g_m(\mathbf{k}; \mathbf{r}, t). \qquad (1.3.5)$$

For point-like sub-sources and free propagation,

$$E_m(\mathbf{k}) = \frac{a_m(\mathbf{k}; \mathbf{r}_m, t_m)}{|\mathbf{r}-\mathbf{r}_m|}\, e^{i\varphi_m(\mathbf{k})} = a_m(\mathbf{k})\, e^{i\varphi_m(\mathbf{k})}$$

$$g_m(\mathbf{k}; \mathbf{r}, t) = e^{-i[\omega(t-t_m)-\mathbf{k}\cdot(\mathbf{r}-\mathbf{r}_m)]}, \qquad (1.3.6)$$

where $a_m(\mathbf{k}; \mathbf{r}_m, t_m)$ is the real-positive amplitude of the $\mathbf{k}$-mode of the mth subfield at its creation point $(\mathbf{r}_m, t_m)$, and $E_m(\mathbf{k}) = a_m(\mathbf{k})\, e^{i\varphi_m(\mathbf{k})}$ is the complex amplitude of the $\mathbf{k}$-mode of the mth subfield at its observation point $(\mathbf{r}, t)$. It is worth noting that when the distance between sub-sources is much smaller than the propagation distance, $a_m(\mathbf{k}; \mathbf{r}_m, t_m) \simeq a_m(\mathbf{k})$. To simplify later calculations and discussions by shortening the notation, we name the phase propagator $g_m(\mathbf{k}; \mathbf{r}, t)$, which propagates the $\mathbf{k}$-mode of the mth subfield from $(\mathbf{r}_{0m}, t_{0m})$ to $(\mathbf{r}, t)$, Green's function.

Simplifying the mathematics to 1-D, we assume a simple setup of a point-like light source at $\mathbf{r}_0 = 0$ contains of a large number of two-level atomic transitions between energy levels E_2 and E_1, and a point-like photodetector at $\mathbf{r}$. The atomic transitions may happen at different times, each labeled as t_{0m}. The observed radiation comes from the point-like source to the point-like photodetector along the 1-D path $r = |\mathbf{r} - \mathbf{r}_0|$. The observed field $E(\mathbf{r}, t)$ is the result of a superposition among a large number of subfields; each is created from a point-like sub-source at time t_{0m}, such as millions of atomic transitions in a point-like discharge tube; the resulting field can be modeled as a superposition of the subfields in terms of the sub-sources and their harmonic modes of frequency ω:

$$E(r, t) = \sum_m \int d\omega \, a_m(\omega) e^{i\varphi_m} e^{-i\omega[(t-t_{0m})-r/c]}, \qquad (1.3.7)$$

where $E_m(\omega) = a_m(\omega) e^{i\varphi_m}$ is the complex amplitude of the Fourier mode, ω, of the mth subfield at its observation point $(\mathbf{r}, t)$. Note, again, that we have treated the superposition of subfields as incoherent, and the superposition of Fourier modes within an atomic transition as coherent. Since the mth subfield is created from the mth atomic transition, φ_m is a function of m (mth atomic transition) only but not a function of ω (energy uncertainty from an atomic transition). Although each Fourier modes may take a different value of $a_m(\omega)$ (real and positive amplitude), all the coherently superposed Fourier modes within the mth atomic transition must have the same φ_m (phase of the complex amplitude). Equation (1.3.7) can be formally written into the following form of Fourier transform:

$$E(r, t) = \sum_m \left\{ \int d\nu \, a_m(\nu) e^{-i\nu\tau_m} \right\} e^{-i(\omega_0\tau_m - \varphi_m)}$$

$$= \sum_m \mathcal{F}_{\tau_m}\{a_m(\nu)\} e^{-i(\omega_0\tau_m - \varphi_m)}, \qquad (1.3.8)$$

where ω_0 is the central frequency of the mth quantized subfield, $\nu \equiv \omega - \omega_0$ is the detuning frequency, $\tau_m \equiv (t - t_{0m}) - r/c = \tau - t_{0m}$, and $\tau \equiv (t - r/c)$.

It is interesting to find that the mathematical representation of each subfield is in the form of a "wavepacket". Each wavepacket has a well-defined envelope, which is the Fourier transform of $a_m(\nu)$, and a harmonic carrier of frequency ω_0 and initial phase φ_m:

$$E_m(r, t) = \mathcal{F}_{\tau_m}\{a_m(\nu)\} e^{-i(\omega_0\tau_m - \varphi_m)}. \qquad (1.3.9)$$

The measured field $E(r, t)$ is the incoherent superposition of a certain number of wave packets, each created independently and randomly from point-like sub-sources at its own creation time with its own initial phase:

$$E(r,t) = \sum_m \mathcal{F}_{\tau_m}\{a_m(\nu)\}\, e^{-i(\omega_0 \tau_m - \varphi_m)}. \qquad (1.3.10)$$

The measured intensity $\langle I(r,t)\rangle$ is thus

$$\langle I(r,t)\rangle = \langle E^*(r,t)E(r,t)\rangle$$

$$= \left\langle \sum_n \mathcal{F}^*_{\tau_n}\{a_n(\nu)\}\, e^{i(\omega_0 \tau_n - \varphi_n)} \sum_m \mathcal{F}_{\tau_m}\{a_m(\nu)\}\, e^{-i(\omega_0 \tau_m - \varphi_m)} \right\rangle$$

$$= \sum_m \left|\mathcal{F}_{\tau_m}\{a_m(\nu)\right|^2, \qquad (1.3.11)$$

where we have considered the incoherent superposition between the randomly created subfields: The only surviving terms in the sum are those with $m = n$. Based on Einstein's granularity, we thus have the following physical picture of the above simple 1-D measurement: A large number of wavepackets excited at space point $\mathbf{r} = 0$ but different time t_{0m} and initial phase φ_m are propagated and superposed at space–time point $(\mathbf{r}, t)$. Equation (1.3.10) is commonly used to model thermal light. We have more discussions about this model in later chapters. It is worth noting that in Eq. (1.3.10), we have two parameters, t_{0m} and φ_m, to specify the randomness of the mth subfield. The random phase relationship between different sub-fields is critical in the calculation of coherence or correlation of incoherent sub-fields. In some measurements, a single parameter, t_{0m} or φ_m, is sufficient to obtain the result of an incoherent superposition. However, in some measurements, we may need to use two parameters.

Einstein's picture of light introduced an additional freedom to the superposition "principle"[1] of the classical EM wave theory: The observed field at a space–time point is not only the result of the superposition among a large number of coherent or incoherent harmonic modes but also the result of the superposition among a large number of coherent or incoherent quantized subfields created from a large number of point-like sub-sources. We have applied Einstein's picture modeling thermal light. In fact, Einstein's picture can be applied to all different types of coherent and incoherent radiations, such as laser light and sunlight.

Assuming a point-like light source at $\mathbf{r}_0 = 0$ and a point-like photodetector at $\mathbf{r}$, the observed radiation propagates from the point-like source to

---

[1] Unlike the quantum theory of light, superposition is not considered a "principle" in the classical electromagnetic wave theory of light; instead, superposition is a mathematical consequence of the Maxwell's wave equation.

the point-like photodetector along the 1-D path $r = |\mathbf{r} - \mathbf{r}_0|$ in vacuum. In Einstein's granularity picture, the observed field $E(\mathbf{r}, t)$ can be represented in the following superposition:

$$E(r, t) = \sum_m \int d\omega \, E_m(\omega) g_m(\omega; r, t)$$

$$\simeq \sum_m \int d\omega \, a_m(\omega) e^{i\varphi_m(\omega)} e^{-i\omega[(t-t_{0m})-r/c]}, \qquad (1.3.12)$$

in terms of the superposition of the subfields and the Fourier modes of the radiation. The expectation value of the measured intensity is thus

$$\langle I(r, t) \rangle = \langle E^*(r, t) E(r, t) \rangle$$

$$= \left\langle \sum_m \int d\omega \, a_m(\omega) e^{-i\varphi_m(\omega)} e^{i\omega(\tau - t_{0m})} \right.$$

$$\times \left. \sum_n \int d\omega' \, a_n(\omega') e^{i\varphi_n(\omega')} e^{-i\omega(\tau - t_{0n})} \right\rangle, \qquad (1.3.13)$$

where $\tau \equiv t - r/c$. Equation (1.3.13) can be written in the following form:

$$\langle I(r, t) \rangle$$

$$= \left\langle \sum_{m=n} \int_{\omega=\omega'} d\omega \, a_m^2(\omega) \right\rangle$$

$$+ \left\langle \sum_{m \neq n} \int_{\omega=\omega'} d\omega \left\{ a_m(\omega) a_n(\omega) e^{-i[\varphi_m(\omega) - \varphi_n(\omega)]} e^{-i\omega(t_{0m} - t_{0n})} \right\} \right\rangle$$

$$+ \left\langle \sum_{m=n} \int_{\omega \neq \omega'} d\omega d\omega' \left\{ a_m(\omega) a_m(\omega') e^{-i[\varphi_m(\omega) - \varphi_m(\omega')]} \right\} e^{i[\omega(\tau - t_{0m}) - \omega'(\tau - t_{0n})]} \right\rangle$$

$$+ \left\langle \sum_{m \neq n} \int_{\omega \neq \omega'} d\omega d\omega' \left\{ a_m(\omega) a_n(\omega') e^{-i[\varphi_m(\omega) - \varphi_n(\omega')]} \right\} e^{i[\omega(\tau - t_{0m}) - \omega'(\tau - t_{0n})]} \right\rangle.$$

$$(1.3.14)$$

Focusing on the coherent and incoherent superposition in terms of the subfields and Fourier modes, we analyze the following four extreme cases:

(I) Incoherent subfields   &   Incoherent Fourier modes;
(II) Coherent subfields   &   Coherent Fourier mode;
(III) Incoherent subfields   &   Coherent Fourier mode;
(IV) Coherent subfields   &   Incoherent Fourier mode;

Equation (1.3.14) indicates two-freedom of superposition in terms of Einstein's subfield, m, and the Fourier mode, ω.

(I) Incoherent subfield and incoherent Fourier modes:

An extreme case of superposition in Eq. (1.3.14) involves incoherently superposed subfields and incoherently superposed Fourier modes. A good example is the classical model of thermal light. Imagine a radiation source in which all its subfields and Fourier modes are created randomly, thus exhibiting random relative phases $\varphi_m(\omega) - \varphi_n(\omega')$. The observed radiation at a space–time point $(\mathbf{r}, t)$ is the result of an incoherent superposition among its incoherent subfields and incoherent Fourier modes. As a result of the incoherent superposition, the only surviving terms in Eq. (1.3.14) are those amplitudes (and their conjugates) that belong to the same sub-source and the same mode of frequency ω, i.e., $m = n$ and $\omega = \omega'$, when taking into account *all possible* values of $\varphi_m(\omega) - \varphi_n(\omega')$, i.e., taking an ensemble average. These surviving terms are known as the "diagonal terms" of the matrix elements

$$\langle\, I(r,t)\,\rangle = \langle E^*(r,t)E(r,t)\rangle = \sum_{m=n} \int_{\omega=\omega'} d\omega\, a_m^2(\omega) \qquad (1.3.15)$$

corresponding to the first term of Eq. (1.3.14). Equation (1.3.15) can also be written in the following form

$$\langle\, I(r,t)\,\rangle = \sum_{m} \int d\omega\, |E_m(\omega)|^2 = \sum_{m} \int d\omega\, I_m(\omega). \qquad (1.3.16)$$

Equation (1.3.16) indicates that the expectation value of the total intensity is the sum of the sub-intensities in terms of the subfields and the Fourier modes. Note that the above evaluation of expectation is only partially completed; we have left out the evaluation for the real-positive amplitudes. The coherence property of light is determined by the relative phases between the sub-radiations.[2] In certain measurements, we may need to take into account *all possible* values of $a_m^2(\omega)$ in terms of the subfields and the Fourier modes.

Examining Eq. (1.3.14), the second, third, and fourth integrals may all contribute to the variation of the intensity δI, if the interference cancelation

---

[2]Note that "sub-radiation" refers a Fourier mode ω within the mth atomic transition. Here, "sub-radiation" is different from "subfield". "Subfield", labeled by m, refers the mth "bundle of ray".

is incomplete. These contributions may vary from time to time causing random fluctuations of the intensity in the neighborhood of its expectation value. The incomplete cancelations are the major contributions to δI compared with the variations of the real-positive amplitude.

(II) Coherent subfields and coherent Fourier modes:

Another extreme case of superposition in Eq. (1.3.14), obviously, involves coherently superposed subfields and coherently superposed Fourier modes. A Q-switched laser pulse is a good example. In this extreme case, all terms of the superposition in Eq. (1.3.14) survive. The expectation function of intensity is thus

$$\langle I(r,t) \rangle = \left\langle \int d\nu \left[\sum_m a_m(\nu) \right] e^{i\nu\tau} \int d\nu' \left[\sum_n a_n(\nu') \right] e^{-i\nu'\tau} \right\rangle$$
$$= \left| \mathcal{F}_\tau \{ A(\nu) \} \right|^2, \tag{1.3.17}$$

where we have defined $A(\nu) = \sum_m a_m(\nu)$ as the total amplitude of the mode of frequency ω. The intensity expectation turns out to be a well-defined pulse in Eq. (1.3.17).

Differing from incoherent subfields, here, we assume a constant phase relationship between the subfields and Fourier modes. The calculated wavepacket, or pulse, is the result of constructive interference or coherent superposition of the subfields and the Fourier modes. In regard to the fluctuations, incoherent field and coherent wavepacket take two extremes. In an incoherent radiation, the subfields may not take all possible relative phases, and in a coherent radiation, the subfields may not take an identical constant phase. The incomplete destructive interference in an incoherent radiation and the incomplete constructive interference in a coherent radiation are the major causes of the intensity fluctuation.

Figure 1.3.1 shows a laser pulse train measured by a fast photodetector. The pulse is a well-defined function of time t deterministically but fluctuates from pulse to pulse in a nondeterministic manner. The expectation value of the intensity can be calculated from Eq. (1.3.14) by taking a constant phase relationship in terms of the subfield and Fourier mode. In a measurement, however, the radiations may not be able to take the same constant phase relationship and thus produce a wavepacket slightly differ from its expectation. Consequently, the measured intensity $I(\mathbf{r},t)$ may fluctuate in the neighborhood of its expectation function $\langle I(\mathbf{r},t) \rangle$ from pulse to pulse randomly in a nondeterministic manor.

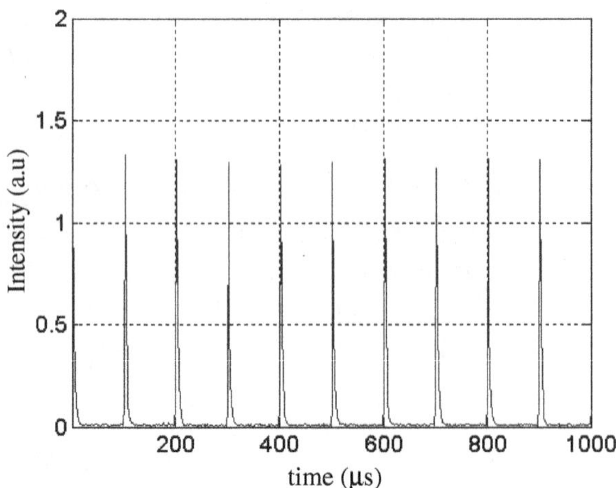

Fig. 1.3.1   Measured laser pulse train. The intensity of each pulse is a function of time t deterministically but fluctuates from pulse to pulse.

(III) Incoherent subfields and coherent Fourier modes.

In the third simplified model, we assume that each sub-source randomly emits its own subfields with random relative phases. The Fourier modes of the mth subfield, however, are coherently excited at time t_{0m}. This model is very close to Einstein's picture of natural light.

Under the assumption of incoherent sub-sources, after the ensemble average by taking into account all possible random phases of the subfields, the only surviving terms in Eq. (1.3.14) are those with $m = n$, which includes the first and third sums:

$$\langle I(r,t) \rangle$$

$$= \int d\omega \, d\omega' \sum_m a_m(\omega) e^{-i\omega t_{0m}} \, a_m(\omega') e^{i\omega' t_{0m}} \, e^{i(\omega-\omega')\tau}$$

$$= \sum_m \left\{ \int d\nu \, [a_m(\nu) e^{-i\nu t_{0m}}] \, e^{i\nu\tau} \right\} \left\{ \int d\nu' \, [a_m(\nu') e^{i\nu' t_{0m}}] \, e^{-i\nu'\tau} \right\}$$

$$= \sum_m \left| \mathcal{F}_{(\tau - t_{0m})} \{ a_m(\nu) \} \right|^2. \tag{1.3.18}$$

The expectation value of intensity, $\langle I(r,t) \rangle$, is the sum of a set of randomly distributed sub-pulses. The mth sub-pulse is the result of a self-interference of the mth wavepacket, or subfield. The expectation value, or expectation function, $\langle I(r,t) \rangle$, is determined by the sum of these well-defined sub-pulses.

It should be emphasized, in the above calculation, that we have "taken into account all possible random phases of the subfields". In fact, a measurement may not be able to "take into account all possible realizations of the field". For instance, if only a limited number of subfields participated the measurement, the result will be different. We examine the following simplified cases:

(1) Single wavepacket

We assume the field is weak enough so only one wavepacket, or one pulse, is involved in the measurement of $\langle I(r,t)\rangle$; we may keep Eq. (1.3.18) as the expectation value, except only one wavepacket left in the sum:

$$\langle I(r,t)\rangle = \left|\mathcal{F}_{(\tau-t_{0m})}\{a(\nu)\}\right|^2. \tag{1.3.19}$$

The expectation value, or expectation function, is the result of a self-interference of a single wavepacket. Obviously, $\langle I(r,t)\rangle$ is a well-defined function of time.

(2) A limited number of wavepackets

If only a few subfields participated the measurement, the measured total field is the result of an incoherent superposition among limited number of wavepackets:

$$E(r,t) = \sum_m e^{-i\omega_0(\tau-t_{0m})}\,\mathcal{F}_{(\tau-t_{0m})}\{a_m(\nu)\}, \tag{1.3.20}$$

for example, the superposition of two wavepackets gives

$$E(r,t) = e^{-i\omega_0(\tau-t_{01})}\,\mathcal{F}_{(\tau-t_{01})}\{a_1(\nu)\} + e^{-i\omega_0(\tau-t_{02})}\,\mathcal{F}_{(\tau-t_{02})}\{a_2(\nu)\}, \tag{1.3.21}$$

where the jth ($j = 1, 2$) wavepacket is centered at time $t = r/c + t_{0j}$, i.e., the jth group of Fourier modes have a common phase at $r = 0$, $t = t_{0j}$. The instantaneous intensity is thus

$$I(r,t) = \left|\mathcal{F}_{(\tau-t_{01})}\{a_1(\nu)\}\right|^2 + \left|\mathcal{F}_{(\tau-t_{02})}\{a_2(\nu)\}\right|^2$$
$$+ 2Re\left[e^{-i\omega_0(t_{01}-t_{02})}\mathcal{F}^*_{(\tau-t_{01})}\{a_1(\nu)\} \cdot \mathcal{F}_{(\tau-t_{02})}\{a_2(\nu)\}\right]. \tag{1.3.22}$$

The first term and second term of Eq. (1.3.22) are the results of the self-interference of wavepacket one and wavepacket two, respectively, and the third term of Eq. (1.3.22) is the interference term between wavepacket one and wavepacket two. The first two terms are deterministic functions of space–time belong to $\langle I(r,t)\rangle$. The contribution of the third term, determined by the relative phase of the wavepacket one and two and their creation times t_{01} and t_{02}, is indeterministic. Its value may vary from pair to pair, from time to time, and from measurement to measurement. We name it intensity fluctuation.

For any number of overlapped–partially overlapped wavepackets, if the measurement cannot take into account all possible random phases, the indeterministic intensity fluctuation $\Delta I(r,t)$ will be observable from the output current of the photodetector:

$$I(r,t) = \sum_m |\mathcal{F}_{(\tau-t_{0m})}\{a_m(\nu)\}|^2$$

$$+ \sum_{m\neq n} [e^{i\omega_0(t_{0n}-t_{0m})}\mathcal{F}^*_{(\tau-t_{0m})}\{a_m(\nu)\} \cdot \mathcal{F}_{(\tau-t_{0n})}\{a_n(\nu)\}]$$

$$= \langle I(r,t)\rangle + \Delta I(r,t). \tag{1.3.23}$$

The second term contributes a random value from time to time or from measurement to measurement in the neighborhood of $\langle I(r,t)\rangle$ as an intensity fluctuation:

$$\Delta I(r,t) = \sum_{m\neq n} [e^{i\omega_0(t_{0n}-t_{0m})}\mathcal{F}^*_{(\tau-t_{0m})}\{a(\nu)\} \cdot \mathcal{F}_{(\tau-t_{0n})}\{a(\nu)\}] \neq 0.$$

$$\tag{1.3.24}$$

(3) A large number of overlapped–partially overlapped wavepackets or pulses

Assume that a large number of overlapped–partially overlapped wavepackets participated the measurement, and the number is large enough for "taking into account all possible random phases of the subfields",

$$\langle I(r,t)\rangle = \sum_m |\mathcal{F}_{(\tau-t_{0m})}\{a_m(\nu)\}|^2$$

$$\cong \int dt_m |\mathcal{F}_{(\tau-t_{0m})}\{a_m(\nu)\}|^2$$

$$= \int d\nu\, d\nu'\, a(\nu)\, a(\nu') \left[\int_\infty dt_m\, e^{i(\nu-\nu')t_m}\right] e^{-i(\nu-\nu')\tau}$$

$$\cong N \int d\nu\, a^2(\nu) \tag{1.3.25}$$

agreeing with the Parseval's theorem. Here, we have assumed that the wavepackets or pulses all have the same function of $\tau - t_{0m}$ and are distributed randomly and continuously so that the summation can be approximated to an integral over t_{0m} from $-\infty$ to $+\infty$. Equation (1.3.25) can be also written in terms of ω:

$$\langle I(r,t)\rangle = \int d\omega \left[N a^2(\omega) \right] = \int d\omega \sum_m a_m^2(\omega) = \int d\omega \sum_m I_m(\omega).$$

It is interesting to find that the expectation value of the intensity of a large number of *randomly* distributed overlapped–partially overlapped wavepackets is the same as that of the radiation with incoherent subfields and incoherent Fourier modes. It should be emphasized that this result is under the approximation of the integral over t_m from $-\infty$ to $+\infty$.[3] This approximation may not apply to certain measurements, for instance, when the integral over t_{0m} is limited within a time window of t_c:

$$\langle I(r,t)\rangle = \int d\nu \, d\nu' \, a(\nu) \, a(\nu') \left[\int_{t_c} dt_m \, e^{i(\nu-\nu')t_{0m}} \right] e^{-i(\nu-\nu')\tau}$$

$$\simeq \int d\nu \, d\nu' \, a(\nu) \, a(\nu') \left[\mathrm{sinc}\, \frac{(\nu-\nu')t_c}{2} \right] e^{i(\nu-\nu')\tau}. \qquad (1.3.26)$$

In this case, $\langle I(r,t)\rangle$ may not be treated as a constant. Note that although the integral over t_{0m} is limited within a time window of t_c, the ensemble average has taken into account all possible random phases of these overlapped–partially overlapped wavepackets, therefore, the sum of Eq. (1.3.24) yields a zero value of ΔI.

Figure 1.3.2 illustrates an experimentally observed intensity, $I(t)$, of an artificial pseudo-thermal light in the photon counting regime by a fast point-like photodetector fixed at a chosen coordinate. The figure reports the counting number, which is proportional to the intensity, at time $t \pm \Delta t/2$ with Δt a chosen short time window in the order of nanosecond. Both $\langle I(t)\rangle$ and $\Delta I(t)$ are observable in this measurement. The intensity fluctuation comes from the incomplete destructive interference between different subfields. In a nanosecond time window, the "weak-light" condition guaranties that only a limited number of subfields contribute to the measurement of $I(\mathbf{r},t)$. It is unlikely that a limited number of subfields could take all possible complex amplitudes. $\Delta I(\mathbf{r},t)$ is therefore observable.

---

[3]Mathematically, integrating over parameter t_m is equivalent to integrating over time t. Physically, the two types of summation correspond to the following two pictures: (1) a measurement at time t deals with a large number of randomly distributed overlapped–partially-overlapped wavepackets and (2) a large number of accumulated measurements happen randomly at different time.

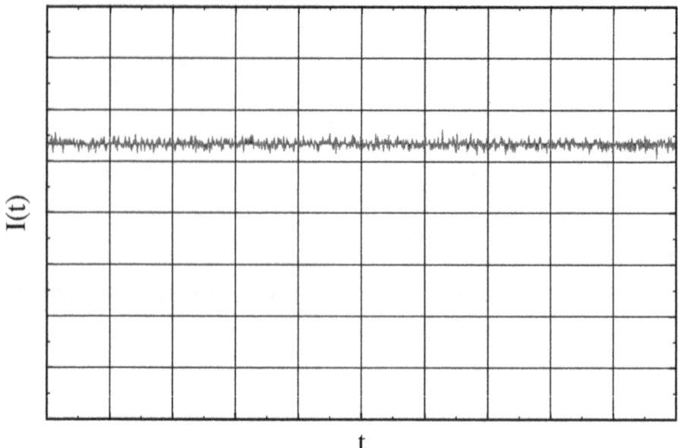

Fig. 1.3.2   A typical measured intensity of pseudo-thermal light by a fast point-like photodetector in nanosecond scale. $I(t)$ fluctuated randomly in the neighborhood of a constant value. The pseudo-thermal light source contains a single-line continuous wave (CW) laser beam and a fast rotating defusing ground glass. The transversely expanded laser beam is scattered by millions of tiny scattering elements of the fast rotating ground glass. These tiny scattering elements play the role of point-like sub-sources. It is reasonable to model this pseudo-thermal light with incoherent subfields and coherent Fourier modes.

(IV) Coherent subfields and incoherent Fourier modes:

We now consider the model of the fourth type of radiation where the sub-sources radiate coherently. The Fourier modes, however, are created randomly with random relative phases. Continuous wave (CW) laser beam with multi-cavity modes is a closer example, except that the cavity modes of the laser are usually discrete rather than continuous. The mechanism of stimulated emission forces all the sub-sources to radiate coherently, however, it may still leave the cavity modes with random relative phases. Since the Fourier modes are assumed incoherent, the only surviving terms in Eq. (1.3.14) are those with $\omega = \omega'$. The expectation value of the intensity is thus

$$\langle I(r,t) \rangle = \int d\omega \sum_{m,n} a_m(\omega)\, a_n(\omega) \tag{1.3.27}$$

which includes to the first and second sums of Eq. (1.3.14). Equation (1.3.27) reflects the incoherent nature of the Fourier modes and the coherent nature of the sub-sources: The measured intensity is the sum of the sub-intensities of the Fourier modes; however, the amplitudes corresponding to the sub-sources add coherently with all cross terms of the sub-sources. Equation (1.3.27) can be further evaluated by assuming $a(\nu)$ to be a well-defined function for all sub-sources:

$$\langle I(r,t) \rangle \cong N^2 \int d\omega\, a^2(\omega) = \int d\nu\, A^2(\omega), \qquad (1.3.28)$$

where $A(\omega) = \sum_m a_m(\omega) \simeq Na(\omega)$.

Comparing with Eq. (1.3.15), we see that $\langle I(r,t) \rangle$ is N times greater. We thus have a "super radiator", a result of coherent superposition or constructive interference in terms of the subfields.

Einstein's picture of light is still in the frame work of Maxwell electromagnetic wave theory, except Einstein introduced the concept of granularity, or quantized subfield to radiation and abandoned the continuum interpretation of Maxwell. Einstein's bundle of ray, light quantum, or subfield is a quantized microscopic realistic entity of electromagnetic wave propagating in space–time. Einstein may have never felt comfortable to this picture. In the last a few years of his life, Einstein posed his students the following question: Suppose a photon of energy $\hbar\omega$ is created from a point source, such as an atomic transition, how big is the photon after propagating one year? This question seems easy to answer for his students. Since the photon is created from a point source, it would propagate in the form of a spherical wave and its wavefront must be a sphere with a diameter of two lightyears after one year propagating. We may find a "consensus" or "common sense" behind the above statement: The total energy of the photon $\hbar\omega$ must be carried by the wavefront of the subfield and thus uniformly distributed on the big sphere. Einstein then asked again the following: Suppose that photon is annihilated by a point-like photon counting detector located on the surface of the big sphere, how long does it take for the energy on the other side of the big sphere to arrive at the detector? Two years? Bohr provided a famous answer to this question: The "wavefunction collapses" instantaneously! Why does the wavefunction need to "collapse"? Bohr did not explain. Nevertheless, Bohr has passed an important message to us: Quantum mechanical picture of photon is different from Einstein's granularity, or EM subfield. Although we still have questions regarding the wave–particle duality of a photon, we have to accept the experimental effect that the energy of the electromagnetic field is quantized in nature. We may have to face the truth that a new theory of radiation with quantization is necessary.

1.4 Field Quantization and the Light Quantum

In blackbody radiation, the atoms on the walls of the cavity box continuously radiate electromagnetic waves into the cavity. In general, there are two fundamental mechanisms governing the physical process of the radiation and determining the physical properties of the radiation field. (1) The Schrödinger equation determines the quantized atomic energy levels. The energy of the created photons, or the frequency of the excited radiations, is determined by the quantized energy levels of the atom, $\hbar\omega = E_2 - E_1$. (2) The excited electromagnetic field must satisfy the Maxwell wave equation which determines the harmonic mode structure and the superposition of the radiation.

In the quantum theory of light, the radiation field is treated as a set of harmonic oscillators. The energy of each mode is quantized in a similar way as that of a harmonic oscillator. To quantize the field, we follow the standard procedure. First, we proceed to link the Hamiltonian of the free electromagnetic field to a set of independent harmonic oscillators. The quantum mechanical results of harmonic oscillators are then adapted to the quantized radiation field. Note, here, that free field means no "sources" or "drains" of the radiation field in the chosen volume of $V = L^3$ that covers the field of interest. The energy of the free field is given by

$$H = \frac{1}{2} \int_V d^3 r \left[\epsilon_0 \mathbf{E}^2(\mathbf{r}, t) + \frac{1}{\mu_0} \mathbf{B}^2(\mathbf{r}, t) \right], \qquad (1.4.1)$$

where V is the total volume of the field of interest. The volume is usually, but not necessarily, treated as a large finite cubic cavity of L^3 to simplify the mathematics. To link the energy of the free field with the Hamiltonian of a set of independent harmonic oscillators, we first need to turn Eq. (1.4.1) into the following form:

$$H = \frac{1}{2} \sum_{\mathbf{k}} [\, p_{\mathbf{k}}^2(t) + \omega^2 q_{\mathbf{k}}^2(t) \,], \qquad (1.4.2)$$

where $q_{\mathbf{k}}(t)$ and $p_{\mathbf{k}}(t)$ are a pair of real canonical variables.

To achieve our goal, we start to construct a classical solution of the vector potential $\mathbf{A}(\mathbf{r}, t)$ of the field. The vector potential of the free electromagnetic field satisfies the EM wave equation

$$\nabla^2 \mathbf{A} - \frac{1}{c^2} \frac{\partial^2 \mathbf{A}}{\partial t^2} = 0 \qquad (1.4.3)$$

with the Coulomb gauge

$$\nabla \cdot \mathbf{A} = 0.$$

The electric and magnetic fields, $\mathbf{E}(\mathbf{r},t)$ and $\mathbf{B}(\mathbf{r},t)$, are thus given in terms of $\mathbf{A}(\mathbf{r},t)$:

$$\mathbf{E}(\mathbf{r},t) = -\frac{\partial}{\partial t}\mathbf{A}(\mathbf{r},t)$$
$$\mathbf{B}(\mathbf{r},t) = \nabla \times \mathbf{A}(\mathbf{r},t). \tag{1.4.4}$$

The most convenient way for analyzing the field structure is to begin with a very large but finite cubic cavity. Applying the periodic boundary condition, we write $\mathbf{A}(\mathbf{r},t)$ in terms of the Fourier expansion of plane-wave modes

$$\mathbf{A}(\mathbf{r},t) = \sum_{\mathbf{k}} \mathbf{A}_{\mathbf{k}}(t)\, e^{i\mathbf{k}\cdot\mathbf{r}} \tag{1.4.5}$$

with

$$\mathbf{A}_{\mathbf{k}}(t) = \mathbf{A}^{*}_{-\mathbf{k}}(t).$$

Where we have introduced a wave vector

$$\mathbf{k} = \left\{ \frac{2\pi l}{L_x},\ \frac{2\pi m}{L_y},\ \frac{2\pi n}{L_z} \right\}, \tag{1.4.6}$$

with

$$l = 0, \pm 1, \pm 2, \ldots$$
$$m = 0, \pm 1, \pm 2, \ldots$$
$$n = 0, \pm 1, \pm 2, \ldots$$

forming a discrete sum of $\mathbf{k}$ in the Fourier expansion of Eq. (1.4.5).

Considering the Coulomb gauge, we have

$$\sum_{\mathbf{k}} \mathbf{k} \cdot \mathbf{A}_{\mathbf{k}}(t)\, e^{i\mathbf{k}\cdot\mathbf{r}} = 0 \tag{1.4.7}$$

for all values of $\mathbf{r}$, which requires that

$$\mathbf{k} \cdot \mathbf{A}_{\mathbf{k}}(t) = 0. \tag{1.4.8}$$

Substituting Eq. (1.4.5) into Eq. (1.4.3), we have

$$\sum_{\mathbf{k}} \left(-k^2 - \frac{1}{c^2} \frac{\partial^2}{\partial t^2} \right) \mathbf{A}_{\mathbf{k}}(t)\, e^{i\mathbf{k}\cdot\mathbf{r}} = 0 \qquad (1.4.9)$$

for all values of $\mathbf{r}$. Thus, we have an equation to determine each of the amplitudes, $\mathbf{A}_{\mathbf{k}}(t)$, of the Fourier expansion

$$\left(\frac{\partial^2}{\partial t^2} + \omega^2 \right) \mathbf{A}_{\mathbf{k}}(t) = 0, \qquad (1.4.10)$$

where $\omega = ck$ is the angular frequency of the mode. A solution of Eq. (1.4.10) is given by

$$\mathbf{A}_{\mathbf{k}}(t) = \hat{\mathbf{e}}_1 \, \mathcal{A}_{\mathbf{k},1}(a^*_{-\mathbf{k},1}\, e^{i\omega t} + a_{\mathbf{k},1}\, e^{-i\omega t}) + \hat{\mathbf{e}}_2 \, \mathcal{A}_{\mathbf{k},2}(a^*_{-\mathbf{k},2}\, e^{i\omega t} + a_{\mathbf{k},2}\, e^{-i\omega t}), \qquad (1.4.11)$$

where $a_{\mathbf{k},s}$ $(a^*_{\mathbf{k},s})$, $s = 1,2$ is the amplitude for the mode $\mathbf{k}$ and the polarization s. $\mathcal{A}_{\mathbf{k},s}$ is determined by the initial conditions of the electromagnetic field. In Eq. (1.4.11), we have assigned two orthogonal polarization, $\hat{\mathbf{e}}_1$ and $\hat{\mathbf{e}}_2$, by considering the transverse condition of Eq. (1.4.8). The unit vectors $\hat{\mathbf{e}}_1$, $\hat{\mathbf{e}}_2$ and the unit vector $\hat{\mathbf{k}}$ in the direction of the wave vector $\mathbf{k}$, together, form a right-hand, orthogonal, Cartesian basis.

To simplify the mathematical expression, we focus on one of the polarizations

$$\mathbf{A}(\mathbf{r}, t) = \sum_{\mathbf{k}} \hat{\mathbf{e}} \, \mathcal{A}_{\mathbf{k}} \left[a_{\mathbf{k}}(t)\, e^{i\mathbf{k}\cdot\mathbf{r}} + a^*_{\mathbf{k}}(t)\, e^{-i\mathbf{k}\cdot\mathbf{r}} \right], \qquad (1.4.12)$$

where $a_{\mathbf{k}}(t) = a_{\mathbf{k}} e^{-i\omega t}$, $a^*_{\mathbf{k}}(t) = a^*_{\mathbf{k}} e^{i\omega t}$. The vector potential $\mathbf{A}(\mathbf{r}, t)$ is written as a superposition of the orthogonal harmonic modes

$$q_{\mathbf{k}}(\mathbf{r}, t) = a_{\mathbf{k}}(t)\, e^{i\mathbf{k}\cdot\mathbf{r}} + a^*_{\mathbf{k}}(t)\, e^{-i\mathbf{k}\cdot\mathbf{r}}. \qquad (1.4.13)$$

These orthogonal (independent) harmonic modes are equivalent to the normal mode of a harmonic oscillator system. In the following, we treat these orthogonal harmonic modes of the electromagnetic field as independent harmonic oscillators and treat $a_{\mathbf{k}}(t)$ $(a^*_{\mathbf{k}}(t))$ as normal mode amplitudes. In terms of the vector potential $\mathbf{A}(\mathbf{r}, t)$ of Eq. (1.4.12), the electric field $\mathbf{E}(\mathbf{r}, t)$ and the magnetic field $\mathbf{B}(\mathbf{r}, t)$ are calculated from Eq. (1.4.4):

$$\mathbf{E}(\mathbf{r}, t) = \sum_{\mathbf{k}} \omega \, \hat{\mathbf{e}} \, \mathcal{A}_{\mathbf{k}} \left[i\, a_{\mathbf{k}}(t)\, e^{i\mathbf{k}\cdot\mathbf{r}} - i\, a^*_{\mathbf{k}}(t)\, e^{-i\mathbf{k}\cdot\mathbf{r}} \right],$$

$$\mathbf{B}(\mathbf{r}, t) = \sum_{\mathbf{k}} (\mathbf{k} \times \hat{\mathbf{e}}) \, \mathcal{A}_{\mathbf{k}} \left[i\, a_{\mathbf{k}}(t)\, e^{i\mathbf{k}\cdot\mathbf{r}} - i\, a^*_{\mathbf{k}}(t)\, e^{-i\mathbf{k}\cdot\mathbf{r}} \right]. \qquad (1.4.14)$$

Now, we examine the energy of the electromagnetic field. We can either treat the electromagnetic field as a system of independent harmonic oscillators or calculate the Hamiltonian of the field from Eq. (1.4.1). Viewing the field as a set of independent harmonic oscillators, the Hamiltonian of the system is readily given in classical mechanics

$$H = 2 \int_V d^3r \sum_{\mathbf{k}} \omega^2 \left| q_{\mathbf{k}}(\mathbf{r}, t) \right|^2 = 2 \sum_{\mathbf{k}} \omega^2 \left| a_{\mathbf{k}}(t) \right|^2, \qquad (1.4.15)$$

where $a_{\mathbf{k}}(t)$ is treated as the normal mode amplitude. To calculate the Hamiltonian from Eq. (1.4.1), substitute Eq. (1.4.14) into Eq. (1.4.1), then the integral gives

$$H = 2 \sum_{\mathbf{k}} |\mathcal{A}_{\mathbf{k}}|^2 \omega^2 \left| a_{\mathbf{k}}(t) \right|^2. \qquad (1.4.16)$$

To be consistent with the result of Eq. (1.4.15), we may take a constant mode distribution of $\mathcal{A}_{\mathbf{k}} = \mathcal{A}_0 = 1/\epsilon_0^{1/2} L^{3/2}$ to "normalize" the energy of the field to $H = 2 \sum_{\mathbf{k}} \omega^2 \left| a_{\mathbf{k}}(t) \right|^2$. It is quite reasonable to treat all the independent and "free" harmonic oscillators equally. Although, in reality, nonconstant mode distribution may have to be taken into consideration, it would not affect the quantization of each independent mode of the electromagnetic field.

Next, we introduce a pair of canonical variables $q_{\mathbf{k}}(t)$ and $p_{\mathbf{k}}(t)$:

$$\begin{aligned} q_{\mathbf{k}}(t) &= a_{\mathbf{k}}(t) + a_{\mathbf{k}}^*(t) \\ p_{\mathbf{k}}(t) &= -i\omega \left[a_{\mathbf{k}}(t) - a_{\mathbf{k}}^*(t) \right]. \end{aligned} \qquad (1.4.17)$$

The Hamiltonian of the field in Eq. (1.4.15) is then rewritten in terms of $q_{\mathbf{k}}(t)$ and $p_{\mathbf{k}}(t)$ as

$$H = \sum_{\mathbf{k}} \frac{1}{2} \left[p_{\mathbf{k}}^2(t) + \omega^2 q_{\mathbf{k}}^2(t) \right] = \sum_{\mathbf{k}} H_{\mathbf{k}}, \qquad (1.4.18)$$

which is recognized as the Hamiltonian of a harmonic oscillator system. Each harmonic oscillator of the system corresponds to a mode of the field specified by wave vector $\mathbf{k}$. In Eq. (1.4.18), $H_{\mathbf{k}}$ indicates the Hamiltonian of the kth harmonic oscillator. The Hamiltonian of a classical harmonic oscillator is allowed to take any nonnegative values because $p_{\mathbf{k}}$ and $q_{\mathbf{k}}$ can take any values in classical theory. It is easy to derive from Eqs. (1.4.17) and (1.4.18) that $q_{\mathbf{k}}(t)$ and $p_{\mathbf{k}}(t)$ satisfy the classical canonic equations of motion:

$$\frac{\partial q_{\mathbf{k}}}{\partial t} = \frac{\partial H}{\partial p_{\mathbf{k}}},$$

$$\frac{\partial p_{\mathbf{k}}}{\partial t} = -\frac{\partial H}{\partial q_{\mathbf{k}}}. \tag{1.4.19}$$

It is also easy to find the results of the following Poisson bracket:

$$\left[q_{\mathbf{k}}(t),\ p_{\mathbf{k}'}(t) \right] = \delta_{\mathbf{kk}'},$$

$$\left[q_{\mathbf{k}}(t),\ q_{\mathbf{k}'}(t) \right] = 0, \tag{1.4.20}$$

$$\left[p_{\mathbf{k}}(t),\ p_{\mathbf{k}'}(t) \right] = 0.$$

Now, we follow the quantum theory of harmonic oscillator to formulate and quantize the electromagnetic field. As a standard procedure, we first replace the canonical variables of $q_{\mathbf{k}}(t)$ and $p_{\mathbf{k}}(t)$ with quantum mechanical canonical conjugate operators $\hat{q}_{\mathbf{k}}(t)$ and $\hat{p}_{\mathbf{k}}(t)$ with the following commutation relations:

$$\left[\hat{q}_{\mathbf{k}}(t),\ \hat{p}_{\mathbf{k}'}(t) \right] = i\hbar\,\delta_{\mathbf{kk}'},$$

$$\left[\hat{q}_{\mathbf{k}}(t),\ \hat{q}_{\mathbf{k}'}(t) \right] = 0, \tag{1.4.21}$$

$$\left[\hat{p}_{\mathbf{k}}(t),\ \hat{p}_{\mathbf{k}'}(t) \right] = 0.$$

The quantum mechanical Hamiltonian of the harmonic oscillator system is thus written in terms of the operators $\hat{q}_{\mathbf{k}}(t)$ and $\hat{p}_{\mathbf{k}}(t)$:

$$\hat{H} = \sum_{\mathbf{k}} \frac{1}{2} \left[\hat{p}_{\mathbf{k}}^2(t) + \omega^2\,\hat{q}_{\mathbf{k}}^2(t) \right] = \sum_{\mathbf{k}} \hat{H}_{\mathbf{k}}, \tag{1.4.22}$$

where $\hat{H}_{\mathbf{k}}$ is the Hamiltonian operator of the $\mathbf{k}$-th harmonic oscillator. Differing from the Hamiltonian, $H_{\mathbf{k}}$, of a classical oscillator in Eq. (1.4.18), which may take any values, $\hat{H}_{\mathbf{k}}$ in Eq. (1.4.22) can only admit quantized energy.

To show the energy quantization of a quantum harmonic oscillator, we introduce a pair of nonHermitian operators to replace the $\hat{p}_{\mathbf{k}}$ and $\hat{q}_{\mathbf{k}}$ in Eq. (1.4.22):

$$\hat{a}_{\mathbf{k}}(t) = \frac{1}{\sqrt{2\hbar\omega}} \left[\omega\,\hat{q}_{\mathbf{k}}(t) + i\hat{p}_{\mathbf{k}}(t) \right],$$

$$\hat{a}_{\mathbf{k}}^{\dagger}(t) = \frac{1}{\sqrt{2\hbar\omega}} \left[\omega\,\hat{q}_{\mathbf{k}}(t) - i\hat{p}_{\mathbf{k}}(t) \right], \tag{1.4.23}$$

and a set of commutation relations derived from Eq. (1.4.21):

$$\begin{aligned}
\left[\hat{a}_{\mathbf{k}}(t),\ \hat{a}_{\mathbf{k}'}^{\dagger}(t)\right] &= \delta_{\mathbf{k}\mathbf{k}'}, \\
\left[\hat{a}_{\mathbf{k}}(t),\ \hat{a}_{\mathbf{k}'}(t)\right] &= 0, \\
\left[\hat{a}_{\mathbf{k}}^{\dagger}(t),\ \hat{a}_{\mathbf{k}'}^{\dagger}(t)\right] &= 0.
\end{aligned} \tag{1.4.24}$$

The Hamiltonian of Eq. (1.4.22) is thus written in terms of $\hat{a}_{\mathbf{k}}(t)$ and $\hat{a}_{\mathbf{k}}^{\dagger}(t)$:

$$\begin{aligned}
\hat{H} &= \frac{1}{2}\sum_{\mathbf{k}} \hbar\omega \left[\hat{a}_{\mathbf{k}}(t)\,\hat{a}_{\mathbf{k}}^{\dagger}(t) + \hat{a}_{\mathbf{k}}^{\dagger}(t)\,\hat{a}_{\mathbf{k}}(t)\right] \\
&= \sum_{\mathbf{k}} \hbar\omega \left[\hat{a}_{\mathbf{k}}^{\dagger}(t)\,\hat{a}_{\mathbf{k}}(t) + \frac{1}{2}\right] \\
&= \sum_{\mathbf{k}} \hbar\omega \left[\hat{n}_{\mathbf{k}} + \frac{1}{2}\right] \\
&= \sum_{\mathbf{k}} \hat{H}_{\mathbf{k}},
\end{aligned} \tag{1.4.25}$$

where we have used the commutation relations of Eq. (1.4.24) and introduced the number operator

$$\hat{n}_{\mathbf{k}} = \hat{a}_{\mathbf{k}}^{\dagger}(t)\,\hat{a}_{\mathbf{k}}(t) \tag{1.4.26}$$

for the mode $\mathbf{k}$. It is clear that $\hat{n}_{\mathbf{k}}$ is Hermitian and is independent of time.

We now show that the Hamiltonian, $\hat{H}_{\mathbf{k}}$, of the $\mathbf{k}$th harmonic oscillator, or mode, in Eq. (1.4.25) only admits quantized energies. To simplify the notation, we drop the subscript $\mathbf{k}$ in the following discussion by considering a single harmonic oscillator, or mode. Assume that $|n\rangle$ is an eigenstate of $\hat{n}$ with eigenvalue of n,

$$\hat{n}\,|n\rangle = n\,|n\rangle, \tag{1.4.27}$$

where n must be real since $\hat{n}$ is Hermitian. It is easy to show that $|n\rangle$ is also an eigenstate of $\hat{H}$ with eigenvalue of $E_n = \hbar\omega(n + 1/2)$:

$$\hat{H}\,|n\rangle = \hbar\omega\left[\hat{n} + \frac{1}{2}\right]|n\rangle = \hbar\omega\left(n + \frac{1}{2}\right)|n\rangle. \tag{1.4.28}$$

We show that the Hamiltonian in Eq. (1.4.28) is quantized because the commutations of Eq. (1.4.24) limit n to integer values. To prove this, we show that $\hat{a}\,|\,n\,\rangle$ is an eigenstate of $\hat{n}$ with eigenvalue of $n-1$ since

$$\hat{n}\,\hat{a}|\,n\,\rangle = \hat{a}\,(n-1)\,|\,n\,\rangle = (n-1)\,\hat{a}\,|\,n\,\rangle, \qquad (1.4.29)$$

where we have used the commutation relations of Eq. (1.4.24). If $\langle\,n\,|\,n\,\rangle = 1$, then $\hat{a}\,|\,n\,\rangle$ has normalization

$$\langle\,n\,|\,\hat{a}^{\dagger}\,\hat{a}\,|\,n\,\rangle = \langle\,n\,|\,\hat{n}\,|\,n\,\rangle = n. \qquad (1.4.30)$$

Therefore,

$$\hat{a}\,|\,n\,\rangle = \sqrt{n}\,|\,n-1\,\rangle, \qquad (1.4.31)$$

where state $|\,n-1\,\rangle$ is normalized $\langle\,n-1\,|\,n-1\,\rangle = 1$.

Similarly, $(\hat{a})^2\,|\,n\,\rangle$ is an eigenstate of $\hat{n}$ with eigenvalue of $n-2$, and

$$(\hat{a})^2\,|\,n\,\rangle = \sqrt{n(n-1)}\,|\,n-2\,\rangle.$$

Repeatedly applying the operator $\hat{a}$, we have all the normalized eigenstates in terms of $|\,n\,\rangle$:

$$(\hat{a})^m\,|\,n\,\rangle = \sqrt{n(n-1)...(n-m+1)}\,|\,n-m\,\rangle. \qquad (1.4.32)$$

Thus, if n is an eigenvalue of the number operator $\hat{n}$, then $n-1$, $n-2$, ... are all eigenvalues of $\hat{n}$. Since

$$n = \langle\,n\,|\,\hat{n}\,|\,n\,\rangle = \langle\,n\,|\,\hat{a}^{\dagger}\,\hat{a}\,|\,n\,\rangle = \langle\,\Phi\,|\,\Phi\,\rangle \geq 0,$$

where $|\,\Phi\,\rangle = \hat{a}\,|\,n\,\rangle$, all the eigenvalues of $\hat{n}$ must be positive. Therefore, n must be an integer and bounded by the lowest eigenvalue of zero. The null eigenstate $|\,0\,\rangle$, or the vacuum state, is defined as

$$\hat{a}\,|\,1\,\rangle = |\,0\,\rangle, \text{ and } \hat{a}\,|\,0\,\rangle = 0. \qquad (1.4.33)$$

The Hermitian operator $\hat{n}$, therefore, takes eigenvalues from a set of integer $0, 1, 2, \ldots$ and is called the number operator for mode $\mathbf{k}$. Regarding Eq. (1.4.32), it is interesting that we can only apply n times the operator $\hat{a}$ to reach the lowest permissible vacuum state $|\,0\,\rangle$ with a nonzero lowest energy $E_0 = \hbar\omega/2$:

$$(\hat{a})^n\,|\,n\,\rangle = \sqrt{n!}\,|\,0\,\rangle. \qquad (1.4.34)$$

So far we have explored two important differences between quantum theory and classical theory regarding the energy of a radiation mode: (1) The energy of a radiation mode is quantized; (2) the lowest energy of a radiation mode, the vacuum state $|\,0\,\rangle$, takes a nonzero value $E_0 = \hbar\omega/2$ which is called the zero-point energy. It should be emphasized that both properties (1) and (2) follow from the commutation relation $[\hat{a}, \hat{a}^\dagger] = 1$. We come back to discuss the physics associated with these two special features later.

Similar to $\hat{a}\,|\,n\,\rangle$, $\hat{a}^\dagger\,|\,n\,\rangle$ is also an eigenstate of $\hat{n}$ with eigenvalue $n+1$ since

$$\hat{n}\,\hat{a}^\dagger|\,n\,\rangle = \hat{a}^\dagger\,(n+1)\,|\,n\,\rangle = (n+1)\,\hat{a}^\dagger\,|\,n\,\rangle, \tag{1.4.35}$$

where we have again used the commutation relations of Eq. (1.4.24). It is easy to show that

$$\hat{a}^\dagger\,|\,n\,\rangle = \sqrt{n+1}\,|\,n+1\,\rangle, \tag{1.4.36}$$

where the state $|\,n+1\,\rangle$ is normalized, $\langle\,n+1\,|\,n+1\,\rangle = 1$. Unlike $\hat{a}$, the eigenvalues of $\hat{a}^\dagger$ are not bounded. By repeatedly applying $\hat{a}^\dagger$ the eigenvalues of $\hat{n}$ go to infinity and therefore are integers from zero to infinity. Using Eq. (1.4.35), we have all the normalized eigenstates in terms of $|\,0\,\rangle$:

$$|\,1\,\rangle = \hat{a}^\dagger\,|\,0\,\rangle,$$

$$|\,2\,\rangle = \frac{1}{\sqrt{2}}\,\hat{a}^\dagger\,|\,1\,\rangle = \frac{1}{\sqrt{2}}\,(\hat{a}^\dagger)^2\,|\,0\,\rangle,$$

$$|\,3\,\rangle = \frac{1}{\sqrt{3}}\,\hat{a}^\dagger\,|\,2\,\rangle = \frac{1}{\sqrt{6}}\,(\hat{a}^\dagger)^3\,|\,0\,\rangle, \tag{1.4.37}$$

$$\vdots$$

$$|\,n\,\rangle = \frac{1}{\sqrt{n!}}\,(\hat{a}^\dagger)^n\,|\,0\,\rangle,$$

where the state $|\,n\,\rangle$ is normalized by means of $\langle\,n\,|\,n\,\rangle = 1$.

By applying $\hat{a}^\dagger$, we have created a set of normalized number states $|\,n_\mathbf{k}\,\rangle$ which are eigenstates of the Hamiltonian $\hat{H}_\mathbf{k}$ for mode $\mathbf{k}$ and for polarization $\hat{\mathbf{e}}_\mathbf{k}$. These energy eigenstates form a complete, orthonormal vector space for characterizing the radiation field. The eigenvalues of the Hamiltonian, $\hat{H}_\mathbf{k}$, are quantized with discrete values of $n\hbar\omega$, or $nh\nu$, in contrast to classical electromagnetic theory where the energy of a radiation mode can have any value. Figure 1.4.1 illustrates the quantized energy levels for a radiation mode of the electromagnetic field.

Connecting $\hat{a}^\dagger$ and $\hat{a}$ with Einstein's concept of an energy bundle, or a photon, $\hat{a}^\dagger$ adds a quantum of energy, $\hbar\omega$, or a photon, to the $\mathbf{k}$th mode of the radiation field. Consequently, it's called the *photon creation operator*. Similarly, $\hat{a}$ subtracts a quantum of energy, $\hbar\omega$, from the $\mathbf{k}$th mode of the radiation field. Therefore, its called the *photon annihilation operator*. The

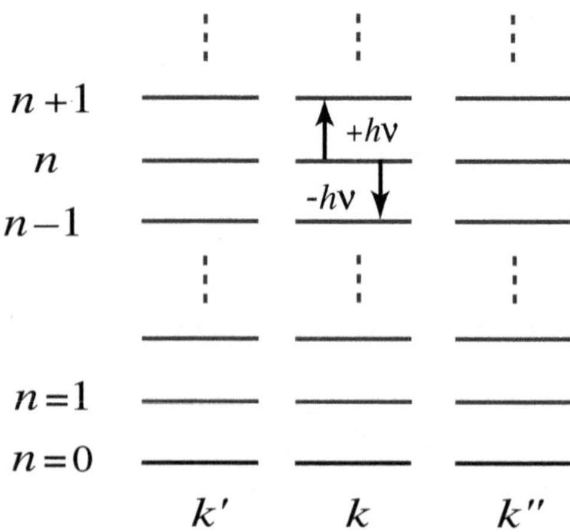

Fig. 1.4.1   Quantized energy levels for a radiation mode of an electromagnetic field. The creation operator $\hat{a}_{\mathbf{k}}^{\dagger}$ adds a quantum of energy $\hbar\omega$, or a photon, to the mode $\mathbf{k}$ to excite the mode to a higher energy level. The annihilation operator $\hat{a}_{\mathbf{k}}$ subtracts the same amount of energy, or annihilates a photon, from the radiation mode. In connection with the concept of photon, the illustrated energy eigenstates of $\hat{H}_{\mathbf{k}}$ are also named as photon number states, or Fock states.

excited state $|\,n\,\rangle$ of a radiation mode contains n quanta of energy along with the zero-point energy, and n is called the *occupation number* or the number of photons. Accordingly, the set of energy eigenstates in Eq. (1.4.37) are called *photon number states*. The physical process of photon creation and annihilation in an atomic transition is addressed later, in the discussions of the quantum state and the field operator.

We are now ready to characterize the quantized radiation field in terms of the energy eigenstates of its Hamiltonian. In general, the state of any radiation field can be described as the superposition of the energy eigenstates,

$$|\,\Psi\,\rangle = \sum_{n} c_n\,|\,n\,\rangle, \qquad (1.4.38)$$

where c_n is the complex amplitude associated with the photon number state, $|\,n\,\rangle$. In principle, we are not restricted to using the energy eigenstates to characterize the state vector of a radiation mode. However, it has advantageous when calculating the time evolution of the radiation field. It is readily shown that an eigenstate of the Hamiltonian $\hat{H}$ simply develops a phase factor $e^{-iE_n(t-t_0)/\hbar}$, from time t_0 to t,

$$| \Psi_n(t) \rangle = e^{-iE_n(t-t_0)/\hbar} | \Psi_n(t_0) \rangle, \qquad (1.4.39)$$

since

$$i\hbar \frac{\partial}{\partial t} | \Psi_n \rangle = \hat{H} | \Psi_n \rangle = E_n | \Psi_n \rangle.$$

With the radiation field quantized, we have simultaneously turned the electromagnetic field into operators in terms of the creation operator and the annihilation operator:

$$\hat{\mathbf{A}}(\mathbf{r}, t) = \sum_{\mathbf{k}} \hat{\mathbf{e}}_k \, \mathcal{A}_{\mathbf{k}} \left[\hat{a}_{\mathbf{k}}(t) \, e^{i\mathbf{k}\cdot\mathbf{r}} + \hat{a}_{\mathbf{k}}^{\dagger}(t) \, e^{-i\mathbf{k}\cdot\mathbf{r}} \right],$$

$$\hat{\mathbf{E}}(\mathbf{r}, t) = \sum_{\mathbf{k}} \hat{\mathbf{e}}_k \, \mathcal{E}_{\mathbf{k}} \left[i \, \hat{a}_{\mathbf{k}}(t) \, e^{i\mathbf{k}\cdot\mathbf{r}} - i \, \hat{a}_{\mathbf{k}}^{\dagger}(t) \, e^{-i\mathbf{k}\cdot\mathbf{r}} \right], \qquad (1.4.40)$$

$$\hat{\mathbf{B}}(\mathbf{r}, t) = \sum_{\mathbf{k}} (\mathbf{k} \times \hat{\mathbf{e}}_k) \, \mathcal{B}_{\mathbf{k}} \left[i \, \hat{a}_{\mathbf{k}}(t) \, e^{i\mathbf{k}\cdot\mathbf{r}} - i \, \hat{a}_{\mathbf{k}}^{\dagger}(t) \, e^{-i\mathbf{k}\cdot\mathbf{r}} \right].$$

We usually write the field operator into two parts:

$$\hat{\mathbf{E}}(\mathbf{r}, t) = \sum_{\mathbf{k}} \hat{\mathbf{e}}_k \, i \, \mathcal{E}_{\mathbf{k}} \, \hat{a}_{\mathbf{k}}(t) \, e^{i\mathbf{k}\cdot\mathbf{r}} + \sum_{\mathbf{k}} \hat{\mathbf{e}}_k (-i) \mathcal{E}_{\mathbf{k}} \, \hat{a}_{\mathbf{k}}^{\dagger}(t) \, e^{-i\mathbf{k}\cdot\mathbf{r}}$$

$$= \hat{\mathbf{E}}^{(+)}(\mathbf{r}, t) + \hat{\mathbf{E}}^{(-)}(\mathbf{r}, t), \qquad (1.4.41)$$

where $\hat{\mathbf{E}}^{(+)}(\mathbf{r}, t)$ and $\hat{\mathbf{E}}^{(-)}(\mathbf{r}, t)$ contain the annihilation and creation operators, respectively.

Note that in the above analysis we have simplified the mathematics by assuming a large but finite cubic cavity and by applying the plane-wave solutions. In certain experimental situations, the plane-wave solutions may need to be generalized for different sizes, shapes, and nature of boundaries. We then replace the plane-wave solutions with the generalized spatial mode functions $\mathbf{u}_k(\mathbf{r})$ in Eq. (1.4.5):

$$\mathbf{A}(\mathbf{r}, t) = A_0 \sum_{\mathbf{k}} \mathbf{A}_{\mathbf{k}}(t) \, \mathbf{u}_k(\mathbf{r}), \qquad (1.4.42)$$

corresponding to the excited frequency ω_k. $\mathbf{u}_k(\mathbf{r})$ must be a solution to the Helmholtz equation

$$\nabla^2 \mathbf{u}_k + \frac{\omega_k^2}{c^2} \mathbf{u}_k = 0 \qquad (1.4.43)$$

subject to the corresponding boundary conditions. We further require that the mode functions form a complete orthonormal set

$$\int dr^3\, \mathbf{u}_l^*(\mathbf{r})\, \mathbf{u}_m(\mathbf{r}) = \delta_{lm}, \qquad (1.4.44)$$

with the condition

$$\nabla \cdot \mathbf{u}_k(\mathbf{r}) = 0.$$

Using the above three equations and the boundary conditions, we find a suitable set of mode functions for the field quantization. The nonplane-wave solutions are used in later chapters.

Bibliography

Dirac P.A.M., *Proc. R. Soc. A* **114**, 243 (1927).
Einstein A., *Ann. Phys.* **17**, 132 (1905).
Hertz H.R., *Ann. Phys.* **267**, 421, 983 (1987).
Jeans J.H., *Phil. Mag.* **10**, 91 (1905).
Landau L.D. and E.M Lifshitz, *Quantum Mechanics*, Pergamon Press, Addison-Wesley, 1958..
Plank M., *Verh. Dtsch. Phys. Ges.* **2**, 202, 237 (1900).
Rayleigh J.W.S., *Phil. Mag.* **49**, 539 (1900); *Nature* **72**, 54, 243 (1905).
Wien W., *Ann. Phys.* **58**, 662 (1896).

Chapter 2

Quantum Theory of Light:
The State of Quantized
Radiation Field and Photon

In this chapter, we study the quantum state of quantized field. To analyze and to solve a problem of quantum optics, we need to know the state of the quantized field, or the state of a photon, or the state of a group of photons. We first study the photon number state and coherent state. Photon number state representation, especially the single photon state representation, and the coherent state representation are widely used to specify the state of a radiation field. The concept of density operator or density matrix is introduced in the process of calculating the expectation value of a quantum observable. We then introduce and distinguish the pure state and mixed state. The quantum states of composite system, especially the state of two-photon field, are studied by introducing a 2-D Hilbert space constructed as the direct or tensor product of the Hilbert spaces of two subsystems. A simple model for the creation of single photon state and multi-photon state is given with details. The concepts of product state, entangled state, and mixed state are introduced and distinguished in the last section of this chapter.

2.1 Photon Number State of Radiation Field

By applying the creation operators, we have generated a set of normalized photon number states which are eigenstates of the Hamiltonian in Eq. (1.4.37) for mode $\mathbf{k}$ and polarization $\hat{\mathbf{e}}_{\mathbf{k}}$. These energy eigenstates form a complete and orthonormal vector space for characterizing the radiation field, or for characterizing the state of light quanta. We now proceed to generalize our discussion to multi-mode radiation. We show that the generalized Fock state of Eq. (2.1.1) is an eigenstate of the multi-mode Hamiltonian in Eq. (1.4.25). In addition, we introduce a useful operator, namely, the total photon number operator, $\hat{n}$, in Eq. (2.1.3).

The generalized multi-mode photon number state, or Fock state of the radiation field is written as

$$| \Psi \rangle = \prod_{\mathbf{k},s} | n_{\mathbf{k},s} \rangle = \prod_{\mathbf{k},s} \frac{1}{\sqrt{n_{\mathbf{k},s}!}} (\hat{a}^{\dagger}_{\mathbf{k},s})^n | 0 \rangle, \qquad (2.1.1)$$

where $\mathbf{k}$ and s indicate the mode and the polarization. Equation (2.1.1) defines the state for all and for each radiation mode and polarization.

It follows immediately that the multi-mode Fock state of Eq. (2.1.1) is an eigenstate of the single-mode photon number operator $\hat{n}_{\mathbf{k},s}$:

$$\hat{n}_{\mathbf{k},s} \left(\prod_{\mathbf{k},s} | n_{\mathbf{k},s} \rangle \right) = n_{\mathbf{k},s} \prod_{\mathbf{k},s} | n_{\mathbf{k},s} \rangle, \qquad (2.1.2)$$

where $n_{\mathbf{k},s}$ is the occupation number of the radiation mode, $\mathbf{k}$, and polarization, s. We now define a total photon number operator, $\hat{n}$, by summing $\hat{n}_{\mathbf{k},s}$ over all the radiation modes, $\mathbf{k}$, and all the polarizations, s:

$$\hat{n} = \sum_{\mathbf{k},s} \hat{n}_{\mathbf{k},s}. \qquad (2.1.3)$$

It is easy to find that the multi-mode Fock state of Eq. (2.1.1) is an eigenstate of the total photon number operator:

$$\hat{n} \left(\prod_{\mathbf{k},s} | n_{\mathbf{k},s} \rangle \right) = \left(\sum_{\mathbf{k},s} n_{\mathbf{k},s} \right) \prod_{\mathbf{k},s} | n_{\mathbf{k},s} \rangle = n \prod_{\mathbf{k},s} | n_{\mathbf{k},s} \rangle, \qquad (2.1.4)$$

where $n = \sum n_{\mathbf{k},s}$ is the total number of photons in the radiation field.

It is easy to show that the multi-mode Fock state of Eq. (2.1.1) is an eigenstate of the total Hamiltonian (multi-mode) in Eq. (1.4.25):

$$\hat{H} | \Psi \rangle = \left[\sum_{\mathbf{k},s} \hbar\omega_{\mathbf{k},s} \left(\hat{n}_{\mathbf{k},s} + \frac{1}{2} \right) \right] \left(\prod_{\mathbf{k},s} | n_{\mathbf{k},s} \rangle \right)$$

$$= \left[\sum_{\mathbf{k},s} \hbar\omega_{\mathbf{k},s} \left(n_{\mathbf{k},s} + \frac{1}{2} \right) \right] \left(\prod_{\mathbf{k},s} | n_{\mathbf{k},s} \rangle \right)$$

$$= E | \Psi \rangle, \qquad (2.1.5)$$

where $E = \sum \hbar\omega_{\mathbf{k},s} (n_{\mathbf{k},s} + \frac{1}{2})$ is the total energy of the radiation.

For convenience, we define a short-hand notation $\{n\}$ and rewrite the multi-mode Fock state of Eq. (2.1.1) as

$$|\{n\}\rangle \equiv \prod_{\mathbf{k},s} |n_{\mathbf{k},s}\rangle. \qquad (2.1.6)$$

Equation (2.1.4) is then rewritten, in short-hand form, as

$$\hat{n}|\{n\}\rangle = n|\{n\}\rangle, \qquad (2.1.7)$$

which simply indicates that the multi-mode Fock state $|\{n\}\rangle$ is an eigenstate of the total number operator, $\hat{n}$, defined in Eq. (2.1.3) with an eigenvalue of n, which represents the total number of photons in the radiation field.

The eigenvalue, n, of the total number operator, is commonly used to name the Fock state, either single-mode or multi-mode, as an n-photon state. For instance, the state

$$|\Psi\rangle = \hat{a}^{\dagger}_{\mathbf{k},s}|0\rangle = |0\rangle \ldots |0\rangle |1_{\mathbf{k},s}\rangle |0\rangle \ldots |0\rangle$$

$$= |0,\ldots 0, 1_{\mathbf{k},s}, 0,\ldots 0\rangle \qquad (2.1.8)$$

is a *single photon* state ($n = 1$) of the mode $\mathbf{k}$ and polarization s. Shorten the notation, it is common to write the above single photon state as $|\Psi\rangle = |1_{\mathbf{k},s}\rangle$, we should not forget the ground states of the other modes. Referring Fig. 1.4.1, $|\Psi\rangle = |1_{\mathbf{k},s}\rangle$ means the occupation number of the mode $(\mathbf{k}, s)$ takes a value of one, while all the other modes stay on their ground state (vacuum) with occupation number of zeros. The states

$$|\Psi\rangle = \hat{a}^{\dagger}_{\mathbf{k},s}\hat{a}^{\dagger}_{\mathbf{k}',s'}|0\rangle = |0,\ldots 0, 1_{\mathbf{k},s}, 0,\ldots 1_{\mathbf{k}',s'}, 0,\ldots 0\rangle \qquad (2.1.9)$$

and

$$|\Psi\rangle = \frac{1}{\sqrt{2}}(\hat{a}^{\dagger}_{\mathbf{k},s})^2|0\rangle = |0,\ldots 0, 2_{\mathbf{k},s}, 0,\ldots 0\rangle \qquad (2.1.10)$$

are both *two-photon states* ($n = 2$) but characterize different physics. The state in Eq. (2.1.9) indicates the excitation of two different radiation modes with occupation numbers $n_{\mathbf{k},s} = 1$ and $n_{\mathbf{k}',s'} = 1$. The state in Eq. (2.1.10) indicates the excitation of a radiation mode with occupation number $n_{\mathbf{k},s} = 2$. Such states in Eqs. (2.1.8), (2.1.9), and (2.1.10) are all known as Fock states, or photon number states. Again, shorten the notation, it is also common to write the two-photon state as $|\Psi\rangle = |1_{\mathbf{k},s}, 1_{\mathbf{k}',s'}\rangle$ and $|\Psi\rangle = |2_{\mathbf{k},s}\rangle$; one should not ignore the vacuum states of the other modes.

In connection with the concept of photon, Eq. (2.1.8) indicates the creation of one light quantum, or a photon, of energy $\hbar\omega_{\mathbf{k},s}$ to the radiation mode

$\mathbf{k}$ and polarization s. Equation (2.1.9) means the creation of two photons with energies $\hbar\omega_{\mathbf{k},s}$ and $\hbar\omega_{\mathbf{k}',s'}$, respectively, to the radiation mode $\mathbf{k}$ — polarization s and the radiation mode $\mathbf{k}'$-polarization s'. Equation (2.1.10) corresponds to the creation of two photons both with energy of $\hbar\omega_{\mathbf{k},s}$ to the same radiation mode $\mathbf{k}$ and polarization s. Equations (2.1.8), (2.1.9), and (2.1.10) are also popular to be considered as the state of a photon with wavevector $\mathbf{k}$ and polarization s, the state of two photons, one with wavevector $\mathbf{k}$ and polarization s and the other one with wavevector $\mathbf{k}'$ and polarization s', and the state of two photons both with wavevector $\mathbf{k}$ and polarization s. The two photons in state Eq. (2.1.10) are "indistinguishable" or "degenerate".

Since Fock states form a complete, orthonormal vector space, following the rules of quantum mechanics, we are able to express any state vector of a radiation field in terms of a superposition of Fock states, or photon number states,

$$|\Psi\rangle = \sum_j c_j |\Psi_j\rangle = \sum_{\{n\}} f(\{n\}) |\{n\}\rangle, \qquad (2.1.11)$$

where $f(\{n\})$ is the normalized probability amplitude to find the radiation field in the state $|\{n\}\rangle$.

As an example, Eq. (2.1.12) represents a single photon state:

$$|\Psi\rangle = \sum_j c_j |\Psi_j\rangle = \sum_{\mathbf{k},s} f(\mathbf{k}, s)\, \hat{a}_{\mathbf{k},s}^\dagger |0\rangle$$

$$= \sum_s \int d\mathbf{k}\, f(\mathbf{k}, s)\, \hat{a}_{\mathbf{k},s}^\dagger |0\rangle, \qquad (2.1.12)$$

where $c_j = f(\mathbf{k}, s) = \langle\Psi_j|\Psi\rangle$ is the normalized probability amplitude for the radiation field to be in the Fock state $|\Psi_j\rangle = \hat{a}_{\mathbf{k},s}^\dagger |0\rangle = |1_{\mathbf{k},s}\rangle$. In the last line of Eq. (2.1.12), we have treated mode $\mathbf{k}$ continuously as usual; we have also formally written the two polarizations, $s = 1, 2$, for each mode $\mathbf{k}$ as a superposition. If the two polarizations have a constant phase relationship, the two polarizations superpose coherently resulting in a pure polarization state, or a vector in the polarization space. However, if the two polarizations are completely independent with random relative phases, the two polarizations superpose incoherently resulting in a mixed polarization state. In this case, we label the state as a mixed state:

$$|\tilde{\Psi}\rangle = \sum_j c_j \,|\,\Psi_j\,\rangle = \sum_{\mathbf{k},s} f(\mathbf{k},s)\,\hat{a}_{\mathbf{k},s}^{\dagger}\,|\,0\,\rangle$$

$$= \sum_s \int d\mathbf{k}\, f(\mathbf{k},s)\,\hat{a}_{\mathbf{k},s}^{\dagger}\,|\,0\,\rangle. \tag{2.1.13}$$

To distinguish the mixed state from pure state, we label pure states as $|\Psi\rangle$ and mixed states as $|\tilde{\Psi}\rangle$. Even if we decide to simplify the mathematics to one polarization by ignoring the sum of s, the state can still be a pure state or a mixed state, depending on the nature of the superposition of $|\Psi_j\rangle = \hat{a}_{\mathbf{k}}^{\dagger}|0\rangle$: A coherent superposition gives a pure state, and an incoherent superposition gives a mixed state. In this section we focus on pure states and leave mixed states until we introduce the concept of density operator.

The single photon state of Eq. (2.1.12) can be generated from a two-level atomic transition. Figure 2.1.1 illustrates a simple model of the process. A two-level atom is placed inside a large but finite size cubic cavity. The upper energy level E_2 has a finite width of $\Delta E_2 \neq 0$. A single photon, or wavepacket, is created from the atomic transition between energy levels E_2 and E_1. The created photon may excite any or excite all permissible states $|\Psi_j\rangle = \hat{a}_{\mathbf{k},s}^{\dagger}|0\rangle$ with probability amplitude $f(\mathbf{k},s) = |f(\mathbf{k},s)|e^{i\varphi(\mathbf{k},s)}$. The space–time behavior of each permissible radiation mode is determined by the boundary condition in solving the Maxwell equations. The probability amplitude distribution is determined by

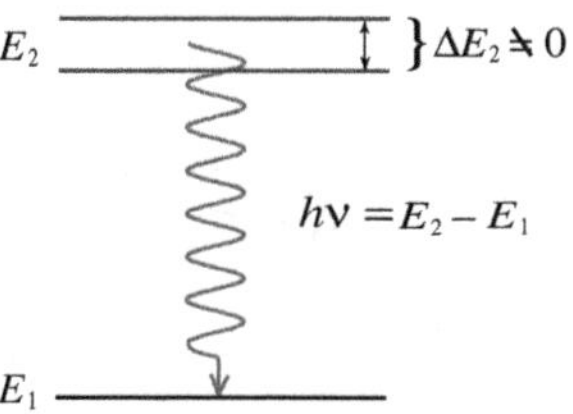

Fig. 2.1.1   Schematic model of a possible atomic transition that generates the single photon state, or wavepacket, characterized by Eq. (2.1.12). A two-level atom is placed inside a large but finite sized cubic cavity. The upper energy level E_2 has a finite width of $\Delta E_2 \neq 0$. The emitted photon, or the radiation field excited by this atomic transition, may be in any, or in all permissible states, of $|\Psi_j\rangle = \hat{a}_{\mathbf{k},s}^{\dagger}|0\rangle$ with probability amplitude $f(\mathbf{k},s)$.

the property of the atomic transition and the property of the cavity, mainly the energy uncertainty ΔE_2. Since all superposed states are generated from the same atomic transition, it is reasonable to consider $\varphi(\mathbf{k}, s) = \varphi_0$. Mathematically, the coherent superposition of a special set of Fock states $|\Psi_j\rangle = |0, \ldots 0, 1_{\mathbf{k},s}, 0, \ldots 0\rangle$ of total occupation number $n = 1$ represents a vector in Hilbert space. Physically, it characters the state of a radiation field that is excited by an atomic transition.

2.2   Coherent State of Radiation Field

Coherent state is defined as the eigenstate of the annihilation operator

$$\hat{a}|\alpha\rangle = \alpha|\alpha\rangle. \tag{2.2.1}$$

It is convenient to write the eigenvalue α in terms of an amplitude and a phase

$$\alpha = a\,e^{i\varphi} \quad \text{with} \quad a = |\alpha|.$$

It is straightforward to obtain an expression of $|\alpha\rangle$ in terms of the number state $|n\rangle$ from Eq. (2.2.1):

$$|\alpha\rangle = e^{-|\alpha|^2/2} \sum_{n=0}^{\infty} \frac{\alpha^n}{\sqrt{n!}} |n\rangle, \tag{2.2.2}$$

by applying

$$\hat{a}|n\rangle = \sqrt{n}\,|n-1\rangle.$$

Since

$$|n\rangle = \frac{1}{\sqrt{n!}}(\hat{a}^{\dagger})^n |0\rangle,$$

Equation (2.2.2) can be written as

$$|\alpha\rangle = e^{-|\alpha|^2/2}\,e^{\alpha\hat{a}^{\dagger}}|0\rangle. \tag{2.2.3}$$

Equation (2.2.3) is thus formally rewritten as

$$|\alpha\rangle = e^{-|\alpha|^2/2}\,e^{\alpha\hat{a}^{\dagger}}e^{-\alpha^*\hat{a}}|0\rangle = e^{\alpha\hat{a}^{\dagger}-\alpha^*\hat{a}}|0\rangle, \tag{2.2.4}$$

since

$$e^{-\alpha^*\hat{a}}|0\rangle = 0,$$

and

$$e^{\hat{A}+\hat{B}} = e^{-[\hat{A},\hat{B}]/2}\, e^{\hat{A}} e^{\hat{B}},$$

where $\hat{A}$ and $\hat{B}$ are any operators that satisfy

$$[[\hat{A},\hat{B}],\hat{A}] = [[\hat{A},\hat{B}],\hat{B}] = 0.$$

Here, we have taken $\hat{A} = \alpha\hat{a}^{\dagger}$ and $\hat{B} = -\alpha^{*}\hat{a}$. By writing $|\alpha\rangle$ in the form of Eq. (2.2.4), we thus introduced a unitary operator, namely, the displacement operator

$$\hat{D}(\alpha) = e^{\alpha\hat{a}^{\dagger} - \alpha^{*}\hat{a}} = e^{-|\alpha|^{2}/2}\, e^{\alpha\hat{a}^{\dagger}} e^{-\alpha^{*}\hat{a}} = e^{|\alpha|^{2}/2}\, e^{-\alpha^{*}\hat{a}} e^{\alpha\hat{a}^{\dagger}},$$

for the purpose of expressing $|\alpha\rangle$ as a unitary transformation of $|0\rangle$:

$$|\alpha\rangle = \hat{D}(\alpha)\,|0\rangle. \tag{2.2.5}$$

Equation (2.2.5) is useful for further theoretical concerns and discussions about coherent state.

Some important properties of the coherent states are listed as follows:

(1) The mean number of photons in the coherent state $|\alpha\rangle$ is $|\alpha|^{2}$,

$$\bar{n} = \langle\hat{n}\rangle = \langle\alpha|\,\hat{n}\,|\alpha\rangle = |\alpha|^{2}, \tag{2.2.6}$$

and the probability of finding n photons in $|\alpha\rangle$ is given by a Poisson distribution:

$$P(n) = \langle n|\alpha\rangle\langle\alpha|n\rangle = \frac{\bar{n}^{n}}{n!}e^{-\bar{n}}. \tag{2.2.7}$$

The photon number distribution of laser light is close to the Poisson distribution. The Poisson distribution is useful for simulating single photon state. For instance, reducing the intensity of a laser field to mean number $\bar{n} = 0.01$, we find that the field has $\sim99\%$ probability to be in its ground state $|0\rangle$; the probabilities of being in $|1\rangle$ and in $|2\rangle$ are $\sim1\%$ and $\sim0.01\%$, respectively. For a photon counting-type experiment with 1% error, the contributions from $|2\rangle$ and higher numbers are ignorable. Taking first-order approximation, the state of the measured field can be approximated as

$$|\Psi\rangle \simeq |0\rangle + \epsilon\,|1\rangle + \epsilon^{2}\dots,$$

where $\epsilon \ll 1$. The Poisson distribution is also useful for simulating number state of $n \gg 1$. For instance, achieving mean number of $\bar{n} = 10^{6}$, its photon number distribution has a peak at $n = 10^{6}$ with $\Delta n \sim 10^{3}$ which is three-orders smaller than $\bar{n}$.

(2) The vector set of coherent states $|\alpha\rangle$ is a complete set in Hilbert space

$$\frac{1}{\pi}\int d^2\alpha\,|\alpha\rangle\langle\alpha| = 1. \tag{2.2.8}$$

The completeness relation of Eq. (2.2.8) indicates that the coherent states can be used as a vector basis for expanding any quantum state of radiation. To prove this, we substitute the number state expansion of the coherent state into the integral, obtaining

$$\int d^2\alpha\,|\alpha\rangle\langle\alpha| = \int d^2\alpha\,e^{-|\alpha|^2}\frac{\alpha^n\,(\alpha^*)^n}{n!}\sum_n |n\rangle\langle n| = \pi\sum_n |n\rangle\langle n|, \tag{2.2.9}$$

where we have applied the result of the following integral[1]

$$\int d^2\alpha\,e^{-|\alpha|^2}\frac{\alpha^n\,(\alpha^*)^n}{n!} = \pi.$$

Since the Fock states $|n\rangle$ form a complete orthonormal set of vector basis, the sum in Eq. (2.2.9) gives the unit operator.

(3) Two coherent states $|\alpha\rangle$ and $|\alpha'\rangle$ are not orthogonal unless $|\alpha - \alpha'| \gg 1$,

$$\langle\alpha|\alpha'\rangle = e^{-\frac{1}{2}(|\alpha|^2 - 2\alpha^*\alpha' + |\alpha'|^2)}, \tag{2.2.10}$$

and

$$|\langle\alpha|\alpha'\rangle|^2 = e^{-|\alpha - \alpha'|^2}. \tag{2.2.11}$$

This means that the vector basis of coherent states is in principle over-complete. However, if the experimental condition achieves $|\alpha - \alpha'| \gg 1$, the set of coherent states can be approximated as orthonormal. Under this approximation, similar to number states, coherent states can be used as an orthonormal vector basis for characterizing radiation field.

(4) Similar to photon number state, we can define a multi-mode coherent state, which is written as a product of single-mode coherent state $|\alpha_{\mathbf{k},s}\rangle$:

$$|\{\alpha\}\rangle = \prod_{\mathbf{k},s} |\alpha_{\mathbf{k},s}\rangle, \tag{2.2.12}$$

---

[1]

$$\int d^2\alpha\,e^{-|\alpha|^2}\,\alpha^m\,(\alpha^*)^n = \int_0^\infty d|\alpha|\,e^{-|\alpha|^2}|\alpha|^{m+n+1}\int_0^{2\pi} d\varphi\,e^{i(m-n)\varphi} = \pi n!\,\delta_{mn}.$$

where $\mathbf{k}$ and s indicate the wavenumber vector and the polarization of the mode, respectively. $|\{\alpha\}\rangle$ is an eigenstate of the annihilation operator with an eigenvalue $\alpha_{\mathbf{k},s}$,

$$\hat{a}_{\mathbf{k},s}\,|\{\alpha\}\rangle = \alpha_{\mathbf{k},s}|\{\alpha\}\rangle. \qquad (2.2.13)$$

We may also define a multi-mode annihilation operator

$$\hat{a} = \sum_{\mathbf{k},s} \hat{a}_{\mathbf{k},s}. \qquad (2.2.14)$$

It is easy to find that the multi-mode coherent state defined in Eq. (2.2.12) is an eigenstate of the multi-mode annihilation operator with eigenvalue $\sum_{\mathbf{k},s}\alpha_{\mathbf{k},s}$:

$$\hat{a}\,|\{\alpha\}\rangle = \left(\sum_{\mathbf{k},s}\hat{a}_{\mathbf{k},s}\right)\prod_{\mathbf{k},s}|\alpha_{\mathbf{k},s}\rangle = \left(\sum_{\mathbf{k},s}\alpha_{\mathbf{k},s}\right)|\{\alpha\}\rangle. \qquad (2.2.15)$$

For example, similar to the number state, we may construct a two-mode coherent state

$$|\Psi\rangle = |0,\ldots,0,\alpha_{\mathbf{k},s},0,\ldots\alpha_{\mathbf{k}',s'},0,\ldots0\rangle, \qquad (2.2.16)$$

which is an eigenstate of $\left(\sum_{\mathbf{k},s}\hat{a}_{\mathbf{k},s}\right)$ with eigenvalue

$$\left(\sum_{\mathbf{k},s}\hat{a}_{\mathbf{k},s}\right)|0,\ldots,0,\alpha_{\mathbf{k},s},0,\ldots\alpha_{\mathbf{k}',s'},0,\ldots\rangle$$

$$= (\alpha_{\mathbf{k},s} + \alpha_{\mathbf{k}',s'})|0,\ldots,0,\alpha_{\mathbf{k},s},0,\ldots\alpha_{\mathbf{k}',s'},0,\ldots\rangle.$$

The eigenvalue is a complex number that has not only a real and positive amplitude but also a phase,

$$\alpha_{\mathbf{k},s} + \alpha_{\mathbf{k}',s'} = |\alpha_{\mathbf{k},s}|e^{i\varphi_{\mathbf{k},s}} + |\alpha_{\mathbf{k}',s'}|e^{i\varphi_{\mathbf{k}',s'}}. \qquad (2.2.17)$$

If the relative phase of the two amplitudes remains constant, the state is a pure state. It will be labeled $|\Psi\rangle$. If the relative phase of the two amplitudes is random, the state is a mixed state. It will be labeled $|\tilde{\Psi}\rangle$.

2.3 Density Operator, Density Matrix and the Expectation Value of an Observable

For a pure state, the expectation value of an observable (operator) is easily calculated:

$$\langle \hat{A} \rangle = \langle \Psi | \hat{A} | \Psi \rangle = \langle \hat{A} \rangle_{\mathrm{QM}} \equiv \langle \hat{A} \rangle_{|\Psi\rangle}. \tag{2.3.1}$$

Unfortunately, we are not always dealing with pure states. In certain measurements, the radiation field is in mixed state and can only be described statistically. In this case, a density operator will be defined to characterize the field. To calculate the expectation value of an observable for mixed state, in addition to the quantum average, a statistical ensemble average is necessary. Density operators help to specify the mixed nature of the radiation field and calculate the expectation value of an observable for mixed states.

Assume a measurement in which we have to deal with radiation in a mixed state. The field can be described by N randomly distributed state vectors $|\Psi_j\rangle$ in Hilbert space, where N can be any integer from two to infinity, $N = 2 \ldots \infty$. The best knowledge we can have about this radiation is that it has a certain probability P_j of being in the state $|\Psi_j\rangle$. The expectation value of an observable $\hat{A}$ is calculated as

$$\langle \hat{A} \rangle = \sum_j P_j \langle \Psi_j | \hat{A} | \Psi_j \rangle = \langle \langle \hat{A} \rangle_{\mathrm{QM}} \rangle_{\mathrm{En}}. \tag{2.3.2}$$

In Eq. (2.3.2), we first calculate the expectation values of $\hat{A}$ for each individual state $|\Psi_j\rangle$ and than evaluate the statistical mean for the ensemble. Applying completeness $\sum_n |n\rangle\langle n| = 1$,

$$\langle \langle \hat{A} \rangle_{\mathrm{QM}} \rangle_{\mathrm{Ensamble}} = \sum_n \sum_j P_j \langle \Psi_j | \hat{A} | n \rangle \langle n | \Psi_j \rangle$$

$$= \sum_n \sum_j P_j \langle n | \Psi_j \rangle \langle \Psi_j | \hat{A} | n \rangle$$

$$= \sum_n \langle n | \hat{\rho} \hat{A} | n \rangle$$

$$= tr\, \hat{\rho} \hat{A} \equiv \langle \hat{A} \rangle_{\hat{\rho}}, \tag{2.3.3}$$

where we have introduced the density operator

$$\hat{\rho} = \sum_j P_j \, |\Psi_j\rangle\langle\Psi_j| \tag{2.3.4}$$

to specify the radiation field in mixed state. By tracing operators $\hat{\rho}\hat{A}$, we obtain the expectation value of an observable from a radiation field in mixed state.

Following the same procedure, we can formally define a density operator for pure state too

$$\hat{\rho} \equiv |\Psi\rangle\langle\Psi|. \tag{2.3.5}$$

The expectation value of an observable (operator) can be calculated accordingly:

$$\langle\hat{A}\rangle = tr\,\hat{\rho}\hat{A} = \langle\Psi|\hat{\rho}\hat{A}\,|\Psi\rangle = \langle\Psi|\Psi\rangle\langle\Psi|\hat{A}|\Psi\rangle = \langle\Psi|\hat{A}|\Psi\rangle. \tag{2.3.6}$$

Since any state $|\Psi_j\rangle$ can be expanded in terms of a chosen vector basis, the density operator $\hat{\rho}$ is also able to be expanded in terms of a chosen vector basis, such as number states and coherent states. For the density operator of a pure state,

$$\hat{\rho} = \sum_m \sum_n |m\rangle\langle m|\Psi\rangle\langle\Psi|n\rangle\langle n| \equiv \sum_m \sum_n \rho_{mn} \, |m\rangle\langle n|, \tag{2.3.7}$$

which defines the density matrix for pure state

$$\rho_{mn} \equiv \langle m|\Psi\rangle\langle\Psi|n\rangle.$$

For the density operator of a mixed state,

$$\hat{\rho} = \sum_m \sum_n \sum_j P_j \, |m\rangle\langle m|\Psi_j\rangle\langle\Psi_j|n\rangle\langle n| \equiv \sum_m \sum_n \rho_{mn} \, |m\rangle\langle n|, \tag{2.3.8}$$

which defines the density matrix for mixed state

$$\rho_{mn} \equiv P_j \, \langle m|\Psi_j\rangle\langle\Psi_j|n\rangle.$$

Some useful properties of the density matrix are listed in the following:

(1) $\rho_{mm} \geq 0$

The diagonal elements of the density matrix are real, and nonnegative. This follows immediately from $\rho_{mm} = \langle m|\Psi_j\rangle\langle\Psi_j|m\rangle = |\langle m|\Psi_j\rangle|^2 \geq 0$.

(2) $\sum_m \rho_{mm} = 1$ or $tr\,\hat{\rho} = 1$

The density matrix is defined in terms of the normalized states, $|\Psi_j\rangle$, and the probability distribution of the field, P_j. This makes the interpretation of the diagonal elements as probabilities valid. It is obvious that $\rho_{mm} < 1$ for a mixed state. Pure state can be treated as a special case of mixed state in which the field is in the state $|\Psi\rangle$ with certainty ($P = 1$).

(3) $\rho_{mn}^* = \rho_{nm}$ or $\hat{\rho}^\dagger = \hat{\rho}$

This means that the density matrix is Hermitian. Since the density matrix is Hermitian it can be diagonalized by a unitary transformation. In this book, we choose photon number state as the basis. Photon number states are the eigen states of the Hamiltonian. The time evolution of photon number states involves phase propagation only in optical measurements. This property is useful for the propagation of the state or the operator. We show in the following a few sections that the density matrix of thermal field only have diagonal elements in the basis of photon number states.

(4) $\hat{\rho}^2 = \hat{\rho}$ for pure state

This property is easy to prove and is useful for distinguishing pure states from mixed states.

We have mentioned earlier that the single photon state of Eq. (2.1.12) can be a pure state or a mixed state. The mixed state is usually represented by the density operator

$$\hat{\rho} = \sum_{\mathbf{k},s} |f(\mathbf{k},s)|^2 \, \hat{a}_{\mathbf{k},s}^\dagger \, |0\rangle\langle 0|\hat{a}_{\mathbf{k},s}. \tag{2.3.9}$$

In certain measurements, we may simplify Eq. (2.3.9) in 1-D:

$$\hat{\rho} = \int d\omega \, |f(\omega)|^2 \, \hat{a}^\dagger(\omega) \, |0\rangle\langle 0| \, \hat{a}(\omega), \tag{2.3.10}$$

where we have also ignored the polarization of the field. In this representation, although we have chosen the same single photon Fock state of $n = 1$

$$\hat{a}^\dagger(\omega)|0\rangle = |0, \ldots 0, 1(\omega), 0, \ldots 0\rangle \tag{2.3.11}$$

as the vector basis of the Hilbert space that was used for specifying a pure state, for the mixed state of Eq. (2.3.10), we do not have enough knowledge on the incoherent superposition between these Fock states. These single photon state vectors are randomly distributed in Hilbert space with random phases. The state of the radiation can only be described statistically. What we know is the probability, $|f(\omega)|^2$, for the radiation field to be in a Fock state of Eq. (2.3.11).

In the following, we introduce another freedom to the single photon state in addition to the wave vector $\mathbf{k}$ and polarization s. Suppose a photodetection device measured $\langle \hat{n} \rangle = 1$ for a weak natural light source. The light source contains a large number of randomly radiated atoms. A photoelectron event can be triggered by a photon that is created from any spontaneous atomic transitions in the source, and we know each atomic transition, such as the mth atomic transition, produces a single photon state of Eq. (2.1.12),

$$|\Psi_m\rangle = \sum_s \int d\mathbf{k}\, f_m(\mathbf{k}, s)\, \hat{a}_m^\dagger(\mathbf{k}, s)\,|\,0\,\rangle, \qquad (2.3.12)$$

as a coherent superposition of Fock states $n = 1$. To simplify the mathematics, we write Eq. (2.3.12) in 1-D:

$$|\Psi_m\rangle = \int d\omega\, f_m(\omega)\, \hat{a}_m^\dagger(\omega)\,|\,0\,\rangle. \qquad (2.3.13)$$

We have enough reason to consider $|\Psi_m\rangle$ a pure state and treat it as a vector in the space of single photon Fock state of Eq. (2.3.11). Mathematically, it is easy to prove $|\Psi_m\rangle$, $m = 1$ to $\sim\infty$, form a complete orthogonal vector basis of a new Hilbert space. The following superposition in the new vector space may represent the state of a radiation excited from these atomic transitions:

$$|\Psi\rangle = \sum_m c_m |\Psi_m\rangle = \sum_m c_m \int d\omega\, f_m(\omega)\, \hat{a}_m^\dagger(\omega)\,|\,0\,\rangle, \qquad (2.3.14)$$

where $c_m = |c_m| e^{i\varphi_m}$ is the normalized complex probability amplitude of observing a photon from the mth atomic transition, or finding the radiation field in state of Eq. (2.3.13). When $\varphi_m =$ constant, the coherent superposition forms a vector in the space of $|\Psi_m\rangle$, $m = 1$ to $\sim\infty$, representing a pure state of the radiation. However, when $\varphi_m =$ random number, the incoherent superposition represents a set of randomly distributed vectors in the space of $|\Psi_m\rangle$, which is usually written as a density operator:

$$\hat{\rho} = \sum_m P_m |\Psi_m\rangle\langle\Psi_m|$$

$$= \sum_m P_m \int\!\!\int d\omega\, d\omega'\, f_m(\omega) f_m^*(\omega')\, \hat{a}_m^\dagger(\omega)|0\rangle\langle 0|\hat{a}_m(\omega'), \qquad (2.3.15)$$

where $P_m = |c_m|^2$ is the probability of observing a photon from the mth atomic transition, or finding the radiation field in the state of Eq. (2.3.13). Equation (2.3.15) indicates a mixed state of the radiation. In general, we usually use Eq. (2.3.14) representing a radiation field in pure state and use Eq. (2.3.15) representing a radiation field in mixed state.

The mean photon number $\langle \hat{n} \rangle$ for the pure state of Eq. (2.3.14) is straightforward,

$$\langle \hat{n} \rangle_{|\Psi\rangle} = \langle \Psi | \, \hat{n} \, | \Psi \rangle = 1. \tag{2.3.16}$$

The mean photon number $\langle \hat{n} \rangle$ for the mixed state of Eq. (2.3.15) is calculated as follows:

$$\langle \hat{n} \rangle_{\hat{\rho}} = tr\, \hat{\rho}\hat{A} = \sum_m P_m \langle \Psi_m | \hat{n} | \Psi_m \rangle = 1. \tag{2.3.17}$$

There should be no surprise that both Eqs. (2.3.14) and (2.3.15) represent single photon states.

To calculate the expectation value of an observable for mixed state, in fact, it is unnecessary to start from tracing operators. For example, we may write a mixed single photon state formally as an incoherent superposition of $|\Psi_m\rangle$:

$$|\tilde{\Psi}\rangle = \sum_m c_m |\Psi_m\rangle = \sum_m |c_m|\, e^{i\varphi_m} \int d\omega\, f_m(\omega)\, \hat{a}_m^\dagger(\omega) \, |0\,\rangle, \tag{2.3.18}$$

where $c_m = |c_m| e^{i\varphi_m}$ is the normalized complex probability amplitude for the system to be in the state $|\Psi_m\rangle$ and φ_m takes a random value. We may name $|\tilde{\Psi}\rangle$ a pseudovector. The expectation value of an arbitrary observable $\langle \hat{A} \rangle$ is calculated accordingly:

$$\langle \langle \hat{A} \rangle_{\mathrm{QM}} \rangle_{\mathrm{En}} = \left\langle \sum_{m,n} |c_m|\, |c_n|\, e^{-i(\varphi_m - \varphi_n)} \langle \Psi_m | \hat{A} | \Psi_n \rangle \right\rangle_{\mathrm{En}}$$

$$= \sum_m P_m \, \langle \Psi_m | \hat{A} | \Psi_m \rangle, \tag{2.3.19}$$

where $P_m = |c_m|^2$, and we have completed the ensemble average by taking into account all possible values of the relative phases of $(\varphi_m - \varphi_n)$. The ensemble average sums all the $m \neq n$ terms to zero, the only surviving terms are the $m = n$ terms. Replacing the arbitrary operator $\hat{A}$ with the photon number operator $\hat{n}$, we obtain

$$\langle\langle\hat{n}\rangle_{\mathrm{QM}}\rangle_{\mathrm{En}} = \left\langle \sum_{m,n} |c_m|\,|c_n|\,e^{-i(\varphi_m-\varphi_n)}\langle\Psi_m|\hat{n}|\Psi_n\rangle \right\rangle_{\mathrm{En}}$$

$$= \sum_m P_m \langle\Psi_m|\hat{n}|\Psi_m\rangle. \tag{2.3.20}$$

2.4   Pure State and Mixed State

The concepts of pure states and mixed states include two important aspects of physics: the state of the individual quantum and the state of the measured ensemble.

(1) In terms of the individual quantum system, a pure state represents a vector in Hilbert space. If the state of a quantum can be described by a vector, the state of the quantum is said to be *pure*. On the contrary, if the state of a quantum cannot be described as a vector but rather a mixture of randomly distributed vectors, the quantum is said to be in a *mixed state*.

(2) In terms of the measured ensemble, if the state of all the measured quanta of the ensemble can be described by the same vector, the state of the ensemble, or the measured systems of quanta, is said to be *homogeneous* or in a statistically *pure state*. If the measured ensemble cannot be described by the same vector, or the measured ensemble has to be described by several, or by many randomly distributed vectors, the ensemble is referred to as a statistical *mixture*.

The second aspect is easy to understand. We usually have more questions about aspect (1). In the following, we give three simple examples: (I) polarization measurement of a single photon state, (II) Schrödinger cat, and (III) general measurement of single photon state. These three simple examples might be helpful for exploring the important physics behind the concept, especially for aspect (1).

Example (I) Polarization measurement of single photon state:

Consider the setup of a polarization measurement of Fig. 2.4.1, which consists of a far-field single photon radiation source of frequency ω, a uniaxial crystal, and two photon counting detectors. The point-like photodetectors D_1 and D_2 are used for measuring the ordinary ray (o-ray) and the extraordinary ray (e-ray) of the uniaxial crystal, respectively.

After a large number of measurements, we find the following: (1) no joint-photodetection event happens, i.e., D_1 and D_2 are never triggered

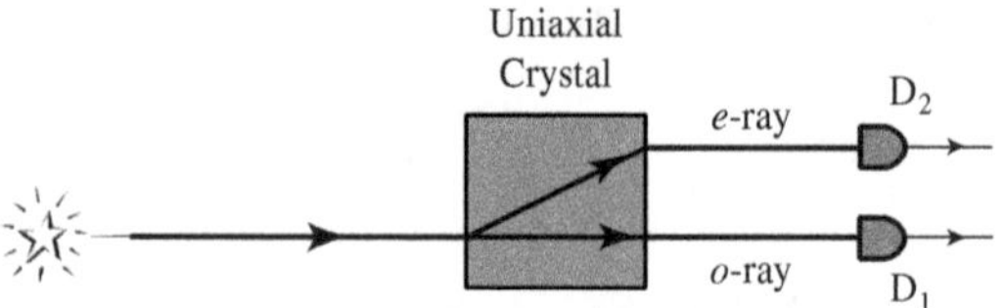

Fig. 2.4.1   Polarization measurement of a photon. The measurement system consists of a uniaxial crystal and two photon counting detectors, D_1 and D_2. D_1 and D_2 are used for measuring the o-ray and the e-ray of the uniaxial crystal, respectively. The o-ray (e-ray) is polarized horizontally (vertically) before any "rotation" of the crystal-detector system.

simultaneously, and (2) the ratio between the single photon counting rates of D_1 and D_2 is 1, i.e., 50% of the photodetection events are registered by D_1 and another 50% of the photodetection events are recorded by D_2. What is the state of the radiation field? From observation (1), the state can be approximated as a single photon state. Observation (2) suggests at least three possible states:

(1) **Possibility one**: The state can be a linearly polarized single photon state with polarization either $45°$ or $-45°$ relative to the polarization of the ordinary ray,

$$|\Psi_{45°}\rangle = \frac{1}{\sqrt{2}} [\hat{a}^\dagger(\omega, \hat{o})|0\rangle + \hat{a}^\dagger(\omega, \hat{e})\,|0\rangle],$$

$$|\Psi_{-45°}\rangle = \frac{1}{\sqrt{2}} [\hat{a}^\dagger(\omega, \hat{o})|0\rangle - \hat{a}^\dagger(\omega, \hat{e})\,|0\rangle], \qquad (2.4.1)$$

where the unit vectors $\hat{o}$ and $\hat{e}$ indicate the polarization of the o-ray and the e-ray, respectively, and the $\pm$ sign indicates that the field is either polarized at $45°$ or $-45°$ relative to the unit vector $\hat{o}$. If the measurements only focus on polarization, we may rewrite Eq. (2.4.1) in the following simplified form:

$$|\Psi_{45°}\rangle = |\hat{o}\rangle\langle\hat{o}|\Psi_{45°}\rangle + |\hat{e}\rangle\langle\hat{e}|\Psi_{45°}\rangle = \frac{1}{\sqrt{2}} [\,|\hat{o}\rangle + |\hat{e}\rangle\,],$$

$$|\Psi_{-45°}\rangle = |\hat{o}\rangle\langle\hat{o}|\Psi_{-45°}\rangle + |\hat{e}\rangle\langle\hat{e}|\Psi_{-45°}\rangle = \frac{1}{\sqrt{2}} [\,|\hat{o}\rangle - |\hat{e}\rangle\,], \qquad (2.4.2)$$

where $|\hat{o}\rangle$ and $|\hat{e}\rangle$ form a complete orthonormal 2-D vector basis. We regard the linear polarization states of $|\Psi_{45°}\rangle$ and $|\Psi_{-45°}\rangle$ as coherent *superposition* of $|\hat{o}\rangle$ and $|\hat{e}\rangle$. The state is a pure state. In the chosen 2-D vector space of

$|\hat{o}\rangle$ and $|\hat{e}\rangle$, the density matrix of the density operator $\hat{\rho} = |\Psi\rangle\langle\Psi|$ is thus

$$\hat{\rho}_{45°} = \begin{pmatrix} 1/2 & 1/2 \\ 1/2 & 1/2 \end{pmatrix}, \quad \hat{\rho}_{-45°} = \begin{pmatrix} 1/2 & -1/2 \\ -1/2 & 1/2 \end{pmatrix}. \tag{2.4.3}$$

(2) **Possibility two**: The state can be a circularly polarized single photon state with right-hand polarization or left-hand polarization:

$$|\Psi_R\rangle = \frac{1}{\sqrt{2}}\left[\hat{a}^\dagger(\omega, \hat{o})|0\rangle + i\,\hat{a}^\dagger(\omega, \hat{e})\,|0\rangle\right],$$

$$|\Psi_L\rangle = \frac{1}{\sqrt{2}}\left[-\hat{a}^\dagger(\omega, \hat{o})|0\rangle + i\,\hat{a}^\dagger(\omega, \hat{e})\,|0\rangle\right], \tag{2.4.4}$$

where, again, the unit vectors $\hat{o}$ and $\hat{e}$ indicate the polarization of the o-ray and the e-ray, respectively. If the measurements only focus on polarization, we may rewrite Eq. (2.4.4) in the following simplified form:

$$|\Psi_R\rangle = |\hat{o}\rangle\langle\hat{o}|\Psi_R\rangle + |\hat{e}\rangle\langle\hat{e}|\Psi_R\rangle = \frac{1}{\sqrt{2}}\left[\,|\hat{o}\rangle + i\,|\hat{e}\rangle\,\right],$$

$$|\Psi_L\rangle = |\hat{o}\rangle\langle\hat{o}|\Psi_L\rangle + |\hat{e}\rangle\langle\hat{e}|\Psi_L\rangle = \frac{1}{\sqrt{2}}\left[-\,|\hat{o}\rangle + i\,|\hat{e}\rangle\,\right]. \tag{2.4.5}$$

We regard the circular polarization states of $|\Psi_R\rangle$ and $|\Psi_L\rangle$ as coherent *superposition* of $|\hat{o}\rangle$ and $|\hat{e}\rangle$. The state is a pure state. In the chosen 2-D vector space of $|\hat{o}\rangle$ and $|\hat{e}\rangle$, the density matrix of the density operator is written as

$$\hat{\rho}_R = \begin{pmatrix} 1/2 & -i/2 \\ i/2 & 1/2 \end{pmatrix}, \quad \hat{\rho}_L = \begin{pmatrix} 1/2 & i/2 \\ -i/2 & 1/2 \end{pmatrix}. \tag{2.4.6}$$

(3) **Possibility three**: The radiation field could be unpolarized. We regard a mixed polarization state in terms of the ordinary ray (o-ray) and the extraordinary ray (e-ray) of the uniaxial crystal:

$$\hat{\rho} = \frac{1}{2}\left[\hat{a}^\dagger(\omega, \hat{o})\,|0\rangle\langle 0\,|\hat{a}(\omega, \hat{o}) + \hat{a}^\dagger(\omega, \hat{e})\,|0\rangle\langle 0\,|\hat{a}(\omega, \hat{e})\,\right], \tag{2.4.7}$$

or in the simplified form

$$\hat{\rho} = P_o\,|\hat{o}\rangle\langle\hat{o}| + P_e\,|\hat{e}\rangle\langle\hat{e}| = \frac{1}{2}\left[\,|\hat{o}\rangle\langle\hat{o}| + |\hat{e}\rangle\langle\hat{e}|\,\right], \tag{2.4.8}$$

where P_o and P_e are the probabilities for the system to be in the polarization states $|\hat{o}\rangle$ and $|\hat{e}\rangle$, respectively. In the chosen vector space in which $|\hat{o}\rangle$ and

$|\hat{e}\rangle$ form a complete orthonormal 2-D vector basis, the matrix form of the density operator is written as

$$\hat{\rho} = \begin{pmatrix} 1/2 & 0 \\ 0 & 1/2 \end{pmatrix}. \tag{2.4.9}$$

It is not difficult to see the above three states (possibilities) yield the same results for the experimental setup of Fig. 2.4.1, i.e., 50%–50% chance to register an photon at D_1 or D_2.

For a more general measurement, we introduce projection operators $\hat{\theta} = |\hat{\theta}\rangle\langle\hat{\theta}|$ and $\hat{\theta}_\perp = |\hat{\theta}_\perp\rangle\langle\hat{\theta}_\perp|$, corresponding to a polarization analyzer oriented at angle θ relative to vector $\hat{o}$. Note that $\hat{o}$ is defined by the uniaxial crystal of Fig. 2.4.1 without any "rotation", see Fig. 2.4.2, in the natural right-hand coordinate system. In the chosen vector space of $|\hat{o}\rangle$ and $|\hat{e}\rangle$, the projection operators $\hat{\theta}$ and $\hat{\theta}_\perp$ are represented by the following matrixes:

$$\hat{\theta} = \begin{pmatrix} \cos^2\theta & \cos\theta\sin\theta \\ \sin\theta\cos\theta & \sin^2\theta \end{pmatrix} \tag{2.4.10}$$

and

$$\hat{\theta}_\perp = \begin{pmatrix} \sin^2\theta & -\sin\theta\cos\theta \\ -\cos\theta\sin\theta & \cos^2\theta \end{pmatrix}, \tag{2.4.11}$$

respectively, where the matrix elements $\langle\hat{i}|\hat{\theta}\rangle\langle\hat{\theta}|\hat{j}\rangle = \cos(\hat{i},\hat{\theta})\cos(\hat{\theta},\hat{j})$ and $\langle\hat{i}|\hat{\theta}_\perp\rangle\langle\hat{\theta}_\perp|\hat{j}\rangle = \cos(\hat{i},\hat{\theta}_\perp)\cos(\hat{\theta}_\perp,\hat{j})$, with $\hat{i},\hat{j} = \hat{o},\hat{e}$, and θ is the angle between the unit vectors $\hat{\theta}$ and $\hat{o}$.

In general, the expectation value of the polarization measurement is calculated by taking the trace of the matrix $\hat{\rho}\hat{\theta}$ and $\hat{\rho}\hat{\theta}_\perp$:

$$\langle\hat{\theta}\rangle = tr\,\hat{\rho}\hat{\theta}, \quad \langle\hat{\theta}_\perp\rangle = tr\,\hat{\rho}\hat{\theta}_\perp. \tag{2.4.12}$$

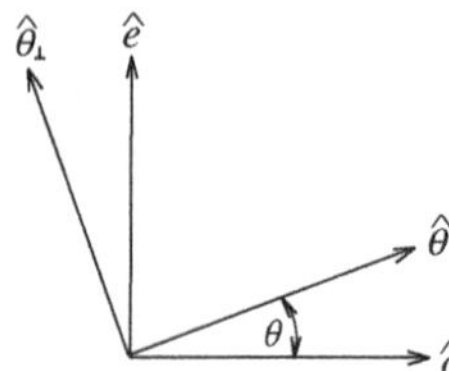

Fig. 2.4.2   A polarization projection operator $\hat{\theta} = |\hat{\theta}\rangle\langle\hat{\theta}|$ corresponds to a polarization analyzer oriented at angle θ relative to vector $\hat{o}$. Note that $\hat{o}$ is defined by the uniaxial crystal in Fig. 2.4.1 without any "rotation".

For pure states,

$$\langle \hat{\theta} \rangle = \langle \Psi | \hat{\theta} \rangle \langle \hat{\theta} | \Psi \rangle = |\langle \hat{\theta} | \Psi \rangle|^2,$$

$$\langle \hat{\theta}_\perp \rangle = \langle \Psi | \hat{\theta}_\perp \rangle \langle \hat{\theta}_\perp | \Psi \rangle = |\langle \hat{\theta}_\perp | \Psi \rangle|^2. \tag{2.4.13}$$

For mixed states,

$$\langle \hat{\theta} \rangle = \sum_j P_j \langle \Psi_j | \hat{\theta} \rangle \langle \hat{\theta} | \Psi_j \rangle = \sum_j P_j |\langle \hat{\theta} | \Psi_j \rangle|^2,$$

$$\langle \hat{\theta}_\perp \rangle = \sum_j P_j \langle \Psi_j | \hat{\theta}_\perp \rangle \langle \hat{\theta}_\perp | \Psi_j \rangle = \sum_j P_j |\langle \hat{\theta}_\perp | \Psi_j \rangle|^2. \tag{2.4.14}$$

In the measurement of Fig. 2.4.1, the projection angle θ of the polarization analyzer (the uniaxial crystal) is chosen to be either $\theta = 0$ (o-ray) ($\theta_\perp = 90°$) or $\theta = 90°$ (e-ray) ($\theta_\perp = 180°$). Definitely, these measurements cannot distinguish the above three different states.

A question naturally arises: Can we distinguish the pure states Eqs. (2.4.1) and (2.4.4) from the mixed state of Eq. (2.4.7), in principle and in practice? For this kind of simple polarization measurement, the answer is positive. For example, one may consider to rotate the polarization analyzer to either $45°$ or $-45°$ directions. If D_1 (D_2) shows a maximum (minimum) counting rate at $45°$ and a minimum (maximum) counting rate at $-45°$, the field is polarized at $45°$:

$$| \Psi \rangle = \hat{a}^\dagger(\omega, 45°) | 0 \rangle \equiv | \widehat{45}^\circ \rangle.$$

If D_1 (D_2) has a minimum (maximum) counting rate at $45°$ and a maximum (minimum) counting rate at $-45°$, the field is polarized at $-45°$:

$$| \Psi_\perp \rangle = \hat{a}^\dagger(\omega, -45°) | 0 \rangle \equiv | -\widehat{45}^\circ \rangle.$$

The $45°$ polarization and $-45°$ polarization are distinguished from the mixed state and the circular polarization. In the mixed state and circular polarization, the count rates of D_1 and D_2 remain constants.

In general, a linearly polarized single photon state of $|\hat{\theta}\rangle$ or $|\hat{\theta}\rangle_\perp$ can be easily distinguished by rotating the polarization analyzer to the directions of θ and $\theta_\perp$. In fact, the states of

$$| \Psi \rangle = \hat{a}^\dagger(\omega, \theta) | 0 \rangle \equiv | \hat{\theta} \rangle \tag{2.4.15}$$

and

$$| \Psi_\perp \rangle = \hat{a}^\dagger(\omega, \theta_\perp) | 0 \rangle \equiv | \hat{\theta}_\perp \rangle \tag{2.4.16}$$

form a new complete orthogonal 2-D vector basis for characterizing any polarization state. For example, the states of horizontal polarization, which was originally labeled as $|\hat{o}\rangle$, and vertical polarization, which was originally labeled as $|\hat{e}\rangle$, can be written as the superposition of $|\hat{\theta}\rangle$ and $|\hat{\theta}_\perp\rangle$:

$$|\Psi_o\rangle = \frac{1}{\sqrt{2}}\left[\hat{a}^\dagger(\omega,\theta)|0\rangle + \hat{a}^\dagger(\omega,\theta_\perp)|0\rangle\right], \qquad (2.4.17)$$

$$|\Psi_e\rangle = \frac{1}{\sqrt{2}}\left[\hat{a}^\dagger(\omega,\theta)|0\rangle - \hat{a}^\dagger(\omega,\theta_\perp)|0\rangle\right],$$

or in the simplified form

$$|\hat{o}\rangle = \frac{1}{\sqrt{2}}\left[|\hat{\theta}\rangle + |\hat{\theta}_\perp\rangle\right], \qquad (2.4.18)$$

$$|\hat{e}\rangle = \frac{1}{\sqrt{2}}\left[|\hat{\theta}\rangle - |\hat{\theta}_\perp\rangle\right].$$

The orthogonal basis vectors θ and $\theta_\perp$ can be used to characterize a mixed state of polarization too:

$$\hat{\rho} = P(\theta)|\hat{\theta}\rangle\langle\hat{\theta}| + P(\theta_\perp)|\hat{\theta}_\perp\rangle\langle\hat{\theta}_\perp|, \qquad (2.4.19)$$

indicating a 50–50% chance to be in the states $|\hat{\theta}\rangle$ and $|\hat{\theta}_\perp\rangle$.

Similarly, if we build a polarization analyzer to distinguish the single photon state of spin $+1$ from that of spin -1, the circularly polarized single photon state with right-hand polarization or left-hand polarization can be distinguished easily. In fact, the states of

$$|\Psi\rangle = \hat{a}^\dagger(\omega,+1)|0\rangle \equiv |\hat{R}\rangle \qquad (2.4.20)$$

and

$$|\Psi_\perp\rangle = \hat{a}^\dagger(\omega,-1)|0\rangle \equiv |\hat{L}\rangle \qquad (2.4.21)$$

form another new complete orthogonal 2-D vector basis for characterizing any polarization state. For example, the states of horizontal polarization, which was originally labeled as $|\hat{o}\rangle$, and vertical polarization, which was originally labeled as $|\hat{e}\rangle$, can be written as the coherent superposition of

$|\hat{R}\rangle$ and $|\hat{L}\rangle$:

$$|\Psi_o\rangle = \frac{1}{\sqrt{2}}\,[\,\hat{a}^\dagger(\omega,+1)\,|\,0\,\rangle - \hat{a}^\dagger(\omega,-1)\,|\,0\,\rangle, \tag{2.4.22}$$

$$|\Psi_e\rangle = \frac{-i}{\sqrt{2}}\,[\,\hat{a}^\dagger(\omega,+1)\,|\,0\,\rangle + \hat{a}^\dagger(\omega,-1)\,|\,0\,\rangle,$$

or in the simplified form

$$|\hat{o}\rangle = \frac{1}{\sqrt{2}}\,[|\,\hat{R}\,\rangle - |\,\hat{L}\,\rangle], \tag{2.4.23}$$

$$|\hat{e}\rangle = \frac{-i}{\sqrt{2}}\,[|\,\hat{R}\,\rangle + |\,\hat{L}\,\rangle].$$

The orthogonal right-hand and left-hand polarization states can be used to characterize a mixed polarization too:

$$\hat{\rho} = P(+1)|\hat{R}\rangle\langle\hat{R}| + P(-1)|\hat{L}\rangle\langle\hat{L}|, \tag{2.4.24}$$

indicating a 50–50% chance to be in the states of $|\hat{R}\rangle$ and $|\hat{L}\rangle$.

A pure state represents a vector in Hilbert space. If we could "rotate" our measurement device to identify the vector and its orthogonal conjugates, it is not surprising to have maximum and minimum counting rates at different "directions" in the chosen Hilbert space. Thus, in principle, it is possible, and "easy", to distinguish a pure state from a mixed state as well as from its orthogonal states, such as the state with the "+" sign and the state with the "−" sign in Eq. (2.4.1), by a simple rotation. If we consider the measurement operation a projection, to distinguish a pure state from mixed states as well as from its orthogonal states, we can project the pure state onto itself, or project it onto its orthogonal states. Mathematically, we may define a projection operator $\hat{P} = |\Psi\rangle\langle\Psi|$, which is nothing but the density operator, to write these projections as

$$\hat{P}\,|\Psi\rangle = |\Psi\rangle\langle\Psi|\Psi\rangle = |\Psi\rangle,$$

$$\hat{P}\,|\Psi_\perp\rangle = |\Psi\rangle\langle\Psi|\Psi_\perp\rangle = 0. \tag{2.4.25}$$

Equation (2.4.25) can be expressed in a general form in terms of the density operator

$$\hat{\rho}^2 = |\Psi\rangle\langle\Psi|\Psi\rangle\langle\Psi| = \hat{\rho}, \tag{2.4.26}$$

where the state $|\Psi\rangle$ (and $|\Psi\rangle_\perp$) is normalized as in Eq. (2.4.1). Any pure state must satisfy Eq. (2.4.26).

It is easy to find that the states $|45°\rangle$ and $|-45°\rangle$ (as well as $|\hat{R}\rangle$ and $|\hat{L}\rangle$) form a complete orthonormal vector basis and satisfy the completeness relation:

$$|\Psi\rangle\langle\Psi| + |\Psi_\perp\rangle\langle\Psi_\perp| = 1. \tag{2.4.27}$$

In measurements other than polarization, however, the situation is not that simple. In many cases, we cannot "rotate" our measurement-projection device in the chosen vector space. We may have to keep the state vector as a superposition of the chosen set of eigenvectors of a known operator, such as the Hamiltonian. In this case, we may be confused again in understanding the physics of a quantum superposition. For example, if one insists that a photon has to be polarized as either an o-ray or an e-ray when passing through a uniaxial crystal, it is very reasonable to ask the following: How could a 45° or −45° polarized photon be an o-ray and an e-ray simultaneously? We know that an energy bundle of $\hbar\omega$, or a photon, can never be divided into parts. A similar question naturally follows: Has the 45° polarized single photon been "which ray or both ray" when passing the uniaxial crystal? This is equivalent to the question we have been asking since the beginning of quantum theory: Has the observed single photon passed "which slit or both slit" in a Young's double-slit interferometer? Regarding this historically long standing problem, the Schrödinger cat may be considered the most vivid cartoon model.

Before turning to the physics of the Schrödinger cat, we need a brief discussion about the measurement of a single photon. The measurement of a single photon does not mean we only measure one photon. In fact, one measurement of one photon cannot provide us any meaningful information about the radiation system. In quantum theory, any meaningful physics has to be learned from a large number of measurements, or from the measurement of an ensemble. There are two types of ensembles in quantum theory: The first type is said to be homogeneous or in a statistically *pure* state. In this case, each of the measured N systems in the ensemble is in the same pure state, and thus the ensemble itself is in a pure state. The second type is referred to as a *mixture*. In this case, the measured ensemble of N systems is not necessarily in the same pure state. The measured N systems can be in several, or in many, different states with certain probabilities, and thus the ensemble itself is inhomogeneous or in a statistically mixed state.

We have two pictures of a pure state of a single photon. The pure state can represent (1) an individual photon or (2) an ensemble of photons.

For example, in the polarization measurement, the pure states of Eq. (2.4.1) with "+" sign

$$|\Psi_{45^\circ}\rangle = \frac{1}{\sqrt{2}}\left[\hat{a}^\dagger(\omega,\hat{o})|0\rangle + \hat{a}^\dagger(\omega,\hat{e})|0\rangle\right]$$

mean (1) each measured photon (individual) is polarized at 45° and (2) all measured photons (ensemble) are polarized at 45°, or simply, the measured radiation field is polarized at 45°. The pure states of Eq. (2.4.1) with "$-$" sign

$$|\Psi_{-45^\circ}\rangle = \frac{1}{\sqrt{2}}\left[\hat{a}^\dagger(\omega,\hat{o})|0\rangle - \hat{a}^\dagger(\omega,\hat{e})|0\rangle\right]$$

mean (1) each measured photon (individual) is polarized at -45° and (2) all measured photons (ensemble) are polarized at -45°, or simply, the measured radiation field is polarized at -45°. Mathematically, the only difference between $|\Psi_{45^\circ}\rangle$ and $|\Psi_{-45^\circ}\rangle$ is the "$\pm$" sign between the orthogonal basis $\hat{a}^\dagger(\omega,\hat{o})|0\rangle$, or $|\hat{o}\rangle$ in the simplified form, and $\hat{a}^\dagger(\omega,\hat{e})|0\rangle$, or $|\hat{e}\rangle$ in the simplified form. The "+" sign means the $|\hat{o}\rangle$ and $|\hat{e}\rangle$ are "in-phase", and the "$-$" sign means $|\hat{o}\rangle$ and $|\hat{e}\rangle$ are "out-of-phase".

In fact, we can formally write any polarization state into the following superposition in the 2-D Hilbert space of single photon Fock state $\hat{a}^\dagger(\omega,\hat{o})|0\rangle$ (or $|\hat{o}\rangle$) and $\hat{a}^\dagger(\omega,\hat{e})|0\rangle$ (or $|\hat{e}\rangle$):

$$|\Psi\rangle = c_o\,\hat{a}^\dagger(\omega,\hat{o})|0\rangle + c_e\,\hat{a}^\dagger(\omega,\hat{e})|0\rangle, \tag{2.4.28}$$

where $c_o = |c_o|e^{i\varphi_o}$ and $c_e = |c_e|e^{i\varphi_e}$ are the probability amplitudes of observing a photon in the state $|\hat{o}\rangle$ and/or in the state $|\hat{e}\rangle$, respectively. It is easy to find, when $|c_o| = |c_e|$ and $\varphi_e - \varphi_o = 0$, $|\Psi\rangle = |\Psi_{45^\circ}\rangle$, while $|c_o| = |c_e|$ and $\varphi_e - \varphi_o = \pi$, $|\Psi\rangle = |\Psi_{-45^\circ}\rangle$. It is also easy to find, when $|c_o| = |c_e|$ and $\varphi_e - \varphi_o = \pi/2$, $|\Psi\rangle = |\Psi_R\rangle$, while $|c_o| = |c_e|$ and $\varphi_e - \varphi_o = -\pi/2$, $|\Psi\rangle = |\Psi_L\rangle$.

Since $|\Psi_{45^\circ}\rangle$ and $|\Psi_{-45^\circ}\rangle$ ($|\Psi_R\rangle$ and $|\Psi_L\rangle$ as well) form a complete orthogonal 2-D Hilbert space, we can also formally write any polarization state into the following superposition in the 2-D Hilbert space of single photon Fock state $\hat{a}^\dagger(\omega,45^\circ)|0\rangle$ (or $|\Psi_{45^\circ}\rangle$) and $\hat{a}^\dagger(\omega,-45^\circ)|0\rangle$ (or $|\Psi_{-45^\circ}\rangle$):

$$|\Psi\rangle = c_{45^\circ}\,\hat{a}^\dagger(\omega,45^\circ)|0\rangle + c_{-45^\circ}\,\hat{a}^\dagger(\omega,-45^\circ)|0\rangle, \tag{2.4.29}$$

where $c_{45^\circ} = |c_{45^\circ}|e^{i\varphi_{45^\circ}}$ and $c_{-45^\circ} = |c_{-45^\circ}|e^{i\varphi_{-45^\circ}}$ are the probability amplitudes of observing a photon in the state $|\Psi_{45^\circ}\rangle$ and/or in the state

$|\Psi_{-45°}\rangle$, respectively. It is easy to find, when $|c_{45°}| = |c_{-45°}|$ and $\varphi_{-45°} - \varphi_{45°} = 0$, $|\Psi\rangle = |\hat{o}\rangle$, while $|c_{45°}| = |c_{-45°}|$ and $\varphi_{-45°} - \varphi_{45°} = \pi$, $|\Psi\rangle = |\hat{e}\rangle$.

In terms of the 2-D Hilbert space of $|\Psi_{45°}\rangle$ and $|\Psi_{-45°}\rangle$, the mixed state of Eq. (2.4.19) may have the following meanings: (1) Half of the measured photons in the ensemble are polarized at $45°$ and the other half of the measured photons in the ensemble are polarized at $-45°$. (2) Each measured photon is polarized neither at $45°$ nor at $-45°$, but a random polarization in the form of an incoherent superposition of $\hat{o}$ and $\hat{e}$. Our early defined pseudovector $|\tilde{\Psi}\rangle$ is a good choice to specify the mixed state of the radiation field:

$$|\tilde{\Psi}\rangle = \frac{e^{i\varphi_{45°}}}{\sqrt{2}}\hat{a}^\dagger(\omega, 45°)|0\rangle + \frac{e^{i\varphi_{-45°}}}{\sqrt{2}}\hat{a}^\dagger(\omega, -45°)|0\rangle \qquad (2.4.30)$$

with $\varphi_{45°}$ = random number and $\varphi_{-45°}$ = random number, or $\varphi_{-45°} - \varphi_{45°}$ = random number. The pseudovector $|\tilde{\Psi}\rangle$ may represent: (1) In the measured ensemble, each photon is polarized with a set of special values of $\varphi_{45°}$ and $\varphi_{-45°}$, and the set takes different values from one to another. (2) An ensemble of randomly created photons with either $|\hat{o}\rangle$ polarization or $|\hat{e}\rangle$ polarization, the ensemble is thus unpolarized.

Example (II) Schrödinger cat:

There has been a famous *cat* in the history of physics, namely, the Schrödinger cat. Schrödinger's cat is in the superposition state of $|\,\text{alive}\,\rangle$ and $|\,\text{dead}\,\rangle$:

$$|\,\text{cat}_+\,\rangle = \frac{1}{\sqrt{2}}\left[\,|\,\text{alive}\,\rangle + |\,\text{dead}\,\rangle\,\right], \qquad (2.4.31)$$

where $|\,\text{alive}\,\rangle$ and $|\,\text{dead}\,\rangle$ are the eigenvectors of the "life status" operator of a cat. Similar to the linear polarization of a radiation field, the vector space defined by the "life status" operator has a dimension of two. The eigenvectors $|\,\text{alive}\,\rangle$ and $|\,\text{dead}\,\rangle$ form a complete, orthonormal vector space. The Schrödinger cat is not in any of the eigenstates but rather a superposition of the eigenstates. Of course, there is no such cat in our everyday life. In the quantum world, however, a Schrödinger cat is allowed to be alive and dead simultaneously. Besides the pure state of Eq. (2.4.31), quantum theory permits another pure state which is orthogonal to the state of Eq. (2.4.31):

$$|\,\text{cat}_-\,\rangle = \frac{1}{\sqrt{2}}\left[\,|\,\text{alive}\,\rangle - |\,\text{dead}\,\rangle\,\right]. \qquad (2.4.32)$$

First, both states $|\,\text{cat}_+\,\rangle$ and $|\,\text{cat}_-\,\rangle$ in Eqs. (2.4.31) and (2.4.32) are pure states. It is easy to show that

$$\hat{\rho}^2_{\text{cat}_+} = \hat{\rho}_{\text{cat}_+}, \qquad \hat{\rho}^2_{\text{cat}_-} = \hat{\rho}_{\text{cat}_-}.$$

Second, the states of $|\,\text{cat}_+\,\rangle$ and $|\,\text{cat}_-\,\rangle$ form a complete orthonormal set of vector basis,

$$|\,\text{cat}_+\,\rangle\langle\,\text{cat}_+\,| + |\,\text{cat}_-\,\rangle\langle\,\text{cat}_-\,| = \hat{1}. \tag{2.4.33}$$

Therefore, any cat states can be written as the superposition of $|\,\text{cat}_+\,\rangle$ and $|\,\text{cat}_-\,\rangle$, such as

$$|\,\text{alive}\,\rangle = \frac{1}{\sqrt{2}}\,[\,|\,\text{cat}_+\,\rangle + |\,\text{cat}_-\,\rangle \tag{2.4.34}$$

$$|\,\text{dead}\,\rangle = \frac{1}{\sqrt{2}}\,[\,|\,\text{cat}_+\,\rangle - |\,\text{cat}_-\,\rangle.$$

If we have a measurement-projection device capable of identifying $|\,\text{cat}_+\,\rangle$ and $|\,\text{cat}_-\,\rangle$, we are going to find that an alive-cat (dead-cat as well) would be a $|\,\text{cat}_+\,\rangle$ (Schrödinger cat) and $|\,\text{cat}_-\,\rangle$ (conjugate Schrödinger cat) simultaneously. Should we ask the following: How could an alive-cat be a Schrödinger cat and its conjugate-cat simultaneously?

Third, the pure states of Eqs. (2.4.31) and (2.4.32) are very different from the mixed state

$$\hat{\rho} = \frac{1}{2}\,[\,|\,\text{alive}\,\rangle\langle\,\text{alive}\,| + |\,\text{dead}\,\rangle\langle\,\text{dead}\,|\,]. \tag{2.4.35}$$

Comparing with the linear polarization measurement in example one, the states of Eqs. (2.4.31) and (2.4.32) are equivalent to the linear polarization states of $45°$ and $-45°$. We have made a similar rotation to achieve Eqs. (2.4.31) and (2.4.32). Theoretically, this is a simple rotation in the 2-D "cat-space". Unfortunately, no one knows how to achieve that rotation in practice. The only measurement we can do is to identify if a cat is alive or dead. If our measurement-projection device can only project a cat state onto the eigenstate $|\,\text{alive}\,\rangle$ or $|\,\text{dead}\,\rangle$, we may never be able to distinguish the pure states of Eqs. (2.4.31) and (2.4.32) from the mixed state of Eq. (2.4.35). Although we do not know how to achieve a rotation from $|\,\text{alive}\,\rangle$ or $|\,\text{dead}\,\rangle$ to $|\,\text{alive}\,\rangle \pm |\,\text{dead}\,\rangle$ in the cat space, it does not mean the nonexistence of this kind of superposition state, especially if we know how to generate such a state.

Similar to the polarization measurement of a single photon, one observation of $|\text{alive}\rangle$ or $|\text{dead}\rangle$ does not give us any meaningful information about the state of the cat. In the theory of Schrödinger's cat, any meaningful physics has to be learned from the measurement of an ensemble of cats. If all measured cats are in the same state, such as that of Eq. (2.4.31) or Eq. (2.4.32), we say that each measured cat is in a pure state and the measured cat-system is in a pure state. If the measured cats are in mixed state of Eq. (2.4.35), it may have the following meaning: (1) Half of the cats in the ensemble are $|\text{alive}\rangle$ and the other half of the cats in the ensemble are $|\text{dead}\rangle$. (2) Each measured cat is neither in $|\text{alive}\rangle$ nor $|\text{dead}\rangle$ but in an incoherent superposition

$$|\tilde{\Psi}\rangle = \frac{e^{i\varphi_{\text{alive}}}}{\sqrt{2}}|\text{alive}\rangle + \frac{e^{i\varphi_{\text{dead}}}}{\sqrt{2}}\hat{a}^\dagger|\text{dead}\rangle \tag{2.4.36}$$

with $\varphi_{\text{alive}} = $ random number and $\varphi_{\text{dead}} = $ random number, or $\varphi_{\text{dead}} - \varphi_{\text{alive}} = $ random number. The pseudovector $|\tilde{\Psi}\rangle$ indicates that each cat is characterized by a set of special values of φ_{alive} and φ_{dead}, and the set takes different values from one cat to another cat. The ensemble of cats is represented by a large number of these randomly distributed vectors in the 2-D cat space.

Example (III) single photon state:

What is the difference between the pure single photon state of Eq. (2.1.12)

$$|\Psi\rangle = \sum_{\mathbf{k},s} f(\mathbf{k}, s)\,\hat{a}^\dagger_{\mathbf{k},s}\,|\,0\,\rangle$$

and the mixed single photon state of Eq. (2.3.9)

$$\hat{\rho} = \sum_{\mathbf{k},s} |f(\mathbf{k}, s)|^2\,\hat{a}^\dagger_{\mathbf{k},s}\,|\,0\,\rangle\langle\,0\,|\,\hat{a}_{\mathbf{k},s}?$$

In terms of the state of a photon, Eq. (2.1.12) represents a pure state by means of a vector in the *multi-dimensional* Hilbert space of single photon Fock state $n = 1$. On the contrary, Eq. (2.3.9) represents a mixed state by means of a set of randomly distributed vectors in the same *multi-dimensional* Hilbert space of single photon Fock state $n = 1$. It should be noted that the Hilbert space of the above single photon states has a dimension of $\gg 2$, mathematically different from that of polarization space and cat space; but the physics is the same in terms of coherent and incoherent superposition.

Equations (2.1.12) and (2.3.9) can be written as a vector

$$|\Psi\rangle = \sum_{\mathbf{k},s} |f(\mathbf{k}, s)| e^{i\varphi_{\mathbf{k},s}}\, \hat{a}^\dagger_{\mathbf{k},s} |0\rangle \qquad (2.4.37)$$

and a pseudovector

$$|\tilde{\Psi}\rangle = \sum_{\mathbf{k},s} |f(\mathbf{k}, s)| e^{i\varphi_{\mathbf{k},s}}\, \hat{a}^\dagger_{\mathbf{k},s} |0\rangle \qquad (2.4.38)$$

respectively, to emphasize the coherent or incoherent superposition. Again, $\varphi_{\mathbf{k},s}$ represents the phase of the probability amplitude for the photon to be in the single photon Fock state $\hat{a}^\dagger_{\mathbf{k},s}|0\rangle$. When $\varphi_{\mathbf{k},s} = $ constant, the coherent superposition among the single photon Fock state $\hat{a}^\dagger_{\mathbf{k},s}|0\rangle$ results in a pure state of Eq. (2.4.37); when $\varphi_{\mathbf{k},s} = $ random number, the incoherent superposition among the single photon Fock state $\hat{a}^\dagger_{\mathbf{k},s}|0\rangle$ results in a mixed state of Eq. (2.4.38).

The pure state of Eq. (2.3.9) must be created from a coherent process in which all radiation modes of $\mathbf{k}$ with polarization s are excited with constant phase and superposed coherently. The probability amplitude for a photon to be in the single photon Fock state $\hat{a}^\dagger_{\mathbf{k},s}|0\rangle$ is $f(\mathbf{k}, s)$, which has a real and positive amplitude $|f(\mathbf{k}, s)|$ and a phase $\varphi_{\mathbf{k},s}$, indicating precise magnitude and phase for all superposed single photon Fock state of $n = 1$. On the contrary, the mixed state Eq. (2.3.9) or Eq. (2.4.38) must be created from a stochastic process in which all radiation modes $\mathbf{k}$ with polarization s are excited with random phases and thus superposed incoherently. The only knowledge we can have about a photon in a mixed state is that it takes probability $|f(\mathbf{k}, s)|^2$ to be in the single photon Fock state $\hat{a}^\dagger_{\mathbf{k},s}|0\rangle$.

Can a simple measurement distinguish the pure state of Eq. (2.1.12) from the mixed state of Eq. (2.3.9) or Eq. (2.4.38)? If we can find a "rotation" in terms of the *multi-dimensional* Hilbert space of single photon Fock state of $n = 1$ to project the pure state of Eq. (2.1.12) onto itself and onto its orthogonal states, certainly, we can. This "rotation" is similar to the rotations we have discussed in earlier discussions on the measurement of polarization and Schrödinger's cat. Do we have such a measurement corresponding to such a "simple rotation"? Unfortunately, we do not.

2.5   Composite System and Two-Photon State of Radiation Field

In certain measurements, such as correction measurements by means of joint detection or coincidence detection of two photodetectors, we need to deal with two-photon states. Two-photon states describe the state of a composite system of two photons. The subsystems may be spatially separated in large distance. Despite the distance between the subsystems, in quantum theory, a composite system composed of two subsystems is described by a Hilbert space constructed as the direct or tensor product of the Hilbert spaces of the two subsystems:

$$H = H_1 \otimes H_2. \tag{2.5.1}$$

If state $|\Psi_1\rangle \in H_1$ and state $|\Psi_2\rangle \in H_2$, then we denote the direct product of these states by $|\Psi\rangle = |\Psi_1\rangle \otimes |\Psi_2\rangle$, or simply $|\Psi\rangle = |\Psi_1\rangle|\Psi_2\rangle$. The inner product on H is defined in terms of the inner product on H_1 and H_2 by

$$\langle\Psi|\Psi'\rangle = \langle\Psi_1|\Psi_1'\rangle\langle\Psi_2|\Psi_2'\rangle. \tag{2.5.2}$$

If $\{|m\rangle\}$ is an orthonormal base of H_1 and $\{|n\rangle\}$ is an orthonormal basis of H_2, then $\{|m\rangle \otimes |n\rangle\}$ or simple $\{|m\rangle|n\rangle\}$ is an orthonormal basis of H. This basis is usually called the Schmidt basis.

In a composite system, the subsystems may be independent, correlated, or entangled. For independent subsystems, the state of the composite system can be written as a product state

$$|\Psi\rangle = \sum_{m,n} c_m\, c_n\, |m\rangle|n\rangle = \sum_m c_m|m\rangle \sum_n c_n\, |n\rangle = |\Psi_1\rangle \otimes |\Psi_2\rangle, \tag{2.5.3}$$

where $c_{mn} = c_m\, c_n$ is factorizable. When the two subsystems are correlated or entangled, in general, $|\Psi\rangle$ cannot be written in the form of Eq. (2.5.3), i.e., c_{mn} is nonfactorizable and consequently the state is nonfactorizable

$$|\Psi\rangle = \sum_{m,n} c_{mn}\, |m\rangle|n\rangle. \tag{2.5.4}$$

We say the two subsystems are in a correlated state or in an entangled state, depending on the form of c_{mn}.

The simplest two-photon states are the Fock states of Eq. (2.1.9). To simplify the mathematics, we rewrite Eq. (2.1.9) in 1-D and ignore the polarization

$$|\Psi\rangle = \hat{a}_1^\dagger(\omega)\, \hat{a}_2^\dagger(\omega')\,|\,0\,\rangle$$

$$= |0,\,\dots\,0,\, 1_1(\omega),\, 0,\,\dots\, 0\rangle|0,\,\dots\, 0,\, 1_2(\omega'),\, 0,\,\dots\, 0\rangle, \tag{2.5.5}$$

where we have also rewritten the state of Eq. (2.1.9) explicitly as that of a composite two subsystems. In general, we may deal with a set of two-photon

Fock states either in a coherent superposition

$$|\Psi\rangle = \sum_{\omega}\sum_{\omega'} f(\omega,\omega')\,\hat{a}_1^\dagger(\omega)\,\hat{a}_2^\dagger(\omega')\,|\,0\,\rangle \tag{2.5.6}$$

or in a statistical mixture

$$\hat{\rho} = \sum_{\omega}\sum_{\omega'} |f(\omega,\omega')|^2\,\hat{a}_1^\dagger(\omega)\,\hat{a}_2^\dagger(\omega')\,|\,0\,\rangle\langle 0|\hat{a}_2(\omega')\,\hat{a}_1(\omega), \tag{2.5.7}$$

where $f(\omega,\omega')$ is the probability amplitude for the quantized field to be in the two-photon Fock state $|0,\dots 0,1_\omega,0,\dots 0\rangle|0,\dots 0,1_{\omega'},0,\dots 0\rangle$. The two-photon probability amplitude $f(\omega,\omega')$ may be factorizable into a product of $f_1(\omega)\times f_2(\omega')$, or nonfactorizable at all. The physical properties of the states are very different between these two cases. If $f(\omega,\omega')$ can be factorized into a product of $f_1(\omega)\times f_2(\omega')$, the state itself is also factorizeable into a product state of two independent single photons. If $f(\omega,\omega')$ is nonfactorizable and consequently the state itself cannot be factorized into a product state, we name the state a correlated or an entangled two-photon state. In terms of the concept of a photon, the product states describe the behavior of two independent photons, the correlated states describe the behavior of two correlated photons, and the entangled states characterize the behavior of an entangled pair of photons.

Example (I) Product state:

Equation (2.5.8) is a two-photon state which can be factorized into a product of two single photon states:

$$|\Psi\rangle = \sum_{\omega}\sum_{\omega'} f(\omega)f'(\omega')\,\hat{a}_1^\dagger(\omega)\,\hat{a}_2^\dagger(\omega')\,|\,0\,\rangle \tag{2.5.8}$$

$$= \sum_{\omega} f(\omega)\,\hat{a}_1^\dagger(\omega)\,|\,0\,\rangle \times \sum_{\omega'} f'(\omega')\,\hat{a}_2^\dagger(\omega')\,|\,0\,\rangle$$

$$= |\Psi_1\rangle|\Psi_2\rangle, \tag{2.5.9}$$

where we have assumed a factorizable amplitude: $f(\omega,\omega') = f(\omega)\times f'(\omega')$.

Example (II) Entangled EPR state:

Equation (2.5.10) is a nonfactorizable two-photon state with total photon number $n = 2$

$$|\Psi\rangle = \Psi_0 \sum_{\omega,\omega'} \delta(\omega+\omega'-\omega_0)\,\hat{a}_1^\dagger(\omega)\,\hat{a}_2^\dagger(\omega')\,|\,0\,\rangle, \tag{2.5.10}$$

where Ψ_0 is a normalization constant and the delta-function and the constant ω_0 together indicate the conservation of energy. In the state of Eq. (2.5.10), a pair of ω and ω' are always excited together and the energy of the pair remains constant, although each ω and ω' may take any value within the superposition. The two-photon state of Eq. (2.5.10) has certain properties that may never be understood classically. For example, in terms of the concept of a photon, the energy of neither photon is determined; however, if one of the photons is measured with a certain value, the energy of the other photon is determined with certainty, despite the distance between the two photons. The state of Eq. (2.5.10) is a coherent superposition of a large number of product substates. While each substate is the product of two individual states of the two photons, the entangled two-photon state, as the result of the coherent superposition, cannot be factorized into a product state of the two photons. These kind of states are named entangled states by Schrödinger following the 1935 paper of Einstein, Podolsky, and Rosen. Quantum entanglement is intensively discussed in later chapters.

Example (III) Number state of $n_\omega = 2$:

Equation (2.5.11) defines another type of two-photon state with total photon number $n = 2$:

$$|\Psi\rangle = \sum_\omega f(\omega)\,[\hat{a}^\dagger(\omega)]^2\,|0\rangle. \tag{2.5.11}$$

Equation (2.5.11) indicates the excitation of one mode with occupation number $n_\omega = 2$.

It should be emphasized that the states in Eqs. (2.5.8), (2.5.10), and (2.5.11) are all pure states. They are very different from the following corresponding mixed states:

$$\hat{\rho} = \sum_{\omega,\omega'} P(\omega)\,P'(\omega')\hat{a}^\dagger(\omega)\,\hat{a}^\dagger(\omega')\,|0\rangle\langle 0|\,\hat{a}(\omega)\,\hat{a}(\omega'),$$

$$\hat{\rho} = \sum_{\omega,\omega'} \delta\,[\omega + \omega' - \omega_0]\,\hat{a}^\dagger(\omega)\,\hat{a}^\dagger(\omega')\,|0\rangle\langle 0|\,\hat{a}(\omega)\,\hat{a}(\omega'),$$

$$\hat{\rho} = \sum_\omega P(\omega)\,[\hat{a}^\dagger(\omega)]^2\,|0\rangle\langle 0|\,[\hat{a}(\omega)]^2.$$

2.6   A Simple Model for the Creation of Single Photon and Multi-Photon State

To simplify the discussion and mathematics, we assume a point light source contains a large number of atoms that are ready for two-level atomic transitions at any time t. For a point source, each atomic transition excites a subfield in the form of a symmetrical spherical wave propagating to all 4π directions. The excited radiations are monitored by a point-like photodetector or a set of independent N point-like photodetectors that are placed at distance, such as a light year, for single photon counting measurement or for joint N-photon counting measurement. We assume the measured light is weak at such a distance.

Although the chance to have a spontaneous emission is very small, there is indeed a small probability for an atom to create a photon whenever the atom decays from its higher energy level E_2 ($\Delta E_2 \neq 0$) down to its ground energy state of E_1. It is reasonable to approximate the mth atomic transition excites a subfield in the following state:

$$|\Psi\rangle = c_0|0\rangle + c_1|\Psi_m\rangle \simeq |0\rangle + \epsilon|\Psi_m\rangle, \tag{2.6.1}$$

where $|c_0| \sim 1$ is the probability amplitude for no-field-excitation and $|c_1| = |\epsilon| \ll 1$ is the probability amplitude for the creation of a photon from the mth atomic transition, and

$$|\Psi_m\rangle = \sum_s \int d\mathbf{k}\, f_m(\mathbf{k}, s)\, \hat{a}_m^\dagger(\mathbf{k}, s)|0\rangle$$

or

$$|\Psi_m\rangle = \int d\omega\, f_m(\omega)\, \hat{a}_m^\dagger(\omega)|0\rangle \quad \text{(1-D, 1-Polarization)}$$

as shown earlier in Eqs. (2.3.12) and (2.3.13). The generalized state of the radiation field that is excited by the light source, which contains such a large number atomic transitions, can be formally written as

$$|\Psi\rangle = \prod_m \{|0\rangle + \epsilon\, c_m\, |\Psi_m\rangle\}$$

$$\simeq |0\rangle + \epsilon \left[\sum_m |c_m|\, e^{i\varphi_m}|\Psi_m\rangle\right]$$

$$+ \epsilon^2 \left[\sum_{m<n} |c_m||c_n|\, e^{i(\varphi_m+\varphi_n)}|\Psi_m\rangle|\Psi_n\rangle\right] + \cdots, \tag{2.6.2}$$

where, again, we have defined a phase factor $e^{i\varphi_m}$ associated with the mth atomic transition. In principle, this phase factor belongs to the amplitude of each excited mode associated with the creation operator. We assume a common phase for each wavepacket, although it can still take arbitrary values from transition to transition. This common phase is determined by a common physical parameter in the creation process, such as the creation time of the mth wavepacket, since $|\epsilon| \ll 1$. In Eq. (2.6.2), we listed the first-order and the second-order approximations on ϵ. For a particular measurement, only a certain lower-order approximations are necessary. In the following, we give four types of approximations corresponding to different types of measurements or different mechanisms of light generation:

(I) Mixed single photon state:

We assume all atomic transitions in the source are radiated randomly and independently with φ_m = random number for all m. In certain measurements, a first-order expansion on ϵ in Eq. (2.6.2) is a good approximation. The higher-order approximations may contribute to the measurement, however, these contributes are usually small enough under the weak light condition. A radiation is said to be at single photon's level when the higher-order approximations are ignorable, and the radiation field can be characterized by the following pseudovector of single photon state:

$$|\tilde{\Psi}\rangle \simeq \sum_m |c_m|\, e^{i\varphi_m}\, |\Psi_m\rangle, \tag{2.6.3}$$

where each coherently superposed wavepacket

$$|\Psi_m\rangle = \int d\omega\, f_m(\omega)\, \hat{a}_m^\dagger(\omega)\, |0\rangle$$

is associated with a random phase φ_m which may take any arbitrary value. The first-order approximation is thus written in the form of an incoherent superposition of a set of single photon state vectors or wavepacket $|\Psi_m\rangle$. The incoherently superposed state of Eq. (2.6.3) can be considered as follows: (1) the state of the measured radiation which may be created from any or from all permissible randomly radiated independent atomic transitions with probability $|c_m|^2$; (2) the state of an ensemble of measured photons, each photon is created from a randomly radiated atomic transition, as an inhomogeneous ensemble.

The expectation value of an operator $\hat{A}$ corresponding to the measurement is calculated from

$$\langle\langle\Psi|\hat{A}|\Psi\rangle\rangle_{\mathrm{En}} = \left\langle \sum_m |c_m| e^{-i\varphi_m} \langle\Psi_m| \hat{A} \sum_n |c_n| e^{i\varphi_n} |\Psi_n\rangle \right\rangle_{\mathrm{En}}$$

$$= \left\langle \sum_{m,n} |c_m||c_n| e^{i(\varphi_n - \varphi_m)} \langle\Psi_m|\hat{A}|\Psi_n\rangle \right\rangle_{\mathrm{En}}$$

$$= \sum_m |c_m|^2 \langle\Psi_m|\hat{A}|\Psi_m\rangle$$

$$= tr\,\hat{\rho}^{(1)} \hat{A}, \tag{2.6.4}$$

where we have defined a density operator to characterize the first-order approximated mixed single photon state

$$\hat{\rho}^{(1)} \equiv \sum_m P_m |\Psi_m\rangle\langle\Psi_m| \tag{2.6.5}$$

with $P_m = |c_m|^2$ the probability for the radiation field to be in the state $|\Psi_m\rangle$. The ensemble average in Eq. (2.6.4) has taken into account all possible values of the relative phases $\varphi_n - \varphi_m$.

Under the above approximation, traditionally, we say the measured field is in a mixed single photon state:

$$\hat{\rho} = \sum_m P_m |\Psi_m\rangle\langle\Psi_m|$$

$$= \sum_m P_m \int d\omega\, d\omega'\, f_m(\omega) f_m^*(\omega')\, \hat{a}_m^\dagger(\omega)\,|0\rangle\langle0|\,\hat{a}_m(\omega'). \tag{2.6.6}$$

If we further assume that all the atomic transitions have the same mode distribution $f_m(\omega) = f(\omega)$ and each mode in the mth wavepacket has a phase $\varphi_m(\omega)$ that is determined by the transition time t_{0m} of the mth atomic transition, $\varphi_m(\omega) = \omega t_{0m}$, and consider a random and continuous distribution on t_{0m}, the sum on m turns into an integral of t_{0m}:

$$\hat{\rho} = \int dt_{0m}\, e^{-i(\omega-\omega')t_{0m}} \int d\omega\, d\omega'\, f(\omega) f^*(\omega')\, \hat{a}^\dagger(\omega)\,|0\rangle\langle0|\,\hat{a}(\omega')$$

$$\cong \int d\omega\, |f(\omega)|^2\, \hat{a}^\dagger(\omega)\,|0\rangle\langle0|\,\hat{a}(\omega), \tag{2.6.7}$$

where the result of the integral on t_{0m} is approximated $\delta(\omega - \omega')$ when taking t_{0m} to ∞.

(II) **Mixed two-photon state:**

Now we consider two independent point-like photodetectors at distance for joint photon counting measurement, i.e., to count the number of events in which each of the two photodetectors observes a photoelectron, respectively, and jointly at space–time $(\mathbf{r}_1, t_1)$ and $(\mathbf{r}_2, t_2)$. For this type of measurement, a second-order approximation is necessary. The third-order and higher-order approximations may also contribute to the measurement, however, these contributions are usually ignorable under weak light condition. In the above given simple model, the second-order approximation, or the two-photon state, can be approximated as a pseudo-vactor:

$$|\tilde{\Psi}\rangle \simeq \sum_{m<n} c_m c_n \, e^{i(\varphi_m + \varphi_n)} |\Psi_m\rangle |\Psi_n\rangle, \tag{2.6.8}$$

where, again,

$$|\Psi_m\rangle = \int d\omega \, f_m(\omega) \, \hat{a}_m^\dagger(\omega) \, |0\rangle.$$

When $\varphi_m + \varphi_n = $ random number for all m and n, the incoherent superposition turns Eq. (2.6.8) into a mixed two-photon state. The incoherently superposed two-photon state can be considered as follows: (1) the state of a jointly measured pair of photons which may be created from any or from all permissible randomly radiated and randomly paired atomic transitions with probability $P_{mn} = |c_m|^2 |c_n|^2$; (2) the state of an ensemble of jointly measured photon pairs, each pair is created from a randomly radiated and randomly paired atomic transitions, as an inhomogeneous ensemble.

In the case of $\varphi_m = $ random number, $\varphi_n = $ random number, and $\varphi_m + \varphi_n = $ random number for all m and n, the incoherent superpositions turn both first-order and second-order approximations into mixed states. Natural light, such as the sunlight, is a typical example of this type of radiation. This kind of light has a popular name of thermal light.

For the mixed two-photon approximation, the expectation value of an operator is calculated from Eq. (2.6.8):

$$\langle\langle \Psi | \hat{A} | \Psi \rangle\rangle_{\text{En}}$$

$$= \left\langle \sum_{m<n} c_m^* c_n^* \, e^{i(\varphi_m + \varphi_n)} \langle\Psi_m| \langle\Psi_n| \hat{A} \sum_{p<q} c_p c_q e^{-i(\varphi_p + \varphi_q)} |\Psi_p\rangle |\Psi_q\rangle \right\rangle_{\text{En}}$$

$$= \left\langle \sum_{m<n,\,p<q} c_m^* c_n^* c_p c_q \, e^{i[(\varphi_m + \varphi_n) - (\varphi_p + \varphi_q)]} \langle\Psi_m| \langle\Psi_n| \hat{A} |\Psi_p\rangle |\Psi_q\rangle \right\rangle_{\text{En}}$$

$$\simeq \sum_{m<n} |c_m|^2 |c_n|^2 \langle \Psi_m | \langle \Psi_n | \hat{A} | \Psi_n \rangle | \Psi_m \rangle$$

$$= tr \, \hat{\rho}^{(2)} \hat{A}, \tag{2.6.9}$$

where we have defined a density operator to characterize the second-order approximated mixed two-photon state:

$$\hat{\rho}^{(2)} = \sum_{m<n} |c_m|^2 |c_n|^2 |\Psi_n \rangle | \Psi_m \rangle \langle \Psi_m | \langle \Psi_n |. \tag{2.6.10}$$

The ensemble average in Eq. (2.6.9) has taken into account all possible values of the relative phases $(\varphi_m + \varphi_n) - (\varphi_p - \varphi_q)$. If we further assume the atoms have the same mode distribution $f(\omega)$ and each mode in the mth wavepacket has a phase $\varphi_m(\omega)$ that is determined by the transition time t_{0m} of the mth atomic transition, $\varphi_m(\omega) = \omega t_{0m}$, and consider a random and continuous distribution on t_{0m} and t_{0n}, similar to the single photon state, the integrals on t_{0m} $(\sim\infty)$ and t_{0n} $(\sim\infty)$ turn the density operator into the following form:

$$\hat{\rho}^{(2)} \cong \int d\omega \, d\omega' \, |f(\omega)|^2 |f(\omega')|^2 \, \hat{a}^\dagger(\omega) \, \hat{a}^\dagger(\omega') \, |0\rangle \langle 0| \, \hat{a}(\omega')\hat{a}(\omega), \tag{2.6.11}$$

where $f(\omega)$ is renormalized.

Equations (2.6.10) and (2.6.11) are both commonly used to represent the mixed two-photon state, such as in the calculation of second-order coherence of thermal light.

(III) **Entangled two-photon state**:

Entangled two-photon state can be created from atomic cascade decay and spontaneous parametric down-conversion (SPDC). The atomic cascade decay creates a pair of entangled photons from one cascade atomic transition. The entangled two-photon state of SPDC is generated from nonlinear optical process. A simplified 1-D entangled two-photon state of SPDC is given as follows:

$$|\Psi\rangle = \Psi_0 \int d\omega_s \, d\omega_i \, \delta(\omega_s + \omega_i - \omega_p) \, \hat{a}^\dagger(\omega_s)\hat{a}^\dagger(\omega_i) \, |0\rangle, \tag{2.6.12}$$

where Ψ_0 is the normalization constant and ω_p is the frequency of the pump laser, which is approximated as a constant. Equation (2.6.12) represents a coherent superposition of infinite number of two-photon product states $\hat{a}^\dagger(\omega_s)\hat{a}^\dagger(\omega_i) |0\rangle$, in which ω_s and ω_i are selected by the energy conservation $\hbar\omega_s + \hbar\omega_i = \hbar\omega_p$. It is this particular coherent superposition results in the nonfactorizable state of the signal–idler photon pair. For historical reason, the created pair of photons are named signal and idler. The signal photon and the idler photon are labeled by frequency ω_s and ω_i, respectively. The entangled two-photon state of the signal–idler photon pair is a pure state; however, the signal photon and the idler photon, respectively, is in mixed single photon state:

$$\hat{\rho}_s = \int d\omega_s\, \hat{a}^\dagger(\omega_s)|0\rangle\langle 0|\hat{a}(\omega_s),$$

$$\hat{\rho}_i = \int d\omega_i\, \hat{a}^\dagger(\omega_i)|0\rangle\langle 0|\hat{a}(\omega_i). \tag{2.6.13}$$

The entangled two-photon state of Eq. (2.6.12) was named by Klyshko "biphoton".

(IV) Coherent radiation:

Assuming a photon creation mechanism that makes all atomic transitions in phase and indistinguishable, the state of Equation (2.6.2) is rewritten in the following form:

$$|\Psi\rangle = \left\{ |0\rangle + \epsilon \int d\omega\, f(\omega)\, \hat{a}^\dagger(\omega)\,|0\rangle \right\}^n, \tag{2.6.14}$$

where n is the number of atomic transitions participate to the coherent excitation of the radiation field. The expansion of Eq. (2.6.14) is thus

$$|\Psi\rangle \simeq |0\rangle + \epsilon\,\frac{n}{1!}|\Psi\rangle + \epsilon^2\,\frac{n(n-1)}{2!}\left(|\Psi\rangle\right)^2 + \epsilon^3\,\frac{n(n-1)(n-2)}{3!}\left(|\Psi\rangle\right)^3 + \cdots$$

$$+ \epsilon^m\,\frac{n(n-1)(n-2)...(n-m+1)}{m!}\left(|\Psi\rangle\right)^m + \cdots, \tag{2.6.15}$$

with

$$|\Psi\rangle = \int d\omega\, f(\omega)\, \hat{a}^\dagger(\omega)\,|0\rangle.$$

In single-mode approximation, the state $|\Psi\rangle$ turns to be

$$|\Psi\rangle \simeq |0\rangle + \epsilon\,\frac{n}{1!}\,\hat{a}^\dagger(\omega)|0\rangle$$

$$+ \epsilon^2\,\frac{n(n-1)}{2!}\,[\hat{a}^\dagger(\omega)]^2|0\rangle + \epsilon^3\,\frac{n(n-1)(n-2)}{3!}\,[\hat{a}^\dagger(\omega)]^3|0\rangle$$

$$+ \epsilon^m\,\frac{n(n-1)(n-2)...(n-m+1)}{m!}\,[\hat{a}^\dagger(\omega)]^m|0\rangle + \cdots. \tag{2.6.16}$$

A laser system has the ability to force all atomic transitions in phase and indistinguishable. Single-mode laser beam is therefore considered in coherent state. A coherent state represents a group of indistinguishable photons.

2.7 Product State, Entangled State and Mixed State of Photon Pairs

In this section, we give a simple discussion about product, entangled and mixed polarization states of a two-photon system.

(1) Product polarization state of two photons

Suppose two independent photons are created from a radiation source at time $t = 0$ with photon-one polarized at $45°$ and photon-two polarized at $-45°$. Due to the independent creation process, or noninteraction between the subsystems, the two created photons can be treated as a photon system of a product state

$$|\Psi\rangle = |45°_1\rangle \times |-45°_2\rangle$$

$$= \frac{1}{2}[|\hat{o}_1\rangle|\hat{o}_2\rangle + |\hat{e}_1\rangle|\hat{e}_2\rangle - |\hat{o}_1\rangle|\hat{e}_2\rangle - |\hat{e}_1\rangle|\hat{o}_2\rangle], \qquad (2.7.1)$$

where $\hat{o}$ and $\hat{e}$ indicate the ordinary ray and the extraordinary ray of the uniaxial crystal that we defined in early section for polarization measurements.

It is not surprising that the two observers at large distances will observe four equal possibilities for the polarization correlation: (1) photon-one and photon-two are both polarized in the $\hat{o}$ direction, (2) photon-one and photon-two are both polarized in the $\hat{e}$ direction, (3) photon-one is polarized in the $\hat{o}$ direction and photon-two is polarized in the $\hat{e}$ direction, and (4) photon-one is polarized in the $\hat{e}$ direction and photon-two is polarized in the $\hat{o}$ direction.

Again, "not surprising" means that there is no difference between the quantum prediction and the classical convention.

(2) Entangled polarization state of two photons

Suppose a pair of photons is created at the radiation source in one of the following states:

$$|\Psi^{(+)}\rangle = \frac{1}{\sqrt{2}}[|\hat{o}_1\rangle|\hat{o}_2\rangle + |\hat{e}_1\rangle|\hat{e}_2\rangle],$$

$$|\Psi^{(-)}\rangle = \frac{1}{\sqrt{2}}[|\hat{o}_1\rangle|\hat{o}_2\rangle - |\hat{e}_1\rangle|\hat{e}_2\rangle],$$

$$|\Phi^{(+)}\rangle = \frac{1}{\sqrt{2}}\left[|\,\hat{o}_1\,\rangle|\,\hat{e}_2\,\rangle + |\,\hat{e}_1\,\rangle|\,\hat{o}_2\,\rangle\right],$$

$$|\Phi^{(-)}\rangle = \frac{1}{\sqrt{2}}\left[|\,\hat{o}_1\,\rangle|\,\hat{e}_2\,\rangle - |\,\hat{e}\,\rangle_1|\,\hat{o}\,\rangle_2\right], \tag{2.7.2}$$

where the state cannot be written in the form of a product $|\Psi\rangle = |\Psi_1\rangle|\Psi_2\rangle$, the state of the photon pair is defined as entangled state. The above four states are named Bell's states. Bell's states form a complete, orthogonal two-photon polarization vector basis. It is worth emphasizing that although Bell's states, i.e., entangled two-photon states, of the pair are pure states, the sigle-photon state of photon one and photon two are both mixed states. The state of a subsystem can easily be found by taking a partial trace of its twin

$$\hat{\rho}_j = \frac{1}{2}\left[|\,\hat{o}_j\,\rangle\langle\,\hat{o}_j\,| + |\,\hat{e}_j\,\rangle\langle\,\hat{e}_j\,|\right], \tag{2.7.3}$$

where $j = 1, 2$. Equation (2.7.3) indicates that the subsystems are unpolarized.

Now, we apply the criterion for free propagation and time evolution to the two-photon system, the state of the photon pair will remain the same as that of Eq. (2.7.2), except for a trivial overall phase factor. Assume two distant observers decide to measure the polarization of the photon pair with two independent polarization analyzers oriented at angles θ_1 and θ_2 relative to the unit vectors $\hat{o}_1$ and $\hat{o}_2$, respectively, as shown in Fig. 2.7.1. Each polarization analyzer has two output ports: θ_j and $\theta_{\perp j}$, $j = 1, 2$. The four possible joint detection events correspond to four joint projections of the

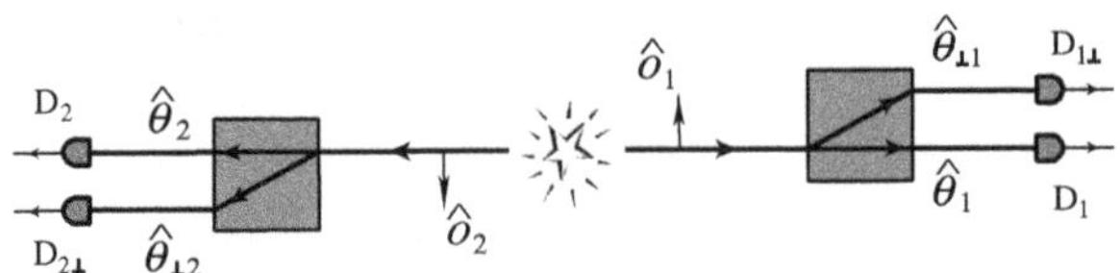

Fig. 2.7.1 Polarization correlation measurement of a photon pair. Two independent polarization analyzers are oriented at angles θ_1 and θ_2 relative to the unit vector $\hat{o}_1$ and $\hat{o}_2$, respectively. Each polarization analyzer has two output ports: $\hat{\theta}_j$ and $\hat{\theta}_{\perp j}$. The four possible joint detection events correspond to four projection operations defined in Eq. (2.7.4). Remark: All unit vectors are defined in the natural right-hand coordinate system.

two-photon state defined as follows:

$$\langle\Psi|\hat{\theta}_1\,\hat{\theta}_2\rangle\langle\hat{\theta}_2\,\hat{\theta}_1|\Psi\rangle = |\langle\hat{\theta}_2\,\hat{\theta}_1|\Psi\rangle|^2,$$

$$\langle\Psi|\hat{\theta}_{\perp 1}\,\hat{\theta}_{\perp 2}\rangle\langle\hat{\theta}_{\perp 2}\,\hat{\theta}_{\perp 1}|\Psi\rangle = |\langle\hat{\theta}_{\perp 2}\,\hat{\theta}_{\perp 1}|\Psi\rangle|^2,$$

$$\langle\Psi|\hat{\theta}_1\,\hat{\theta}_{\perp 2}\rangle\langle\hat{\theta}_{\perp 2}\,\hat{\theta}_1|\Psi\rangle = |\langle\hat{\theta}_{\perp 2}\,\hat{\theta}_1|\Psi\rangle|^2,$$

$$\langle\Psi|\hat{\theta}_{\perp 1}\,\hat{\theta}_2\rangle\langle\hat{\theta}_2\,\hat{\theta}_{\perp 1}|\Psi\rangle = |\langle\hat{\theta}_2\,\hat{\theta}_{\perp 1}|\Psi\rangle|^2, \tag{2.7.4}$$

where we have written $|\hat{\theta}_1\rangle\,|\hat{\theta}_2\rangle$ as $|\hat{\theta}_1\,\hat{\theta}_2\rangle$. Bell's states have the following polarization correlation:

	$|\Psi^{(+)}\rangle$	$|\Psi^{(-)}\rangle$	$|\Phi^{(+)}\rangle$	$|\Phi^{(-)}\rangle$
$\langle\hat{\theta}_2\,\hat{\theta}_1|$	$\cos^2(\theta_1-\theta_2)$	$\cos^2(\theta_1+\theta_2)$	$\sin^2(\theta_1+\theta_2)$	$\sin^2(\theta_1-\theta_2)$
$\langle\hat{\theta}_{\perp 2}\,\hat{\theta}_{\perp 1}|$	$\cos^2(\theta_1-\theta_2)$	$\cos^2(\theta_1+\theta_2)$	$\sin^2(\theta_1+\theta_2)$	$\sin^2(\theta_1-\theta_2)$
$\langle\hat{\theta}_{\perp 2}\,\hat{\theta}_1|$	$\sin^2(\theta_1-\theta_2)$	$\sin^2(\theta_1+\theta_2)$	$\cos^2(\theta_1+\theta_2)$	$\cos^2(\theta_1-\theta_2)$
$\langle\hat{\theta}_2\,\hat{\theta}_{\perp 1}|$	$\sin^2(\theta_1-\theta_2)$	$\sin^2(\theta_1+\theta_2)$	$\cos^2(\theta_1+\theta_2)$	$\cos^2(\theta_1-\theta_2)$

$$\tag{2.7.5}$$

In Eq. (2.7.5), we have ignored the common constant of $1/2$.

The correlation functions in Eq. (2.7.5) specify the probabilities for the photodetector at output port θ_1 (θ_2) to register a photon while its twin is registered at output port θ_2 (θ_1), for each of Bell's states. It is easy to see that for any value of θ_1 (θ_2) there is a unique orientation, θ_2 (θ_1), for which we expect a maximum probability of $1/2$ to register the pair simultaneously, while the pair is prepared in any one of the Bell's states. This is truly a surprise. For each photon pair, coming from the two-photon source, each polarization analyzer has a 50–50% chance to register its photon at θ_j or $\theta_{\perp j}$, as if the photon is unpolarized. However, whenever observer one (two) finds a photon polarized along θ_1 (θ_2) or $\theta_{\perp 1}$ ($\theta_{\perp 2}$), the photon measured by observer two (one) must be polarized at a unique orientation θ_2 (θ_1) or $\theta_{\perp 2}$ ($\theta_{\perp 1}$). Surprisingly, after a large number of measurements, the two distant observers conclude that although the polarization of neither photon one nor photon two is defined, if one of the photons is measured with a polarization along a certain orientation, the polarization of its twin is uniquely determined, despite the distance between the two observers. Remember, (1) there is no interaction between the two photons during their propagation and there is no communication between the two observers either; (2) the ensemble is prepared neither with photon pairs of $|\hat{\theta}_1\rangle$ & $|\hat{\theta}_2\rangle$ and $|\hat{\theta}_{\perp 1}\rangle$ & $|\hat{\theta}_{\perp 2}\rangle$, nor with photon pairs of $|\hat{\theta}_1\rangle$ & $|\hat{\theta}_{\perp 2}\rangle$ and $|\hat{\theta}_{\perp 1}\rangle$ & $|\hat{\theta}_2\rangle$, except when $\hat{\theta} = \hat{o}$, or $\hat{\theta} = \hat{e}$.

Here, again, a "surprise" means that the quantum predication is different from the classical convention and it seems impossible to understand it under the framework of classical physics.

(3) **Mixed state**

It is very clear that Bell's states in Eq. (2.7.2) are very different from the following mixed states

$$\hat{\rho}^{(S)} = \frac{1}{2} \left[\, |\,\hat{o}_1\,\rangle\langle\,\hat{o}_2\,| + |\,\hat{e}_1\,\rangle\langle\,\hat{e}_2\,| \,\right],$$

$$\hat{\rho}^{(A)} = \frac{1}{2} \left[\, |\,\hat{o}_1\,\rangle\langle\,\hat{e}_2\,| + |\,\hat{e}_1\,\rangle\langle\,\hat{o}_2\,| \,\right], \tag{2.7.6}$$

where the ensemble is prepared with one-half of the photon–pairs with polarization $\hat{o}_1$ & $\hat{o}_2$ ($\hat{o}_1$ & $\hat{e}_2$) and the other one-half photon–pairs with polarization $\hat{e}_1$ & $\hat{e}_2$ ($\hat{e}_1$ & $\hat{o}_2$) for $\hat{\rho}^{(S)}$ ($\hat{\rho}^{(A)}$). Although in $\hat{\rho}^{(S)}$ ($\hat{\rho}^{(A)}$) the two distant observers will receive half of the pairs in $\hat{o}_1$ & $\hat{o}_2$ ($\hat{o}_1$ & $\hat{e}_2$) and another half of the pairs in $\hat{e}_1$ & $\hat{e}_2$ ($\hat{e}_1$ & $\hat{o}_2$), neither $\hat{\rho}^{(S)}$ nor $\hat{\rho}^{(A)}$ can achieve any correlation similar to that in Eq. (2.7.5), if θ_1 and θ_2 are chosen at orientations other then $\hat{o}$ and $\hat{e}$.

Bibliography

Baym G., *Lectures on Quantum Mechanics*, Benjamin/Cummings, 1981.

Glauber R.J., *Phys. Rev.* **130**, 2529 (1963); *Phys. Rev.* **131**, 2766 (1963).

Loudon R., *The Quantum Theory of Light*, Oxford Science, 2000.

Mandel L. and Wolf E., *Optical Coherence and Quantum Optics*, Cambridge, London, 1995.

Scully M.O., and Zubairy M.S., *Quantum Optics*, Cambridge University Press, Cambridge, 1997.

Chapter 3

Measurement of Light: Photon Counting and Multi-Photon Joint-Detection

In the Maxwell continuum wave theory of light, a photodetector measures the intensity of the EM wave. The output current of the photodetector is proportional to the intensity of input light. Since the introduction of the concept of photon, or the concept of field quantization, physicists started to premeditate single-photon detection theoretically and experimentally. Not until the 1950s, did we have the ability to observe a single light quantum, not theoretically but experimentally. Is this ability important? In fact, it is important not only experimentally but also theoretically. As we know, for a certain period of time, we used to consider a quantum state of radiation only meaningful in terms of an ensemble. For example, in the early experiments of Young's double-slit interferometer, the interference was judged by human eyes. Human eyes do not work well at weak light conditions. In terms of the measurement of an ensemble, the interpretation of Young's interference is reasonable and simple: 50% of the photons passed slit-A and another 50% passed slit-B; the photons that passed slit-A "meet" with the photons that passed slit-B at different locations in the observation plane. Due to the slight differences in their optical paths, at certain locations, the radiations superposed constructively, while at other locations, the field superposed destructively. A sinusoidal modulation is thus observable. However, we found this interpretation may not be appropriate when the experimental physicists improved the sensitivity of their photodetector to the single-photon level. Almost everyone was surprised that the interference was still observable in the observation plane, and the sinusoidal modulation observed from single-photon was exactly the same as before. A single-photon counting detector is able to count photons one by one. How does a photon pass the double-slit then? Does it pass one

77

slit or both slits? One may not care which slit a photon passes, but we do need to understand the measurement process: How does a photoelectron or a pair of photoelectrons physically relate to the quantized radiation field? Nowadays, we may easily find a statement like "... the observed photon pair is in an entangled state ... ". What does it mean?

3.1 Measurement of Einstein's Bundle of Ray

Einstein's granularity picture of light is still in the framework of Maxwell's electromagnetic wave theory, except that Einstein introduced the concept of bundle of ray, or quantized subfield, to radiation and abandoned the continuum interpretation of Maxwell. Einstein's bundle of ray, light quantum, or subfield is a realistic quantized microscopic entity of electromagnetic wavepacket propagating in space–time. In the framework of electromagnetic field theory, Einstein introduced an additional freedom of superposition in terms of quantized subfields: the superposition of a large number of coherent or incoherent bundle of rays, or quantized subfields, in addition to the superposition of a large number of coherent or incoherent harmonic radiation modes. In Einstein's granularity picture, we have written the radiation field measured at coordinate $(\mathbf{r}, t)$ as the result of a superposition among a large number of subfields, each created from a point-like sub-source at $(\mathbf{r}_m, t_m)$,

$$E(\mathbf{r}, t) = \sum_m E_m(\mathbf{r}, t) = \sum_m E_m(\mathbf{r}_m, t_m)\, g_m(\mathbf{r}, t),$$

where $E_m(\mathbf{r}_m, t_m)$ labels the subfield emitted from the mth sub-source at coordinate $(\mathbf{r}_m, t_m)$, and $g_m(\mathbf{r}, t)$ represents the field propagator or Green's function that propagates the mth subfield from coordinate $(\mathbf{r}_m, t_m)$ to coordinate $(\mathbf{r}, t)$. If there exist more than one different yet indistinguishable paths or alternatives for the mth subfield to propagate from $(\mathbf{r}_m, t_m)$ to $(\mathbf{r}, t)$, $g_m(\mathbf{r}, t)$ must be written as a superposition

$$g_m(\mathbf{r}, t) = \frac{1}{\sqrt{N}} \sum_{s=1}^{N} g_{ms}(\mathbf{r}, t),$$

where $g_m(\mathbf{r}, t)$ is normalized, $g_{ms}(\mathbf{r}, t)$ indicates the sth path of the mth subfield.

(I) **Measurement of** $\langle I(\mathbf{r}, t) \rangle$

Assume that a point-like photodetector D, placed at coordinate $\mathbf{r}$, measures the intensity of a radiation from a natural light source at time t, the expectation value of its measured intensity at $(\mathbf{r}, t)$ is

$$\langle I(\mathbf{r}, t) \rangle = \langle E^*(\mathbf{r}, t) E(\mathbf{r}, t) \rangle$$

$$= \left\langle \sum_m E_m^*(\mathbf{r}, t) \sum_n E_n(\mathbf{r}, t) \right\rangle$$

$$= \left\langle \sum_m \left| E_m(\mathbf{r}, t) \right|^2 \right\rangle + \left\langle \sum_{m \neq n} E_m^*(\mathbf{r}, t) E_n(\mathbf{r}, t) \right\rangle$$

$$= \sum_m \left| E_m(\mathbf{r}, t) \right|^2 + 0$$

$$= \sum_m I_m(\mathbf{r}, t). \tag{3.1.1}$$

In Eq. (3.1.1), the ensemble average makes the second term vanishing. Equation (3.1.1) connects the expectation of the measured intensity, which is proportional to the mean current of the photodetector, $\langle i(\mathbf{r}, t) \rangle \propto \langle I(\mathbf{r}, t) \rangle$, with Einstein's concept of subfield or "bundle of ray". Equation (3.1.1) indicates that $\langle I(\mathbf{r}, t) \rangle$ is the sum of subintensities, $\sum_m I_m(\mathbf{r}, t)$, and each $I_m(\mathbf{r}, t)$ is the result of the mth subfield interfering with the mth subfield itself. The interference between different subfields, corresponding to the $m \neq n$ term, vanishes while taking into account all possible random phases of the subfields,

$$\langle \Delta I(\mathbf{r}, t) \rangle = \left\langle \sum_{m \neq n} E_m^*(\mathbf{r}, t) \, E_n(\mathbf{r}, t) \right\rangle = 0. \tag{3.1.2}$$

However, if the measurement is under weak light condition, or if the measurement happens within a short time window in which only a few photons or a limited number of photons contribute to the measurement, so that the measurement is unable to take into account all possible random phases of the subfields, the $m \neq n$ term may not vanish and is observable from the measurement of a photodetector. In this case, the $m \neq n$ terms contribute to the intensity a fluctuation:

$$\Delta I(\mathbf{r}, t) = \sum_{m \neq n} E_m^*(\mathbf{r}, t) E_n(\mathbf{r}, t) \neq 0. \tag{3.1.3}$$

Equation (3.1.3) indicates $\Delta I(\mathbf{r}, t)$ is the sum of a set of interferences between different subfields. Due to the random phases associated with the subfields, $\Delta I(\mathbf{r}, t)$ takes random positive or negative values in the neighborhood of $\langle I(\mathbf{r},t)\rangle$ from time to time or from measurement to measurement. $\Delta I(\mathbf{r},t)$ is usually considered "noise" in the measurement of $\langle I(\mathbf{r},t)\rangle$.

(II) Measurement of $\langle I(\mathbf{r}_1,t_1)I(\mathbf{r}_2,t_2)\rangle$

Now, we set up a correlation measurement of thermal field by means of joint photodetection between two point-like photodetectors D_1 and D_2 placed at coordinates $\mathbf{r}_1$ and $\mathbf{r}_2$, respectively. Figure 3.1.1 schematically illustrates such a measurement. The output currents of D_1 and D_2 are amplified by two individual amplifiers, respectively. The output currents, $i_1(t)$ and $i_2(t)$, of the two amplifiers are multiplied by an electronic linear multiplier or RF mixer at time t. The correlation of $i_1(t)$ and $i_2(t)$ is proportional to the correlation of the intensities measured by D_1 and D_2 at early times t_1 and t_2, respectively, $\langle i_1(t)i_2(t)\rangle \propto \langle I(\mathbf{r}_1,t_1)I(\mathbf{r}_2,t_2)\rangle$, where $t_1 = t - \tau_1^e$, $t_2 = t - \tau_2^e$ are the times of the two photodetection events of D_1 and D_2. τ_1^e and τ_2^e are the electronic time delays, including the delays of the detectors, the amplifiers, and the interconnection or delay-line cables. The expectation value of $\langle I(\mathbf{r}_1,t_1)I(\mathbf{r}_2,t_2)\rangle$ is calculated in Einstein's picture as follows.

$$\langle I(\mathbf{r}_1,t_1)I(\mathbf{r}_2,t_2)\rangle = \langle E^*(\mathbf{r}_1,t_1)E(\mathbf{r}_1,t_1)E^*(\mathbf{r}_2,t_2)E(\mathbf{r}_2,t_2)\rangle$$

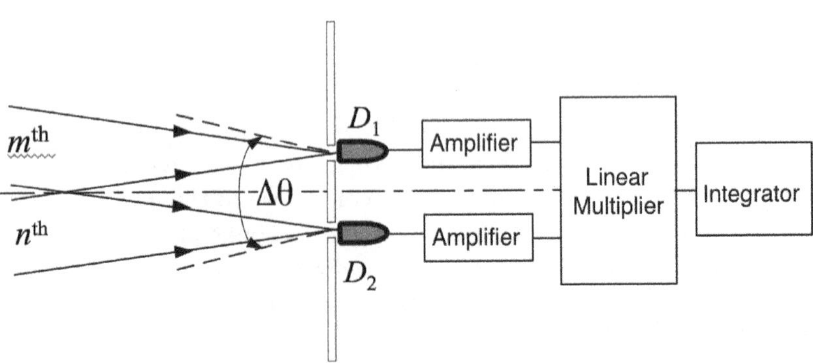

Fig. 3.1.1 Schematic setup for the intensity correlation measurement of $\langle I(\mathbf{r}_1,t_1) I(\mathbf{r}_2,t_2)\rangle$ by two point-like photodetectors placed at coordinates $\mathbf{r}_1,t_1$ and $\mathbf{r}_2,t_2$, respectively. An electronic linear multiplier, or RF mixer, is employed to multiply the two currents $i_1(t)$ and $i_2(t)$ at time t. Note that the times of the two photodetection events t_1 and t_2, respectively, as well as the relative time delay $t_1 - t_2 = \tau_1^e - \tau_2^e$ are all defined by the electronics.

$$= \left\langle \sum_m E_m^*(\mathbf{r}_1, t_1) \sum_n E_n(\mathbf{r}_1, t_1) \sum_p E_p^*(\mathbf{r}_2, t_2) \sum_q E_q(\mathbf{r}_2, t_2) \right\rangle$$

$$= \sum_m E_m^*(\mathbf{r}_1, t_1) \, E_m(\mathbf{r}_1, t_1) \sum_n E_n^*(\mathbf{r}_2, t_2) \, E_n(\mathbf{r}_2, t_2)$$

$$+ \sum_{m \neq n} E_m^*(\mathbf{r}_1, t_1) \, E_n(\mathbf{r}_1, t_1) \, E_n^*(\mathbf{r}_2, t_2) \, E_m(\mathbf{r}_2, t_2)$$

$$= \sum_{m \neq n} \left| \frac{1}{\sqrt{2}} \left[E_m(\mathbf{r}_1, t_1) \, E_n(\mathbf{r}_2, t_2) + E_n(\mathbf{r}_1, t_1) \, E_m(\mathbf{r}_2, t_2) \right] \right|^2$$

$$= \langle I(\mathbf{r}_1, t_1) \rangle \langle I(\mathbf{r}_2, t_2) \rangle + \langle \Delta I(\mathbf{r}_1, t_1) \Delta I(\mathbf{r}_2, t_2) \rangle, \tag{3.1.4}$$

where $m, n, p,$ and q label the randomly created and randomly distributed subfields and their corresponding point-like sub-sources. Considering the random phases of the subfields, as the result of destructive interference cancelation when taking into account all possible phases of the subfields, the only surviving terms from the ensemble sum are the following two groups: (1) $m = q, n = p$, (2) $m = p, n = q$. $\langle I(\mathbf{r}_1, t_1) I(\mathbf{r}_2, t_2) \rangle$ can be written into the form of Eq. (3.1.4). The first terms of Eq. (3.1.4) is a product of two mean intensities measured by D_1 and D_2, respectively. The second term of Eq. (3.1.4) is the intensity fluctuation correlation measured by D_1 and D_2 jointly. For randomly created thermal radiations, the first term usually contributes a constant to $\langle I(\mathbf{r}_1, t_1) I(\mathbf{r}_2, t_2) \rangle$, and the second term contributes a nontrivial function of space–time $(\mathbf{r}_1, t_1)$ and $(\mathbf{r}_2, t_2)$ to $\langle I(\mathbf{r}_1, t_1) I(\mathbf{r}_2, t_2) \rangle$ in terms of either temporal delay or spatial separation between the two photodetection events of D_1 and D_2. Mathematically, it is straightforward to find that Eq. (3.1.4) indicates an interference phenomenon: A randomly created and randomly paired subfields, or photons in thermal state, interfere with the pair itself. Physically, it is easy to find that the mth–nth pair of subfields or photons has two different yet indistinguishable alternatives to produce a joint photodetection event: (1) the mth subfield, or photon, is annihilated at D_1 while the nth subfield, or photon, is annihilated at D_2; (2) the nth subfield, or photon, is annihilated at D_1 while the mth subfield, or photon, is annihilated at D_2. We name this interference or superposition "two-photon interference".

It might be necessary to emphasize that the intensity fluctuation correlation $\langle \Delta I(\mathbf{r}_1, t_1) \Delta I(\mathbf{r}_2, t_2) \rangle$, corresponding to the cross term of Eq. (3.1.4), is the result of two-photon interference, but not the intrinsic statistical correlation of the thermal fields. We cannot ignore that the thermal field is created from stochastic processes in a complete random manner.

(III) **Measurement of** $\langle \Delta I(\mathbf{r}_1, t_1) \Delta I(\mathbf{r}_2, t_2) \rangle$

It is also easy to find from Eq. (3.1.4) the two-photon, or two-subfield, interference-induced intensity fluctuation correlation of randomly created and randomly distributed thermal radiation is observable when

$$\langle \Delta I(\mathbf{r}_1, t_1) \Delta I(\mathbf{r}_2, t_2) \rangle$$

$$= \sum_{m \neq n} E_m^*(\mathbf{r}_1, t_1) \, E_n(\mathbf{r}_1, t_1) \, E_n^*(\mathbf{r}_2, t_2) \, E_m(\mathbf{r}_2, t_2) \neq 0. \qquad (3.1.5)$$

Indeed, it is the cross-interference term of the two-subfield or two-photon superposition of Eq. (3.1.4) that results in the nontrivial intensity fluctuation correlation.

How does a randomly created thermal radiation have its nontrivial intensity fluctuation correlation in terms of the temporal delay or spatial separation of the measured fields at $(\mathbf{r}_1, t_1)$ and $(\mathbf{r}_2, t_2)$? To estimate the correlation, we may rewrite Eq. (3.1.5) in terms of the superposition of a large number of subfields and Fourier modes. To simplify the math, we calculate the temporal correlation only by taking the Fourier transform in 1D as usual:

$$\langle \Delta I(r_1, t_1) \Delta I(r_2, t_2) \rangle$$

$$= \sum_{m \neq n} \int d\omega \, E_m^*(\omega) e^{i(\omega t_1 - k r_1)} \int d\omega' E_n(\omega') e^{-i(\omega' t_1 - k' r_1)}$$

$$\times \int d\omega'' \, E_n^*(\omega'') e^{i(\omega'' t_2 - k'' r_2)} \int d\omega'' E_m(\omega''') e^{-i(\omega''' t_2 - k''' r_2)}$$

$$= \sum_{m \neq n} \mathcal{F}_{\tau_1}^* \{ E_m(\nu) \} \, \mathcal{F}_{\tau_1} \{ E_n(\nu) \} \, \mathcal{F}_{\tau_2}^* \{ E_n(\nu) \} \, \mathcal{F}_{\tau_2} \{ E_m(\nu) \} \qquad (3.1.6)$$

indicating the product of two wavepackets and their conjugates. This means the two-photon interference of the m-nth pair of wavepackets will contribute a nonzero value to $\langle \Delta I(r_1, t_1) \Delta I(r_2, t_2) \rangle$ when the two wavepackts "overlap" in space–time τ_1 and τ_2. We have more detailed discussions on the measurement of intensity fluctuation correlation in later chapters.

3.2   Time-Dependent Perturbation Theory

In certain problems of practice, the Schrödinger equation may not be solved with exact solutions, but only with a good approximation. In quantum optics, we usually face a problem in finding the eigenstates and the eigenvalues of a Hamiltonian in the form of $H_0 + V(t)$, where $V(t)$ is time-dependent and H_0 has exact solutions. Photodetection is typically formulated as time-dependent perturbation. We have a very brief review of the subject. A model and calculation of the photon counting process will follow as an exercise of time-dependent perturbation theory.

Suppose we shine light on an atomic system at time t_0 and ask what are the chances that the light ionizes the atoms or releases one photoelectron, two photoelectrons, three photoelectrons, ..., or N photoelectrons from the atomic system. Alternatively, the problem may be stated in a more general way: We begin with a system of Hamiltonian H_0 and then add a time-dependent perturbation $V(t)$ at time t_0. What changes of the eigenstates and eigenvalues of the system occurred due to the perturbation?

It is assumed that, before time t_0, the system is in state $|\Psi^{(0)}\rangle$ which is an exact solution of the Schrödinger equation

$$i\hbar \frac{\partial}{\partial t} |\Psi^{(0)}\rangle = H_0 |\Psi^{(0)}\rangle \quad \text{for } t < t_0 \tag{3.2.1}$$

and the time-dependent perturbation $V(t)$ is applied at t_0. To find the new state $|\Psi^{(t)}\rangle$ of the system we need to solve the Schrödinger equation

$$i\hbar \frac{\partial}{\partial t} |\Psi^{(t)}\rangle = [H_0 + V(t)] |\Psi^{(t)}\rangle \quad \text{for } t > t_0 \tag{3.2.2}$$

subject to the boundary condition $|\Psi^{(0)}\rangle = |\Psi^{(t)}\rangle$ at $t = t_0$. Although we may not be able to find exact solutions for Eq. (3.2.2), a good approximation of $|\Psi^{(t)}\rangle$ is permissible in terms of the superposition

$$|\Psi^{(t)}\rangle = \sum_j c_j |j\rangle, \tag{3.2.3}$$

where c_j is the probability amplitude for the system staying in the jth eigenstate, $|j\rangle$, of H_0, and $|j\rangle$ forms a complete orthonormal vector basis. Furthermore, if $V(t)$ is much smaller than H_0, only a few lower-order terms of the expansion are necessary for a good approximation. In this case, $V(t)$ is treated as a perturbation. In the following discussion, we assume that

$V(t) \ll H_0$. Thus, $|\Psi^{(t)}\rangle$ can be approximated as

$$|\Psi^{(t)}\rangle = e^{-iH_0 t/\hbar} |\Psi(t)\rangle, \tag{3.2.4}$$

where $e^{-iH_0 t/\hbar}$ dominates the time dependence of $|\Psi^{(t)}\rangle$. It is easy to see that this term contributes to the jth eigenstate of H_0 a phase factor $e^{-iE_j t/\hbar} = e^{-i\omega_j t}$.

Substituting Eq. (3.2.4) into Eq. (3.2.2), we obtain the Schrödinger equation that $|\Psi(t)\rangle$ must obey

$$i\hbar \frac{\partial}{\partial t} |\Psi(t)\rangle = H_I(t) |\Psi(t)\rangle, \tag{3.2.5}$$

where

$$H_I(t) = e^{iH_0 t/\hbar} V(t) e^{-iH_0 t/\hbar}. \tag{3.2.6}$$

The state $|\Psi(t)\rangle$ and the operator $H_I(t)$ in Eq. (3.2.5) are said to be in the interaction representation. In quantum mechanics, Eq. (3.2.6) defines the interaction representation for any operator.

Equation (3.2.5) can be formally integrated as

$$|\Psi(t)\rangle = \left[e^{\frac{1}{i\hbar} \int_{t_0}^{t} dt' \, H_I(t')} \right]_T |\Psi(t_0)\rangle, \tag{3.2.7}$$

where T stands for time order. The first three leading terms of the state, i.e., the zero-order, the first-order and the second-order changes of the state are thus

$$|\Psi(t)\rangle \cong |\Psi(t)\rangle^{(0)} + |\Psi(t)\rangle^{(1)} + |\Psi(t)\rangle^{(2)}$$

$$= |\Psi(t_0)\rangle - \frac{i}{\hbar} \int_{t_0}^{t} dt' \, H_I(t') \, |\Psi(t_0)\rangle$$

$$+ \frac{1}{(i\hbar)^2} \int_{t_0}^{t} dt' \int_{t_0}^{t'} dt'' \, H_I(t') \, H_I(t'') \, |\Psi(t_0)\rangle. \tag{3.2.8}$$

The notation $|\Psi(t)\rangle^{(s)}$, $s = 0, 1, 2 \ldots$, indicates the sth order changes of the state. It should be emphasized that the operators in $|\Psi(t)\rangle^{(2)}$ are in time order. Care has to be taken when the operators do not commute and, in general, this is the more likely case.

It is interesting to find, term by term, the physical correspondence of Eq. (3.2.8), in connection with the photodetection process. The zero-order term $|\Psi(t)\rangle^{(0)}$, of course, is the no-action term. The state of the atomic system and the state of the field have no changes. The second term,

$|\Psi(t)\rangle^{(1)}$, and the third term, $|\Psi(t)\rangle^{(2)}$, in Eq. (3.2.8) are the first-order and the second-order changes of the state corresponding to the annihilation of one photon and two photons from the field, respectively.

3.3   Measurement of Light: Photon Counting

In classical theory, a photodetector measures the intensity of a light wave and thus relates to the electromagnetic fields in a phenomenological way. It is quantum theory that provides a successful solution to the physical process of atomic transition: photon annihilation. Although a detailed study of the theory is not the goal of the following discussion, a brief review of the concepts is necessary for understanding the physics of photon counting detection and the measurement of quantum correlations. The following review follows Glauber's theory of photodetection.

We start by defining an idealized photon counting detector as a point-like detector. One may imagine it as an atom with all concerned transition probabilities in terms of the frequencies of the light field. H_0 has two parts:

$$H_0 = H_0^A + H_0^F, \tag{3.3.1}$$

where H_0^A and H_0^F are the Hamiltonian for the free atom and the free field, with complete sets of orthonormal eigenstates $|j\rangle$ and $|n\rangle$, respectively. To simplify the analysis, we assume the atom is in its ground state $|0\rangle$ before the interaction. The initial state of the field, $|\Psi_F^{(0)}\rangle$, will be treated more generally as a superposition, in terms of the eigenstates, $|n\rangle$, of $H_0^F = \hbar\omega\hat{N}$, where $\hat{N}$ is the number operator. To simplify the discussion, we restrict ourselves to single mode. The result can be easily generalized to the multi-mode case. Assume that the atom is ionized from its ground state due to the dipole–field interaction, $V(t) = -\hat{\mu} \cdot \mathbf{E}^{(+)}$, where $\hat{\mu}$ is the dipole moment of the atom. The interaction Hamiltonian is thus

$$H_I(t) = e^{iH_0 t/\hbar}\, V(t)\, e^{-iH_0 t/\hbar}$$

$$= -e^{iH_0^A t/\hbar}\, \hat{\mu}\, e^{-iH_0^A t/\hbar}\, e^{iH_0^F t/\hbar}\, E^{(+)}\, e^{-iH_0^F t/\hbar}, \tag{3.3.2}$$

where we have simplified the problem to 1D. Following Eq. (3.2.8), the state $|\Psi(t)\rangle$ in first-order approximation is thus

$$|\Psi(t)\rangle \cong |0\rangle|\Psi_F^{(0)}\rangle - \frac{i}{\hbar}\int_{t_0}^t dt'\, e^{iH_0 t'/\hbar}\, (-\hat{\mu}\, E^{(+)})\, e^{-iH_0 t'/\hbar}\, |0\rangle|\Psi_F^{(0)}\rangle. \tag{3.3.3}$$

The total probability of ionizing the atom from its ground state to a set of possible final states by absorbing a photon from the field is thus

$$
P = \sum_f |\langle f | \Psi(t) \rangle|^2 \cong \frac{1}{\hbar^2} \int_{t_0}^{t} dt' \int_{t_0}^{t'} dt''
$$

$$
\times \sum_{j,n} \langle 0 | e^{-iH_0^A t''/\hbar} \, \hat{\mu}^* \, e^{iH_0^A t''/\hbar} | j \rangle \langle j | e^{iH_0^A t'/\hbar} \, \hat{\mu} \, e^{-iH_0^A t'/\hbar} | 0 \rangle
$$

$$
\times \langle \Psi_F^{(0)} | e^{-iH_0^F t''/\hbar} E^{(-)}(t'') e^{iH_0^F t''/\hbar} | n \rangle
$$

$$
\times \langle n | e^{iH_0^F t'/\hbar} E^{(+)}(t') e^{-iH_0^F t'/\hbar} | \Psi_F^{(0)} \rangle
$$

$$
\cong \frac{1}{\hbar^2} \int_{t_0}^{t} dt' \int_{t_0}^{t'} dt'' \sum_{j,n} e^{-i\left(\frac{E_j - E_0}{\hbar} - \omega\right)(t'' - t')} \langle 0 | \hat{\mu}^* | j \rangle \langle j | \hat{\mu} | 0 \rangle
$$

$$
\times \langle \Psi_F^{(0)} | E^{(-)}(t'') | n \rangle \langle n | E^{(+)}(t') | \Psi_F^{(0)} \rangle. \tag{3.3.4}
$$

To complete the calculation, we first consider summing Eq. (3.3.4) over all final states. Suppose $\langle j | \hat{\mu} | 0 \rangle$ and $\langle n | E^{(+)} | \Psi_F^{(0)} \rangle$ do not change violently within the group of final states. The sum turns into an integral over the energy E_j with a certain density distribution function. We further assume a smooth density distribution function. The total transition probability is thus proportional to

$$
P \propto \int_{t_0}^{t} dt' \langle \Psi_F^{(0)} | E^{(-)}(t') E^{(+)}(t') | \Psi_F^{(0)} \rangle, \tag{3.3.5}
$$

where we have used the completeness relation

$$
\sum_m | m \rangle \langle m | = 1 \tag{3.3.6}
$$

and the mathematical approximation for broadband integral

$$
\int dE_j \, e^{-i\left(\frac{E_j - E_0}{\hbar} - \omega\right)(t'' - t')} \sim \delta(t'' - t'). \tag{3.3.7}
$$

The transition rate of ionizing the atom by absorbing a photon from the field at coordinate $\mathbf{r}$ between time t and $t + dt$ is thus

$$
G^{(1)}(\mathbf{r}, t) \propto \langle \Psi_F^{(0)} | E^{(-)}(\mathbf{r}, t) E^{(+)}(\mathbf{r}, t) | \Psi_F^{(0)} \rangle. \tag{3.3.8}
$$

In other words, we may conclude that the probability of annihilating a photon from the radiation field of state $|\Psi\rangle$, at space–time point $(\mathbf{r}, t)$, is proportional to

$$G^{(1)}(\mathbf{r}, t) = \langle\Psi| E^{(-)}(\mathbf{r}, t)\, E^{(+)}(\mathbf{r}, t)|\Psi\rangle = \left| E^{(+)}(\mathbf{r}, t)|\Psi\rangle \right|^2.$$

Note that a pure state has been used in the above discussion. Equation (3.3.8) can be easily generalized to a mixed state

$$G^{(1)}(\mathbf{r}, t) = \sum_j P_j \langle\Psi_j| E^{(-)}(\mathbf{r}, t)\, E^{(+)}(\mathbf{r}, t)|\Psi_j\rangle$$

$$= Tr\,[\,\hat{\rho}\, E^{(-)}(\mathbf{r}, t)\, E^{(+)}(\mathbf{r}, t)\,], \tag{3.3.9}$$

where P_j is the probability of being in the jth state $|\Psi_j\rangle$, and

$$\hat{\rho} = \sum_j P_j\, |\Psi_j\rangle\langle\Psi_j| \tag{3.3.10}$$

is the density operator representing the mixed state of the radiation field or the system of photons. Note that we have dropped the superscript from the state $|\Psi_j^{(0)}\rangle$.

If the mixed state can be formally written into an incoherent superposition

$$|\Psi\rangle = \sum_j e^{i\varphi_j}|\Psi_j\rangle,$$

where φ_j is a random number. The probability of annihilating a photon from the radiation field can be calculated, in general, from

$$G^{(1)}(\mathbf{r}, t) = \langle\langle\Psi| E^{(-)}(\mathbf{r}, t)\, E^{(+)}(\mathbf{r}, t)|\Psi\rangle\rangle_{\mathrm{En}}. \tag{3.3.11}$$

$G^{(1)}(\mathbf{r}, t)$ is a function of $(\mathbf{r}, t)$ and is often called the first-order coherence function of the measured radiation field.

The time-averaged photon counting rate is thus

$$R \propto \frac{1}{T}\int_0^T dt\, G^{(1)}(\mathbf{r}, t) = \int_T dt\, \langle\, E^{(-)}(\mathbf{r}, t)\, E^{(+)}(\mathbf{r}, t)\,\rangle, \tag{3.3.12}$$

where T is the total time of averaging.

3.4   Measurement of Light: Joint Detection of Photons

In the following, we calculate the probability of the joint detection of photon pairs by two independent point-like photon counting detectors, such as two independent atoms. To simplify the analysis, we assume the two independent atoms are both in their ground state, $|0_1\rangle$ and $|0_2\rangle$, before the interaction, where the subscripts label the individual atoms, or the point-like photon counting detectors. The Hamiltonian H_0 has three parts:

$$H_0 = H_{01}^A + H_{02}^A + H_0^F, \tag{3.4.1}$$

where H_{0j}^A, $j = 1, 2$, is the Hamiltonian of the jth free atom. Assume that the two atoms are both ionized from their ground state due to the dipole–field interaction $V_1(t) = -\hat{\mu}_1 \cdot \mathbf{E}_1^{(+)}$ and $V_2(t) = -\hat{\mu}_2 \cdot \mathbf{E}_2^{(+)}$, where $\hat{\mu}_j$, $j = 1, 2$ is the dipole moment of the jth atom. The interaction Hamiltonian is thus

$$\begin{aligned}
H_I(t) &= e^{iH_0 t/\hbar} \left[V_1(t) + V_2(t)\right] e^{-iH_0 t/\hbar} \\
&= -e^{iH_{01}^A t/\hbar}\, \hat{\mu}_1\, e^{-iH_{01}^A t/\hbar}\, e^{iH_0^F t/\hbar}\, E_1^{(+)}\, e^{-iH_0^F t/\hbar} \\
&\quad - e^{iH_{02}^A t/\hbar}\, \hat{\mu}_2\, e^{-iH_{02}^A t/\hbar}\, e^{iH_0^F t/\hbar}\, E_2^{(+)}\, e^{-iH_0^F t/\hbar} \\
&= H_{I1}(t) + H_{I2}(t). \tag{3.4.2}
\end{aligned}$$

Following Eq. (3.2.8), the first-order change of the state, $|\Psi(t)\rangle^{(1)}$, is thus

$$|\Psi(t)\rangle^{(1)} = -\frac{i}{\hbar} \int_{t_0}^{t} dt'\, H_{I1}(t')\, |\Psi(t_0)\rangle - \frac{i}{\hbar} \int_{t_0}^{t} dt'\, H_{I2}(t')\, |\Psi(t_0)\rangle.$$

The two terms correspond to the single-photon transition of atoms one and two, respectively. The second-order change of the state, $|\Psi(t)\rangle^{(2)}$, has four terms:

$$\begin{aligned}
|\Psi(t)\rangle^{(2)} = {}&\frac{1}{(i\hbar)^2} \int_{t_0}^{t} dt' \int_{t_0}^{t'} dt''\, H_{I1}(t')\, H_{I1}(t'')\, |\Psi(t_0)\rangle \\
&+ \frac{1}{(i\hbar)^2} \int_{t_0}^{t} dt' \int_{t_0}^{t'} dt''\, H_{I2}(t')\, H_{I2}(t'')\, |\Psi(t_0)\rangle \\
&+ \frac{1}{(i\hbar)^2} \int_{t_0}^{t} dt' \int_{t_0}^{t'} dt''\, H_{I1}(t')\, H_{I2}(t'')\, |\Psi(t_0)\rangle \\
&+ \frac{1}{(i\hbar)^2} \int_{t_0}^{t} dt' \int_{t_0}^{t'} dt''\, H_{I2}(t')\, H_{I1}(t'')\, |\Psi(t_0)\rangle.
\end{aligned}$$

All four terms correspond to the joint annihilation of two photons from the field. The first two terms correspond to the two-photon transition associated with an atom that contributes to the single-detector counting rate. For the joint detection events of two photodetectors, we are interested in the last two terms of this group. Assume that the coincidence detection circuit only selects one of these joint detection events by time ordering. The total probability of jointly ionizing atom one and atom two from their ground states to a set of possible final states is thus

$$\sum_f |\langle f | \Psi(t) \rangle|^2$$

$$\cong \frac{1}{\hbar^4} \int_{t_0}^{t_1} dt_1' \int_{t_0}^{t_1'} dt_1'' \int_{t_0}^{t_2} dt_2' \int_{t_0}^{t_2'} dt_2''$$

$$\times \sum_{j_1, j_2, n,} \langle 0_1 | e^{-iH_{01}^A t_1''/\hbar} \hat{\mu}_1^* e^{iH_{01}^A t_1''/\hbar} | j_1 \rangle \langle j_1 | e^{iH_{01}^A t_1'/\hbar} \hat{\mu}_1 e^{-iH_{01}^A t_1'/\hbar} | 0_1 \rangle$$

$$\times \langle 0_2 | e^{-iH_{02}^A t_2''/\hbar} \hat{\mu}_2^* e^{iH_{02}^A t_2''/\hbar} | j_2 \rangle \langle j_2 | e^{iH_{02}^A t_2'/\hbar} \hat{\mu}_2 e^{-iH_{02}^A t_2'/\hbar} | 0_2 \rangle$$

$$\times \langle \Psi_F^{(0)} | e^{-iH_{01}^F t_1''/\hbar} e^{-iH_{02}^F t_2''/\hbar} E_1^{(-)}(t_1'') E_2^{(-)}(t_2'') e^{iH_{02}^F t_2''/\hbar} e^{iH_{01}^F t_1''/\hbar} | n \rangle$$

$$\times \langle n | e^{iH_{01}^F t_1'/\hbar} e^{iH_{02}^F t_2'/\hbar} E_2^{(+)}(t_2') E_1^{(+)}(t_1') e^{-iH_{02}^F t_2'/\hbar} e^{-iH_{01}^F t_1'/\hbar} | \Psi_F^{(0)} \rangle.$$

$$(3.4.3)$$

Similar to the single detection case, to complete the calculation, we first consider summing Eq. (3.4.3) over all final states and then apply the completeness relation and the delta function approximation. The total joint transition probability is thus proportional to

$$P \propto \int_{t_0}^{t_1} dt_1' \int_{t_0}^{t_2} dt_2' \langle \Psi_F^{(0)} | E_1^{(-)}(t_1') E_2^{(-)}(t_2') E_2^{(+)}(t_2') E_1^{(+)}(t_1') | \Psi_F^{(0)} \rangle.$$

$$(3.4.4)$$

The joint transition rate of ionizing the two atoms by absorbing a photon from the field at coordinate $\mathbf{r}_1$ between time t_1 and $t_1 + dt_1$ and another photon from the field at coordinate $\mathbf{r}_2$ between time t_2 and $t_2 + dt_2$ is thus

$$G^{(2)}(\mathbf{r}_1, t_1; \mathbf{r}_2, t_2)$$

$$\propto \langle \Psi_F^{(0)} | E_1^{(-)}(\mathbf{r}_1, t_1) E_2^{(-)}(\mathbf{r}_2, t_2) E_2^{(+)}(\mathbf{r}_2, t_2) E_1^{(+)}(\mathbf{r}_1, t_1) | \Psi_F^{(0)} \rangle.$$

$$(3.4.5)$$

In other words, we may conclude that the probability of annihilating a pair of photons from the radiation field of state $|\Psi\rangle$, at space–time point $(\mathbf{r}_1, t_1)$

and $(\mathbf{r}_2, t_2)$, is proportional to

$$G^{(2)}(\mathbf{r}_1, t_1; \mathbf{r}_2, t_2) = \langle \Psi | E_1^{(-)}(\mathbf{r}_1, t_1)\, E_2^{(-)}(\mathbf{r}_2, t_2)\, E_2^{(+)}(\mathbf{r}_2, t_2)\, E_1^{(+)}(\mathbf{r}_1, t_1)\, | \Psi \rangle.$$

Note again that we have been using a pure state in the above calculation. The result of Eq. (3.4.5) can easily be generalized to a mixed state

$$
\begin{aligned}
G^{(2)}&(\mathbf{r}_1, t_1; \mathbf{r}_2, t_2) \\
&\propto \sum_j P_j \, \langle \Psi_j | E^{(-)}(\mathbf{r}_1, t_1)\, E^{(-)}(\mathbf{r}_2, t_2)\, E^{(+)}(\mathbf{r}_2, t_2)\, E^{(+)}(\mathbf{r}_1, t_1) | \Psi_j \rangle \\
&= tr\,[\,\hat{\rho}\, E^{(-)}(\mathbf{r}_1, t_1)\, E^{(-)}(\mathbf{r}_2, t_2)\, E^{(+)}(\mathbf{r}_2, t_2)\, E^{(+)}(\mathbf{r}_1, t_1)\,],
\end{aligned}
\qquad (3.4.6)
$$

where P_j is the probability of being in the jth initial pure state $|\Psi_j\rangle$ and $\hat{\rho}$ is the density operator representing the mixed state of the radiation field or the system of photons, as defined earlier.

If the mixed state can be formally written into an incoherent superposition

$$|\widetilde{\Psi}\rangle = \sum_j e^{i\varphi_j}|\Psi_j\rangle \quad \text{or} \quad |\widetilde{\Psi}\rangle = \sum_{j,k} e^{i(\varphi_j + \varphi_k)}|\Psi_j\rangle|\Psi_k\rangle,$$

where $(\varphi_j + \varphi_k)$ is a random number. The probability of annihilating a photon from the radiation field can be calculated, in general, as follows:

$$
\begin{aligned}
G^{(2)}&(\mathbf{r}_1, t_1; \mathbf{r}_2, t_2) \\
&= \big\langle \langle \widetilde{\Psi} |\, E^{(-)}(\mathbf{r}_1, t_1)\, E^{(-)}(\mathbf{r}_2, t_2)\, E^{(+)}(\mathbf{r}_2, t_2)\, E^{(+)}(\mathbf{r}_1, t_1)\, |\widetilde{\Psi}\rangle \big\rangle_{\mathrm{En}}.
\end{aligned}
$$
$$(3.4.7)$$

$G^{(2)}(\mathbf{r}_1, t_1; \mathbf{r}_2, t_2)$ is a function of $(\mathbf{r}_1, t_1)$ and $(\mathbf{r}_2, t_2)$, typically a function of relative temporal delay and spatial separation between the two photodetection events, and is often called the second-order coherence function of the measured radiation field.

The averaged joint detection counting rate is thus

$$
\begin{aligned}
R &\propto \int_T dt_1 \int_T dt_2\, G^{(2)}(\mathbf{r}_1, t_1; \mathbf{r}_2, t_2) \\
&= \int_T dt_1 \int_T dt_2\, \langle\, E^{(-)}(\mathbf{r}_1, t_1)\, E^{(-)}(\mathbf{r}_2, t_2)\, E^{(+)}(\mathbf{r}_2, t_2)\, E^{(+)}(\mathbf{r}_1, t_1)\,\rangle,
\end{aligned}
$$
$$(3.4.8)$$

where T is the total time of averaging.

In the above equations, $|\Psi\rangle$ is the state of the field at t_0, which is the time when the perturbation $V(t)$ is applied, i.e., at the very beginning of the photodetection event. In typical optical measurements, $|\Psi\rangle$, is often evaluated from the state $|\Psi(0)\rangle$ that is prepared at the radiation source. The state $|\Psi(0)\rangle$ of the field that is excited at the radiation source is usually known from early measurements or predictable through the study of the radiation mechanism of the light source. This is a typical quantum mechanical dynamical problem: We are interested in knowing the expectation value of an operator $\hat{A}$ at time t

$$\langle\,\hat{A}\,\rangle_t = \langle\Psi(t)|\,\hat{A}\,|\Psi(t)\rangle, \tag{3.4.9}$$

with the knowledge of the quantum state of the system at an early time, such as $|\Psi(0)\rangle$ at time $t = 0$. Following quantum mechanics, we can either develop the state (Schrödinger picture) or the operator (Heisenberg picture) in time to calculate the expectation value of the operator $\hat{A} = E^{(-)}(\mathbf{r}_1, t_1)\, E^{(-)}(\mathbf{r}_2, t_2)\, E^{(+)}(\mathbf{r}_2, t_2)\, E^{(+)}(\mathbf{r}_1, t_1)$.

3.5 Quantized Subfield and Effective Wavefunction of Photon

Einstein's bundle of ray, or subfield, is a quantized microscopic realistic entity of electromagnetic wavepacket propagating in space–time. This concept is still in the frame work of Maxwell electromagnetic wave theory, except that Einstein introduced a granularity picture and abandoned the continuum interpretation of Maxwell. It is interesting to find that an "effective wavefunction", $\psi_m(\mathbf{r}, t)$, of the mth photon, or the mth group of indistinguishable photons, among an ensemble of photons, or groups of indistinguishable photons, can be defined from the quantum photodetection theory, precisely, from the calculations of $G^{(1)}(\mathbf{r}, t)$ and $G^{(2)}(\mathbf{r}_1, t_1; \mathbf{r}_2, t_2)$. It is also interesting to find that $\psi_m(\mathbf{r}, t)$ has the same mathematical representation and plays the same role as that of Einstein's subfield $E_m(\mathbf{r}, t)$. It should be emphasized even before introducing the concept, however, that $\psi_m(\mathbf{r}, t)$ represents the probability amplitude for the mth photon, or the mth group of indistinguishable photons, to produce a photodetection event at space–time $(\mathbf{r}, t)$, which is different in nature from a quantized realistic entity of electromagnetic wavepacket. In the following, we introduce the "effective wavefunction" of a photon, or a group of indistinguishable photons, in terms of the single-photon state representation and the coherent state representation.

(I) **Single-photon state representation:**

Assuming a point-like weak natural light source at coordinate $\mathbf{r} = 0$, and a far-field measurement of a point-like photodetector at coordinate $\mathbf{r}$. In the light source, a large number of randomly distributed spontaneous atomic transitions excite a large number of subfields. Although the chance of having a spontaneous emission is very small, there is indeed a tiny probability for an atom to create a photon whenever the atom decays from its higher energy level E_2 down to its lower energy state of E_1. We have approximated the state of the mth subfield that is excited by the mth atomic transition as

$$|\Psi\rangle \simeq c_0|0\rangle + c_1|\Psi_m\rangle \simeq |0\rangle + \epsilon|\Psi_m\rangle,$$

where $|c_0| \sim 1$ is the probability amplitude for no-field-excitation and $|c_1| = |\epsilon| \ll 1$ is the probability amplitude for the creation of a photon from the mth atomic transition. In a 1D approximation,

$$|\Psi_m\rangle = e^{i\varphi_m} \int d\omega\, a_m(\omega)\, \hat{a}_m^\dagger(\omega)\,|0\rangle,$$

we have intentionally written the complex amplitudes $a_m(\omega)e^{i\varphi_m(\omega)}$, which is determined by the creation and annihilation operators, into the state and assumed that all radiation modes ω excited by the mth atomic transition have a common phase $\varphi_m(\omega) = \varphi_m$. The generalized state of the radiation field, created from such a large number atomic transitions, can be formally written as

$$|\widetilde{\Psi}\rangle = \prod_m \left\{|0\rangle + \epsilon\, c_m\,|\Psi_m\rangle\right\}$$

$$\simeq |0\rangle + \epsilon\left[\sum_m c_m\,|\Psi_m\rangle\right] + \epsilon^2\left[\sum_{m<n} c_m\,c_n|\Psi_m\rangle|\Psi_n\rangle\right] + \cdots, \qquad (3.5.1)$$

where, again, we have assumed a common complex amplitude for each wavepacket, although it may take arbitrary random phases from transition to transition. Since $|\epsilon| \ll 1$, in Eq. (3.5.1), we have listed the first-order and the second-order approximations on ϵ only.

Considering a simple measurement in which a point-like photodetector is facing a point-like thermal light source, the field operator at $(\mathbf{r}, t)$ can be approximated in 1D:

$$\hat{E}^{(+)}(\mathbf{r}, t) = \sum_p \int d\omega\, \hat{a}_p(\omega)\, g_p(\omega; \mathbf{r}, t),$$

$$\hat{E}^{(-)}(\mathbf{r}, t) = \sum_q \int d\omega'\, \hat{a}_q^\dagger(\omega')\, g_q^*(\omega'; \mathbf{r}, t), \qquad (3.5.2)$$

where $g_p(\omega; \mathbf{r}, t)$ is Green's function that propagates the ω-mode of the pth subfield from the point-like source to space–time coordinate $(\mathbf{r}, t)$. Following Glauber's theory, the probability to produce a photoelectron event at $(\mathbf{r}, t)$ is proportional to $G^{(1)}(\mathbf{r}, t)$:

$$G^{(1)}(\mathbf{r}, t)$$

$$= \langle\langle \widetilde{\Psi}|E^{(-)}(\mathbf{r}, t)E^{(+)}(\mathbf{r}, t)|\widetilde{\Psi}\rangle\rangle_{\mathrm{En}}$$

$$= \left\langle \langle \widetilde{\Psi}| \left[\sum_p \int d\omega\, \hat{a}_p^\dagger(\omega)\, g_p^*(\omega; \mathbf{r}, t) \right] \left[\sum_q \int d\omega'\, \hat{a}_q(\omega') g_q(\omega'; \mathbf{r}, t) \right] |\widetilde{\Psi}\rangle \right\rangle_{\mathrm{En}}$$

$$= \left\langle \langle \widetilde{\Psi}| \left[\sum_p \int d\omega\, \hat{a}_p^\dagger(\omega)\, g_p^*(\omega; \mathbf{r}, t) \right] |0\rangle \right.$$

$$\left. \times \langle 0| \left[\sum_q \int d\omega'\, \hat{a}_q(\omega') g_q(\omega'; \mathbf{r}, t) \right] |\widetilde{\Psi}\rangle \right\rangle_{\mathrm{En}}$$

$$= \left\langle \left[\sum_m \psi_m^*(\mathbf{r}, t) \right] \left[\sum_n \psi_n(\mathbf{r}, t) \right] \right\rangle_{\mathrm{En}}$$

$$= \left\langle \sum_m |\psi_m(\mathbf{r}, t)|^2 \right\rangle_{\mathrm{En}} + \left\langle \sum_{m \neq n} \psi_m^*(\mathbf{r}, t)\psi_n(\mathbf{r}, t) \right\rangle_{\mathrm{En}}$$

$$\propto \langle n(\mathbf{r}, t) \rangle, \tag{3.5.3}$$

where we have applied the completeness relation

$$\sum_l |l\rangle\langle l| = 1. \tag{3.5.4}$$

In Eq. (3.5.3), we have defined the effective wavefunction of the mth photon,

$$\psi_m(\mathbf{r}, t) \equiv \langle 0|E^{(+)}(\mathbf{r}, t)|\Psi_m\rangle = \int d\omega\, a_m(\omega)e^{i\varphi_m}\, g_m(\omega; \mathbf{r}, t). \tag{3.5.5}$$

where, again, we have assumed a common phase φ_m for the mth atomic transition, and we have also absorbed $|c_m|$ into $a_m(\omega)$ as usual. In the point-to-point propagation, or the far-field approximation, $\psi_m(\mathbf{r}, t)$ can be written as

$$\psi_m(\mathbf{r}, t) = \left[\int d\nu\, a_m(\nu)e^{-i\nu\tau_m} \right] e^{-i(\omega_0\tau_m - \varphi_m)}$$

$$= \mathcal{F}_{\tau_m}\{a_m(\nu)\}\, e^{-i(\omega_0\tau_m - \varphi_m)}. \tag{3.5.6}$$

It is interesting to find that the effective wavefunction $\psi_m(\mathbf{r}, t)$ has the same mathematical representation and plays the same role as that of Einstein's subfield $E_m(\mathbf{r}, t)$.

It is easy to find that $\langle \Delta n(\mathbf{r}, t) \rangle = 0$ for thermal state:

$$\langle \Delta n(\mathbf{r}, t) \rangle = \left\langle \sum_{m \neq n} \psi_m^*(\mathbf{r}, t) \psi_n(\mathbf{r}, t) \right\rangle_{En}$$

$$= \left\langle \sum_{m \neq n} e^{-i(\varphi_m - \varphi_n)} \mathcal{F}_\tau^* \{a_m(\nu)\} \mathcal{F}_\tau \{a_n(\nu)\} \right\rangle_{En} = 0$$

while taking into account all possible values of $(\varphi_m - \varphi_n)$, agreeing with that of Einstein's picture.

(II) Coherent state representation:

We have assumed a weak thermal radiation in the above calculation, in which only the lower order of Eq. (3.5.1) is taken into account. In the case of arbitrary brightness, especially in bright light condition, we may need to keep all orders of the expansion. In this case, quantum coherent state, representing a group of indistinguishable photons, is a better choice. In the coherent state representation, the state of a radiation field, in general, can be written as

$$|\Psi\rangle = \prod_m |\Psi_m\rangle = \prod_m |\{\alpha_m\}\rangle = \prod_{m,\mathbf{k}} |\alpha_m(\mathbf{k})\rangle. \qquad (3.5.7)$$

If the radiation field consists of a large number of randomly created and randomly distributed subfields, we may rewrite Eq. (3.5.7) in the form of a pseudovector:

$$|\tilde{\Psi}\rangle = \prod_m |\Psi_m\rangle = \prod_m |\{\alpha_m\}\rangle = \prod_{m,\mathbf{k}} |\alpha_m(\mathbf{k})\rangle. \qquad (3.5.8)$$

In the above state, a subfield may represent a radiation at single-photon's level, or represent a group of indistinguishable photons with $|\alpha| \gg 1$. In Eqs. (3.5.7) and (3.5.8), m labels the mth subfield, and $\mathbf{k}$ labels the $\mathbf{k}$th mode. Each subfield may consist of a large number of radiation modes of $\mathbf{k}$. $|\alpha_m(\mathbf{k})\rangle$ is an eigenstate of the annihilation operator with an eigenvalue $\alpha_m(\mathbf{k})$,

$$\hat{a}_m(\mathbf{k})|\alpha_m(\mathbf{k})\rangle = \alpha_m(\mathbf{k})|\alpha_m(\mathbf{k})\rangle, \qquad (3.5.9)$$

where $\alpha_m(\mathbf{k})$ takes a complex value, $\alpha_m(\mathbf{k}) = a_m(\mathbf{k})e^{i\varphi_m(\mathbf{k})}$, $\varphi_m(\mathbf{k})$ can be a random number or a constant. We thus have,

$$\hat{a}_m(\mathbf{k})|\Psi\rangle = \alpha_m(\mathbf{k})|\Psi\rangle. \qquad (3.5.10)$$

Considering the same simple measurement with the same field operators of Eq. (3.5.2), we have

$$\langle\Psi|\hat{E}^{(-)}(\mathbf{r},t)|\Psi\rangle = \langle\Psi|\sum_m \int d\mathbf{k}\,\hat{a}_m^\dagger(\mathbf{k})\,g_m^*(\mathbf{k};\mathbf{r},t)|\Psi\rangle$$

$$= \sum_m \int d\mathbf{k}\,\alpha_m^*(\mathbf{k})\,g_m^*(\mathbf{k};\mathbf{r},t)]$$

$$\langle\Psi|\hat{E}^{(+)}(\mathbf{r},t)|\Psi\rangle = \langle\Psi|\left[\sum_n \int d\mathbf{k}\,\hat{a}_n(\mathbf{k})\,g_n(\mathbf{k};\mathbf{r},t)\right]|\Psi\rangle$$

$$= \sum_n \int d\mathbf{k}\,\alpha_n(\mathbf{k})\,g_n(\mathbf{k};\mathbf{r},t). \tag{3.5.11}$$

Following Glauber's theory, the probability to produce a photoelectron event at $(\mathbf{r},t)$ from a radiation field is proportional to $G^{(1)}(\mathbf{r},t) = \langle\langle E^{(-)}(\mathbf{r},t)E^{(+)}(\mathbf{r},t)\rangle_{\mathrm{QM}}\rangle_{\mathrm{En}}$. Substituting the quantum state and the field operators into $G^{(1)}(\mathbf{r},t)$, we have

$$G^{(1)}(\mathbf{r},t) = \langle\langle\Psi|E^{(-)}(\mathbf{r},t)E^{(+)}(\mathbf{r},t)|\Psi\rangle\rangle_{\mathrm{En}}$$

$$= \left\langle\langle\Psi|\left[\sum_m \int d\mathbf{k}\,\hat{a}_m^\dagger(\mathbf{k})\,g_m^*(\mathbf{k};\mathbf{r},t)\right]\right.$$

$$\left.\times\left[\sum_n \int d\mathbf{k}\,\hat{a}_n(\mathbf{k})g_n(\mathbf{k};\mathbf{r},t)\right]|\Psi\rangle\right\rangle_{\mathrm{En}}$$

$$= \left\langle\langle\Psi|\left[\sum_m \int d\mathbf{k}\,\hat{a}_m^\dagger(\mathbf{k})\,g_m^*(\mathbf{k};\mathbf{r},t)\right]|\Psi\rangle\right.$$

$$\left.\times\langle\Psi|\left[\sum_n \int d\mathbf{k}\,\hat{a}_n(\mathbf{k})g_n(\mathbf{k};\mathbf{r},t)\right]|\Psi\rangle\right\rangle_{\mathrm{En}}$$

$$= \left\langle\left[\sum_m \psi_m^*(\mathbf{r},t)\right]\left[\sum_n \psi_n(\mathbf{r},t)\right]\right\rangle_{\mathrm{En}}$$

$$= \left\langle\sum_m |\psi_m(\mathbf{r},t)|^2\right\rangle_{\mathrm{En}} + \left\langle\sum_{m\neq n}\psi_m^*(\mathbf{r},t)\psi_n(\mathbf{r},t)\right\rangle_{\mathrm{En}}$$

$$\propto \langle n(\mathbf{r},t)\rangle, \tag{3.5.12}$$

where we have defined $\psi_m(\mathbf{r}, t)$ the effective wavefunction of the mth subfield:

$$\psi_m(\mathbf{r}, t) \equiv \langle \Psi_m | E^{(+)}(\mathbf{r}, t) | \Psi_m \rangle = \int d\mathbf{k}\, \alpha_m(\mathbf{k}) e^{i\varphi_m}\, g_m(\mathbf{k}; \mathbf{r}, t). \quad (3.5.13)$$

In Eq. (3.5.12) we approximate the photon number fluctuation $\langle \Delta n(\mathbf{r}, t) \rangle = 0$ by taking into account all possible random values of $(\varphi_m - \varphi_n)$ in the ensemble average.

It should be emphasized that although the quantum mechanical concept of effective wavefunction has the same mathematical representation and plays the same role as Einstein's concept of subfield, the physical means of the two are different in nature: Einstein's subfield $E_m(\mathbf{r}, t)$ is a quantized real entity of electromagnetic field. The quantum mechanical effective wavefunction $\psi_m(\mathbf{r}, t)$ is the probability amplitude for the mth subfield to produce a photodetection event at space–time $(\mathbf{r}, t)$.

If there exist two different yet indistinguishable alternatives, such as in Young's double-slit interferometer, to produce a photodetection event at space–time coordinate $(\mathbf{r}, t)$, the effective wavefunction is calculated as follows:

$$\begin{aligned}
\psi_m(\mathbf{r}, t) &= \int d\mathbf{k}\, a_m(\mathbf{k}) \frac{1}{\sqrt{2}} \big[g_{mA}(\mathbf{k}; \mathbf{r}, t) + g_{mB}(\mathbf{k}; \mathbf{r}, t) \big] \\
&= \frac{1}{\sqrt{2}} \left[\int d\mathbf{k}\, a_m(\mathbf{k}) g_{mA}(\mathbf{k}; \mathbf{r}, t) + \int d\mathbf{k}\, a_m(\mathbf{k}) g_{mB}(\mathbf{k}; \mathbf{r}, t) \right] \\
&= \frac{1}{\sqrt{2}} \big[\psi_{mA}(\mathbf{r}, t) + \psi_{mB}(\mathbf{r}, t) \big]
\end{aligned} \quad (3.5.14)$$

which is, again, the same superposition as that in Einstein's picture except that the electromagnetic subfield $E_m(\mathbf{r}, t)$ is replaced with the effective wavefunction $\psi_m(\mathbf{r}, t)$.

3.6   Joint Measurement of Composite Radiation Systems

In the following, we give two simple calculations of joint measurements on composite radiation systems and introduce the "effective wave function" of a pair of photons in the single-photon state representation, or of a pair of groups of indistinguishable photons in the coherent state representation.

(I) Single-photon state representation.

Following Glauber's theory, the probability to produce a joint photodetection event at space–time coordinates $(\mathbf{r}_1, t_1)$ and $(\mathbf{r}_2, t_2)$ by a radiation field in thermal state is calculated as follows:

$$
\begin{aligned}
G^{(2)}&(\mathbf{r}_1, t_1; \mathbf{r}_2, t_2) \\
&= \left\langle \left\langle E^{(-)}(\mathbf{r}_1, t_1) E^{(-)}(\mathbf{r}_2, t_2) E^{(+)}(\mathbf{r}_2, t_2) E^{(+)}(\mathbf{r}_1, t_1) \right\rangle_{\mathrm{QM}} \right\rangle_{\mathrm{En}} \\
&= \left\langle \langle \widetilde{\Psi} | \sum_m E_m^{(-)}(\mathbf{r}_1, t_1) \sum_n E_n^{(-)}(\mathbf{r}_2, t_2) | 0 \rangle \right. \\
&\qquad\qquad \left. \times \langle 0 | \sum_q E_q^{(+)}(\mathbf{r}_2, t_2) \sum_p E_p^{(+)}(\mathbf{r}_1, t_1) | \widetilde{\Psi} \rangle \right\rangle_{\mathrm{En}} \\
&= \left\langle \sum_m \psi_m^*(\mathbf{r}_1, t_1) \sum_n \psi_n^*(\mathbf{r}_2, t_2) \sum_p \psi_p(\mathbf{r}_2, t_2) \sum_q \psi_q(\mathbf{r}_1, t_1) \right\rangle_{\mathrm{En}} \\
&= \sum_m \psi_m^*(\mathbf{r}_1, t_1) \psi_m(\mathbf{r}_1, t_1) \sum_n \psi_n^*(\mathbf{r}_2, t_2) \psi_n(\mathbf{r}_2, t_2) \\
&\qquad + \sum_{m \neq n} \psi_m^*(\mathbf{r}_1, t_1) \psi_n(\mathbf{r}_1, t_1) \psi_n^*(\mathbf{r}_2, t_2) \psi_m(\mathbf{r}_2, t_2) \\
&= \sum_{m,n} \left| \frac{1}{\sqrt{2}} [\psi_m(\mathbf{r}_1, t_1) \psi_n(\mathbf{r}_2, t_2) + \psi_m(\mathbf{r}_2, t_2) \psi_n(\mathbf{r}_1, t_1)] \right|^2 \\
&= \langle n(\mathbf{r}_1, t_1) \rangle \langle n(\mathbf{r}_2, t_2) \rangle + \langle \Delta n(\mathbf{r}_1, t_1) \Delta n(\mathbf{r}_2, t_2) \rangle.
\end{aligned}
\tag{3.6.1}
$$

Here, we define the effective wave function of the mth–nth pair of randomly generated and randomly distributed photons in the single-photon state representation:

$$
\psi_{mn}(\mathbf{r}_1, t_1; \mathbf{r}_2, t_2) \equiv \langle 0 | E^{(+)}(\mathbf{r}_2, t_2) E^{(+)}(\mathbf{r}_1, t_1) | \Psi_{mn} \rangle,
\tag{3.6.2}
$$

with

$$
|\Psi_{mn}\rangle = |\Psi_m\rangle |\Psi_n\rangle
$$

indicating a product of two single-photon effective wavefunctions.

$G^{(2)}(\mathbf{r}_1, t_1; \mathbf{r}_2, t_2)$, often called the second-order coherence function, is the result of the superposition of two different yet indistinguishable two-photon amplitudes: (1) the mth photon is annihilated at $(\mathbf{r}_1, t_1)$ while the

nth photon is annihilated at $(\mathbf{r}_2, t_2)$; (2) the mth photon is annihilated at $(\mathbf{r}_2, t_2)$ while the nth photon is annihilated at $(\mathbf{r}_1, t_1)$. We name this superposition two-photon interference: two randomly created and randomly paired photons in thermal state interfering with the pair itself. Examining the cross-interference term of Eq. (3.6.1) which contributes to the photon number fluctuation correlation,

$$\langle \Delta n(\mathbf{r}_1, t_1) \Delta n(\mathbf{r}_2, t_2) \rangle = \sum_{m \neq n} \psi_m^*(\mathbf{r}_1, t_1) \psi_n(\mathbf{r}_1, t_1) \psi_n^*(\mathbf{r}_2, t_2) \psi_m(\mathbf{r}_2, t_2),$$

$$(3.6.3)$$

we may find that this term is the same as that we have calculated in the previous sections except that Einstein's subfield is replaced by the effective wavefunction.

(II) Coherent state representation:

In the case of arbitrary brightness, especially in bright light condition, coherent state representation is a better choice. As we have discussed in previous section, in the coherent state representation, the state of a radiation field can be written as that in Eq. (3.5.7)

$$|\Psi\rangle = \prod_m |\Psi_m\rangle = \prod_m |\{\alpha_m\}\rangle = \prod_{m,\mathbf{k}} |\alpha_m(\mathbf{k})\rangle,$$

where m labels the mth subfield, and $\mathbf{k}$ labels the kth mode, $|\alpha_m(\mathbf{k})\rangle$, again, is an eigenstate of the annihilation operator

$$\hat{a}_m(\mathbf{k})|\alpha_m(\mathbf{k})\rangle = \alpha_m(\mathbf{k})|\alpha_m(\mathbf{k})\rangle.$$

We thus have

$$\hat{a}_m(\mathbf{k})|\Psi\rangle = \alpha_m(\mathbf{k})|\Psi\rangle, \qquad (3.6.4)$$

where $\alpha_m(\mathbf{k}) = a_m(\mathbf{k})e^{i\varphi_m(\mathbf{k})}$ takes a complex value. If the subfields or identical-photon groups are created randomly with random relative phases, then $\varphi_m(\mathbf{k})$ will be treated as a random number that can take all possible random values. In this case, $|\Psi\rangle$ will be replaced by $|\tilde{\Psi}\rangle$ as a pseudovector.

Considering the same simple measurement of thermal field with the same field operators of Eq. (3.5.2), following Glauber's theory, the probability to produce a joint photodetection event at space–time coordinates

$(\mathbf{r}_1, t_1)$ and $(\mathbf{r}_2, t_2)$ is calculated as follows:

$$G^{(2)}(\mathbf{r}_1, t_1; \mathbf{r}_2, t_2)$$

$$= \left\langle \langle E^{(-)}(\mathbf{r}_1, t_1) E^{(-)}(\mathbf{r}_2, t_2) E^{(+)}(\mathbf{r}_2, t_2) E^{(+)}(\mathbf{r}_1, t_1) \rangle_{\text{QM}} \right\rangle_{\text{En}}$$

$$= \left\langle \langle \Psi | \sum_m E_m^{(-)}(\mathbf{r}_1, t_1) \sum_n E_n^{(-)}(\mathbf{r}_2, t_2) \right.$$

$$\left. \times |\Psi\rangle\langle\Psi| \sum_q E_q^{(+)}(\mathbf{r}_2, t_2) \sum_p E_p^{(+)}(\mathbf{r}_1, t_1) | \Psi \rangle \right\rangle_{\text{En}}$$

$$= \left\langle \sum_m \psi_m^*(\mathbf{r}_1, t_1) \sum_n \psi_n^*(\mathbf{r}_2, t_2) \sum_p \psi_p(\mathbf{r}_2, t_2) \sum_q \psi_q(\mathbf{r}_1, t_1) \right\rangle_{\text{En}}$$

$$= \sum_m \psi_m^*(\mathbf{r}_1, t_1) \psi_m(\mathbf{r}_1, t_1) \sum_n \psi_n^*(\mathbf{r}_2, t_2) \psi_n(\mathbf{r}_2, t_2)$$

$$+ \sum_{m \neq n} \psi_m^*(\mathbf{r}_1, t_1) \psi_n(\mathbf{r}_1, t_1) \psi_n^*(\mathbf{r}_2, t_2) \psi_m(\mathbf{r}_2, t_2)$$

$$= \sum_{m,n} \left| \frac{1}{\sqrt{2}} [\psi_m(\mathbf{r}_1, t_1) \psi_n(\mathbf{r}_2, t_2) + \psi_m(\mathbf{r}_2, t_2) \psi_n(\mathbf{r}_1, t_1)] \right|^2$$

$$= \langle n(\mathbf{r}_1, t_1) \rangle \langle n(\mathbf{r}_2, t_2) \rangle + \langle \Delta n(\mathbf{r}_1, t_1) \Delta n(\mathbf{r}_2, t_2) \rangle. \tag{3.6.5}$$

Here, we have taken all possible random phases of the subfields or identical-photon groups. We have also defined the effective wave function of the m-nth pair of randomly created and randomly distributed identical-photon groups:

$$\psi_{mn}(\mathbf{r}_1, t_1; \mathbf{r}_2, t_2) \equiv \langle \Psi_{mn} | E^{(+)}(\mathbf{r}_2, t_2) E^{(+)}(\mathbf{r}_1, t_1) | \Psi_{mn} \rangle \tag{3.6.6}$$

with

$$|\Psi_{mn}\rangle = |\Psi_m\rangle |\Psi_n\rangle,$$

indicating a product of two effective wavefunctions of identical-photon groups.

$G^{(2)}(\mathbf{r}_1, t_1; \mathbf{r}_2, t_2)$ of Eq. (3.6.5), often called the second-order coherence function, is the result of a superposition of two different yet indistinguishable quantum amplitudes: (1) one or more indistinguishable photons from the mth subfield triggers D_1 at (r_1, t_1) while one or more indistinguishable photons from the nth subfield triggers D_2 at (r_2, t_2); and (2) one or more indistinguishable photons from the mth subfield triggers D_2 at (r_2, t_2) while one or more indistinguishable photons from the nth subfield triggers D_1 at (r_1, t_1), indicating a randomly paired two groups of indistinguishable photons interfering with the pair itself. Similar to the intensity fluctuation correlation we calculated in the Einstein picture, the photon number fluctuation correlation, expressed by the second term in the last line of Eq. (3.6.5), corresponds to the interference cross terms of the above superposition. In other words, the photon number fluctuation correlation is produced by the self-interference of a pair of randomly created and randomly paired identical-photon groups.

Bibliography

Glauber R.J., *Phys. Rev.* **130**, 2529 (1963); *Phys. Rev.* **131**, 2766 (1963).
Scully M.O. and Zubairy M.S., *Quantum Optics*, Cambridge University Press, 1997.

Chapter 4

First-Order Coherence of Light: A Photon Interferes With the Photon Itself

The concept of first-order coherence was introduced in classical theory of light much earlier than that of the quantum theory. The degree of first-order coherence is defined to quantify the interference between temporally delayed or spatially separated electromagnetic waves. The superposed radiations are defined as first-order coherent if the interference fringes exhibit 100% modulation or first-order incoherent if no interference fringes are observable. The radiation fields are considered as partial coherent if the modulation is less than 100% but greater than zero. The higher the degree of first-order coherence, the higher interference visibility we could observe.

Although it is named "coherence" and is an intrinsic property of the radiation itself, either temporal coherence or spatial coherence, the concept is very different from the coherence property of light that we have discussed in previous chapters. For example, radiation produced by lasers is considered coherent light, and natural light from distant stars is called incoherent thermal fields. However, under different experimental conditions, both laser radiation and thermal fields can be either first-order coherent or first-order incoherent. For a certain spectral bandwidth $\Delta\omega$ and spatial frequency $\Delta\vec{\kappa}$ or Δk_x (Δk_y), the interference observed in an interferometer is the same for laser light and thermal light. The measurement cannot distinguish a laser beam from thermal radiation by means of the degree of first-order coherence. Although laser radiation is considered coherent, it does not mean we are able to observe interference for a temporal delay beyond certain limit. Similarly, thermal light, such as the sunlight, is named incoherent radiation, but this does not prevent us from observing interference fringes under the "white-light" condition, i.e., in the neighborhood of equal optical paths of an interferometer. One should pay special attention to this.

The goal of this chapter is to probe the quantum nature of optical coherence, which has been especially successful in dealing with the coherent phenomena of light at quantum level. In fact, although no field quantization was involved, we implicitly introduced the quantum concepts of radiation by means of the granularity picture of light in the very beginning of this book. In Einstein's picture of light, the large number of sub-sources, either radiate independently or coherently, are connected with a large number of atomic transitions. The very basic contribution to the radiation field from each sub-source is therefore the creation of a photon. In the bright light condition, a measurement may involve the creation and annihilation of a large number of photons. The individual behavior of a photon may not be that significant in the final observation. However, it does not prevent us from asking very simple questions: What would happen if a photon and only one photon were present in the observation? Can a photon interfere with itself? In a "single-photon interferometer", does a photon passes "both paths" or "which path"? See Fig. 4.0.1. In fact, even in the "bright light" condition, we may find that the interference is the result of a subfield interfering with the subfield itself. Connecting Einstein's bundle of ray, or subfield, with the concept of photon or a group of indistinguishable photons, we would have the same questions. Quantum theory of coherence deals with this kind of physics and gives its quantitative predictions and qualitative interpretations according to the principles of quantum mechanics. In the light of new technology, we are now able to study these problems experimentally. In certain aspects, we may say that quantum theory of coherence studies the interference of a photon with itself, or a group of indistinguishable photons interference with the group itself. In this regard, the first-order quantum coherence studies and quantifies the self-interference of a photon or a group of indistinguishable photons. The degree of first-order quantum coherence is introduced in terms of the superposition of quantum amplitudes, in the form of self-superposition of either Einstein's subfield or the effective wavefunctions of a photon or a group of indistinguishable photons.

In this chapter, we start from a simple Young's double-pinhole interferometer to show the observed first-order interference fringes is the result of a photon, or a group of indistinguishable photons, interferes with the photon or the group of indistinguishable photons itself. We then introduce the first-order coherence of light by defining the classical mutual-coherence

Fig. 4.0.1 On one hand, in quantum theory of light, a photon can never be divided into parts; on the other hand, we have never lost interference at single-photon's level. In fact, according to quantum theory, interference is a single-photon phenomenon. In Dirac's language: "... photon ... only interferes with itself."

function $\Gamma^{(1)}(\mathbf{r}_1, t_1; \mathbf{r}_2, t_2)$ and the normalized complex degree of first-order classical coherence $\gamma^{(1)}(\mathbf{r}_1, t_1; \mathbf{r}_2, t_2)$. The first-order quantum coherence $G^{(1)}(\mathbf{r}_1, t_1; \mathbf{r}_2, t_2)$ and the normalized degree of first-order quantum coherence $g^{(1)}(\mathbf{r}_1, t_1; \mathbf{r}_2, t_2)$ are subsequently introduced and defined.

4.1 First-Order Interference of Light: Einstein's Picture

A Young's double-pinhole interference experiment is shown in Fig. 4.1.1. The light source is a bright distant star, composed of a large number of point-like sub-sources, which can be treated as either a point-like source or an extended source with a certain angular size. A large number of subfields, each created from a sub-sources at $(\mathbf{r}_m, t_m)$, propagate and pass though the

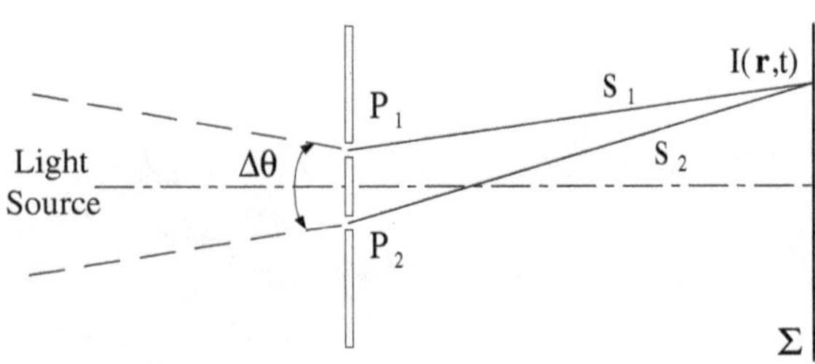

Fig. 4.1.1   Schematic of an Young's double-pinhole interference experiment. The interference pattern is observed by a photodetector array, such as a CCD or a CMOS, or by scanning a point-like photodetector on the observation plane Σ.

upper pinhole and the lower pinhole at $(\mathbf{r}_1, t_1)$ and $(\mathbf{r}_2, t_2)$, respectively. Observation is made by scanning a point-like photodetector on the far-field observation plane Σ, or by a photodetector array such as a CCD or a CMOS. The pinholes and the observation plan are arranged symmetrically with respect to the optical axis, as shown in Fig. 4.1.1.

The measured intensity $\langle I(\mathbf{r}, t) \rangle$ is the result of a superposition between $E_1(\mathbf{r}, t)$ which passed through the upper pinhole P_1 and $E_2(\mathbf{r}, t)$ which passed through the lower pinhole P_2, at space–time $(\mathbf{r}, t)$ of the observation plane Σ:

$$\begin{aligned}
\langle I(\mathbf{r}, t) \rangle &= \langle |E_1(\mathbf{r}, t) + E_2(\mathbf{r}, t)|^2 \rangle \\
&= \langle |E_1(\mathbf{r}, t)|^2 \rangle + \langle |E_2(\mathbf{r}, t)|^2 \rangle + \langle E_1^*(\mathbf{r}, t) E_2(\mathbf{r}, t) \rangle \\
&\quad + \langle E_1(\mathbf{r}, t) E_2^*(\mathbf{r}, t) \rangle.
\end{aligned} \tag{4.1.1}$$

It is easy to find that the first two terms of Eq. (4.1.1) are the constant intensities of the fields that passed through P_1 and P_2, respectively. It is the last two terms of Eq. (4.1.1) that produce the sinusoidal fringes. We need to prove the last two cross-interference terms of Eq. (4.1.1) are the result of a photon, or a group of indistinguishable photons, interferes with the photon, or the group of indistinguishable photons, itself.

In Einstein's granularity picture, the thermal field $E(\mathbf{r}, t)$ at space–time $(\mathbf{r}, t)$ is the result of an incoherent superposition among a large number of subfields randomly created from a large number of independent point-like sub-sources and propagated to $(\mathbf{r}, t)$:

$$E_j(\mathbf{r}, t) = \sum_m E_{jm}(\mathbf{r}, t). \tag{4.1.2}$$

The interference fringe observable on the observation plane Σ is therefore

$$\langle I(\mathbf{r},t)\rangle = \langle |E_1(\mathbf{r},t) + E_2(\mathbf{r},t)|^2\rangle = \left\langle \left|\sum_m E_{1m}(\mathbf{r},t) + \sum_n E_{2n}(\mathbf{r},t)\right|^2\right\rangle.$$

$$(4.1.3)$$

The cross-interference terms are calculated as follows:

$$\langle E_1^*(\mathbf{r},t)E_2(\mathbf{r},t)\rangle$$

$$= \left\langle \sum_m E_{1m}^*(\mathbf{r},t) \sum_n E_{2n}(\mathbf{r},t)\right\rangle$$

$$= \left\langle \sum_{m=n} E_{1m}^*(\mathbf{r},t)\, E_{2n}(\mathbf{r},t)\right\rangle + \left\langle \sum_{m\neq n} E_{1m}^*(\mathbf{r},t)\, E_{2n}(\mathbf{r},t)\right\rangle$$

$$= \left\langle \sum_m E_{1m}^*(\mathbf{r},t)\, E_{2m}(\mathbf{r},t)\right\rangle. \qquad (4.1.4)$$

Due to the random relative phases between the mth and nth subfields, the sum of $m \neq n$ vanishes when taking into account all possible realization of the field, i.e., the ensemble average. Taking our previous results of the intensity calculation in Einstein granularity picture, we thus have

$$\langle I(\mathbf{r},t)\rangle = \sum_m \big|E_{1m}(\mathbf{r},t) + E_{2m}(\mathbf{r},t)]\big|^2. \qquad (4.1.5)$$

This result indicates that *the only observable interference is the interference in which the mth subfield interferes with the mth subfield itself.* Interference between two different subfields ($m \neq n$) becomes unobservable due to the ensemble average.

In Einstein's granularity picture of light, consider an incoherent superposition of the randomly created subfields but a coherent superposition of the Fourier modes in each subfield. The set of coherent Fourier modes of the mth subfield produces the mth wavepacket:

$$E_{jm}(\mathbf{r},t) = \int d\omega\, E_m(\omega)\, g_m(\omega; \mathbf{r}_j, t_j)\, g_j(\omega; \mathbf{r}, t)$$

$$\simeq \int d\omega\, E_m(\omega)\, e^{-i\omega[(t_j-t_{0m})-r_{mj}/c]}\, e^{-i\omega[(t-t_j)-s_j/c]}$$

$$= \int d\omega\, E_m(\omega)\, e^{-i\omega[(t_j-t_{0m})-r_{mj}/c]}$$

$$= \mathcal{F}_{\tau_{jm}}\{a_m(\nu)\}\, e^{-i\omega_0\tau_{jm}}, \qquad (4.1.6)$$

where $g_m(\omega; \mathbf{r}_j, t_j)$, $j = 1, 2$, is Green's function that propagates the mth wavepacket from the mth sub-source to $(\mathbf{r}_j, t_j)$; $g(\mathbf{r}_j, t_j; \mathbf{r}, t)$ is Green's function that propagates the wavepacket from $(\mathbf{r}_j, t_j)$ to $(\mathbf{r}, t)$. Simplifying the notation of Fourier transform, here, we define a temporal parameter $\tau_{jm} \equiv (t_j - t_{0m}) - r_{mj}/c$ with $t_j \equiv t - s_j/c$. t_j is defined by the observation time t, corresponding to the time the wavepacket passes through pinhole j. τ_{jm} is also written as $\tau_{jm} = \tau_j - t_{0m}$ with $\tau_j \equiv t_j - r_{mj}/c$.

Accordingly, $\langle I(\mathbf{r}, t) \rangle$, the expectation function of the measured intensity on the observation plane, is the sum of a large number of individual sub-interferences; each is the result of an individual subfield or wavepacket interfering with the subfield or wavepacket itself:

$$\langle I(\mathbf{r}, t) \rangle \simeq \sum_m \left| \frac{1}{\sqrt{2}} \left[\mathcal{F}_{\tau_{1m}}\{a(\nu)\} e^{-i\omega_0 \tau_{1m}} + \mathcal{F}_{\tau_{2m}}\{a(\nu)\} e^{-i\omega_0 \tau_{2m}} \right] \right|^2.$$

$$(4.1.7)$$

Due to the common relative delay $\tau = (s_1 - s_2)/c$, all sub-interference fringes contain the same sinusoidal modulation. The sub-interference fringes add at the observation plane as a function of $\tau = (s_2 - s_1)/c$.

4.2 First-Order Interference of Light: Quantum Theory

The schematic setup of Young's double-pinhole interference experiment is the same as that shown in Fig. 4.1.1, except the measurement device is a single-photon counting detector array, either a CCD or a CMOS working in photon counting mode, or a point-like photon counting detector scannable on the observation plane. Following the photodetection theory introduced in Chapter 3, the probability of observing a photodetection event from a point-like photodetector by annihilating a photon at space-time point $(\mathbf{r}, t)$ is

$$P(\mathbf{r}, t) \propto \langle \hat{E}^{(-)}(\mathbf{r}, t) \hat{E}^{(+)}(\mathbf{r}, t) \rangle = \left\langle \langle \hat{E}^{(-)}(\mathbf{r}, t)\, \hat{E}^{(+)}(\mathbf{r}, t) \rangle_{\mathrm{QM}} \right\rangle_{\mathrm{En}},$$

where $\hat{E}^{(-)}$ and $\hat{E}^{(+)}$ are the negative frequency and the positive frequency field operators, and the expectation value is calculated by averaging the field operators over the quantum state which may be a pure state or a mixed state, depending on the light source. In the case of mixed state, an additional ensemble average is necessary. Calculating the interference pattern of Young's double-pinhole interferometer in Heisenberg picture, we need to (1) model the created radiation and determine the quantum state of the radiation field and (2) propagate the operators from each point-like sub-source to each point of the observation plane.

To be consistent with our previous discussion, we assume a radiation in thermal state. The light source consists of a large number of independent point-like sub-sources, and each creates a photon or a group of indistinguishable photons at time t_m from coordinate $\mathbf{r}_m$ in a randomly manner. We present the radiation field either in single-photon state representation or in coherent state representation.

(I) $\langle n(\mathbf{r},t)\rangle$: Thermal state in single-photon state representation

In this section, we consider the randomly created and randomly distributed photons in mixed single-photon state:

$$|\tilde{\Psi}\rangle = \sum_m c_m \int d\omega\, f_m(\omega)\, \hat{a}_m^\dagger(\omega)\, |0\rangle = \sum_m c_m |\Psi_m\rangle, \qquad (4.2.1)$$

where m labels the mth sub-source, c_m is the complex amplitude of the mth single-photon state, and $f_m(\omega)$, mainly determined by the mth atomic transition, is the spectral distribution function of the mth single-photon state. To simplify the mathematics, we assume equal chance of photon creation at each sub-source, $|c_m| = $ constant, however, c_m contains a random phase. Equation (4.2.1) represents an incoherent superposition of a set of single-photon states.

Next, we propagate the field operator from each point-like sub-source to the measurement point of space–time $(\mathbf{r},t)$:

$$
\begin{aligned}
\hat{E}^{(+)}(\mathbf{r},t) &= \sum_m \int d\omega\, \hat{a}_m(\omega)\, g_m(\omega;\mathbf{r}_1,t_1)\, g_1(\omega;\mathbf{r},t) \\
&\quad + \sum_n \int d\omega'\, \hat{a}_n(\omega')\, g_n(\omega';\mathbf{r}_2,t_2)\, g_2(\omega';\mathbf{r},t) \\
&\simeq \sum_m \int d\omega\, \hat{a}_m(\omega)\, e^{-i\omega\tau_{1m}} + \sum_n \int d\omega'\, \hat{a}_n(\omega')\, e^{-i\omega'\tau_{2n}} \\
&\equiv \sum_m \hat{E}_{1m}^{(+)}(\mathbf{r},t) + \sum_n \hat{E}_{2n}^{(+)}(\mathbf{r},t) \\
&\equiv \hat{E}_1^{(+)}(\mathbf{r},t) + \hat{E}_2^{(+)}(\mathbf{r},t), \qquad\qquad (4.2.2)
\end{aligned}
$$

where $g_m(\omega;\mathbf{r}_j,t_j)$, $j = 1,2$, is Green's function that propagates the field from the mth sub-source to $(\mathbf{r}_j,t_j)$; $g(\mathbf{r}_j,t_j;\mathbf{r},t)$ is Green's function that propagates the field from $(\mathbf{r}_j,t_j)$ to the jth pinhole at $(\mathbf{r},t)$; again,

$\tau_{jm} \equiv (t_j - t_{0m}) - r_{mj}/c$ with $t_j \equiv t - s_j/c$. t_j is defined by the observation time t, corresponding to the time that the wavepacket passes through pinhole j. τ_{jm} is also written as $\tau_{jm} = \tau_j - t_{0m}$ with $\tau_j \equiv t_j - r_{mj}/c$.

The interference observable on the observation plane Σ is thus

$$\langle \hat{n}(\mathbf{r},t) \rangle = \langle \hat{E}^{(-)}(\mathbf{r},t)\hat{E}^{(+)}(\mathbf{r},t) \rangle$$

$$= \langle\, [\hat{E}_1^{(-)}(\mathbf{r},t) + \hat{E}_2^{(-)}(\mathbf{r},t)][\hat{E}_1^{(+)}(\mathbf{r},t) + \hat{E}_2^{(+)}(\mathbf{r},t)] \,\rangle$$

$$= \langle \hat{E}_1^{(-)}(\mathbf{r},t)\hat{E}_1^{(+)}(\mathbf{r},t) \rangle + \langle \hat{E}_2^{(-)}(\mathbf{r},t)\hat{E}_2^{(+)}(\mathbf{r},t)| \rangle$$

$$+ \langle \hat{E}_1^{(-)}(\mathbf{r},t)\hat{E}_2^{(+)}(\mathbf{r},t) \rangle + \langle \hat{E}_2^{(-)}(\mathbf{r},t)\hat{E}_1^{(+)}(\mathbf{r},t) \rangle. \qquad (4.2.3)$$

It is easy to find that the first two terms of Eq. (4.2.3), $\langle \hat{E}_1^{(-)}(\mathbf{r},t)\hat{E}_1^{(+)}(\mathbf{r},t) \rangle$ and $\langle \hat{E}_2^{(-)}(\mathbf{r},t)\hat{E}_2^{(+)}(\mathbf{r},t) \rangle$, represent the mean number of photons that passed through pinholes P_1 and P_2, respectively. It is the last two terms of Eq. (4.2.3) that produce the sinusoidal fringes. We need to prove the last two cross-interference terms of Eq. (4.2.3) are the result of a photon, or a group of indistinguishable photons, interferes with the photon, or the group of indistinguishable photons, itself.

Before doing that, we rewrite Eq. (4.2.3) in terms of the effective wave function of a photon or a group of indistinguishable photons. The four terms in Eq. (4.2.3) can be written in the form of $\langle \hat{E}_j^{(-)}(\mathbf{r},t)\hat{E}_k^{(+)}(\mathbf{r},t) \rangle$, where $j = 1, 2$, $k = 1, 2$:

$$\langle \hat{E}_j^{(-)}(\mathbf{r},t)\hat{E}_k^{(+)}(\mathbf{r},t) \rangle = \left\langle \langle \Psi| \sum_m \hat{E}_{jm}^{(-)}(\mathbf{r},t) \sum_n \hat{E}_{kn}^{(+)}(\mathbf{r},t)|\Psi\rangle \right\rangle_{En}$$

$$= \left\langle \langle \Psi| \sum_m \hat{E}_{jm}^{(-)}(\mathbf{r},t) \left[\sum_l |l\rangle\langle l| \right] \sum_n \hat{E}_{kn}^{(+)}(\mathbf{r},t)\,|\Psi\rangle \right\rangle_{En}$$

$$= \left\langle \langle \Psi| \sum_m \hat{E}_{jm}^{(-)}(\mathbf{r},t)\,|0\rangle\langle 0| \sum_n \hat{E}_{kn}^{(+)}(\mathbf{r},t)\,|\Psi\rangle \right\rangle_{En}$$

$$= \left\langle \left[\sum_m \psi_{jm}^{*}(\mathbf{r},t) \right] \left[\sum_n \psi_{kn}(\mathbf{r},t) \right] \right\rangle_{En}, \qquad (4.2.4)$$

where the effective wavefunction of a photon or a group of indistinguishable photons that passed through the kth pinhole is defined as

$$\psi_{kn}(\mathbf{r},t) \equiv \langle 0| \hat{E}_{kn}^{(+)}(\mathbf{r},t)|\Psi\rangle. \qquad (4.2.5)$$

We also have

$$\langle \hat{E}_j^{(-)}(\mathbf{r},t)\hat{E}_k^{(+)}(\mathbf{r},t)\rangle = \langle\langle\Psi|\hat{E}_j^{(-)}(\mathbf{r},t)\hat{E}_k^{(+)}(\mathbf{r},t)|\Psi\rangle\rangle_{\mathrm{En}}$$

$$= \left\langle \langle\Psi|\hat{E}_j^{(-)}(\mathbf{r},t)\left[\sum_l |l\rangle\langle l|\right]\hat{E}_k^{(+)}(\mathbf{r},t)|\Psi\rangle \right\rangle_{\mathrm{En}}$$

$$= \langle\langle\Psi|\hat{E}_j^{(-)}(\mathbf{r},t)|0\rangle\langle 0|\hat{E}_k^{(+)}(\mathbf{r},t)|\Psi\rangle\rangle_{\mathrm{En}}$$

$$= \langle \Psi_j^*(\mathbf{r},t)\,\Psi_k(\mathbf{r},t)\rangle_{\mathrm{En}}, \tag{4.2.6}$$

where the effective wavefunction of the field that passed through the kth pinhole is defined as

$$\Psi_k(\mathbf{r},t) \equiv \langle 0|\hat{E}_k^{(+)}(\mathbf{r},t)|\Psi\rangle. \tag{4.2.7}$$

It is easy to find that

$$\Psi_k(\mathbf{r},t) = \sum_n \psi_{kn}(\mathbf{r},t). \tag{4.2.8}$$

Substituting the field operator and the state into Eq. (4.2.7), we have

$$\Psi_k(\mathbf{r},t) = \sum_n \mathcal{F}_{\tau_{kn}}\{f_n(\nu)\}\,e^{-i\omega_0\tau_{kn}}. \tag{4.2.9}$$

The cross-interference terms is calculated as follows:

$$\langle \hat{E}_1^{(-)}(\mathbf{r},t)\hat{E}_2^{(+)}(\mathbf{r},t)\rangle$$

$$= \left\langle \sum_m \mathcal{F}_{\tau_{1m}}^*\{f_m(\nu)\}\,e^{i\omega_0\tau_{1m}} \sum_n \mathcal{F}_{\tau_{2n}}\{f_n(\nu)\}\,e^{-i\omega_0\tau_{2n}} \right\rangle$$

$$= \left\langle \sum_{m=n} e^{i\omega_0(r_{m1}-r_{m2})}\,\mathcal{F}_{\tau_{1m}}^*\{f_m(\nu)\}\,\mathcal{F}_{\tau_{2n}}\{f_n(\nu)\} \right\rangle e^{i\omega_0(s_1-s_2)/c}$$

$$+ \left\langle \sum_{m\neq n} e^{i\omega_0(t_n-t_m)}\,e^{i\omega_0(r_{m1}-r_{m2})} \right.$$

$$\left. \times \mathcal{F}_{\tau_{1m}}^*\{f_m(\nu)\}\,\mathcal{F}_{\tau_{2n}}\{f_n(\nu)\} \right\rangle e^{i\omega_0(s_1-s_2)/c}$$

$$= \left\langle \sum_m e^{i\omega_0(r_{m1}-r_{m2})}\,\mathcal{F}_{\tau_{1m}}^*\{f_m(\nu)\}\,\mathcal{F}_{\tau_{2m}}\{f_m(\nu)\} \right\rangle e^{i\omega_0(s_1-s_2)/c}, \tag{4.2.10}$$

where we have assumed that all wavepackets have the same carrier frequency and also defined $\tau_{jm} = (t_j - t_m) - r_{mj}/c$ and $t_j = t - s_j/c$ $j = 1, 2$. The second sum ($m \neq n$) vanishes when taking into account all possible random phases and/or random values of $t_m - t_n$. This result indicates that *the only observable interference is the interference in which the mth photon interferes with the mth photon itself*. Interference between two different photons $m \neq n$ becomes unobservable due to the incoherent ensemble average.

Accordingly, the measured mean number of photons $\langle n(\mathbf{r}, t) \rangle$, as an interference pattern on the observation plane, is the result of the sum of a large number of individual sub-interferences, and each is produced by the self-interference of a photon — the mth photon interferes with the mth photon itself:

$$\langle \hat{n}(\mathbf{r}, t) \rangle \propto \sum_m \left| \frac{1}{\sqrt{2}} [\psi_{1m}^*(\mathbf{r}, t) + \psi_{2m}^*(\mathbf{r}, t)] \right|^2$$

$$= \sum_m \left| \frac{1}{\sqrt{2}} [\mathcal{F}_{\tau_{1m}}\{f_m(\nu)\} e^{-i\omega_0 \tau_{1m}} + \mathcal{F}_{\tau_{2m}}\{f_m(\nu)\} e^{-i\omega_0 \tau_{2m}}] \right|^2.$$

$$(4.2.11)$$

This result is consistent with that calculated in Einstein's picture. It is not difficult to find that the effective wave function of a photon or a group of indistinguishable photons in the single-photon state representation plays the same role as Einstein's quantized subfield.

(II) $\langle \hat{n}(\mathbf{r}, t) \rangle$: Thermal state in coherent state representation

Now, we calculate interference fringes in coherent state representation. The state of a thermal field in coherent state representation has been introduced in Eq. (3.5.7):

$$|\tilde{\Psi}\rangle = \prod_m \prod_\omega |\alpha_m(\omega)\rangle, \qquad (4.2.12)$$

where m labels the mth sub-source. Equation (4.2.12) represents a mixed coherent state. The mixed coherent state is defined in Chapter 2. In Eq. (4.2.12), we explicitly separate the indices m and ω as usual. In the following calculations, we assume a large number of independent random radiating and randomly distributed sub-sources, each of which produces a group of indistinguishable photons with coherent Fourier modes.

The propagation of field operators is the same as in the single-photon state representation.

We need to introduce the concept of effective wavefunction in the coherent state representation. Similar to what we did in the single-photon state representation, we introduce the effective wavefunction form the calculation of $\langle \hat{E}_j^{(-)}(\mathbf{r},t)\hat{E}_k^{(+)}(\mathbf{r},t)\rangle$:

$$\langle \hat{E}_j^{(-)}(\mathbf{r},t)\hat{E}_k^{(+)}(\mathbf{r},t)\rangle = \left\langle \langle \Psi | \sum_m \hat{E}_{jm}^{(-)}(\mathbf{r},t) \sum_n \hat{E}_{kn}^{(+)}(\mathbf{r},t)|\Psi\rangle \right\rangle_{\mathrm{En}}$$

$$= \left\langle \langle \Psi | \sum_m \hat{E}_{jm}^{(-)}(\mathbf{r},t) |\Psi\rangle\langle\Psi| \sum_n \hat{E}_{kn}^{(+)}(\mathbf{r},t) |\Psi\rangle \right\rangle_{\mathrm{En}}$$

$$= \left\langle \left[\sum_m \psi_{jm}^*(\mathbf{r},t)\right]\left[\sum_n \psi_{kn}(\mathbf{r},t)\right] \right\rangle_{\mathrm{En}}, \qquad (4.2.13)$$

where the effective wavefunction of a photon or a group of indistinguishable photons that passed through the kth pinhole is defined as

$$\psi_{kn}(\mathbf{r},t) \equiv \langle \Psi | \hat{E}_{kn}^{(+)}(\mathbf{r},t)|\Psi\rangle. \qquad (4.2.14)$$

We also have

$$\langle \hat{E}_j^{(-)}(\mathbf{r},t)\hat{E}_k^{(+)}(\mathbf{r},t)\rangle = \langle\langle\Psi|\hat{E}_j^{(-)}(\mathbf{r},t)\hat{E}_k^{(+)}(\mathbf{r},t)|\Psi\rangle\rangle_{\mathrm{En}}$$

$$= \langle\langle\Psi|\hat{E}_j^{(-)}(\mathbf{r},t)|\Psi\rangle\langle\Psi|\hat{E}_k^{(+)}(\mathbf{r},t)|\Psi\rangle\rangle_{\mathrm{En}}$$

$$= \langle \Psi_j^*(\mathbf{r},t)\,\Psi_k(\mathbf{r},t)\rangle_{\mathrm{En}}, \qquad (4.2.15)$$

where the effective wavefunction of the field that passed through the kth pinhole is defined as

$$\Psi_k(\mathbf{r},t) \equiv \langle \Psi | \hat{E}_k^{(+)}(\mathbf{r},t)|\Psi\rangle. \qquad (4.2.16)$$

It is easy to find that

$$\Psi_k(\mathbf{r},t) = \sum_n \psi_{kn}(\mathbf{r},t). \qquad (4.2.17)$$

Substituting the field operator and the state into Eq. (4.2.16), we have

$$\Psi_k(\mathbf{r},t) = \sum_m \int d\omega\, \alpha_m(\omega)\, e^{-i\omega[(t_k-t_m)-r_{mk}/c]]\rangle}$$

$$= \sum_m \mathcal{F}_{\tau_{km}}\{\alpha_m(\nu)\}\, e^{-i\omega_0 \tau_{km}}. \qquad (4.2.18)$$

The cross-interference terms is calculated as follows:

$$\langle \hat{E}_1^{(-)}(\mathbf{r},t)\hat{E}_2^{(+)}(\mathbf{r},t)\rangle$$

$$= \left\langle \sum_m \mathcal{F}_{\tau_{1m}}^* \{\alpha_m(\nu)\}\, e^{i\omega_0 \tau_{1m}} \sum_n \mathcal{F}_{\tau_{2n}}\{\alpha(\nu)\}\, e^{-i\omega_0 \tau_{2n}} \right\rangle$$

$$= \left\langle \sum_{m=n} e^{i\omega_0(r_{m1}-r_{m2})}\, \mathcal{F}_{\tau_{1m}}^*\{\alpha_m(\nu)\}\, \mathcal{F}_{\tau_{2n}}\{\alpha_n(\nu)\} \right\rangle e^{i\omega_0(s_1-s_2)/c}$$

$$+ \left\langle \sum_{m \neq n} e^{i\omega_0(t_n-t_m)}\, e^{i\omega_0(r_{m1}-r_{m2})} \right.$$

$$\left. \times\, \mathcal{F}_{\tau_{1m}}^*\{\alpha_m(\nu)\}\, \mathcal{F}_{\tau_{2n}}\{\alpha_n(\nu)\} \right\rangle e^{i\omega_0(s_1-s_2)/c}$$

$$= \left\langle \sum_m e^{i\omega_0(r_{m1}-r_{m2})}\, \mathcal{F}_{\tau_{1m}}^*\{\alpha_m(\nu)\}\, \mathcal{F}_{\tau_{2m}}\{\alpha_m(\nu)\} \right\rangle e^{i\omega_0(s_1-s_2)/c},$$

$$(4.2.19)$$

where, again, we have assumed that all wavepackets have the same carrier frequency. The second sum ($m \neq n$) vanishes when taking into account all possible random phases and/or random values of $t_m - t_n$. This result indicates that *the only observable interference is the interference in which the mth photon or the mth group indistinguishable photons interfere with the mth photon or the mth group of photons*. Interference between two different photons or two different groups of indistinguishable photons ($m \neq n$) becomes unobservable due to the ensemble average.

Accordingly, the measured mean number of photons $\langle n(\mathbf{r},t)\rangle$, as an interference pattern on the observation plane, is the result of the sum of a large number of individual sub-interferences, and each is produced by the self-interference of a photon or a group of indistinguishable photons — the mth photon, or the mth group of indistinguishable photons, interferes with the mth photon, or the mth group of indistinguishable photons, itself:

$$\langle \hat{n}(\mathbf{r},t)\rangle \propto \sum_m \left| \frac{1}{\sqrt{2}} [\psi_{1m}^*(\mathbf{r},t) + \psi_{2m}^*(\mathbf{r},t)] \right|^2$$

$$= \sum_m \left| \frac{1}{\sqrt{2}} [\mathcal{F}_{\tau_{1m}}\{\alpha_m(\nu)\} e^{-i\omega_0 \tau_{1m}} + \mathcal{F}_{\tau_{2m}}\{\alpha_m(\nu)\} e^{-i\omega_0 \tau_{2m}}] \right|^2.$$

$$(4.2.20)$$

This result is consistent with the results calculated from Einstein's picture and the single-photon state representation. It is not difficult to find that the effective wave function of a photon or a group of indistinguishable photons in the coherent state representation plays the same role as Einstein's quantized subfield.

4.3   First-Order Coherence of Light

In this section, we introduce the concept of first-order coherence of light. The degree of first-order coherence is defined to quantify the interference between temporally delayed and/or spatially separated radiations, considered as either classical waves or light quanta. In an Young's double-pinhole interferometer, the radiation created from the source is selected by the upper pinhole and the lower pinhole and superposed on the observation plane Σ, as shown in Fig. 4.1.1. The questions are as follows: (1) Are the two temporally delayed and/or spatially separated radiations always able to produce observable interference fringes? (2) How to quantify the quality of their interference? The first-order coherence theory of light is aimed to address these questions. The superposed radiations are defined as first-order coherent if the interference fringes exhibit 100% modulation or first-order incoherent if no interference fringes are observable. The radiation fields are considered as partial coherent if the modulation is less than 100% but greater than zero. The higher the degree of first-order coherence, the higher interference visibility we are able to observe.

(I) First-order coherence of light: Electromagnetic theory

The concept of first-order coherence was introduced in classical theory much earlier than quantum theory. Part (I) of our introduction to first-order coherence is in the framework of electromagnetic wave theory of light. To simplify the discussion and mathematics, we assume a far-field point-like radiation source located at $\mathbf{r} = 0$ of Young's double-pinhole interferometer shown in Fig. 4.1.1. The expectation value of the intensity, $\langle I(\mathbf{r}, t) \rangle$, measured at space-time point $(\mathbf{r}, t)$ of the observation plan Σ is

$$\langle I(\mathbf{r}, t) \rangle = \langle |E(\mathbf{r}, t)|^2 \rangle$$

$$= \langle |E(\mathbf{r}_1, t_1)\, g_1(\mathbf{r}, t) + E(\mathbf{r}_2, t_2)\, g_2(\mathbf{r}, t)|^2 \rangle$$

$$= \left\langle \left| E\left(\mathbf{r}_1, t - \frac{s_1}{c}\right) + E\left(\mathbf{r}_2, t - \frac{s_2}{c}\right) \right|^2 \right\rangle$$

$$= \left\langle \left| E(\mathbf{r}_1, t_1) + E(\mathbf{r}_2, t_2) \right|^2 \right\rangle$$

$$= \left\langle E^*(\mathbf{r}_1, t_1) E(\mathbf{r}_1, t_1) \right\rangle + \left\langle E^*(\mathbf{r}_2, t_2) E(\mathbf{r}_2, t_2) \right\rangle$$

$$+ \left\langle E^*(\mathbf{r}_1, t_1) E(\mathbf{r}_2, t_2) \right\rangle + \left\langle E(\mathbf{r}_1, t_1) E^*(\mathbf{r}_2, t_2) \right\rangle, \qquad (4.3.1)$$

where $E(\mathbf{r}_1, t_1)$ and $E(\mathbf{r}_2, t_2)$ are the electromagnetic fields that pass through P_1 and P_2 at times t_1 and t_2, and $t_1 \equiv t - s_1/c$ and $t_2 \equiv t - s_2/c$ are the "earlier" times relative to the measurement time t; $g_j(\mathbf{r}, t)$, $j = 1, 2$, is Green's function that propagates the field from the jth pinhole to $(\mathbf{r}, t)$ along the jth optical path. The early times t_1 and t_2 are defined by the observation time t and the optical paths s_1 and s_2: $(t - t_1) = s_1/c$ and $(t - t_2) = s_2/c$. Note that Eq. (4.3.1) adopted the important conclusion about t_1 and t_2 from previous sections. Similarly, we recognized that the first two terms in Eq. (4.3.1) correspond to the light intensity passing through the upper and lower pinholes at space–time points $(\mathbf{r}_1, t_1)$ and $(\mathbf{r}_2, t_2)$, respectively. The third and fourth terms, which involve the cross product of the fields at space–time point $(\mathbf{r}_1, t_1)$ [or $(\mathbf{r}_2, t_2)$] with its conjugate at space–time point $(\mathbf{r}_2, t_2)$ [or $(\mathbf{r}_1, t_1)$], give rise to an interference pattern at the observation plane and results in a sinusoidal modulation of the photocurrent as a function of the position of the scanning photodetector or the coordinate of the pixels of the CCD.

We break $\left\langle I(\mathbf{r}, t) \right\rangle$ of Eq. (4.3.1) into two groups:

(1) $\langle E^*(\mathbf{r}_1, t_1) E(\mathbf{r}_1, t_1) \rangle$ and $\langle E^*(\mathbf{r}_2, t_2) E(\mathbf{r}_2, t_2) \rangle$,

(2) $\langle E^*(\mathbf{r}_1, t_1) E(\mathbf{r}_2, t_2) \rangle$ and $\langle E(\mathbf{r}_1, t_1) E^*(\mathbf{r}_2, t_2) \rangle$.

We define the first group as the self-coherence function:

$$\Gamma^{(1)}(\mathbf{r}_1, t_1; \mathbf{r}_1, t_1) \equiv \langle E^*(\mathbf{r}_1, t_1) E(\mathbf{r}_1, t_1) \rangle,$$

$$\Gamma^{(1)}(\mathbf{r}_2, t_2; \mathbf{r}_2, t_2) \equiv \langle E^*(\mathbf{r}_2, t_2) E(\mathbf{r}_2, t_2) \rangle \qquad (4.3.2)$$

and the second group as the mutual-coherence function:

$$\Gamma^{(1)}(\mathbf{r}_1, t_1; \mathbf{r}_2, t_2) \equiv \langle E^*(\mathbf{r}_1, t_1) E(\mathbf{r}_2, t_2) \rangle,$$

$$\Gamma^{(1)}(\mathbf{r}_2, t_2; \mathbf{r}_1, t_1) \equiv \langle E(\mathbf{r}_1, t_1) E^*(\mathbf{r}_2, t_2) \rangle. \qquad (4.3.3)$$

It is obvious that

$$\Gamma^{(1)}(\mathbf{r}_1, t_1; \mathbf{r}_2, t_2) = \Gamma^{(1)*}(\mathbf{r}_2, t_2; \mathbf{r}_1, t_1). \qquad (4.3.4)$$

In connection with the concepts of classical statistics, the self-coherence function of Eq. (4.3.2) and the mutual-coherence function of Eq. (4.3.3) are recognized as the self-correlation function and the cross-correlation function, respectively, of the fields. Physically, $\Gamma(\mathbf{r}_1, t_1; \mathbf{r}_2, t_2)$ and $\Gamma(\mathbf{r}_2, t_2; \mathbf{r}_1, t_1)$ determine the visibility of the interference, which is quantified in the following. The self-coherence function $\Gamma(\mathbf{r}_j, t_j; \mathbf{r}_j, t_j)$, $j = 1, 2$, defined in Eq. (4.3.2) represents the expectation value, or expectation function of intensity which has been discussed in Chapter 1. Applying the mutual-coherence function and the self-coherence function, the expectation value of $I(\mathbf{r}, t)$ in Eq. (4.3.1) is written as

$$\langle I(\mathbf{r}, t) \rangle = \Gamma_{11}^{(1)} + \Gamma_{22}^{(1)} + \Gamma_{12}^{(1)} + \Gamma_{21}^{(1)}, \tag{4.3.5}$$

where we have used the shortened notation:

$$\Gamma_{11}^{(1)} = \Gamma^{(1)}(\mathbf{r}_1, t_1; \mathbf{r}_1, t_1), \quad \Gamma_{22}^{(1)} = \Gamma^{(1)}(\mathbf{r}_2, t_2; \mathbf{r}_2, t_2),$$

$$\Gamma_{12}^{(1)} = \Gamma^{(1)}(\mathbf{r}_1, t_1; \mathbf{r}_2, t_2), \quad \Gamma_{21}^{(1)} = \Gamma^{(1)}(\mathbf{r}_2, t_2; \mathbf{r}_1, t_1).$$

Now, we introduce the normalized complex degree of first-order coherence by writing Eq. 4.3.15 in the following form:

$$\langle I(\mathbf{r}, t) \rangle = \Gamma_{11}^{(1)} + \Gamma_{22}^{(1)} + 2\sqrt{\Gamma_{11}^{(1)} \Gamma_{22}^{(1)}} \, Re \, \gamma_{12}^{(1)}$$

$$= \langle I_1 \rangle + \langle I_2 \rangle + 2\sqrt{\langle I_1 \rangle \langle I_2 \rangle} \, |\gamma_{12}^{(1)}| \cos(\omega\tau), \tag{4.3.6}$$

where $\tau = \tau_1 - \tau_2 = (t_1 - t_2) - (r_1 - r_2)/c = (s_2 - s_1)/c + (r_2 - r_1)/c$, $(r_1 - r_2)/c$ is the relative time delay between the optical paths from the light source to pinholes P_1 and P_2; the function of $\cos(\omega\tau)$ is adapted from previous section. In Eq. (4.3.6), we have defined the normalized complex degree of first-order coherence:

$$\gamma^{(1)}(\mathbf{r}_1, t_1; \mathbf{r}_2, t_2) \equiv \frac{\Gamma^{(1)}(\mathbf{r}_1, t_1; \mathbf{r}_2, t_2)}{\left[\Gamma^{(1)}(\mathbf{r}_1, t_1; \mathbf{r}_1, t_1)\right]^{\frac{1}{2}} \left[\Gamma^{(1)}(\mathbf{r}_2, t_2; \mathbf{r}_2, t_2)\right]^{\frac{1}{2}}} \tag{4.3.7}$$

$$= \frac{\langle E^*(\mathbf{r}_1, t_1) E(\mathbf{r}_2, t_2) \rangle}{\left[\langle |E(\mathbf{r}_1, t_1)|^2 \rangle\right]^{\frac{1}{2}} \left[\langle |E(\mathbf{r}_2, t_2)|^2 \rangle\right]^{\frac{1}{2}}},$$

for $\Gamma(\mathbf{r}_1, t_1; \mathbf{r}_1, t_1) \neq 0$ and $\Gamma(\mathbf{r}_2, t_2; \mathbf{r}_2, t_2) \neq 0$. It is easy to find that

$$0 \leq |\gamma_{12}^{(1)}| \leq 1. \tag{4.3.8}$$

$|\gamma_{12}^{(1)}|$ is thus related to the visibility of the interference fringe modulation

$$V \equiv \frac{I_{\text{MAX}} - I_{\text{MIN}}}{I_{\text{MAX}} + I_{\text{MIN}}} = 2\frac{\sqrt{\langle I_1 \rangle \langle I_2 \rangle}}{\langle I_1 \rangle + \langle I_2 \rangle} |\gamma_{12}^{(1)}|. \qquad (4.3.9)$$

If $\langle I_1 \rangle = \langle I_2 \rangle$, which is perhaps the most common arrangement for an optimized interferometer, $|\gamma_{12}|$ is identical to the visibility of the interference modulation:

$$V = |\gamma_{12}^{(1)}|.$$

In this case, the expectation function of the intensity on the observation plane is thus

$$\langle I(\mathbf{r}, t) \rangle = I_0 \left[1 + V \cos(\omega\tau) \right]. \qquad (4.3.10)$$

The radiation fields at space–time points $(\mathbf{r}_1, t_1)$ and $(\mathbf{r}_2, t_2)$ are named first-order coherent, partially coherent, and incoherent in terms of the value of $|\gamma_{12}^{(1)}|$:

$$
\begin{array}{lcl}
\text{1st-order Coherent} & \text{if} & |\gamma^{(1)}(\mathbf{r}_1, t_1; \mathbf{r}_2, t_2)| = 1, \\
\text{1st-order Partially Coherent} & \text{if} & 0 < |\gamma^{(1)}(\mathbf{r}_1, t_1; \mathbf{r}_2, t_2)| < 1, \\
\text{1st-order Incoherent} & \text{if} & |\gamma^{(1)}(\mathbf{r}_1, t_1; \mathbf{r}_2, t_2)| = 0.
\end{array}
$$

(II) First-order coherence of light: Quantum theory

Part (II) of our introduction to first-order coherence is in the framework of quantum theory of light. We start from Young's double-pinhole interferometer shown in Fig. 4.1.1. To simplify the discussion and mathematics, we assume a far-field point-like radiation source located at $\mathbf{r} = 0$. Similar to the previous discussion, the interference observable on the observation plane Σ is

$$
\begin{aligned}
\langle \hat{n}(\mathbf{r}, t) \rangle &= \left\langle \left[\hat{E}^{(-)}(\mathbf{r}_1, t_1) + \hat{E}^{(-)}(\mathbf{r}_1, t_1) \right]\left[\hat{E}^{(+)}(\mathbf{r}_2, t_2) + \hat{E}^{(+)}(\mathbf{r}_2, t_2) \right] \right\rangle \\
&= \left\langle \hat{E}^{(-)}(\mathbf{r}_1, t_1)\hat{E}^{(+)}(\mathbf{r}_1, t_1) \right\rangle + \left\langle \hat{E}^{(-)}(\mathbf{r}_2, t_2)\hat{E}^{(+)}(\mathbf{r}_2, t_2) \right\rangle \\
&\quad + \left\langle \hat{E}^{(-)}(\mathbf{r}_1, t_1)\hat{E}^{(+)}(\mathbf{r}_2, t_2) \right\rangle + \left\langle \hat{E}^{(-)}(\mathbf{r}_2, t_2)\hat{E}^{(+)}(\mathbf{r}_1, t_1) \right\rangle.
\end{aligned}
$$
$$(4.3.11)$$

It is easy to find that the first two terms of Eq. (4.3.11), $\langle \hat{E}^{(-)}(\mathbf{r}_1, t_1)\hat{E}^{(+)}(\mathbf{r}_1, t_1) \rangle$ and $\langle \hat{E}^{(-)}(\mathbf{r}_2, t_2)\hat{E}^{(+)}(\mathbf{r}_2, t_2) \rangle$, represent the mean number of photons that passed through pinhole P_1 and P_2, respectively. It is the last two terms of Eq. (4.3.11) that produce the sinusoidal fringes.

We break $\langle n(\mathbf{r}, t) \rangle$ of Eq. (4.3.11) into two groups:

(1) $\langle \hat{E}^{(-)}(\mathbf{r}_1, t_1) \hat{E}^{(+)}(\mathbf{r}_1, t_1) \rangle$ and $\langle \hat{E}^{(-)}(\mathbf{r}_2, t_2) \hat{E}^{(+)}(\mathbf{r}_2, t_2) | \rangle$,

(2) $\langle \hat{E}^{(-)}(\mathbf{r}_1, t_1) \hat{E}^{(+)}(\mathbf{r}_2, t_2) \rangle$ and $\langle \hat{E}^{(-)}(\mathbf{r}_2, t_2) \hat{E}^{(+)}(\mathbf{r}_1, t_1) \rangle$.

We define the first group as the self-coherence function:

$$G^{(1)}(\mathbf{r}_1, t_1; \mathbf{r}_1, t_1) \equiv \langle \hat{E}^{(-)}(\mathbf{r}_1, t_1) \hat{E}^{(+)}(\mathbf{r}_1, t_1) \rangle,$$

$$G^{(1)}(\mathbf{r}_2, t_2; \mathbf{r}_2, t_2) \equiv \langle \hat{E}^{(-)}(\mathbf{r}_2, t_2) \hat{E}^{(+)}(\mathbf{r}_2, t_2) \rangle \tag{4.3.12}$$

and the second group as the mutual-coherence function:

$$G^{(1)}(\mathbf{r}_1, t_1; \mathbf{r}_2, t_2) \equiv \langle \hat{E}^{(-)}(\mathbf{r}_1, t_1) \hat{E}^{(+)}(\mathbf{r}_2, t_2) \rangle,$$

$$G^{(1)}(\mathbf{r}_2, t_2; \mathbf{r}_1, t_1) \equiv \langle \hat{E}^{(-)}(\mathbf{r}_2, t_2) \hat{E}^{(+)}(\mathbf{r}_1, t_1) \rangle. \tag{4.3.13}$$

It is obvious that

$$G^{(1)}(\mathbf{r}_1, t_1; \mathbf{r}_2, t_2) = G^{(1)*}(\mathbf{r}_2, t_2; \mathbf{r}_1, t_1). \tag{4.3.14}$$

In connection with the concepts of quantum statistics, the self-coherence function of Eq. (4.3.12) and the mutual-coherence function of Eq. (4.3.13) are recognized as the self-correlation function and the cross-correlation function, respectively, of the quantized fields. Physically, $G^{(1)}(\mathbf{r}_1, t_1; \mathbf{r}_2, t_2)$ determines the visibility of the interference, which is quantified in the following. The self-coherence function $G^{(1)}(\mathbf{r}_j, t_j; \mathbf{r}_j, t_j)$, $j = 1, 2$, defined in Eq. (4.3.12) represents the expectation value, or expectation function of the photon number operator, or simply the mean number of photons that passed through the jth pinhole at $(\mathbf{r}_j, t_j)$. Applying the mutual-coherence function and the self-coherence function, the expectation value of $\hat{n}(\mathbf{r}, t)$ in Eq. (4.3.11) is written as

$$\langle \hat{n}(\mathbf{r}, t) \rangle = G^{(1)}_{11} + G^{(1)}_{22} + G^{(1)}_{12} + G^{(1)}_{21}, \tag{4.3.15}$$

where we have used the shortened notation

$$G^{(1)}_{11} = G^{(1)}(\mathbf{r}_1, t_1; \mathbf{r}_1, t_1), \quad G^{(1)}_{22} = G^{(1)}(\mathbf{r}_2, t_2; \mathbf{r}_2, t_2),$$

$$G^{(1)}_{12} = G^{(1)}(\mathbf{r}_1, t_1; \mathbf{r}_2, t_2), \quad G^{(1)}_{21} = G^{(1)}(\mathbf{r}_2, t_2; \mathbf{r}_1, t_1).$$

Now, we introduce the normalized complex degree of first-order coherence by writing Eq. (4.3.11) in the following form:

$$\langle \hat{n}(\mathbf{r}, t) \rangle = G^{(1)}_{11} + G^{(1)}_{22} + 2\sqrt{G^{(1)}_{11} G^{(1)}_{22}} \, Re \, g^{(1)}_{12}$$

$$= \langle \hat{n}_1 \rangle + \langle \hat{n}_2 \rangle + 2\sqrt{\langle \hat{n}_1 \rangle \langle \hat{n}_2 \rangle} \, |g^{(1)}_{12}| \cos(\omega\tau), \tag{4.3.16}$$

where $\tau = \tau_1 - \tau_2 = (t_1 - t_2) - (r_1 - r_2)/c = (s_2 - s_1)/c + (r_2 - r_1)/c$, $(r_1 - r_2)/c$ is the relative time delay between the optical paths from the light source to pinholes P_1 and P_2; the function of $\cos(\omega\tau)$ is adapted from previous section. In Eq. (4.3.16), we have defined the normalized complex degree of first-order coherence:

$$
\begin{aligned}
g^{(1)}&(\mathbf{r}_1, t_1; \mathbf{r}_2, t_2) \\[2mm]
&\equiv \frac{G^{(1)}(\mathbf{r}_1, t_1; \mathbf{r}_2, t_2)}{\left[G^{(1)}(\mathbf{r}_1, t_1; \mathbf{r}_1, t_1)\right]^{\frac{1}{2}} \left[G^{(1)}(\mathbf{r}_2, t_2; \mathbf{r}_2, t_2)\right]^{\frac{1}{2}}} \\[3mm]
&= \frac{\langle \hat{E}^{(-)}(\mathbf{r}_1, t_1)\hat{E}^{(+)}(\mathbf{r}_2, t_2)\rangle}{\left[\langle \hat{E}^{(-)}(\mathbf{r}_1, t_1)\hat{E}^{(+)}(\mathbf{r}_1, t_1)\rangle\right]^{\frac{1}{2}} \left[\langle \hat{E}^{(-)}(\mathbf{r}_2, t_2)\hat{E}^{(+)}(\mathbf{r}_2, t_2)\rangle\right]^{\frac{1}{2}}}
\end{aligned}
\tag{4.3.17}
$$

for $G(\mathbf{r}_1, t_1; \mathbf{r}_1, t_1) \neq 0$ and $G(\mathbf{r}_2, t_2; \mathbf{r}_2, t_2) \neq 0$. It is easy to find that

$$
0 \leq |g_{12}^{(1)}| \leq 1.
\tag{4.3.18}
$$

$|g_{12}^{(1)}|$ is thus related to the visibility of the interference fringe modulation

$$
V \equiv \frac{n_{\text{MAX}} - n_{\text{MIN}}}{n_{\text{MAX}} + n_{\text{MIN}}} = 2\frac{\sqrt{\langle \hat{n}_1\rangle\langle \hat{n}_2\rangle}}{\langle \hat{n}_1\rangle + \langle \hat{n}_2\rangle} |g_{12}^{(1)}|.
\tag{4.3.19}
$$

If $\langle \hat{n}_1\rangle = \langle \hat{n}_2\rangle$, which is perhaps the most common arrangement for an optimized interferometer, $|g_{12}|$ is identical to the visibility of the interference modulation:

$$
V = |g_{12}^{(1)}|.
$$

In this case, the mean photon number measured on the observation plane is thus

$$
\langle \hat{n}(\mathbf{r}, t)\rangle = n_0\left[1 + V\cos(\omega\tau)\right].
\tag{4.3.20}
$$

4.4   First-Order Temporal Coherence: Einstein's Picture

In this section, we focus on the first-order temporal coherence of light. To simplify the mathematics and the discussion, we assume a far-field point-like radiation source at $\mathbf{r} = 0$ for Young's double-slit experimental setup of Fig. 4.4.1.

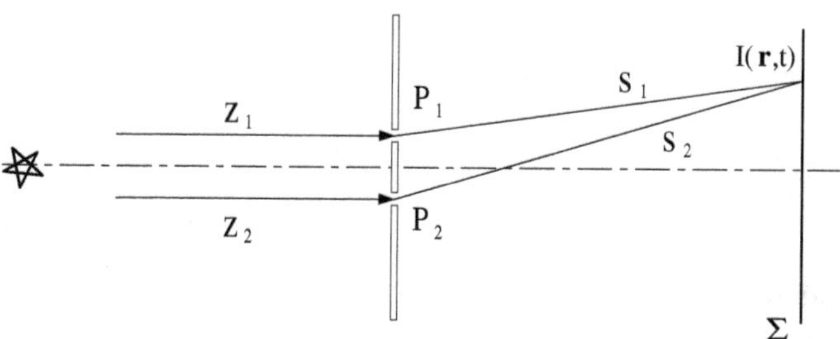

Fig. 4.4.1   Schematic of Young's double-slit interference experiment which measures the degree of first-order coherence of incoherent or coherent light. To simplify the discussion, we focus on the measurement of first-order temporal coherence by assuming a far-field point-like radiation source at $\mathbf{r} = 0$. We have $r_1 \simeq z_1$, $r_2 \simeq z_2$, and $z_1 = z_2$.

(I) $\Gamma^{(1)}(\mathbf{r}_1, t_1; \mathbf{r}_2, t_2)$: Incoherent subfields and incoherent Fourier modes

Consider a point-like light source which consists of a large number of independent point sub-sources that radiate randomly. The measured radiation coming from the source is an incoherent superposition of a large number of such incoherent subfields and incoherent Fourier modes. This is a typical classical model for natural radiation, such as radiation from distant stars. Each atomic transition is treated as a sub-source, and each atomic transition produces a Fourier mode of single frequency at a random time with random phase. Compared with the broadband of natural light, the finite line-width of the atomic transitions is ignorable.

Determining its degree of first-order coherence, we need to calculate the mutual-coherence function of $E(\mathbf{r}_1, t_1)$ and $E(\mathbf{r}_2, t_2)$:

$$\Gamma^{(1)}(\mathbf{r}_1, t_1; \mathbf{r}_2, t_2)$$

$$= \left\langle \sum_m E_m^*(\mathbf{r}_1, t_1) \sum_n E_n(\mathbf{r}_2, t_2) \right\rangle$$

$$= \left\langle \sum_m \int d\omega \, a_m(\omega) e^{-i\varphi_m(\omega)} \, e^{i\omega\tau_{1m}} \sum_n \int d\omega' \, a_n(\omega') e^{i\varphi_n(\omega')} \, e^{-i\omega'\tau_{2n}} \right\rangle$$

$$\simeq \sum_m \int d\omega \, a_m^2(\omega) e^{i\omega(\tau_1 - \tau_2)}$$

$$\simeq \mathcal{F}_\tau \left\{ \sum_m a_m^2(\nu) \right\} e^{i\omega_0\tau}, \tag{4.4.1}$$

where the expectation value calculation or ensemble average has taken into account all possible realizations of the random phases of the subfields and the Fourier modes. In Eq. (4.4.1), again, we have defined $\tau_1 = t_1 - z_1/c$, $\tau_2 = t_2 - z_2/c$, and $\tau = \tau_1 - \tau_2 = (s_2 - s_1)/c + (z_2 - z_1)/c$. In the case of $z_1 = z_2$, see Fig. 4.4.1, we have $\tau = (s_2 - s_1)/c$. In the simplified situation in which all the subfields hold identical spectrum distribution $a^2(\omega)$, it is easy to find that the optical delay of τ and the spectral bandwidth $\Delta\omega$ are restricted by the relation of $\Delta\omega\,\tau = 2\pi$ for a nonzero mutual-coherence function of $\Gamma_{12}^{(1)}$ and consequently for an observable interference.

$\Gamma_{12}^{(1)}$ of the above classical field is a function of the temporal delay $\tau = \tau_1 - \tau_2 = (s_2 - s_1)/c + (z_2 - z_1)/c$, which implies that the temporal correlation function of a thermal field is invariant under the displacements of temporal variables, i.e., invariant for time t. This is the characteristic of stationary fields. For stationary fields, the temporal mutual-correlation function $\Gamma^{(1)}(\mathbf{r}_1, t_1; \mathbf{r}_2, t_2)$ is usually written as $\Gamma_{12}^{(1)}(\tau)$. Compared with the expectation value of the intensity of thermal radiation in Eq. (1.3.15), we find $\langle I(\mathbf{r}_1, t_1)\rangle = \Gamma_{11}^{(1)}(0)$ and $\langle I(\mathbf{r}_2, t_2)\rangle = \Gamma_{22}^{(1)}(0)$. The expectation function of the intensity on the observation plane of Young's double-slit experiment illustrated in Fig. 4.1.1 is thus

$$\langle I(\mathbf{r}, t)\rangle = \Gamma_{11}^{(1)}(0) + \Gamma_{22}^{(1)}(0) + 2\,Re\,\Gamma_{12}^{(1)}(\tau)$$

$$= \Gamma_{11}^{(1)}(0) + \Gamma_{22}^{(1)}(0) + 2\sqrt{\Gamma_{11}^{(1)}(0)\,\Gamma_{22}^{(1)}(0)}\;Re\,\gamma_{12}^{(1)}(\tau). \qquad (4.4.2)$$

Substituting the early calculated $\Gamma_{11}^{(1)}(0)$ $[\Gamma_{22}^{(1)}(0)]$ and $\Gamma_{12}^{(1)}(\tau)$ $[\Gamma_{21}^{(1)}(\tau)]$ into Eq. (4.4.2), we find that the maximum interference occurs in the neighborhood of $\tau = 0$. The interference starts to be invisible at $\tau = 2\pi/\Delta\omega$. In Eq. (4.4.2), the normalized complex degree of first-order temporal coherence is

$$\gamma_{12}^{(1)}(\tau) \equiv \frac{\Gamma_{12}^{(1)}(\tau)}{[\,\Gamma_{11}^{(1)}(0)\,]^{\frac{1}{2}}\,[\,\Gamma_{22}^{(1)}(0)\,]^{\frac{1}{2}}} = \frac{\mathcal{F}_\tau\left\{\sum_m a_m^2(\nu)\right\}e^{i\omega_0\tau}}{\sqrt{I_1\,I_2}}, \qquad (4.4.3)$$

where I_j, $j = 1, 2$, is the mean intensity passed through P_j. The time delay

$$\tau = \frac{2\pi}{\Delta\omega} \equiv \tau_c \qquad (4.4.4)$$

is defined as the coherence time of the field and consequently $c\tau_c$ is defined as the coherence length of the field (longitudinal). The fields with separation

τ are named temporally coherent when $|\gamma^{(1)}(\tau)| = 1$, corresponding to $\tau \ll \tau_c$; temporally partially coherent when $0 < |\gamma(\tau)| < 1$, corresponding to $0 < \tau < \tau_c$; and temporally incoherent when $|\gamma^{(1)}(\tau)| = 0$, corresponding to $\tau \geq \tau_c$. Therefore, when we loosely say that "a thermal radiation source has a coherence time τ_c or a coherence length $c\tau_c$" we mean that first-order interference is observable between temporally delayed radiations with a temporal separation of $\tau < \tau_c$.

So far, we have calculated the degree of first-order temporal coherence of the classical field and concluded that its temporal coherence is a function of the temporal delay between the superposed radiations: The larger the delay, the lower the interference visibility. But why? In the model of "incoherent subfields and incoherent Fourier modes", the radiations are just a set of incoherent collection of *continues waves*, or "infinite long" sinusoidal oscillations! What is the physical reason behind this?

The observed interference fringes, i.e., the measured intensity, $\langle I(\mathbf{r}, t) \rangle$, on the observation plane as a function of $(\mathbf{r}, t)$ can be calculated as follows:

$$\langle I(\mathbf{r}, t) \rangle = \left\langle \sum_m \int d\omega\, E_m^*(\omega; \mathbf{r}, t) \sum_n \int d\omega'\, E_n(\omega'; \mathbf{r}, t) \right\rangle$$

$$= \sum_m \int d\omega\, I_m(\omega; \mathbf{r}, t)$$

$$= \sum_m \int d\omega \left| \frac{1}{\sqrt{2}} \left[E_m(\omega; z_1, t_1) + E_m(\omega; z_2, t_2) \right] \right|^2$$

$$= \sum_m \int d\omega\, I_{m0}(\omega) \left[1 + \cos(\omega\,\tau) \right], \tag{4.4.5}$$

where we explicitly write $1/\sqrt{2}$ for later discussions. Equation (4.4.5) indicates that the observed interference pattern is the sum of a large number of individual sub-interference patterns, $I_m(\omega; \mathbf{r}, t)$; each is produced by a Fourier mode of ω associated with a sub-source of mth. If each individual interference fringe is produced by self-interference of a continuous wave sinusoidal oscillation, then why is the temporal coherence a function of the time delay?

In fact, the physics is quite simple: In the neighborhood of $\tau = 0$, these individual sub-patterns coincide so that the intensity modulation can be easily observed ($V(0) \cong 100\%$). When the detector moves away from $\tau = 0$, i.e., the value of $|\tau|$ increases, the relative phase shifts between the sub-patterns of different Fourier modes of ω increase. The spread of these sub-patterns smoothes the light intensity distribution on the observation plane, and the interference visibility is then reduced from 100% to 0, which results in a constant distribution of intensity on the observation plane. The interference visibility, $V(\tau)$, is determined by the maximum relative phase shift $\Delta\omega\tau$, where $\Delta\omega = \omega_{\max} - \omega_{\min}$ is the bandwidth of the field. If $\Delta\omega\tau \ll 2\pi$, i.e., the relative phase shifts are not large enough to produce notable separation between the sub-patterns of different ω for the value of τ, then the interference modulation will be observable. When the relative phase shifts increase, however, the sub-patterns of these Fourier modes are significantly separated at the value of τ. The interference modulation visibility is then reduced to zero. In other words, for a certain spectral bandwidth of $\Delta\omega$, if the time delay $\tau \ll 2\pi/\Delta\omega$, the fields at t and $t + \tau$ are considered temporally coherent by definition. Otherwise, the fields are considered temporally incoherent by definition, which means no interference is observable on the observation plane.

(II) $\Gamma^{(1)}(\mathbf{r}_1, t_1; \mathbf{r}_2, t_2)$: Incoherent subfields and coherent Fourier modes

Consider a point-like light source contains of a large number of independent and randomly radiating point-like sub-sources; each sub-source creates a large set of coherent Fourier modes. The coherent superposition of the Fourier modes results in a wavepacket. The measurement of the point-like detector at space–time $(\mathbf{r}, t)$ may annihilate a wavepacket, two wavepackets, or a large number of wavepackets. The wavepackets may have the same carrier frequency ω_0, or different carrier frequencies ω_{0j}, $j = 1, 2, \ldots \infty$.

(A) A single subfield or wavepacket

Now, we turn our discussion into an interesting and controversial topic of quantum optics: interference of a single-photon or a quantized single subfield in Einstein's picture. We assume the measurement involves only a single subfield or wavepacket at a time period of $\Delta t > \tau_c$. We usually consider this kind of weak light at single-photon level. The question is as follows: Do we observe interference? In classical optics, even if a wavepacket or subfield only carries energy of $\hbar\omega$, this wavepacket can still be divided into two[1] by taking path s_1 (passing through pinhole P_1) and/or path

---

[1] We may name it partial-wavepacket or partial-subfield to distinguish it from Einstein's quantized subfield that carries quantized energy of $\hbar\omega$.

s_2 (passing through pinhole P_2), and then to be superposed at coordinate $(\mathbf{r}, t)$. In the measurement of a single subfield, the statistical concept of classical first-order coherence is not defined. We start our discussion from the calculation of the possible interference fringe produced by the mth subfield.

The instantaneous intensity $I(\mathbf{r}, t)$ is the result of a superposition of two partial-wavepackets: one passing through pinhole P_1 and another passing through pinhole P_2:

$$I(\mathbf{r}, t) = \left| \frac{1}{\sqrt{2}} \left[\mathcal{F}_{\tau_{1m}}\{a(\nu)\} e^{-i\omega_0 \tau_{1m}} + \mathcal{F}_{\tau_{2m}}\{a(\nu)\} e^{-i\omega_0 \tau_{2m}} \right] \right|^2$$

$$= \frac{1}{2} \left\{ \left| \mathcal{F}_{\tau_{1m}}\{a(\nu)\} \right|^2 + \left| \mathcal{F}_{\tau_{2m}}\{a(\nu)\} \right|^2 \right.$$

$$\left. + 2 \operatorname{Re} \left[\mathcal{F}^*_{\tau_{1m}}\{a(\nu)\} \mathcal{F}_{\tau_{2m}}\{a(\nu)\} e^{i\omega_0 \tau} \right] \right\}. \tag{4.4.6}$$

It is easy to see from Eq. (4.4.6) that an interference pattern is potentially observable when the two partial wavepackets overlap at the observation points $(\mathbf{r}, t)$. The maximum interference occurs in the neighborhood of $\tau = \tau_1 - \tau_2 \sim 0$ when the two partial-wavepackets completely overlap. For $\tau > 2\pi/\Delta\omega$, the two Fourier transforms cannot take nonzero values simultaneously, the interference cross term of Eq. (4.4.6) vanishes, and consequently no interference is observable when the temporal delay is greater than the temporal width of the wavepacket. Obviously, the coherence time of a single-photon wavepacket is $\tau_c = 2\pi/\Delta\omega$, or concisely, the coherence time of a single-photon wavepacket equals the temporal width of the wavepacket.

On one hand, it is interesting to find that *one subfield or photon produces an interference pattern* of $I(\mathbf{r}, t)$ based on classical superposition. On the other hand, we are facing the following serious question: How to measure the interference pattern? Suppose we place a 2-D photodetector, CCD or CMOS, in the observation plane; a photon can only produce one photoelectron from a pixel of the CCD or CMOS. The registration of one pixel cannot provide any meaningful information about the interference. A timely accumulative measurement of photon-counting is always necessary in this case. After a timely accumulative counting on a large number of subfields or wavepackets, one by one, we find the accumulated charges in each pixel of the CCD or CMOS, which is proportional to the number of photons registered at coordinate $(\mathbf{r})$, within the accumulative time period of ΔT:

$$Q(\mathbf{r}) = \int_{\Delta T} dt\, I(\mathbf{r}, t)$$

$$\propto \sum_m \left| \mathcal{F}_{\tau_{1m}}\{a(\nu)\} \right|^2 + \sum_m \left| \mathcal{F}_{\tau_{2m}}\{a(\nu)\} \right|^2$$

$$+ 2\,\mathrm{Re} \sum_m \left[\mathcal{F}^*_{\tau_{1m}}\{a(\nu)\} \mathcal{F}_{\tau_{2m}}\{a(\nu)\}\, e^{i\omega_0\tau} \right]. \tag{4.4.7}$$

The interference pattern is observable from $Q(\mathbf{r})$ after a timely accumulative measurement of photon counting. When the accumulative time is long enough, we may define the first-order temporal coherence $\Gamma^{(1)}(\tau)$ of a single wavepacket as follows:

$$\Gamma^{(1)}(\tau) = \left\langle \sum_m E^*_m(\mathbf{r}_1, t_1)\, E_m(\mathbf{r}_2, t_2) \right\rangle$$

$$= \sum_m \left[\mathcal{F}^*_{\tau_{1m}}\{a(\nu)\} \mathcal{F}_{\tau_{2m}}\{a(\nu)\}\, e^{i\omega_0\tau} \right]$$

$$\simeq \left[\int_\infty dt_m\, \mathcal{F}^*_{\tau_{1m}}\{a(\nu)\} \mathcal{F}_{\tau_{2m}}\{a(\nu)\} \right] e^{i\omega_0\tau}$$

$$= \mathcal{F}_\tau\{a^2(\nu)\} e^{i\omega_0\tau}, \tag{4.4.8}$$

where all constants associated with the integral have been absorbed into the Fourier transform, as usual, and $\tau = \tau_1 - \tau_2 = (s_2 - s_1)/c + (z_2 - z_1)/c$. In Eq. (4.5.9), the ensemble average has taken into account all possible realizations of the fields. The first-order temporal coherence $\Gamma^{(1)}(\tau)$ takes a nonzero value when $\tau < 2\pi/\Delta\omega$. The coherence time of a single subfield is therefore defined as $\tau_c = 2\pi/\Delta\omega$.

Another serious question then naturally arises: What do we mean *each wavepacket produces an interference-pattern*? In classical optics, this means the energy of the subfield $\hbar\omega$ is distributed on the entire interference pattern and the expected amount of energy crossing a unit area per unit time, precisely the Poynting vector or the intensity, of the electromagnetic field, is modulated sinusoidally as a function of the space–time coordinate $(\mathbf{r}, t)$. There is no problem in classical theory to have an interference pattern produced by the mth subfield in terms of an intensity distribution $I_m(\mathbf{r}, t)$ in space–time, even if the mth wavepacket only carries the energy $\hbar\omega$ of a single photon. However, definitely, this is not the picture corresponding to the above photon-counting measurement. The energy $\hbar\omega$ of a subfield must be localized within one pixel of the photodetector array. If the energy $\hbar\omega$ of a subfield is distributed on the entire interference pattern, when a photoelectron is excited from a pixel of the photodetector array at $(\mathbf{r}, t)$ by taking an amount of energy $\hbar\omega$ from the electromagnetic field, how much time does it take for the energy from other coordinates of the interference pattern to arrive at that pixel? Remember, Einstein had asked a similar question to his students based on his two light-year big wavepacket.[2] Although the distance between two points of an interference pattern may not be as big as a light-year, we must answer the same question.

From the measurement point of view, the interference pattern produced by a subfield means nothing but a probability distribution function $P(\mathbf{r}, t)$: the probability for a subfield, or a photon, to produce a photodetection event at space–time coordinate $(\mathbf{r}, t)$. In an accumulative measurement involving a large number of N wavepackets, there will be $NP(\mathbf{r})$ photodetection events occurring at coordinate $\mathbf{r}$, corresponding to the time averaged intensity $\langle I(\mathbf{r}, t)\rangle_{\Delta T}$.

(B) Two independent subfields or wavepackets with the same carrier frequency ω_0

Assuming two subfields, mth and nth, created from two independent and randomly radiated atomic transitions with the same carrier frequency ω_0, are superposed through an Young's double-pinhole interferometer. The mth wavepacket passes the upper pinhole and propagates along path s_1, and the nth wavepacket passes the lower pinhole and propagates along path s_2. Similar to the case of a single subfield, the statistical concept of classical first-order coherence of two independent subfields or wave packets is not defined. We start our discussion from the calculation of the possible interference produced by two independent subfields or wave packets. The instantaneous intensity $I(\mathbf{r}, t)$ in the observation plane is thus

$$
\begin{aligned}
I(\mathbf{r}, t) &= \left| \mathcal{F}_{\tau_{1m}}\{a_m(\nu)\}\, e^{-i\omega_0\tau_{1m}} + \mathcal{F}_{\tau_{2n}}\{a_n(\nu)\}\, e^{-i\omega_0\tau_{2n}} \right|^2 \\
&= \left| \mathcal{F}_{\tau_{1m}}\{a_m(\nu)\} \right|^2 + \left| \mathcal{F}_{\tau_{2n}}\{a_n(\nu)\} \right|^2 \\
&\quad + 2\,\mathrm{Re}\left[\mathcal{F}^*_{\tau_{1m}}\{a_m(\nu)\} \mathcal{F}_{\tau_{2n}}\{a_2(\nu)\}\, e^{i\omega_0[(t_n - t_m) + \tau]} \right],
\end{aligned} \qquad (4.4.9)
$$

where $\tau_{1m} = \tau_1 - t_m$ and $\tau_{2n} = \tau_2 - t_n$; t_m and t_n are the initial creation times of the two wavepackets at the mth and the nth independent

---

[2]Einstein's story is given at the end of Section 2.3.

sub-sources. It is easy to see from Eq. (4.4.9) that an instantaneous interference pattern is potentially observable when the two wavepackets are overlapped at $(\mathbf{r}, t)$. Again, we are facing a serous problem: How to measure the two wavepacket produced interference pattern? Suppose we place a 2-D photodetector, CCD or CMOS, in the observation plane; two photons can only produce two photoelectrons from two pixels of the CCD or CMOS. The registration of two pixels cannot provide any meaningful information about the interference. Similar to the single-photon case, a timely accumulative measurement of photon-counting is always necessary. In a timely accumulative measurement, each pair of wavepacket may produce two photoelectrons at a time t to charge two electronic integrators. From time to time, or from wavepacket-pair to wavepacket-pair, the photocurrents keep charging the integrators until achieving an observable level:

$$Q(\mathbf{r}) = \int_{\Delta T} dt\, I(\mathbf{r}, t)$$

$$\propto \sum_m \left| \mathcal{F}_{\tau_{1m}} \{a_m(\nu)\} \right|^2 + \sum_n \left| \mathcal{F}_{\tau_{2n}} \{a_n(\nu)\} \right|^2$$

$$+ 2\mathrm{Re} \sum_{m \neq n} e^{i\omega_0(t_n - t_m)} \left[\mathcal{F}^*_{\tau_{1m}} \{a_m(\nu)\} \mathcal{F}_{\tau_{2n}} \{a_2(\nu)\} e^{i\omega_0 \tau} \right]. \quad (4.4.10)$$

The interference fringes gradually disappear while the number of photo-electron pairs accumulates due to the random radiation times of t_m and t_n. In this case, the first-order temporal mutual-coherence function of two independent single-photon wavepackets is expected to be

$$\Gamma^{(1)}(\tau) = \sum_{m,n} e^{i\omega_0(t_n - t_m)} \left[\mathcal{F}^*_{\tau_{1m}} \{a(\nu)\} \mathcal{F}_{\tau_{2n}} \{a(\nu)\} e^{i\omega_0 \tau} \right] \simeq 0. \quad (4.4.11)$$

$\Gamma^{(1)}(\tau) \simeq 0$ is the result of the random relative phases $\omega_0(t_n - t_m)$: Each m–nth pair of wavepackets produces an "instantaneous" interference pattern with a particular phase of $\omega_0[(t_n - t_m) + \tau]$; due to the randomness of t_m and t_n from wavepacket-pair to wavepacket-pair, the averaged sinusoidal function of $\cos\{\omega_0[(t_n - t_m) + \tau]\}$ results in a value of zero. Based on Eq. (4.4.11), we may conclude that the interference between two randomly created wavepackets with the same carrier frequency is practically nonobservable; at least it has no contribution to the expectation value of $\Gamma^{(1)}(\mathbf{r}_1, \mathbf{r}_2)$ which takes account all possible values of $(t_n - t_m)$.

However, (1) if we can force the two "independent" sub-sources radiating in phase, i.e., achieving $(t_n - t_m) = $ constant, these identical "instantaneous" interferences would add together accumulatively produce an observable interference pattern from the CCD or CMOS. (2) If the measurement cannot take all possible values of $(t_n - t_m)$, the interference between two randomly created wavepackets with the same carrier frequency may contribute to the measurement of intensity fluctuations and is practically observable in the measurement of intensity fluctuations.

(C) Two independent subfields or wavepackets with different carrier frequencies $\omega_{0j} \neq \omega_{0k}$

Assuming two subfields, mth and nth, created from two independent and randomly radiated atomic transitions with different carrier frequencies $\omega_{0j} \neq \omega_{0k}$, are superposed through an Young's double-pinhole interferometer for interference observation. The mth wavepacket passes the upper pinhole and propagates along path s_1, and the nth wavepacket passes the lower pinhole and propagates along path s_2. Similar to the above two cases we have discussed, we start our discussion from the calculation of the possible interference produced by two independent subfields or wave packets with different carrier frequencies. The instantaneous intensity $I(\mathbf{r}, t)$ in the observation plane is

$$
\begin{aligned}
I(\mathbf{r}, t) &= \left| \mathcal{F}_{\tau_{1m}}\{a_m(\nu)\} e^{-i\omega_{0j}\tau_{1m}} + \mathcal{F}_{\tau_{2n}}\{a_n(\nu)\} e^{-i\omega_{0k}\tau_{2n}} \right|^2 \\
&= \left| \mathcal{F}_{\tau_{1m}}\{a_m(\nu)\} \right|^2 + \left| \mathcal{F}_{\tau_{2n}}\{a_n(\nu)\} \right|^2 \qquad (4.4.12) \\
&\quad + 2\mathrm{Re}\left[\mathcal{F}^*_{\tau_{1m}}\{a_m(\nu)\} \mathcal{F}_{\tau_{2n}}\{a_n(\nu)\} \, e^{i[(\omega_{0k}t_{0n} - \omega_{0j}t_{0m}) + (\omega_{0j}\tau_1 - \omega_{0k}\tau_2)]} \right].
\end{aligned}
$$

It is not difficult to see from Eq. (4.4.12) that an instantaneous interference beat of $\cos[(\omega_{0j} - \omega_{0k})t - (\omega_{0j}s_1/c - \omega_{0k}s_2/c)]$ is potentially observable when the two wavepackets are overlapped at $(\mathbf{r}, t)$.

Similar to the two cases we have discussed above, after a large number of photoelectron accumulation, the first-order temporal mutual-coherence function of two independent single-photon wavepackets with different carrier frequencies is expected to be

$$
\Gamma^{(1)}(\tau) = \left\langle \sum_{m,n} \mathcal{F}^*_{\tau_{1m}}\{a_m(\nu)\} \mathcal{F}_{\tau_{2n}}\{a_2(\nu)\} \, e^{i[(\omega_{0k}t_{0n} - \omega_{0j}t_{0m}) + (\omega_{0j}\tau_1 - \omega_{0k}\tau_2)]} \right\rangle
$$

$$
\simeq 0, \qquad\qquad\qquad\qquad\qquad\qquad\qquad\qquad\qquad\qquad\qquad (4.4.13)
$$

where we have taken into account all possible values of $(t_n - t_m)$. Repeating the same argument, we conclude the following: (1) $\Gamma^{(1)}(\tau) \simeq 0$ is the result of random values of $(t_n - t_m)$; (2) if we force the two "independent" sub-sources radiating in phase, i.e., achieving $(t_n - t_m) = \text{constant}$, these identical "instantaneous" beats would add together accumulatively produce an observable beat; (3) if the measurement cannot take all possible values of $(t_n - t_m)$, the interference beat between two randomly created wavepackets with different carrier frequencies may contribute to the measurement of intensity fluctuations and is practically observable in the measurement of intensity fluctuations.

(D) Thermal light, consisting of a large number of overlapped and partially overlapped wavepackets with the same carrier frequency

In Einstein's picture, the radiation field $E(\mathbf{r}, t)$ at space–time $(\mathbf{r}, t)$ is the result of a superposition among a large number subfields, or wavepackets, randomly created from a large number of independent point-like sub-sources, such as billions of independent atomic transitions. We label the subfield or wavepacket associated with the mth sub-source with parameter t_m, indicating the initial creation time of the wavepacket at the mth sub-source. Each photodetection event at space–time point $(\mathbf{r}, t)$ involves a large number of randomly created and randomly distributed overlapped–partially overlapped wavepackets. The referred earlier fields $E(\mathbf{r}_1, t_1)$ and $E(\mathbf{r}_2, t_2)$ have the same nature in terms of the superposition of the subfields. Substituting the formally integrated wavepackets of $E^*(\mathbf{r}_1, t_1)$ and $E(\mathbf{r}_2, t_2)$, into $\langle E^*(\mathbf{r}_1, t_1) E(\mathbf{r}_2, t_2) \rangle$ the mutual-coherence function $\Gamma^{(1)}(\mathbf{r}_1, t_1; \mathbf{r}_2, t_2)$ is calculated as

$$\Gamma^{(1)}(\mathbf{r}_1, t_1; \mathbf{r}_2, t_2)$$

$$= \left\langle \sum_{m=1}^{M} e^{i\omega_0 \tau_{m1}} \mathcal{F}^*_{\tau_{m1}}\{a(\nu)\} \sum_{n=1}^{N} e^{-i\omega_0 \tau_{n2}} \mathcal{F}_{\tau_{n2}}\{a(\nu)\} \right\rangle$$

$$= \left\langle \sum_{m=n} e^{i\omega_0 \tau} \mathcal{F}^*_{\tau_{m1}}\{a(\nu)\} \mathcal{F}_{\tau_{m2}}\{a(\nu)\} \right\rangle$$

$$+ \left\langle \sum_{m \neq n} e^{i\omega_0 [(t_n - t_m) + \tau]} \mathcal{F}^*_{\tau_{m1}}\{a(\nu)\} \mathcal{F}_{\tau_{n2}}\{a(\nu)\} \right\rangle, \qquad (4.4.14)$$

where we have assumed all wavepackets have the same carrier frequency, and m and n label the mth and nth wavepacket. We have also defined

$\tau_{m1} = (t-t_m)-s_1/c-z_1/c = \tau_1-t_m$ and $\tau_{n2} = (t-t_n)-s_2/c-z_2/c = \tau_2-t_n$, $\tau = \tau_1 - \tau_2 = (s_2-s_1)/c + (z_2-z_1)/c$. It is convenient to break up the sum into two groups. If we consider a random distribution of the subfield or wavepackets, i.e., stochastically radiated atomic transitions with random relative phases, or at arbitrary initial creation times of t_m and t_n, the second group ($m \neq n$) of the sum vanishes when taking into account all possible relative phases or all possible values of $t_m - t_n$ in the superposition. This result indicates that *the only observable interference is the interference in which the mth wavepacket interfering with the mth wavepacket itself.* Interference between two different wavepackets becomes unobservable after the ensemble average.

Accordingly, $\langle I(\mathbf{r},t) \rangle$, the expectation function of the intensity in the observation plane, is the sum of a large number of individual sub-interferences; each is the result of an individual wavepacket interfering with the wavepacket itself:

$$\langle I(\mathbf{r},t) \rangle \simeq \sum_m \left| \frac{1}{\sqrt{2}} \left[\mathcal{F}_{\tau_{m1}}\{a(\nu)\} e^{-i\omega_0 \tau_{m1}} + \mathcal{F}_{\tau_{m2}}\{a(\nu)\} e^{-i\omega_0 \tau_{m2}} \right] \right|^2.$$

$$(4.4.15)$$

Due to the common relative delay $\tau = \tau_1 - \tau_2$, all the sub-interference patterns comprise the same sinusoidal modulation. The identical sub-interference patterns add at the observation plane.

Now, we approximate the sum into an integral of t_m, similar to what we have done earlier. The temporal mutual-coherence function $\Gamma^{(1)}(\tau)$ is calculated as

$$\Gamma^{(1)}(\tau) = \left\langle \sum_m \mathcal{F}^*_{\tau_{m1}}\{a(\nu)\}\, \mathcal{F}_{\tau_{m2}}\{a(\nu)\} \right\rangle e^{i\omega_0 \tau}$$

$$\simeq \left[\int_\infty dt_m\, \mathcal{F}^*_{(\tau_1-t_m)}\{a(\nu)\}\, \mathcal{F}_{(\tau_2-t_m)}\{a(\nu)\} \right] e^{i\omega_0 \tau}$$

$$\simeq \mathcal{F}_\tau\{a^2(\nu)\}\, e^{i\omega_0 \tau}, \qquad\qquad (4.4.16)$$

where all constants associated with the integral have been absorbed into the Fourier transform function, as usual, and $\tau = \tau_1 - \tau_2 = (s_2-s_1)/c + (z_2-z_1)/c$.

The mutual-coherence function, and consequently the interference visibility, is quantitatively determined by how much the two wavepackets, $\mathcal{F}_{\tau_{1m}}\{a(\nu)\}$ and its delayed conjugate $\mathcal{F}^*_{\tau_{2m}}\{a(\nu)\}$, are overlapped in space–time. The integral, or the convolution, has a maximum value when $\tau = \tau_1 - \tau_2 = 0$. The radiation fields become first-order incoherent when $\tau \geq 2\pi/\Delta\omega$. Again,

$$\tau_c = 2\pi/\Delta\omega$$

is called the coherence time of the field, which is nothing but the temporal width of the wavepackets.

The integral over t_m in Eq. (4.4.16) is mathematically equivalent to a time integral and, consequently, an autocorrelation or a self-convolution of the wavepacket:

$$\Gamma^{(1)}(\tau) = e^{-i\omega_0\tau} \int_\infty dt\, \mathcal{F}^*_{\tau_1}\{E(\nu)\}\, \mathcal{F}_{\tau_2}\{E(\nu)\} = \int_\infty dt'\, E^*(t')\, E(t' - \tau).$$

$$(4.4.17)$$

Applying the Wiener–Khinchin theorem, we have

$$\Gamma^{(1)}(\tau) = \int_\infty dt\, E^*(t)\, E(t - \tau) = \int_\infty d\nu\, |E(\nu)|^2\, e^{-i\nu\tau}.$$

(E) Natural light, consisting of a large number of overlapped and partially overlapped wavepackets with the same carrier frequency and different carrier frequencies

We model a natural radiation, such as sunlight, a superposition among a large number subfields, or wavepackets, randomly created from a large number of independent point-like sub-sources, such as billions of independent atomic transitions. The wavepackets may come from identical atomic transitions with the same carrier frequency, or come from distinctive atomic transitions with different carrier frequencies. We label the subfield or wavepacket associated with the mth sub-source with parameter t_{0m}, indicating the initial creation time of the wavepacket at the mth sub-source, and carrier frequency ω_{0j}. Each photodetection event at space–time point $(\mathbf{r}, t)$ is the result of a superposition of a large number of randomly created and randomly distributed overlapped–partially overlapped wavepackets. Based on earlier discussions, the only observable interferences is the self-interference of the wavepacket. Interference between two different wavepackets becomes unobservable after the ensemble average.

The temporal mutual-coherence function $\Gamma^{(1)}(\tau)$ is calculated as

$$\Gamma^{(1)}(\tau) \simeq \left\langle \left[\sum_j \sum_m A^*(\omega_{0j}) \, \mathcal{F}^*_{\tau_{1m}}\{a(\nu)\} e^{i\omega_{0j}\tau_{1m}} \right] \right.$$

$$\left. \times \left[\sum_k \sum_n A(\omega_{0k}) \, \mathcal{F}_{\tau_{2n}}\{a(\nu)\} e^{-i\omega_{0k}\tau_{2n}} \right] \right\rangle$$

$$\simeq \sum_{j=k} \sum_{m=n} |A(\omega_{0j})|^2 \, e^{i\omega_{0j}\tau} \, \mathcal{F}^*_{\tau_{1m}}\{a(\nu)\} \, \mathcal{F}_{\tau_{2m}}\{a(\nu)\}$$

$$\simeq \mathcal{F}_\tau\{|A(\nu_0)|^2\} \, \mathcal{F}_\tau\{a^2(\nu)\} \, e^{i\bar{\omega}_0\tau}, \tag{4.4.18}$$

where $A_j(\omega)$ is the complex amplitude of the jth carrier wave, $\bar{\omega}_0$ is the average carrier frequency, and $\nu_0 = \omega_0 - \bar{\omega}_0$. In Eq. (4.4.18), we approximate a continuous spectrum $|A(\omega_0)|^2$ of the carrier wave. Since the spectrum of the carrier waves is usually much broader than that of each atomic transition, the above calculated first-order temporal coherence of natural light is mainly determined by the spectral width of the carrier waves. The result of Eq. (4.4.18) is consistent with that of Eq. (4.4.1), where the radiation is modeled as "incoherent subfields and incoherent Fourier modes" in Einstein's granularity picture of light.

4.5 First-Order Temporal Coherence: Quantum Theory

This section introduces the quantum theory of first-order temporal coherence. We calculate the first-order temporal coherence functions of different radiations in the single-photon state representation and the coherent state representation.

(I) The first-order temporal coherence of a single photon

Considering the same Young's double-pinhole interferometer in Fig. 4.4.1, we assume a weak far-field point-like light source. The radiation is so weak that only one photon is observable at the interferometer within the detection time of the photon counting detector. We usually call this weak light at the single-photon level. The statistical concept of first-order coherence of a photon is not defined in classical theory of statistics.

Do we have a defined mutual-coherence function of a single-photon? Does a photon have defined coherence time or coherence length?

In quantum theory, the first-order coherence is defined if a photon is in pure state. For example, quantum theory defines a first-order coherence function $G^{(1)}(z_1, t_1; z_2, t_2) = \langle\Psi|\hat{E}^{(-)}(z_1, t_1)\,\hat{E}^{(+)}(z_2, t_2)|\Psi\rangle$ if the state of the measured field can be approximated as

$$|\Psi\rangle = \int d\omega\, f(\omega)\,\hat{a}^\dagger(\omega)|0\rangle. \tag{4.5.1}$$

Equation (4.5.1) represents a single-photon state in the form of coherent superposition of Fock states $\hat{a}^\dagger(\omega)|0\rangle = |\ldots 0, 1_\omega, \ldots, 0, \ldots\rangle$ with $f(\omega)$ the normalized complex probability amplitude of the Fock state $|1(\omega)\rangle$. The first-order temporal coherence function of a single-photon is easily calculated by submitting the field operators

$$\hat{E}^{(-)}(z_j, t_j) = \int d\omega\, \hat{a}^\dagger(\omega)\, g_m^*(\omega; z_j, t_j),$$

$$\hat{E}^{(+)}(z_k, t_k) = \int d\omega'\, \hat{a}(\omega')\, g_m(\omega'; z_k, t_k) \tag{4.5.2}$$

and the pure state of Eq. (4.5.1) into $G^{(1)}(z_1, t_1; z_2, t_2)$. In the above field operators, $g_m(\omega; z_j, t_j)$ is Green's function that propagates the field operator from (z_j, t_j) to (z_m, t_m). Here, we assume the measured single-photon is created from the mth sub-source at $(z_m = 0, t_m)$. The quantum mutual-coherence function $G^{(1)}(z_1, t_1; z_2, t_2)$ is defined and calculated as follows:

$$G^{(1)}(z_1, t_1; z_2, t_2)$$

$$= \langle\Psi|\,\hat{E}^{(-)}(z_1, t_1)\,\hat{E}^{(+)}(z_2, t_2)\,|\Psi\rangle$$

$$= \sum_n \langle\Psi|\hat{E}^{(-)}(z_1, t_1)|n\rangle\langle n|\hat{E}^{(+)}(z_2, t_2)|\Psi\rangle$$

$$= \langle\Psi|\hat{E}^{(-)}(z_1, t_1)|0\rangle\langle 0|\hat{E}^{(+)}(z_2, t_2)|\Psi\rangle$$

$$= \left[\int d\omega\, f^*(\omega)\,\langle 0\,|\hat{a}(\omega)\right]\left[\int d\omega''\,\hat{a}^\dagger(\omega'')\,g_m^*(\omega; z_j, t_j)\right]|0\rangle$$

$$\times \langle 0|\left[\int d\omega'''\,\hat{a}(\omega''')\,g_m(\omega; z_j, t_j)\right]\left[\int d\omega'\,f(\omega')\,\hat{a}^\dagger(\omega')|0\rangle\right]$$

$$\simeq \left[\mathcal{F}_{\tau_{1m}}^*\{f(\nu)\}e^{i\omega_0\tau_{1m}}\right]\left[\mathcal{F}_{\tau_{2m}}\{f(\nu)\}e^{-i\omega_0\tau_{2m}}\right]$$

$$= \Psi^*(z_1, t_1)\,\Psi(z_2, t_2). \tag{4.5.3}$$

For a pure state of single-photon wavepacket, $G^{(1)}(z_1, t_1; z_2, t_2)$ cannot be written as a stationary function of the optical delay in general; however, it can be easily written as a function of two time-dependent wavepackets, or effective wavefunctions, $\Psi^*(z_1, t_1)$ and $\Psi(z_2, t_2)$ with

$$\Psi(z_j, t_j) = \langle 0|\hat{E}^{(+)}(z_j, t_j)|\Psi\rangle = \mathcal{F}_{\tau_{jm}}\{f(\nu)\}e^{-i\omega_0 \tau_{jm}}. \tag{4.5.4}$$

It is clear that the first-order temporal coherence of a single-photon is determined by the overlapping or nonoverlapping of the effective wavefunctions. The first-order quantum coherence time is obviously the temporal width of the single-photon wavepacket $\tau_c = 2\pi/\Delta\omega$.

Similar to the mutual-coherence function $G^{(1)}(z_1, t_1; z_2, t_2)$, the self-correlation functions $G^{(1)}(z_1, t_1; z_1, t_1)$ and $G^{(1)}(z_2, t_2; z_2, t_2)$ can be expressed as the function of the time dependent wavepackets $\Psi(z_1, t_1)$ and $\Psi(z_2, t_2)$ with their conjugates:

$$G^{(1)}(z_1, t_1; z_1, t_1) = \left| \mathcal{F}_{\tau_{1m}}\{f(\nu)\} \right|^2 = \left| \Psi(z_1, t_1) \right|^2,$$

$$G^{(1)}(z_2, t_2; z_2, t_2) = \left| \mathcal{F}_{\tau_{2m}}\{f(\nu)\} \right|^2 = \left| \Psi(z_2, t_2) \right|^2. \tag{4.5.5}$$

The probability of having a photodetection event (a "click") at space–time point $(\mathbf{r}, t)$ is thus written in terms of the superposition of the effective wavefunctions $\Psi(z_1, t_1)$ and $\Psi(z_2, t_2)$:

$$P(\mathbf{r}, t) \propto \left| \Psi(\mathbf{r}, t) \right|^2 = \left| \frac{1}{\sqrt{2}}[\Psi(z_1, t_1) + \Psi(z_2, t_2)] \right|^2$$

$$= \frac{1}{2}\left[|\Psi(z_1, t_1)|^2 + |\Psi(z_2, t_2)|^2 + \Psi^*(z_1, t_1)\Psi(z_2, t_2) \right.$$

$$\left. + \Psi(z_1, t_1)\Psi^*(z_2, t_2) \right]. \tag{4.5.6}$$

Although no wavefunction is defined for a photon in quantum mechanics, it is interesting to see that the effective wavefunction $\Psi(\mathbf{r}, t)$ plays the role of wavefunction in Eq. (4.5.6). The effective wavefunction represents the probability amplitude of observing a photon at space–time point $(\mathbf{r}, t)$, or the probability amplitude of having a photodetection event at space–time point $(\mathbf{r}, t)$. In the above experiment, the effective wavefunction $\Psi(\mathbf{r}, t)$ has two different yet indistinguishable[3] amplitudes $\Psi(z_1, t_1)$ and

---

[3]Note, here "indistinguishable" refers to a photodetection event at space–time point $(\mathbf{r}, t)$.

$\Psi(z_2, t_2)$, namely the probability amplitude of observing a photon at (z_1, t_1) and the probability amplitude of observing a photon at (z_2, t_2), respectively. Each amplitude is in the form of a wavepacket in space–time. The observed effective wavefunction $\Psi(\mathbf{r}, t)$ is the result of the superposition of two effective wavefunctions or two dynamical wavepackets in space–time. The effective wavefunction reflects the wave–particle nature of a photon. The normalized degree of first-order temporal coherence function is thus defined in terms of the effective wavefunction of a photon in pure state:

$$g^{(1)}(z_1, t_1; z_2, t_2) = \frac{\Psi^*(z_1, t_1)\, \Psi(z_2, t_2)}{\left[\,|\, \Psi(z_1, t_1)\,|^2\,\right]^{\frac{1}{2}} \left[\,|\, \Psi(z_2, t_2)\,|^2\,\right]^{\frac{1}{2}}}. \tag{4.5.7}$$

It is interesting to find that one photon produces an interference pattern. Can we observe a single-photon produced interference fringe? Obviously impossible, a single-photon can only create one photoelectron. Similar to Einstein's picture, a timely accumulative measurement of photon-counting is always necessary in this case. After a timely accumulative counting on a large number of photons, one by one, the counting rate of a photon counting detector, which is proportional to the number of photons registered at coordinate $(\mathbf{r})$, within the accumulative time period of ΔT is

$$R(\mathbf{r}) = \int_{\Delta T} dt\, N\, P(\mathbf{r}, t)$$

$$\propto \sum_m \left|\mathcal{F}_{\tau_{1m}}\{f(\nu)\}\right|^2 + \sum_m \left|\mathcal{F}_{\tau_{2m}}\{f(\nu)\}\right|^2$$

$$+ 2\,\mathrm{Re} \sum_m \left[\mathcal{F}^*_{\tau_{1m}}\{f(\nu)\}\mathcal{F}_{\tau_{2m}}\{f(\nu)\}\, e^{i\omega_0\tau}\right]. \tag{4.5.8}$$

The interference pattern is observable from the counting rate of the photon counting detector, $R(\mathbf{r})$, after a timely accumulative measurement. When the accumulative time is long enough, we may define the first-order temporal coherence $\Gamma^{(1)}(\tau)$ of a single-photon wavepacket

$$\Gamma^{(1)}(\tau) = \sum_m \left[\mathcal{F}^*_{\tau_{1m}}\{f(\nu)\}\mathcal{F}_{\tau_{2m}}\{f(\nu)\}\, e^{i\omega_0\tau}\right]$$

$$\simeq \left[\int_\infty dt_{0m}\, \mathcal{F}^*_{\tau_{1m}}\{f(\nu)\}\mathcal{F}_{\tau_{2m}}\{f(\nu)\}\right] e^{i\omega_0\tau}$$

$$= \mathcal{F}_\tau\{f^2(\nu)\} e^{i\omega_0\tau}, \tag{4.5.9}$$

where $\tau = \tau_1 - \tau_2 = (s_2 - s_1)/c + (z_2 - z_1)/c$ as usual. The first-order temporal coherence $\Gamma^{(1)}(\tau)$ takes nonzero value when $\tau < 2\pi/\Delta\omega$. The coherence time of a single photon is therefore defined as $\tau_c = 2\pi/\Delta\omega$.

(II) $G^{(1)}(\mathbf{r}_1, t_1; \mathbf{r}_2, t_2)$: Thermal state in single-photon state representation

In this section, we study the first-order coherence of thermal field. Considering the same Young's double-pinhole interferometer of Fig. 4.4.1, we assume a far-field point-like light source containing a large number of independent and randomly radiated atoms that are ready for two-level atomic transitions. Taking the lowest-order contribution to a photodetection event, the first-order temporal mutual-coherence function and self-coherence function can be calculated from

$$G^{(1)}(z_j, t_j; z_k, t_k) = \left\langle \langle \tilde{\Psi} | \hat{E}^{(-)}(z_j, t_j) \hat{E}^{(+)}(z_k, t_k) | \tilde{\Psi} \rangle \right\rangle_{\mathrm{En}}, \qquad (4.5.10)$$

where $j = 1, 2$, $k = 1, 2$, and

$$|\tilde{\Psi}\rangle = \sum_m c_m \int d\omega \, f_m(\omega) \, \hat{a}_m^\dagger(\omega) \, |0\rangle = \sum_m c_m |\Psi_m\rangle,$$

representing an incoherently superposition of single-photon states. Note, c_m is the complex probability amplitude of the mth state that contains a random phase. To simplify the calculation, we assume a constant $|c_m|^2$ that is physically reasonable for a randomly radiated thermal light source.

$G^{(1)}(z_j, t_j; z_k, t_k)$ can be also calculated from the density operator:

$$G^{(1)}(z_j, t_j; z_k, t_k) = \mathrm{tr}\, \hat{\rho}^{(1)} \hat{E}^{(-)}(z_j, t_j) \hat{E}^{(+)}(z_k, t_k) \qquad (4.5.11)$$

with

$$\hat{\rho}^{(1)} = \sum_m P_m \, |\Psi_m\rangle\langle\Psi_m|.$$

The only difference between Eqs. (4.5.10) and (4.5.11) is that Eq. (4.5.11) calculated ensemble average before quantum expectation. In Eq. (4.5.11), ensemble average has been completed in the process of calculating the density operator. To have a better view on the quantum nature of light, the following excises is based on Eq. (4.5.10), which completes the ensemble average after the quantum expectation.

Similar to our early discussion, the field operators $\hat{E}^{(-)}(z_j, t_j)$ and $\hat{E}^{(+)}(z_k, t_k)$ can be approximated as

$$\hat{E}^{(-)}(z_j, t_j) = \sum_m \int d\omega\, \hat{a}_m^\dagger(\omega)\, g_m^*(\omega; z_j, t_j) = \sum_m \int d\omega\, \hat{a}_m^\dagger(\omega)\, e^{i\omega \tau_{jm}},$$

$$\hat{E}^{(+)}(z_k, t_k) = \sum_n \int d\omega'\, \hat{a}_n(\omega')\, g_n(\omega'; z_k, t_k) = \sum_n \int d\omega'\, \hat{a}_n(\omega')\, e^{-i\omega' \tau_{kn}},$$

$$(4.5.12)$$

where $g_m(\omega; z_j, t_j)$ is Green's function that propagates the ω mode of the mth subfield from (z_j, t_j) to (z_m, t_m). In Eq. (4.5.12), we consider the mth subfield which is created from the mth distinguishable subfield.

Following Eq. (4.5.10), the mutual-coherence function $G^{(1)}(z_1, t_1; z_2, t_2)$ is calculated as

$$G^{(1)}(z_1, t_1; z_2, t_2)$$

$$= \left\langle \sum_m c_m^* \langle \Psi_m | \hat{E}^{(-)}(z_1, t_1)\, \hat{E}^{(+)}(z_2, t_2) \sum_n c_n | \Psi_n \rangle \right\rangle_{\text{En}}$$

$$= \left\langle \sum_l \left\langle \sum_m c_m^* \langle \Psi_m | \hat{E}^{(-)}(z_1, t_1) | l \rangle \langle l | \hat{E}^{(+)}(z_2, t_2) \sum_n c_n | \Psi_n \rangle \right\rangle \right\rangle_{\text{En}}$$

$$= \left\langle \sum_m c_m^* \langle \Psi_m | \hat{E}^{(-)}(z_1, t_1) | 0 \rangle \langle 0 | \hat{E}^{(+)}(z_2, t_2) \sum_n c_n | \Psi_n \rangle \right\rangle_{\text{En}}$$

$$= \left\langle \sum_m c_m^* \psi_m^*(z_1, t_1) \sum_n c_n \psi_n(z_2, t_2) \right\rangle_{\text{En}}$$

$$= \left\langle \Psi^*(z_1, t_1) \Psi(z_2, t_2) \right\rangle_{\text{En}}, \qquad (4.5.13)$$

where we have defined an effective wavefunction for the nth photon:

$$\psi_n(z_j, t_j) \equiv \langle 0 | \hat{E}^{(+)}(z_j, t_j) | \Psi_n \rangle. \qquad (4.5.14)$$

Obviously, the effective wavefunction of the photon system is the result of an incoherent superposition of the single-photon effective wavefunctions:

$$\Psi(z_j, t_j) = \sum_n c_n\, \psi_n(z_j, t_j). \qquad (4.5.15)$$

The above superposition is consistent with Einstein's model of thermal field. The effective wavefunction of a single-photon plays the same role as that of

Einstein's subfield. In fact, we could simply replace Einstein's subfield with the effective wavefunction of single-photon in all early analysis for thermal radiation and obtain the same results.

Substituting the effective wavefunctions into Eq. (4.5.13), the quantum mutual-coherence function $G^{(1)}(z_1, t_1; z_2, t_2)$ turns to be

$$G^{(1)}(z_1, t_1; z_2, t_2)$$

$$\propto \left\langle \sum_{m,n} c_m^* c_n e^{-i\omega_0(t_{0m}-t_{0n})} e^{i\omega_0\tau} \mathcal{F}_{\tau_{1m}}^* \{f_m(\nu)\} \mathcal{F}_{\tau_{2n}} \{f_n(\nu)\} \right\rangle_{En}$$

$$\simeq \sum_m |c_m|^2 \mathcal{F}_{\tau_{1m}}^* \{f_m(\nu)\} \mathcal{F}_{\tau_{2m}} \{f_m(\nu)\} e^{i\omega_0\tau}$$

$$= \sum_m |c_m|^2 \psi_m^*(z_1, t_1)\psi_m(z_2, t_2). \tag{4.5.16}$$

Equation (4.5.16) indicates that the first-order temporal coherence is determined by the overlapping and nonoverlapping of the effective wavefunction of each single-photon and its conjugate. The physical picture behind the first-order quantum coherence of thermal field, which consists a large number of randomly distributed and randomly created wavepackets, is similar to that of single-photon wavepacket. This suggests that the first-order coherence time of thermal field should also be determined by the temporal width of each single-photon wavepackets $2\pi/\Delta\omega_m$, where $\Delta\omega_m$ is the spectrum of the mth single-photon wavepacket. Assuming identical spectrum for all measured randomly radiated and randomly distributed photons, the sum of m can be approximated as the following convolution yielding a mutual-coherence function of $G^{(1)}(\tau)$ which is a function of the optical delay τ:

$$G^{(1)}(\tau) = \sum_m |c_m|^2 \psi_m^*(z_1, t_1)\psi_m(z_2, t_2)$$

$$\simeq e^{i\omega_0\tau} \int dt_{0m} \, \mathcal{F}_{\tau_{1m}}^* \{f(\nu)\} \mathcal{F}_{\tau_{2m}} \{f(\nu)\}$$

$$\propto \mathcal{F}_\tau \{a^2(\nu)\} \, e^{i\omega_0\tau}. \tag{4.5.17}$$

This result is constant with that of Eq. (4.4.16). The first-order temporal coherence function and coherence time of the thermal field calculated from quantum theory are the same as those calculated from Einstein's granularity picture.

The probability of observing a photoelectron event from the point-like photon counting detector at space–time coordinate $(\mathbf{r}, t)$ is therefore

$$P(\mathbf{r}, t) \propto G_{11}^{(1)} + G_{22}^{(1)} + G_{12}^{(1)} + G_{21}^{(1)}$$

$$\simeq \sum_m \left| \frac{1}{\sqrt{2}} \left[\psi_m(z_1, t_1) + \psi_m(z_2, t_2) \right] \right|^2. \tag{4.5.18}$$

Equation (4.5.18) indicates that for thermal field, or a radiation system that consists of a large number of randomly distributed and randomly created photons, "...photon... only interferes with itself. Interference between two different photons never occurs". The physical picture is very clear: Each independently and randomly created photon may produce an interference pattern by means of a probability distribution function of that photon in space–time. The sum of these individual single-photon interferences yields the final observed interference. Replacing the effective wavefunctions $\Psi_n(z_j, t_j)$ in Eq. (4.5.18) with their Fourier transform representation as wavepackets, we obtain

$$P(\mathbf{r}, t) \propto \sum_n \left[\left| \mathcal{F}_{\tau_{1m}}\{f(\nu)\} \right|^2 + \left| \mathcal{F}_{\tau_{2m}}\{f(\nu)\} \right|^2 \right.$$

$$\left. + 2\mathrm{Re}\, e^{i\omega_0 \tau} \mathcal{F}_{\tau_{1m}}^*\{f(\nu)\} \mathcal{F}_{\tau_{2m}}\{f_n(\nu)\} \right]$$

$$\simeq \int dt_{0m}\, \frac{1}{2} \left[\left| \mathcal{F}_{\tau_{1m}}\{f(\nu)\} \right|^2 + \left| \mathcal{F}_{\tau_{2m}}\{f(\nu)\} \right|^2 \right.$$

$$\left. + 2\mathrm{Re}\, e^{i\omega_0 \tau} \mathcal{F}_{\tau_{1m}}^*\{f(\nu)\} \mathcal{F}_{\tau_{2m}}\{f_n(\nu)\} \right]$$

$$\propto 1 + \mathcal{F}_\tau\{|f(\nu)|^2\} \cos \omega_0 \tau, \tag{4.5.19}$$

where we have assumed a random and continuous distribution of t_{0m} with $|c_m|^2 =$ constant. The integral over t_{0m} is mathematically equivalent to a time integral. The results of the integrals are easily obtained by applying the Parseval theorem and the Wiener–Khinchin theorem as we have done earlier.

(III) $G^{(1)}(\mathbf{r}_1, t_1; \mathbf{r}_2, t_2)$: Thermal state in coherent state representation

Now, we calculate the first-order temporal coherence function, $G^{(1)}(\mathbf{r}_1, t_1; \mathbf{r}_2, t_2)$ or $G^{(1)}(\tau)$, of thermal field in the coherent state representation. The mixed state of thermal field in the coherence state representation has been introduced in Eq. (3.5.7):

$$|\tilde{\Psi}\rangle = \prod_m \prod_\omega |\alpha_m(\omega)\rangle,$$

where m labels the mth sub-source; again, we separate the sub-indices of m and ω. In the following, we assume a large number of independent random radiating and randomly distributed sub-sources, each producing a group of indistinguishable photons with coherent Fourier modes.

The field operator is the same as that in the single-photon state representation:

$$\hat{E}^{(-)}(z_j, t_j) = \sum_m \int d\omega\, \hat{a}_m^\dagger(\omega)\, g_m^*(\omega; z_j, t_j) = \sum_m \int d\omega\, \hat{a}_m^\dagger(\omega)\, e^{i\omega\tau_{jm}},$$

$$\hat{E}^{(+)}(z_k, t_k) = \sum_n \int d\omega'\, \hat{a}_n(\omega')\, g_n(\omega'; z_k, t_k) = \sum_n \int d\omega'\, \hat{a}_n(\omega')\, e^{-i\omega'\tau_{kn}}.$$

Following Eqs. (4.5.12), the first-order temporal mutual-coherence function of the measured thermal field is thus

$$G^{(1)}(z_1, t_1; z_2, t_2) = \left\langle \langle \tilde{\Psi}| \hat{E}^{(-)}(z_1, t_1) \hat{E}^{(+)}(z_2, t_2) |\tilde{\Psi}\rangle \right\rangle_{\mathrm{En}}$$

$$= \left\langle \sum_m \psi_m^*(z_1, t_1) \sum_n \psi_n(z_2, t_2) \right\rangle_{\mathrm{En}}. \qquad (4.5.20)$$

Here, we have used

$$\langle \Psi | \hat{E}^{(+)}(z_j, t_j) | \Psi \rangle = \langle \Psi | \sum_m \int d\omega\, \hat{a}_m(\omega)\, g_m(\omega; z_j, t_j) | \Psi \rangle$$

$$= \sum_m \int d\omega\, \alpha_m(\omega)\, g_m(\omega; z_j, t_j)$$

$$= \sum_m \psi_m(z_j, t_j) \qquad (4.5.21)$$

and defined the effective wavefunction of the mth group of indistinguishable photons

$$\psi_m(z_j, t_j) = \int d\omega\, \alpha_m(\omega)\, g_m(\omega; z_j, t_j) = \int d\omega\, \alpha_m(\omega)\, e^{-i\omega\tau_{jm}}. \qquad (4.5.22)$$

For a natural thermal light source, we regard a coherent superposition of the Fourier modes radiated from a sub-source and incoherent superposition of the radiation radiated from different sub-sources. In this case,

$$G^{(1)}(\tau) = \left\langle \sum_m \int d\omega\, \alpha_m^*(\omega)\, e^{i\omega\tau_{1m}} \sum_n \int d\omega'\, \alpha_n(\omega')\, e^{-i\omega'\tau_{2n}} \right\rangle_{\text{En}}$$

$$= \sum_m \int d\omega\, \alpha_m^*(\omega)\, e^{i\omega\tau_{1m}} \int d\omega'\, \alpha_m(\omega')\, e^{-i\omega'\tau_{2m}}$$

$$= \sum_m \mathcal{F}_{\tau_{1m}}^* \{\alpha_m^*(\nu)\} \mathcal{F}_{\tau_{2m}} \{\alpha_m(\nu)\}\, e^{i\omega_0\tau}$$

$$\simeq \int dt_{0m}\, \mathcal{F}_{\tau_{1m}}^* \{\alpha_m^*(\nu)\} \mathcal{F}_{\tau_{2m}} \{\alpha_m(\nu)\}\, e^{i\omega_0\tau}$$

$$= \mathcal{F}_\tau \{\alpha_m^2(\nu)\}\, e^{i\omega_0\tau}. \tag{4.5.23}$$

This result agrees with our early result that was derived from Einstein's granularity picture of thermal field, except replacing the subfields with the effective wavefunction. Obviously, the first-order coherence time of thermal field calculated from coherence state representation is $\tau_c = 2\pi/\Delta\omega$.

4.6   First-Order Spatial Coherence: Einstein's Picture

The finite bandwidth of the temporal spectrum $\Delta\omega$ is not the only factor determining the degree of first-order coherence and the visibility of first-order interference. The finite bandwidth of the spatial frequency $\Delta\vec{\kappa}$ is another important factor we have to take into account regarding the visibility of interference and the degree of first-order coherence observed from an Young's double-pinhole interferometer. This concern leads to the concepts of spatial coherence and the degree of first-order spatial coherence. The spatial coherence of light is directly related to the transverse dimension, or the angular size of the light source.

(I) Spatial coherence: Thermal radiation source

In the following discussion, we assume a Young's double-pinhole interferometer facing a distant star of finite angular size. The disk-like star, which consists of a large number of randomly radiated and randomly distributed independent point-like sub-sources of radiation, has an angular diameter of $\Delta\theta$ relative to the interferometer. To simplify the mathematics and to focus on the physics of spatial coherence, we assume each of the mth subfield radiated from each of the mth point-like sub-sources of the distant star, $E_m(\mathbf{r}, t)$, monochromatic. From the view point of the Young's double-pinhole interferometer, which is schematically illustrated in Fig. 4.6.1, each

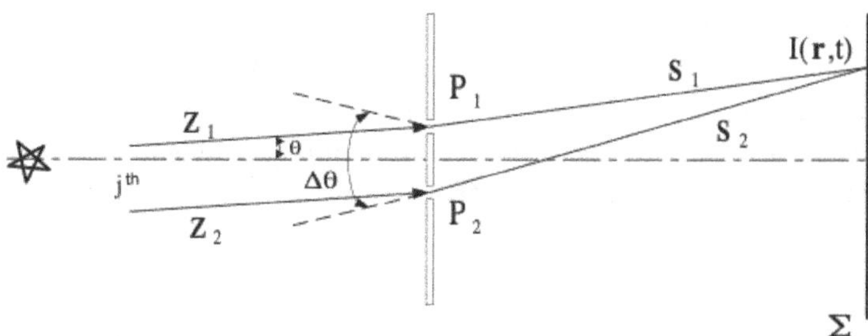

Fig. 4.6.1   Schematic of Young's double-slit interference experiment. The disk-like light source is a distant star of finite angular size ($\Delta\theta \neq 0$) consisting of a large number of randomly radiated and randomly distributed point-like sub-sources or atomic transitions. The spacing between the upper pinhole P_1 and the lower pinhole P_2 is b. To simplify the mathematics, we restrict our calculations in 1-D.

point-like sub-source is identified by an angular coordinate θ (1-D), and each subfield radiated from a sub-source has a different transverse wavevector along the x-axis, $k_x \sim k\theta$ (1-D). The fields $E(\mathbf{r}_1, t_1)$ at pinhole P_1 and $E(\mathbf{r}_2, t_2)$ at pinhole P_2 are treated as a superposition of all the randomly radiated and randomly distributed subfields from the entire surface of the distant star.

The mutual-coherence function $\Gamma^{(1)}(\mathbf{r}_1, t_1; \mathbf{r}_2, t_2)$ is then written as

$$\Gamma^{(1)}(\mathbf{r}_1, t_1; \mathbf{r}_2, t_2)$$

$$= \left\langle \sum_{m=1}^{M} E_m^*(\mathbf{r}_1, t_1) \sum_{n=1}^{N} E_n(\mathbf{r}_2, t_2) \right\rangle \tag{4.6.1}$$

$$= \left\langle \sum_{m=n} E_m^*(\mathbf{r}_1, t_1) E_m(\mathbf{r}_2, t_2) \right\rangle + \left\langle \sum_{m\neq n} E_m^*(\mathbf{r}_1, t_1) E_n(\mathbf{r}_2, t_2) \right\rangle,$$

where m and n label each of the independent contributions of a point-like sub-source. It is convenient to break up the sum into two groups. The second group ($m \neq n$) of the sum vanishes in the expectation calculation when taking into account all possible values of the random relative phase differences. $\Gamma^{(1)}(\mathbf{r}_1, t_1; \mathbf{r}_2, t_2)$ becomes

$$\Gamma^{(1)}(\mathbf{r}_1, t_1; \mathbf{r}_2, t_2) = \sum_{m=1}^{M} E_m^*(\mathbf{r}_1, t_1)\, E_m(\mathbf{r}_2, t_2) \tag{4.6.2}$$

after ensemble averaging. Equation (4.6.2) indicates a summation of a large number of individual interferences each is the result of a self-interference: a subfield interfering with the subfield itself.

We can transfer Eq. (4.6.2) into a simple integral by assuming a random distribution of the sub-sources on the distant star

$$\Gamma^{(1)}(\mathbf{r}_1, t_1; \mathbf{r}_2, t_2)$$

$$= \sum_{m=1}^{M} E_{1m}^* \, e^{i[\omega(t-s_1/c)-kz_{m1}]} \, E_{2m} e^{-i[\omega(t-s_2/c)-kz_{m2}]}$$

$$= e^{i\omega\tau} \sum_{m=1}^{M} a_{1m} a_{2m} \, e^{ik(z_{m2}-z_{m1})}$$

$$\cong e^{i\omega\tau} \int_{-\frac{\Delta\theta}{2}}^{\frac{\Delta\theta}{2}} \left(\frac{I_0}{\Delta\theta}\right) d\theta \, e^{ik\theta b}$$

$$= G_0^{(1)} \, \mathrm{sinc}\left(\frac{\pi b \Delta\theta}{\lambda}\right) e^{i\omega\tau}, \tag{4.6.3}$$

where $G_0^{(1)}$ absorbed all constants and $\tau = (s_2 - s_1)/c$. To simply the mathematics, we have restricted our calculation in 1-D, and the integral is taken over the entire angular diameter of the star along the x direction from $-\Delta\theta/2$ to $\Delta\theta/2$.

The normalized degree of first-order spatial coherence of the two fields at the upper and the lower pinholes is thus

$$\gamma^{(1)}(\mathbf{r}_1, t_1; \mathbf{r}_2, t_2) = \mathrm{sinc}\left(\frac{\pi b \Delta\theta}{\lambda}\right) e^{i\omega\tau} = \mathrm{sinc}\left(\Delta k_x \, b\right) e^{i\omega\tau}. \tag{4.6.4}$$

Here, $\gamma^{(1)}(\mathbf{r}_1, t_1; \mathbf{r}_2, t_2)$ is a function of the separation b between the upper and the lower pinholes and the angular size $\Delta\theta$ of the distant star. The two fields at the upper and lower pinholes are said to be spatially coherent when $b \ll \lambda/\Delta\theta$, or $b \ll 2\pi/\Delta k_x$, ($|\gamma^{(1)}(\mathbf{r}_1, t_1; \mathbf{r}_2, t_2)| \cong 1$), and the two fields are said to be spatially incoherent when $b \geq \lambda/\Delta\theta$, or $b \geq 2\pi/\Delta k_x$ ($|\gamma^{(1)}(\mathbf{r}_1, t_1; \mathbf{r}_2, t_2)| \cong 0$). For a point source, $\Delta\theta \sim 0$, and consequently $|\gamma_{12}| \cong 1$ for any value of b. This means the radiation fields excited by a point radiation source are spatially coherent despite the spatial separation between the fields.

The interference pattern on the observation plane can be written as

$$\langle I(\mathbf{r}, t) \rangle = I_0 \left[1 + \mathrm{sinc}\left(\frac{\pi b \Delta\theta}{\lambda}\right) \cos\left(\omega\tau\right) \right], \tag{4.6.5}$$

with an interference visibility

$$V = |\gamma_{12}^{(1)}| = \text{sinc}\left(\frac{\pi b \Delta\theta}{\lambda}\right) = \text{sinc}\left(\Delta k_x\, b\right). \tag{4.6.6}$$

Note we have simplified the calculation of the $\gamma^{(1)}$ function by assuming a monochromatic plane wave of single wavelength λ so that $V = |\gamma_{12}^{(1)}|$ is independent of τ_s.

We may find the following physical picture useful for understanding the concept of spatial coherence. In Fig. 4.6.1, we have assumed a large number, $M \sim \infty$, of randomly radiated and randomly distributed point-like sub-sources. The subfields created from these sub-sources are all emitted with random relative phases. The only observable interferences are the self-interferences of these subfields. The cross-interference between subfields excited from different sub-sources cancels completely while taking into account all possible values of the relative phases between subfields. The measured intensity on the observation plane is thus

$$\begin{aligned}
\langle I(\mathbf{r}, t)\rangle &= \sum_{m=1}^{M} I_m(\mathbf{r}, t) \\
&= \sum_{m=1}^{M} \left|\frac{1}{\sqrt{2}}\left[E_m(\mathbf{r}_1, t_1) + E_m(\mathbf{r}_2, t_2)\right]\right|^2 \\
&= \sum_{m=1}^{M} I_{m0}\left\{1 + \cos[\omega\tau + k(z_{m2} - z_{m1})]\right\} \\
&\simeq \sum_{m=1}^{M} I_{m0}\left[1 + \cos(\omega\tau + k\theta_m b)\right]. \tag{4.6.7}
\end{aligned}$$

We may consider that each subfield from a point sub-source on the distant star, identified by angle θ_m, produces a Young's double-slit sinusoidal interference pattern on the observation plane. The phase shift of each interference pattern, relative to the one that is produced by the subfields generated from the sub-sources at $\theta = 0$, is determined by the value of $k(z_{2m} - z_{1m}) \sim k\theta_m b \sim k_x b$. The maximum relative phase separation between these individual patterns is $k(\Delta\theta b)$, or $\Delta k_x b$, corresponding to the phase shift between the interference patterns excited by the sub-sources of $\Delta\theta$ and $-\Delta\theta$. Therefore, when $2\pi\Delta\theta b/\lambda \ll 2\pi$, i.e., $b \ll \lambda/\Delta\theta$, or $b \ll 2\pi/\Delta k_x$, the relative phase shifts are not large enough to produce noticeable separation between the sub-interference patterns and so the interference modulation is observable. However, when the relative phase shifts increase to a certain value of $2\pi\Delta\theta b/\lambda \sim 2\pi$, i.e., $b \sim \lambda/\Delta\theta$ or $b \sim 2\pi/\Delta k_x$, the individual patterns become significantly separated. The spread of the patterns smoothes the light intensity distribution on the observation plane so the interference pattern can no longer be identified, and the interference modulation visibility is reduced from 100% to 0. Based on the above observation, we introduce the concept of spatial coherence of the field. For a given light source with angular size $\Delta\theta$, the fields $E(\mathbf{r}_1, t)$ and $E(\mathbf{r}_2, t)$ are considered as spatially coherent if their transverse spatial separation is less than

$$l_c = \lambda/\Delta\theta = 2\pi/\Delta k_x, \qquad (4.6.8)$$

where l_c is defined as the transverse coherence length of the radiation. Any fields with spatial separation beyond l_c are considered as spatially incoherent, which implies no observable first-order interference from a double-slit interferometer. The degree of first-order spatial coherence, defined in Eq. (4.3.7) and calculated in Eq. (4.6.4), is a quantitative measure of Young's double-slit interference for a radiation source of finite angular diameter $\Delta\theta$.

Taking up an early suggestion by Fizeau, Michelson designed a stellar interferometer based on the mechanism of Young's double-pinhole interference. One important application of the Michelson stellar interferometer is the measurement of the angular size of a distant star or the angular separation between distant double stars. The principle and operation of this stellar interferometer is quite simple. What one needs to do is to manipulate the separation between the two pinholes from $b = 0$, point by point, to a critical value $b = l_c$. If one can make an accurate judgment at the critical value of $b = l_c$ at which the interference pattern becomes invisible, this value of l_c can be used to estimate of the angular size of the distant star. According to Eq. (4.6.5), the double-slit interferometer starts to lose its interference at $l_c = \lambda/\Delta\theta$. The angular size of the distant star, $\Delta\theta = \lambda/l_c$, is thus measured with certain accuracy. Of course, making an accurate judgment is never easy, as there are too many physical parameters that contribute to the instability of an interference pattern.

(II) A single-photon wavepacket interferes with the wavepacket itself

Now, again, we turn our discussion into an interesting and controversy topic of quantum optics: interference of a single-photon wavepacket. We assume the light source is so weak that only a single subfield or wavepacket of $\hbar\omega$

is received at the double-pinhole interferometer within a time period of the observation. The observed single-photon wavepacket may be created from any point-like sub-sources that are randomly distributed on the surface of the distant star. The questions are as follows: (1) Do we observe interference from the double-slit interferometer? (2) If we do, how to measure it?

Based on our previous discussions in this section, the answer to question (1) is positive:

$$I(\mathbf{r}, t) = \left| \frac{1}{\sqrt{2}} \left[E_m(\mathbf{r}_1, t_1) + E_m(\mathbf{r}_2, t_2) \right] \right|^2$$

$$= I_{m0} \left\{ 1 + \cos[\omega\tau + k(z_{m2} - z_{m1})] \right\}$$

$$\simeq I_{m0} \left[1 + \cos(\omega\tau + k\theta b) \right]. \tag{4.6.9}$$

Again, in classical theory, although the mth subfield only carries energy of $\hbar\omega$, this wavepacket can be divided into two by taking path s_1 (passing through pinhole P_1) and/or path s_2 (passing through pinhole P_2), and then to be superposed at coordinate $(\mathbf{r}, t)$. The classical intensity distribution on the observation plane is the result of a superposition of two partial wavepackets: one passing through pinhole P_1 and another passing through pinhole P_2.

The answer to question (2) is the same as our previous answer in this section: photon counting. Placing a modern photon-counting photodetector array in the observation plane, one wavepacket or subfield can only excite one photoelectron from one element of the photodetector array, which may not give us any meaningful information about the interference. However, after a timely accumulative counting on a large number of subfields or wavepackets, one by one, we find the accumulated charges in each photoelement, which is proportional to the number of photons registered at coordinate $(\mathbf{r})$, within the accumulative time period of ΔT:

$$Q(\mathbf{r}) \propto \int_{\Delta T} dt\, I(\mathbf{r}, t) \propto n(\mathbf{r}) \propto \langle I(\mathbf{r}, t) \rangle_{\Delta T}$$

$$= I_0 \left[1 + \mathrm{sinc}\left(\frac{\pi b \Delta\theta}{\lambda} \right) \cos(\omega\tau) \right]. \tag{4.6.10}$$

The interference pattern is observable from $Q(\mathbf{r})$ or $n(\mathbf{r})$ after a timely accumulative photon counting.

We are then facing the same serious question: How long does it take for the energy from other part of the interference pattern to arrive at the excited single CCD pixal? In the classical picture of a subfield, its energy of $\hbar\omega$ should be uniformly distributed on the entire interference pattern. If we insist the classical picture of a photon, we would be easily tripped to the above question as Einstein did to himself and to his students.

From the view point of quantum theory, the entire interference pattern "produced" by a subfield of $\hbar\omega$ means nothing but a probability distribution function $P(\mathbf{r}, t)$: the probability for a subfield, or a photon, to produce a photodetection event at space–time coordinate $(\mathbf{r}, t)$. In an accumulative measurement involving a large number of M photons, there will be $MP(\mathbf{r})$ photodetection events occurring at coordinate $\mathbf{r}$, corresponding to the time averaged intensity $\langle I(\mathbf{r}, t)\rangle_{\Delta T}$.

We are ready to introduce the concept of quantum first-order coherence of light.

4.7   First-Order Spatial Coherence: Quantum Theory

Similar to the discussion of first-order spatial classical coherence, we assume a double-pinhole interferometer facing a weak distant star of finite angular size. The distant star, which consists of a large number of randomly radiated and randomly distributed independent atomic transitions, has an angular diameter of $\Delta\theta$ relative to the interferometer, as shown in Fig. 4.6.1. The observed radiation field is in thermal state at single-photon's level. A photon counting CCD or CMOS or a scannable point-like photon counting detector is placed on the observation plane of the interferometer to measure the interference. In this section, we introduce the concept of first-order spatial quantum coherence through a simple calculation based on the single-photon state representation.

The thermal state in single-photon state representation can be written as

$$|\tilde{\Psi}\rangle \simeq \sum_m c_m |\Psi_m\rangle = \sum_m c_m \int d\mathbf{k}\, f_m(\mathbf{k})\, \hat{a}_m^\dagger(\mathbf{k})|0\rangle \tag{4.7.1}$$

which represents an incoherent superposition of a set of single-photon states, where c_m has a random phase. To simplify the mathematics, we assume $|c_m| = 1$.

The field operators are similar to that of Eq. (4.5.12), except approximated into a more general form of 3-D:

$$\hat{E}^{(+)}(\mathbf{r}, t) = \sum_m \int d\mathbf{k}\, \hat{a}_m(\mathbf{k})\, g_m(\mathbf{k}; \mathbf{r}, t),$$

$$\hat{E}^{(-)}(\mathbf{r}, t) = \sum_m \int d\mathbf{k}\, \hat{a}_m^\dagger(\mathbf{k})\, g_m^*(\mathbf{k}; \mathbf{r}, t), \tag{4.7.2}$$

where $g_m(\mathbf{k}; \mathbf{r}, t)$ is Green's function that propagates the $\mathbf{k}$-mode of the mth subfield from the mth atomic transition to space–time coordinate $(\mathbf{r}, t)$.

The first-order quantum mutual-coherence function and self-coherence functions, respectively, are calculated as follows:

$$G^{(1)}(\mathbf{r}_1, t_1; \mathbf{r}_2, t_2)$$

$$= \left\langle \sum_m c_m \langle \Psi_m | \hat{E}^{(-)}(\mathbf{r}_1, t_1) \hat{E}^{(+)}(\mathbf{r}_2, t_2) \sum_n c_n | \Psi_n \rangle \right\rangle_{En}$$

$$= \left\langle \sum_m c_m \, \psi_m^*(\mathbf{r}_1, t_1) \sum_n c_n \, \psi_n(\mathbf{r}_2, t_2) \right\rangle$$

$$= \sum_m \psi_m^*(\mathbf{r}_1, t_1) \psi_m(\mathbf{r}_2, t_2) \tag{4.7.3}$$

and

$$G^{(1)}(\mathbf{r}_j, t_j; \mathbf{r}_j, t_j)$$

$$= \left\langle \sum_m c_m \langle \Psi_m | \hat{E}^{(-)}(\mathbf{r}_j, t_j) \hat{E}^{(+)}(\mathbf{r}_j, t_j) \sum_n c_n | \Psi_n \rangle \right\rangle_{En}$$

$$= \left\langle \sum_m c_m \, \psi_m^*(\mathbf{r}_j, t_j) \sum_n c_n \, \psi_n(\mathbf{r}_j, t_j) \right\rangle$$

$$= \sum_m |\psi_m(\mathbf{r}_j, t_j)|^2, \tag{4.7.4}$$

where $\psi_m(\mathbf{r}_j, t_j)$, $j = 1, 2$, is defined as the effective wavefunction of the mth photon:

$$\psi_m(\mathbf{r}_j, t_j) = \int d\mathbf{k} \, f_m(\mathbf{k}) \, g_m(\mathbf{k}; r_j, t_j). \tag{4.7.5}$$

In a 1-D approximation, replacing the effective wavefunctions with Einstein's subfields, we find $G^{(1)}(\mathbf{r}_1, t_1; \mathbf{r}_2, t_2)$ is the same as $\Gamma^{(1)}(\mathbf{r}_1, t_1; \mathbf{r}_2, t_2)$ that has been calculated earlier. We thus have

$$G^{(1)}(\mathbf{r}_1, t_1; \mathbf{r}_2, t_2) = G_0^{(1)} \, \text{sinc}\left(\frac{\pi b \Delta\theta}{\lambda}\right) e^{i\omega\tau}, \tag{4.7.6}$$

where $\tau = (s_2 - s_1)/c$, and the normalized quantum degree of first-order spatial coherence is

$$g^{(1)}(\mathbf{r}_1, t_1; \mathbf{r}_2, t_2) = \text{sinc}\left(\frac{\pi b \Delta\theta}{\lambda}\right) e^{i\omega\tau} = \text{sinc}\left(\Delta k_x \, b\right) e^{i\omega\tau}. \tag{4.7.7}$$

The probability for a single photon to produce a photoelectron event at space–time coordinate $(\mathbf{r}, t)$ is thus a sinusoidal function of $\tau = (s_2 - s_1)/c$:

$$P(\mathbf{r}, t) \propto 1 + \mathrm{sinc}\left(\frac{\pi b \Delta\theta}{\lambda}\right) \cos(\omega\tau), \tag{4.7.8}$$

with modulation visibility

$$V = |g_{12}^{(1)}| = \mathrm{sinc}\left(\frac{\pi b \Delta\theta}{\lambda}\right) = \mathrm{sinc}(\Delta k_x\, b), \tag{4.7.9}$$

which is determined mainly by the angular diameter $\Delta\theta$ of the light source. Similar to the classical first-order spatial coherence, we define $\Delta l_c \equiv \lambda/\Delta\theta$ the first-order spatial quantum coherence length of the thermal field. Obviously, the interference is unobservable when the spatial separation of the two pinholes is greater than the first-order spatial quantum coherence length $b > \Delta l_c$.

Note, similar to that of the first-order spatial classical coherence in Section 6.3, to simplify the mathematics, we have approximated a monochromatic thermal field.

After a timely accumulative counting on a large number of photons, one by one, we find the number of photons registered in each photoelement of the CCD or CMOS, or registered in the scannable point-like photodetector on the observation plane, is modulated sinusoidally as a function of $\tau_s = (s_2 - s_1)/c$ with modulation visibility $\mathrm{sinc}(\pi b \Delta\theta/\lambda)$:

$$n(\mathbf{r}) \propto n_0 \left[1 + \mathrm{sinc}\left(\frac{\pi b \Delta\theta}{\lambda}\right) \cos(\omega\tau)\right], \tag{4.7.10}$$

where n_0 is the mean number of photons accumulatively received by Young's double-pinhole interferometer. The interference pattern on the observation plane is observable from $n(\mathbf{r})$ after a timely accumulative photon counting. Again, the visibility of the interference modulation is determined by the angular diameter $\Delta\theta$ of the light source. The interference achieves maximum visibility for a point-like source when $\Delta\theta \sim 0$ and vanishes when the separation of the two pinholes becomes greater then the first-order spatial quantum coherence length $b > \Delta l_c$.

4.8 First-Order Coherence of Laser Beam

In this section, we discuss first-order coherence $\Gamma^{(1)}(\mathbf{r}_1, t_1; \mathbf{r}_2, t_2)$ of radiations of lasers.

(I) **First-order temporal coherence of a laser pulse**

A single laser pulse with enough energy is able to produce an observable interference pattern on the observation plane of an Young's double-pinhole interferometer. A modern 2-D photodetector array, CCD or CMOS, is able to monitor the "one-shot" interference pattern $I(\mathbf{r}, t)$ of a laser pulse.

We have shown in our early discussions that the mutual-coherence function $\Gamma(\mathbf{r}_1, t_1; \mathbf{r}_2, t_2)$ can be calculated from the classical fields $E(\mathbf{r}_1, t_1)$ and $E(\mathbf{r}_2, t_2)$ passing through the upper pinhole P_1 and the lower pinhole P_2, respectively, at early times $t_1 = t - s_1/c$ and $t_2 = t - s_2/c$, where t refers the measurement time of the photodetector. The mutual-coherence function $\langle E^*(\mathbf{r}_1, t_1) E(\mathbf{r}_2, t_2) \rangle$ is found to be

$$\Gamma^{(1)}(\tau_1, \tau_2) = \langle E^*(\mathbf{r}_1, t_1) E(\mathbf{r}_2, t_2) \rangle = \mathcal{F}^*_{\tau_1}\{E(\nu)\}\, \mathcal{F}_{\tau_2}\{E(\nu)\}\, e^{-i\omega_0 \tau} \tag{4.8.1}$$

by *taking into account all possible realizations of the fields* $E(\mathbf{r}_1, t_1)$ and $E(\mathbf{r}_2, t_2)$,

In Einstein's picture of light, *taking into account all possible realizations of the fields* means *taking into account all possible phases and amplitudes of the subfields*. We have shown that a laser pulse is the result of coherent superposition among a large number of subfields and Fourier modes:

$$\begin{aligned}
\Gamma^{(1)}_{12}(\tau_1, \tau_2) &= \left\langle \sum_m E^*_m(\mathbf{r}_1, t_1)\rangle \sum_n E_n(\mathbf{r}_2, t_2) \right\rangle \\
&= \sum_m \mathcal{F}^*_{\tau_1}\{a_m(\nu)\} \sum_n \mathcal{F}_{\tau_2}\{a_n(\nu)\}\, e^{i\omega_0 \tau} \\
&= \mathcal{F}^*_{\tau_1}\{A(\nu)\} \mathcal{F}_{\tau_2}\{A(\nu)\}\, e^{i\omega_0 \tau},
\end{aligned} \tag{4.8.2}$$

where $A(\nu) = \sum_m a(\nu)$. The expectation function is calculated by taking into account a constant relative phase for all subfields created from a large number of coherently radiated sub-sources. Similar to the wavepacket at single-photon level, the degree of first-order coherence is determined by the overlapping–nonoverlapping of the pulses. At the neighborhood of $\tau \sim 0$, $\Gamma^{(1)}(\tau_1, \tau_2)$ achieves its maximum value; we consider the fields $E(\mathbf{r}_1, t_1)$ and $E(\mathbf{r}_2, t_2)$ first-order coherent; however, when $\tau > 2\pi/\Delta\omega$, the two pulses cannot have nonzero values simultaneously, $\Gamma^{(1)}(\tau_1, \tau_2) = 0$, we define the fields of $E(\mathbf{r}_1, t_1)$ and $E(\mathbf{r}_2, t_2)$ first-order incoherent, although the laser pulse itself is considered as coherent radiation.

From Eq. (4.8.2), we find that $\Gamma^{(1)}(\mathbf{r}_1, t_1; \mathbf{r}_2, t_2)$ indicates a nonstationary field, which is consistent with the nature of a laser pulse. Due to the nonstationary nature, we need to pay attention to the normalized degree of first-order coherence function $\gamma_{12}(\tau_1, \tau_2)$:

$$\gamma_{12}^{(1)}(\tau_1, \tau_2) = \frac{\mathcal{F}_{\tau_1}^*\{A(\nu)\}\mathcal{F}_{\tau_2}\{A(\nu)\}\,e^{i\omega_0\tau}}{\sqrt{\left|\mathcal{F}_{\tau_1}\{A(\nu)\}\right|^2 \left|\mathcal{F}_{\tau_2}\{A(\nu)\}\right|^2}}, \qquad (4.8.3)$$

under the condition of

$$\left|\mathcal{F}_{\tau_1}\{A(\nu)\}\right|^2 \left|\mathcal{F}_{\tau_2}\{A(\nu)\}\right|^2 \neq 0.$$

One may find that Eq. (4.8.3) leads to $|\gamma_{12}^{(1)}(\tau_1, \tau_2)| = 1$ for real functions, even if the two pulses only slightly overlap. This is because the product of $\Gamma_{11}^{(1)}(\tau_1)$ and $\Gamma_{22}^{(1)}(\tau_2)$, which is used for normalization, behaves the same as that of $\Gamma_{12}^{(1)}(\tau_1, \tau_2)$. Examine Eq. (4.3.9); we find the visibility of the interference fringe modulation is very different from the degree of fist-order coherence $|\gamma_{12}^{(1)}(\tau_1, \tau_2)|$. In this case, the interference visibility has to be estimated from its definition of Eq. (4.3.9).

(II) First-order temporal coherence of CW laser beam

Continuous wave (CW) laser radiation may contain a single cavity mode, a few cavity modes, or a large number of cavity modes, each centered at frequency ω_{0j} with finite bandwidth $\Delta\omega_j$, where j labels the jth cavity mode. Each cavity mode can be treated as either a coherent wavepacket of temporal width $2\pi/\Delta\omega_j$ or a single frequency mode of $\Delta\omega_j \sim 0$. In principle, our earlier treatments of $\Gamma^{(1)}(\mathbf{r}_1, t_1; \mathbf{r}_2, t_2)$ may apply to CW laser radiation, except each wavepacket has a different center frequency ω_{0j}.

In the following, we estimate the first-order coherence function of the radiations of a CW laser by approximating the fields $E(\mathbf{r}_1, t_1)$ and $E(\mathbf{r}_2, t_2)$ as a set of wavepackets each centered at ω_{0j} with spectral bandwidth $\Delta\omega_j$:

$$\Gamma^{(1)}(\mathbf{r}_1, t_1; \mathbf{r}_2, t_2)$$

$$= \left\langle \sum_j \mathcal{F}_{\tau_{1j}}^*\{A_{1j}(\nu)\}e^{i\omega_{0j}(\tau_{1j})} \sum_k \mathcal{F}_{\tau_{2k}}\{A_{2k}(\nu)\}e^{-i\omega_{0k}(\tau_{2k})} \right\rangle$$

$$= \left\langle \sum_{j=k} \mathcal{F}_{\tau_{1j}}^*\{A_{1j}(\nu)\}\mathcal{F}_{\tau_{2j}}\{A_{2j}(\nu)\}\,e^{i\omega_{0j}\tau} \right\rangle \qquad (4.8.4)$$

$$+ \left\langle \sum_{j\neq k} \mathcal{F}_{\tau_{1j}}^*\{A_{1j}(\nu)\}\mathcal{F}_{\tau_{2k}}\{A_{2k}(\nu)\}\,e^{i[\omega_{0j}(\tau_{1j})-\omega_{0k}(\tau_{2k})]} \right\rangle,$$

where $A_{1j}(\nu)$ and $A_{2k}(\nu)$, respectively, with $\nu = \omega_j - \omega_{0j}$, are the amplitude distribution function of the jth cavity mode along path one and path two of the interferometer.

The result of Eq. (4.8.4) depends on the coherent or incoherent relation between the cavity modes. For incoherent cavity modes, the second term in the summation vanishes, the observable sinusoidal modulation comes from the self-interferences of each cavity mode with itself:

$$\Gamma^{(1)}(\mathbf{r}_1, t_1; \mathbf{r}_2, t_2) = \sum_j \mathcal{F}^*_{\tau_{1j}}\{A_{1j}(\nu)\} \mathcal{F}_{\tau_{2j}}\{A_2(\nu)\} e^{i\omega_{0j}\tau}$$

$$= \sum_j \Gamma^{(1)}(\tau_{1j}, \tau_{2j}). \tag{4.8.5}$$

It is interesting to find that each cavity mode of a CW laser behaves like an individual wavepacket or laser pulse. The CW laser beam is the result of incoherent superposition of these "long" wavepackets. The mutual-coherence function of a CW laser beam is the sum of the mutual-coherence functions of individual cavity mode. Considering the narrow spectrum of the laser cavity mode, $\Gamma^{(1)}(\tau_{1j}, \tau_{2j})$ is effectively a constant of $\tau_1 - \tau_2$. It is also clear that the first-order beats between cavity modes are unobservable from a CW laser source. For coherent cavity modes, such as that in a mode-locked laser, both first term and the second term of the summation have to be taken into account. In this case, the beating frequencies between cavity modes are observable.

(III) First-order temporal coherence of two independent laser pulses

In Einstein's granularity picture of light, each laser pulse or wavepacket may consist of a large number of identical and coherently superposed sub-wavepackets. Ensemble average is physically meaningful for the measurement of the interference between a pair of laser pulses. Taking into account all possible realizations of the fields $E(\mathbf{r}_1, t_1)$ and $E(\mathbf{r}_2, t_2)$ from the j–kth pair of independent laser wavepackets, the mutual-coherence function $\langle E^*(\mathbf{r}_1, t_1) E(\mathbf{r}_2, t_2) \rangle$ is found to be

$$\Gamma^{(1)}(\mathbf{r}_1, t_1; \mathbf{r}_2, t_2) = \mathcal{F}^*_{\tau_{1j}}\{A_j(\nu)\} \mathcal{F}_{\tau_{2k}}\{A_k(\nu)\} e^{i\omega_0[(t_k - t_j) + \tau]}, \tag{4.8.6}$$

where we have assumed the two independent laser pulses have the same central frequency ω_0; again, $A_j(\nu)$ is the amplitude for the mode ω of the jth laser wavepacket, and t_j and t_k are the initial creation times of the two individual laser pulses. Note, here we have used j and k to label the jth and the kth laser pulses to distinguish the mth and the nth quantized subfield in Einstein's picture. It is easy to see that within a selected pair of laser wavepackets $(t_k - t_j)$ holds a well-defined value. If the measurement is completed within a pair of laser pulses, the interference will be observable. In fact, the interference between two independent lasers was experimentally demonstrated by Mandel *et al.* during the years 1960s–1970s after the invention of the laser.[4]

What will happen if the measurement involves a large number of individual pairs of wavepackets accumulatively? Can we still observe interference? As we have discussed earlier, in this kind of measurement, the finally measured interference pattern on the observation plane is the sum of a large number of sub-interference patterns, each produced by a pair of wavepackets:

$$\langle I(\mathbf{r}, t) \rangle \simeq \sum_{j,k} \left\{ \left| \mathcal{F}_{\tau_{1j}} \{ A_j(\nu) \} \right|^2 + \left| \mathcal{F}_{\tau_{2k}} \{ A_k(\nu) \} \right|^2 \right.$$

$$\left. + 2 \operatorname{Re} \mathcal{F}^*_{\tau_{1j}} \{ A_1(\nu) \} \mathcal{F}_{\tau_{2k}} \{ A_2(\nu) \} \, e^{i\omega_0 [(t_k - t_j) + \tau]} \right\}, \qquad (4.8.7)$$

where subscript indexes j and k label the j–kth laser wavepacket pair. If the laser wavepacket pairs are "phase-locked", where "phase-lock" means forcing the two lasers to generate their wavepackets at $t_k - t_j = \text{constant}$ for all j–k pairs to have the relative phase $\omega_0 [(t_k - t_j) + \tau = \omega_0 [\text{constant} + \tau]$, which is independent of j–k, the sub-interference-patterns would be identical from pulse pair to pulse pair, and consequently, the timely accumulative observation would be the sum of these identical sub-interference patterns. In this case, the timely accumulative interference, which involves a large number of phase-locked laser pulses, is observable.

(IV) **First-order spatial coherence of laser beam**

The condition of having an observable Young's double-pinhole interference pattern for a coherent laser beam is obvious: The two pinholes must be illuminated by the radiation simultaneously. We have discussed the propagation of spatially coherent radiation, such as a TEM_{00} laser beam, in Sections 2.3 and 3.2. The coherent radiation propagates in a collimated manner with diffraction limited diverging angle $\Delta\vartheta = \lambda/b$ in 1-D, or

---

[4]These experiments stimulated a great deal of attention on a fundamental issue: Can interference take place between two different photons? The debate was partially provoked from a statement of Dirac: "...photon... only interferes with itself. Interference between two different photons never occurs".

$\Delta\vartheta = 1.22\lambda/D$ in 2-D, where D is the diameter of the source. If the spatially coherent radiation is regarded as a wavepacket in the transverse dimension, the above condition indicates that the distance between the two pinholes must be less than the transverse width of the spatial wavepacket. In other words, the spatial separation between the fields $E(\mathbf{r}_1, t_1)$ and $E(\mathbf{r}_2, t_2)$ must be within the spatial coherence of the field, which is the transverse width of the spatial wavepacket in this case, in order to have observable interference.

4.9   Measurement of First-Order Coherence of Light

We have emphasized earlier that in the quantum theory of light, a measurement of a photon or a wavepacket cannot provide us any meaningful knowledge of physics, except a record of an photoelectron event. A time accumulative measurement, which involves the measurement of a large number of photons, is always necessary. This type of measurement is named photon counting. An idealized point-like photon counting detector and associated counting circuit, namely a photon counting device, counts how many photoelectron event occur within a chosen time interval at a chosen spatial coordinate of the photodetector. Basically, it provides a time-averaged counting rate of photoelectrons:

$$R_d = \int_T dt\, \langle\, \hat{E}^{(-)}(\mathbf{r}, t)\hat{E}^{(+)}(\mathbf{r}, t)\,\rangle, \tag{4.9.1}$$

where T is the accumulative time interval of the photon counting. In Young's double-pinhole interferometer, a photon counting device measures the time averaged first-order coherence function by means of an interference pattern:

$$R_d = \int_T dt\, \big[G^{(1)}(\mathbf{r}_1, t_1; \mathbf{r}_1, t_1) + G^{(1)}(\mathbf{r}_2, t_2; \mathbf{r}_2, t_2)$$

$$+ G^{(1)}(\mathbf{r}_1, t_1; \mathbf{r}_2, t_2) + G^{(1)}(\mathbf{r}_2, t_2; \mathbf{r}_1, t_1)\big]. \tag{4.9.2}$$

We have learned that time average and ensemble average are equivalent for stationary field. Time averaging may not have any effect on the first-order coherence function of a stationary field; however, a time average may not be trivial for time dependent effective wavefunction, or the first-order coherence function. What do we learn about the first-order coherence of a single-photon state after the time average?

Example I: Pure single-photon state

The following discussion focuses on the first-order mutual-coherence function or cross-correlation function $G^{(1)}(\mathbf{r}_1, t_1; \mathbf{r}_2, t_2)$. In the case of single-photon pure state, the first-order mutual-coherence function is a product of two wavepackets in space–time: $G^{(1)}(\mathbf{r}_1, t_1; \mathbf{r}_2, t_2) = \Psi^*(\mathbf{r}_1, t_1)\,\Psi(\mathbf{r}_2, t_2)$. In general, it cannot be written as a time independent function of $t_1 - t_2$.

The time averaged $G^{(1)}(\mathbf{r}_1, t_1; \mathbf{r}_2, t_2)$ of the single-photon pure state, which is measurable in an Young's double-pinhole interferometer, can be formally calculated as

$$\int_T dt\,\Psi^*(z_1, t_1)\,\Psi(z_2, t_2)$$

$$\simeq \int_{T\sim\infty} dt\, e^{i(\omega_0 t_1 - k_0 z_1)} \mathcal{F}^*_{\tau_1}\{f(\nu)\}\, e^{-i(\omega_0 t_2 - k_0 z_2)} \mathcal{F}_{\tau_2}\{f(\nu)\}.$$

$$= G_0^{(1)}\,\mathcal{F}_\tau\{|f(\nu)|^2\}\, e^{-i\omega_0\tau}, \tag{4.9.3}$$

where, again, $\nu = \omega - \omega_0$, ω_0 is the center frequency of the spectrum, and $\tau = [(z_2 + s_2) - (z_1 + s_1)]/c$ for arbitrary z_1 and z_2.

Substituting Eq. (4.9.3) into Eq. (4.9.2) and assuming equal chance for a photon to pass the upper and the lower pinholes, the time-averaged counting rate of the photon counting detector at a chosen coordinate is thus a function of τ:

$$R_d(\tau) = R_0\left[1 + \mathcal{F}_\tau\{|f(\nu)|^2\}\,\cos\omega_0\tau\right], \tag{4.9.4}$$

where R_0 is a constant that is calculated from the time integral of the self-correlation function. Figure 4.9.1 shows a measured result of a photon

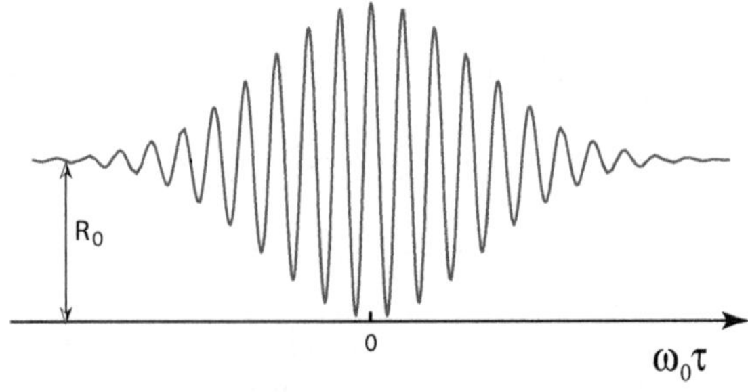

Fig. 4.9.1   Counting rate of the photon counting detector as a function of τ in Young's double-pinhole experiment. The observed interference measures the time averaged first-order coherence. In this observation, a Gaussian spectrum is assumed.

counting detector. Its time-averaged counting rate is a function of τ. This is the only and all information we can directly observe from the measurement.

Example II: Mixed single-photon state

For a mixed single-photon state, the calculation of the time-averaged first-order mutual-coherence function is straightforward:

$$
\begin{aligned}
\int_T & dt\; G^{(1)}(z_1, t_1; z_2, t_2) \\
&= \int_T dt \left\langle \langle \tilde{\Psi}| \, \hat{E}^{(-)}(z_1, t_1) \hat{E}^{(+)}(z_2, t_2) |\tilde{\Psi}\rangle \right\rangle_{En} \\
&= \int_T dt\, tr\left[\hat{\rho}\, \hat{E}^{(-)}(z_1, t_1) \hat{E}^{(+)}(z_2, t_2) \right] \\
&= \int_T dt \sum_n P_n\, G_n^{(1)}(z_1, t_1; z_2, t_2) \\
&\simeq G_0^{(1)}\, \mathcal{F}_\tau\{ |f(\nu)|^2 \}\, e^{-i\omega_0 \tau}.
\end{aligned}
\tag{4.9.5}
$$

It is clearly shown that the time averaging is equivalent to the statistical ensemble averaging in this case. Although after ensemble averaging, time averaging becomes a trivial calculation in this example, a time accumulative integral is always necessary for a photon counting measurement. The observed interference pattern is thus the same as that of Example (I)

$$
R_d(\tau) = R_0 \left[1 + \mathcal{F}_\tau\{ |f(\nu)|^2 \}\, \cos\omega_0 \tau \right].
$$

A time-averaged measurement of first-order coherence cannot distinguish a pure state from mixed state. This point has been emphasized in early chapters from the classical point of view.

Bibliography

Born M. and Wolf E., *Principle of Optics*, Cambridge, 2002.

Glauber R.J., *Phys. Rev.* **130**, 2529 (1963); *Phys. Rev.* **131**, 2766 (1963).

Loudon R., *The Quantum Theory of Light*, Oxford Science, Oxford, 2000.

Scully M.O. and Zubairy M.S., *Quantum Optics*, Cambridge University Press, Cambirdge, 1997.

Chapter 5

Second-Order Coherence of Light:
A Pair of Photons Interferes
with the Pair Itself

In the history of optical science, we were satisfied with the measurement of mean intensity $\langle I(\mathbf{r}, t)$ of light, namely the measurement of first-order coherence of light, for optical studies and observations, until the middle of 20th century when two astrophysicists Hanbury Brown and Twiss (HBT) introduced their "intensity interferometer". Unlike all other optical interferometers, HBT uses two independent photodetectors to measure the intensity correlation of light, $\langle I(\mathbf{r}_1, t_1) I(\mathbf{r}_2, t_2) \rangle$ as a function of the temporal delay or spatial separation between the two measured radiations, namely the second-order correlation or the second-order coherence of light. Whether it is named "second-order correlation" or "second-order coherence" refers to the same physical phenomenon. "Correlation" emphasizes its statistical nature, while "coherence" emphasizes its interference nature. In general, the second-order correlation function measures the probability of observing a joint photodetection at space–time coordinates ($\mathbf{r}_1, t_1$ and $\mathbf{r}_2, t_2$); the second-order coherence function reveals the visibility of an interference in which a pair of fields, photons, or groups of indistinguishable photons, interferences with the pair itself.

Figure 5.0.1 is a schematic illustration of a HBT interferometer. Comparing with Fig. 4.6.1, we can easily find the differences between a classic Young's double-slit interferometer and a HBT interferometer. The classic Young's double-slit interferometer measures the first-order coherence of fields $E(\mathbf{r}_1, t_1)$ and $E(\mathbf{r}_2, t_2)$, i.e., the first-order coherence function $\Gamma^{(1)}(\mathbf{r}_1, t_1; \mathbf{r}_2, t_2)$ or $G^{(1)}(\mathbf{r}_1, t_1; \mathbf{r}_2, t_2)$ as well as the normalized degree of first-order coherence $\gamma^{(1)}(\mathbf{r}_1, t_1; \mathbf{r}_2, t_2)$ or $g^{(1)}(\mathbf{r}_1, t_1; \mathbf{r}_2, t_2)$ by

157

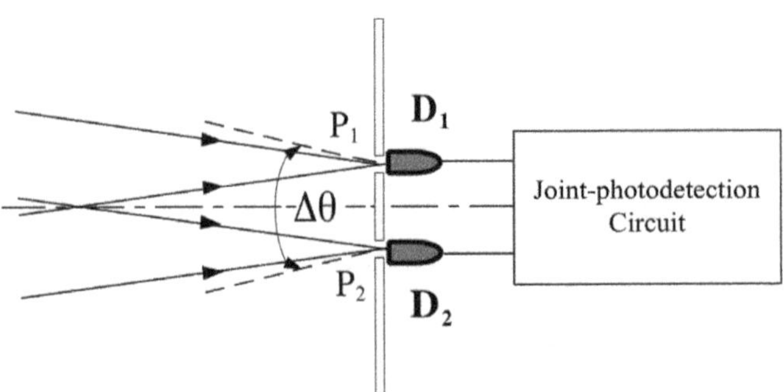

Fig. 5.0.1 Schematic setup of a HBT interferometer. The interferometer is similar to the classic Young's double-pinhole interferometer and the Michelson stellar interferometer, except two photodetectors, either in analog mode or in photon counting mode, are placed behind the pinholes P_1 and P_2, respectively, at space–time coordinates $(\mathbf{r}_1, t_1)$ and $(\mathbf{r}_2, t_2)$. The joint-photodetection integrator or counter can be either a current–current correlator (analog mode), which gives the value of $\langle i_1(t) i_2(t) \rangle \propto \langle I(\mathbf{r}_1, t_1) I(\mathbf{r}_2, t_2) \rangle \propto \langle n(\mathbf{r}_1, t_1) n(\mathbf{r}_2, t_2) \rangle$, or a photon-counting coincidence circuit (photon counting mode), which gives the value of $\langle n(\mathbf{r}_1, t_1) n(\mathbf{r}_2, t_2) \rangle \propto \langle I(\mathbf{r}_1, t_1) I(\mathbf{r}_2, t_2) \rangle$ as a function of either the temporal delay or the spatial separation of the two measured fields or photons.

means of the measurement of mean intensity (analog) or mean photon number (photon-counting) at space–time coordinate $(\mathbf{r}, t)$. In a classic Young's double-slit interferometer, the fields $E(\mathbf{r}_1, t_1)$ and $E(\mathbf{r}_2, t_2)$ are superposed and measured at a chosen space coordinate $\mathbf{r}$ at a later time t, $t = t_1 + s_1/c = t_2 + s_2/c$, by a photodetector. Neither $\Gamma^{(1)}(\mathbf{r}_1, t_1; \mathbf{r}_2, t_2)$ $[\gamma^{(1)}(\mathbf{r}_1, t_1; \mathbf{r}_2, t_2)]$ nor $G^{(1)}(\mathbf{r}_1, t_1; \mathbf{r}_2, t_2)$ $[g^{(1)}(\mathbf{r}_1, t_1; \mathbf{r}_2, t_2)]$ is directly measured at space–time points $(\mathbf{r}_1, t_1)$ and $(\mathbf{r}_2, t_2)$. A photodetection event can never happen at two different space–time coordinates. However, the second-order coherence or correlation function $\Gamma^{(2)}(\mathbf{r}_1, t_1; \mathbf{r}_2, t_2)$ that is defined and calculated based on the electromagnetic theory of light, and $G^{(2)}(\mathbf{r}_1, t_1; \mathbf{r}_2, t_2)$ that is defined and calculated based on the quantum theory of light, as well as the normalized degree of second-order coherence $\gamma^{(2)}(\mathbf{r}_1, t_1; \mathbf{r}_2, t_2)$ and $g^{(2)}(\mathbf{r}_1, t_1; \mathbf{r}_2, t_2)$, are measured directly by two independent photodetectors, either in the analog mode or in the photon counting mode, at space–time points $(\mathbf{r}_1, t_1)$ and $(\mathbf{r}_2, t_2)$, respectively, and jointly.

Historically, the study of second-order coherence or correlation started from the measurement of natural light. Natural light is radiated from natural light sources, such as the sun or distant stars. Laser was invented in the early 1960s. Laser field can be approximated as coherent state. The concept of coherent state and the study of second-order coherence of coherent state were introduced by Glauber at that time. Not until the 1980s, a biphoton state, i.e., entangled two photon state of an signal–idler photon pair generated from spontaneous parametric down-conversion (SPDC), was introduced to the measurement of Bell state and the testing of Bell inequalities. The concept of biphoton was theoretically introduced by Klyshko. The concept of biphoton interference was experimentally introduced by Alley and Shih. The second-order correlation of entangled state of SPDC is very peculiar and unusual. Its nonlocal spatial-temporal correlation, anti-correlation, and polarization correlation begun to attract people's attention. From the perspective of quantum mechanics, whether it is the second-order correlation or the second-order anti-correlation of the thermal field or the entangled state, it is the result of two-photon interference: A pair of photons, whether in a thermal state or in an entangled state, interferes with the pair itself.

In this chapter, first, we define the second-order coherence or correlation function $\Gamma^{(2)}(\mathbf{r}_1, t_1; \mathbf{r}_2, t_2)$, which is formulated from the electromagnetic theory of light, including Einstein's quantized granularity picture of light, and $G^{(2)}(\mathbf{r}_1, t_1; \mathbf{r}_2, t_2)$, which is formulated from the quantum theory of light, in the process of introducing the Hanbury Brown and Twiss interferometer (HBT interferometer). We then give a few simple calculations and discussions on the degree of second-order coherence for incoherent and coherent light as well as for entangled photon pairs. In the process of calculating $G^{(2)}(\mathbf{r}_1, t_1; \mathbf{r}_2, t_2)$, we introduce the concept of two-photon effective wavefunction $\Psi(\mathbf{r}_1, t_1; \mathbf{r}_2, t_2)$ of a pair of jointly detected photons or two groups of jointly detected indistinguishable photons and compare the two-photon effective wavefunctions of product state and entangled state. At the end of this chapter, we extend the concept of second-order coherence to Nth-order coherence $(N > 2)$ and analyze a few simple measurements on the third-order temporal and spatial coherence of thermal field.

5.1 Second-Order Coherence: Maxwell's Continuum EM-Wave Theory

Assume two analog mode photodetectors, D_1 and D_2 in Fig. 5.1.1, are used to measure the instantaneous intensities of light at $(\mathbf{r}_1, t_1)$ and $(\mathbf{r}_2, t_2)$, respectively. The output currents of the two photodetectors are amplified by two independent Radio Frequency (RF) amplifiers. An electronic linear multiplier, or RF mixer, is used to multiply the two currents $i_1(t)$ and $i_2(t)$ at time t. The output voltage of the linear multiplier is proportional to

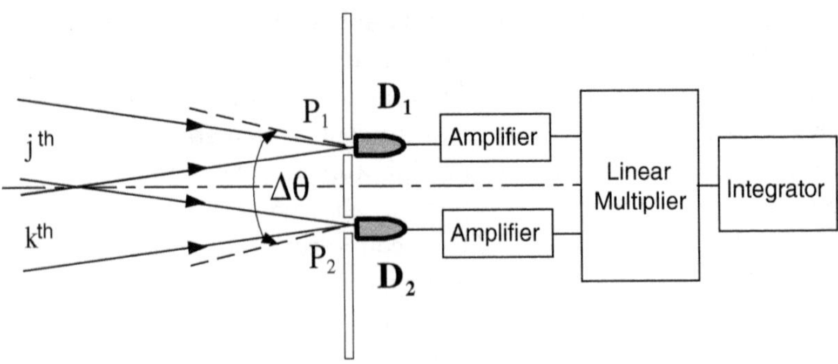

Fig. 5.1.1 Schematic setup for the measurement of second-order coherence function of $\Gamma^{(2)}(\mathbf{r}_1, t_1; \mathbf{r}_2, t_2)$ that is defined as the intensity correlation of $\langle I(\mathbf{r}_1, t_1) I(\mathbf{r}_2, t_2) \rangle$, and is proportional to the photon number correlation $\langle n(\mathbf{r}_1, t_1) n(\mathbf{r}_2, t_2) \rangle$, in terms of either temporal delay or spatial separation between the measured intensities of $I(\mathbf{r}_1, t_1) \propto n(\mathbf{r}_1, t_1)$ and $I(\mathbf{r}_2, t_2) \propto n(\mathbf{r}_2, t_2)$. Two analog photodetectors, D_1 and D_2, are placed behind the pinholes for joint-detection of the radiations. An electronic linear multiplier, or RF mixer, is followed to multiply the two currents $i_1(t)$ and $i_2(t)$ at time t for joint measurement of D_1 and D_2 and gives the value of $\langle i_1(t) i_2(t) \rangle \propto \langle I(\mathbf{r}_1, t_1) I(\mathbf{r}_2, t_2) \rangle \propto \langle n(\mathbf{r}_1, t_1) n(\mathbf{r}_2, t_2) \rangle$.

$i_1(t) \times i_2(t)$, which is proportional to the jointly measured instantaneous intensities, $I(\mathbf{r}_1, t_1) \times I(\mathbf{r}_1, t_1)$, by the two analog photodetectors:

$$V_{12}(t) \propto i_1(t) \times i_2(t)$$

$$\propto I(\mathbf{r}_1, t_1) I(\mathbf{r}_2, t_2)$$

$$= E^*(\mathbf{r}_1, t_1) E(\mathbf{r}_1, t_1) E^*(\mathbf{r}_2, t_2) E(\mathbf{r}_2, t_2), \qquad (5.1.1)$$

where $V_{12}(t)$ is the output voltage of the linear multiplier, $i_1(t)$ and $i_2(t)$ are the electronically amplified photocurrent of D_1 and D_2, respectively, at time t of the multiplication, and $t_1 = t - \tau_1^e$, $t_2 = t - \tau_2^e$ are the early times that are defined by the electronic time delays τ_1^e and τ_2^e, including the delays of the detectors, the amplifiers, and the adjustable delay-line cables. The above statements have assumed an idealized correlation measurement by neglecting "unavoidable" time averages on t_1, t_2, and t: implying *instantaneous* responses of the photodetectors and the associated electronics, including the linear multiplier. The instantaneous intensities $I(\mathbf{r}_1, t_1)$ and $I(\mathbf{r}_2, t_2)$ are identified by the electronic delays. Note that t_1 and t_2 of the two photodetection events as well as the relative time delay $t_1 - t_2 = \tau_1^e - \tau_2^e$ are all defined by the electronics in this setup.

The second-order coherence or correlation function $\Gamma^{(2)}(\mathbf{r}_1, t_1; \mathbf{r}_2, t_2)$, in terms of either temporal delay or spatial separation between the measured intensities of $I(\mathbf{r}_1, t_1)$ and $I(\mathbf{r}_2, t_2)$, is defined as follows in the the electromagnetic field theory of light:

$$\Gamma^{(2)}(\mathbf{r}_1, t_1; \mathbf{r}_2, t_2) = \langle I(\mathbf{r}_1, t_1)\, I(\mathbf{r}_2, t_2) \rangle$$

$$= \langle E^*(\mathbf{r}_1, t_1)E(\mathbf{r}_1, t_1)E^*(\mathbf{r}_2, t_2)E(\mathbf{r}_2, t_2) \rangle, \qquad (5.1.2)$$

and the normalized degree of second-order coherence is defined as

$$\gamma^{(2)}(\mathbf{r}_1, t_1; \mathbf{r}_2, t_2) = \frac{\langle E^*(\mathbf{r}_1, t_1)E(\mathbf{r}_1, t_1)\, E^*(\mathbf{r}_2, t_2)E(\mathbf{r}_2, t_2) \rangle}{\langle E^*(\mathbf{r}_1, t_1)E(\mathbf{r}_1, t_1) \rangle \langle E^*(\mathbf{r}_2, t_2)E(\mathbf{r}_2, t_2) \rangle}, \qquad (5.1.3)$$

where the expectation or ensemble average $\langle \ldots \rangle$ denotes, again, *taking into account all possible realizations of the field.*

We now introduce the concept of second-order correlation and coherence through a classic calculation of $\Gamma^{(2)}(\mathbf{r}_1, t_1; \mathbf{r}_2, t_2)$ for thermal radiation in Maxwell's continuous picture of light. To simplify the mathematics, we consider a 1-D joint measurement of two point-like photodetectors D_1 and D_2, and a point-like thermal source:

$$\Gamma^{(2)}(r_1, t_1; r_2, t_2)$$

$$= \left\langle \int d\omega\, E^*(\omega)\, e^{i\omega\tau_1} \int d\omega'\, E(\omega')\, e^{-i\omega'\tau_1} \right.$$

$$\left. \times \int d\omega''\, E^*(\omega'')\, e^{i\omega''\tau_2} \int d\omega'''\, E(\omega''')\, e^{-i\omega'''\tau_2} \right\rangle$$

$$= \left\langle \int d\omega d\omega'\, d\omega''\, d\omega'''\, E_1^*(\omega)E(\omega')E^*(\omega'')E_2(\omega''')e^{i(\omega-\omega')\tau_1}\, e^{i(\omega''-\omega''')\tau_2} \right\rangle$$

$$= \int_{\omega=\omega'} d\omega\, |E(\omega)|^2 \int_{\omega''=\omega'''} d\omega''\, |E(\omega'')|^2$$

$$+ \int_{\omega=\omega'''} d\omega\, E^*(\omega)E(\omega)\, e^{i\omega(\tau_1-\tau_2)} \int_{\omega'=\omega''} d\omega'\, E^*(\omega')E(\omega')\, e^{-i\omega'(\tau_1-\tau_2)}$$

$$= \Gamma_{11}^{(1)}\Gamma_{22}^{(1)} + \Gamma_{12}^{(1)}\Gamma_{21}^{(1)}, \qquad (5.1.4)$$

where the ensemble average has taken account all possible realizations of the fields, i.e., all possible random phases of the plane waves. Note, in the last lines of Eq. (5.1.4), we connect the survival terms from the ensemble average to the first-order coherence functions of the field:

$$\Gamma_{11}^{(1)} \equiv \int_{\omega=\omega'} d\omega \, |E(\omega)|^2 = \langle I_1 \rangle = \Gamma_0,$$

$$\Gamma_{22}^{(1)} \equiv \int_{\omega''=\omega'''} d\omega'' \, |E(\omega'')|^2 = \langle I_2 \rangle = \Gamma_0,$$

$$\Gamma_{12}^{(1)} \equiv \int_{\omega=\omega'''} d\omega \, E^*(\omega) E(\omega) \, e^{i\omega(\tau_1 - \tau_2)}$$

$$\simeq \Gamma_0 \, \mathcal{F}_\tau \{ a^2(\nu) \} e^{i\omega_0 \tau},$$

$$\Gamma_{21}^{(1)} \equiv \int_{\omega'=\omega''} d\omega' \, E^*(\omega') E(\omega') \, e^{-i\omega'(\tau_1 - \tau_2)}$$

$$\simeq \Gamma_0 \, \mathcal{F}_\tau^* \{ a^2(\nu) \} e^{-i\omega_0 \tau}, \tag{5.1.5}$$

where Γ_0 is a constant and $\tau \equiv \tau_1 - \tau_2$. The second-order temporal correlation function of thermal field is thus approximated as follows:

$$\gamma^{(2)}(r_1, t_1; r_2, t_2) \simeq 1 + \left| \mathcal{F}_\tau \{ a^2(\nu) \} \right|^2 = \gamma^{(2)}(\tau). \tag{5.1.6}$$

Since the second-order correlation function of thermal field is a function of $\tau \equiv \tau_1 - \tau_2$, $\gamma^{(2)}(r_1, t_1; r_2, t_2)$ is measurable accumulatively in time. This property is important for the measurement of weak light, especially at single-photon level.

(I) $\Gamma^{(2)}(r_1, t_1; r_2, t_2)$: A measure of the second-order correlation of radiation fields

We first discuss the statistical correlation nature of $\Gamma^{(2)}(r_1, t_1; r_2, t_2)$. What do we observe if there is no correlation between the two measured intensities at (r_1, t_1) and (r_2, t_2)? Obviously, we should have

$$\langle I(r_1, t_1) I((r_2, t_2)) \rangle = \langle I(r_1, t_1) \rangle \langle I((r_2, t_2)) \rangle = \Gamma_{11}^{(1)} \Gamma_{22}^{(1)}, \tag{5.1.7}$$

which is usually called trivial second-order correlation. It is easy to find from Eq. (5.1.4) that the nontrivial correlation, if there is any, must be produced by the second term: $\Gamma_{12}^{(1)} \Gamma_{21}^{(1)} = |\Gamma_{12}^{(1)}|^2$. The value of $|\Gamma_{12}^{(1)}|^2$ for different temporal delays can be easily estimated from the Fourier transform:

$$\gamma^{(2)}(\tau) = \begin{cases} 1 & |\tau| \geqslant 2\pi/\Delta\omega \\ 1 \text{ to } 2 & |\tau| < 2\pi/\Delta\omega \, . \\ 2 & \tau = 0 \end{cases}$$

We thus have the following conclusions:

(1) When $|\tau| \geqslant 2\pi/\Delta\omega$, the second term $\Gamma_{12}^{(1)}\Gamma_{21}^{(1)} = |\Gamma_{12}^{(1)}|^2$ in Eq. (5.1.4) vanishes. $\Gamma^{(2)}(r_1, t_1; r_2, t_2) = \Gamma_{11}^{(1)}\Gamma_{22}^{(1)}$, corresponding to the product of two mean values of the measured intensities at (r_1, t_1) and (r_2, t_2). In this case, $\gamma^{(2)}(\tau)$ takes the value $\gamma^{(2)}(\tau) = 1$. Statistically, a factorizable $\Gamma^{(2)}(\mathbf{r}_1, t_1; \mathbf{r}_2, t_2)$ with $\gamma^{(2)} = 1$ means the two measured intensities $I(\mathbf{r}_1, t_1)$ and $I(\mathbf{r}_2, t_2)$ are independent with no correlation. The radiation fields measured at (r_1, t_1) and (r_2, t_2) are considered second-order uncorrelated fields.

(2) When $\tau < 2\pi/\Delta\omega$, the second term $\Gamma_{12}^{(1)}\Gamma_{21}^{(1)} = |\Gamma_{12}^{(1)}|^2$ in Eq. (5.1.4) survives. $\Gamma^{(2)}(r_1, t_1; r_2, t_2)$ is no longer the product of the two mean values of the measured intensities at (r_1, t_1) and (r_2, t_2). In this case, $\gamma^{(2)}(\tau)$ takes the value $2 > \gamma^{(2)}(\tau) > 1$. Statistically, a nonfactorizable $\Gamma^{(2)}(\mathbf{r}_1, t_1; \mathbf{r}_2, t_2)$ with $2 > \gamma^{(2)} > 1$ means that the two measured intensities $I(\mathbf{r}_1, t_1)$ and $I(\mathbf{r}_2, t_2)$ are no longer independent of each other. The radiation fields measured at (r_1, t_1) and (r_2, t_2) are considered second-order partially correlated fields.

(3) When $\tau = 0$, the second term $\Gamma_{12}^{(1)}\Gamma_{21}^{(1)} = |\Gamma_{12}^{(1)}|^2$ in Eq. (5.1.4) reaches its maximum value of one. In this case, $\gamma^{(2)}(\tau)$ takes the value $\gamma^{(2)}(\tau) = 2$. Statistically, this means the two measured intensities $I(\mathbf{r}_1, t_1)$ and $I(\mathbf{r}_2, t_2)$ are correlated. The radiation fields measured at (r_1, t_1) and (r_2, t_2) are named second-order correlated fields.

The trivial second-order correlation is thus define with

$$\gamma^{(2)}(\mathbf{r}_1, t_1; \mathbf{r}_2, t_2) = 1. \tag{5.1.8}$$

The nontrivial second-order correlation and anti-correlation, respectively, are defined with

$$\gamma^{(2)}(\mathbf{r}_1, t_1; \mathbf{r}_2, t_2) > 1 \tag{5.1.9}$$

and

$$\gamma^{(2)}(\mathbf{r}_1, t_1; \mathbf{r}_2, t_2) < 1. \tag{5.1.10}$$

Traditionally, the nontrivial contribution of $\Gamma_{12}^{(1)}\Gamma_{21}^{(1)}$ to the second-order correlation is considered to be the "intensity fluctuation correlation" of the measured radiation, since $I(\mathbf{r}_j, t_j) = \bar{I}(\mathbf{r}_j, t_j) + \Delta I(\mathbf{r}_j, t_j)$,

$$\begin{aligned}
\Gamma^{(2)}(\mathbf{r}_1, t_1; \mathbf{r}_2, t_2) &= \langle [\bar{I}(\mathbf{r}_1, t_1) + \Delta I(\mathbf{r}_1, t_1)] \, [\bar{I}(\mathbf{r}_2, t_2) + \Delta I(\mathbf{r}_2, t_2)] \rangle \\
&= \langle I(\mathbf{r}_1, t_1) \rangle \langle I(\mathbf{r}_2, t_2) \rangle + \langle \Delta I(\mathbf{r}_1, t_1 \Delta I(\mathbf{r}_2, t_2) \rangle \\
&= \Gamma_{11}^{(1)}\Gamma_{22}^{(1)} + \Gamma_{12}^{(1)}\Gamma_{21}^{(1)}.
\end{aligned} \tag{5.1.11}$$

A question naturally arises: Where does the intensity fluctuation correlation come from? Conventional theory considers that the intensity fluctuation correlation is caused by the intrinsic statistical fluctuations of the radiation field carried from the light source. Thermal fields have historically been called "correlated fields", perhaps with "photon bunching". Obviously this is incorrect, since we all know that thermal fields are created randomly in a stochastic process. After 70 years of pondering since HBT, we now have enough reason to conclude that the observed intensity fluctuation correlations are the result of second-order interference. In other words, $\Gamma^{(2)}(r_1, t_1; r_2, t_2)$ is a measure of the second-order coherence of the radiation field.

(II) $\Gamma^{(2)}(r_1, t_1; r_2, t_2)$: A measure of the second-order coherence of radiation fields

Next, we answer the following question: Why do we also name $\Gamma^{(2)}(r_1, r_1; r_2, r_2)$ as the second-order coherence function? To answer this question, we rewrite Eq. (5.1.4) as follows:

$$
\Gamma^{(2)}(r_1, t_1; r_2, t_2)
$$

$$
= \left\langle \int d\omega\, E^*(\omega)\, e^{i\omega\tau_1} \int d\omega'\, E(\omega')\, e^{-i\omega'\tau_1} \right.
$$

$$
\left. \times \int d\omega''\, E^*(\omega'')\, e^{i\omega''\tau_2} \int d\omega'''\, E(\omega''')\, e^{-i\omega'''\tau_2} \right\rangle
$$

$$
= \left\langle \int d\omega d\omega'\, d\omega''\, d\omega'''\, E^*(\omega)E(\omega')E^*(\omega'')E(\omega''')e^{i(\omega-\omega')\tau_1}\, e^{i(\omega''-\omega''')\tau_2} \right\rangle
$$

$$
= \int_{\omega=\omega'} d\omega\, |E(\omega)|^2 \int_{\omega''=\omega'''} d\omega''\, |E(\omega'')|^2
$$

$$
+ \int_{\omega=\omega'''} d\omega\, E^*(\omega)E(\omega)\, e^{i\omega(\tau_1-\tau_2)} \int_{\omega'=\omega''} d\omega'\, E^*(\omega')E(\omega')\, e^{-i\omega'(\tau_1-\tau_2)}
$$

$$
= \int d\omega\, d\omega'\, \frac{1}{2}\left| E(\omega)e^{-i\omega\tau_1} E(\omega')e^{-i\omega'\tau_2} + E(\omega)e^{-i\omega\tau_2} E(\omega')e^{-i\omega'\tau_1} \right|^2.
$$

$$
(5.1.12)
$$

The last line of Eq. (5.1.12) contains the sum of a large number of superpositions between two possible indistinguishable measurements: (1) radiation of frequency ω is detected at coordinate (r_1, t_1) while radiation of frequency ω' is detected at coordinate (r_2, t_2); (2) radiation of frequency ω is detected at coordinate (r_2, t_2) while radiation of frequency ω' is detected at coordinate (r_1, t_1), indicating an interference phenomenon — a pair of plane waves interferences with the pair itself. The cross-interference term is connected to the above estimated first-order coherence functions $\Gamma_{12}^{(1)}$ and $\Gamma_{21}^{(1)}$:

$$\int d\omega \, d\omega' \, E^*(\omega)e^{i\omega\tau_1} E^*(\omega')e^{i\omega'\tau_2} E(\omega)e^{-i\omega\tau_2} E(\omega')e^{-i\omega'\tau_1}$$

$$= \left[\int d\omega \, E^*(\omega)E(\omega) \, e^{i\omega(\tau_1-\tau_2)} \right] \left[\int d\omega' \, E^*(\omega')E(\omega') \, e^{-i\omega'(\tau_1-\tau_2)} \right]$$

$$= \Gamma_{12}^{(1)}\Gamma_{21}^{(1)}. \tag{5.1.13}$$

It is easy to estimate the contribution of the second-order interference, if there is any, must be produced by the cross-interference term of Eq. (5.1.12), $\Gamma_{12}^{(1)}\Gamma_{21}^{(1)} = |\Gamma_{12}^{(1)}|^2$, which can be easily calculated from the Fourier transform:

$$\gamma^{(2)}(\tau) = \begin{cases} 1 & |\tau| \geqslant 2\pi/\Delta\omega \\ 1 \text{ to } 2 & |\tau| < 2\pi/\Delta\omega \\ 2 & \tau = 0 \end{cases}.$$

We thus have the following conclusions:

(1) When $|\tau| \geqslant 2\pi/\Delta\omega$, the cross-interference term $\Gamma_{12}^{(1)}\Gamma_{21}^{(1)} = |\Gamma_{12}^{(1)}|^2$ in Eq. (5.1.12) vanishes. In this case, $\gamma^{(2)}(\tau)$ takes the value $\gamma^{(2)}(\tau) = 1$. From the view point of interference, $\gamma^{(2)} = 1$ means no observable second-order interference. The radiation fields measured at (r_1, t_1) and (r_2, t_2) are considered second-order incoherent.

(2) When $\tau < 2\pi/\Delta\omega$, the cross-interference term $\Gamma_{12}^{(1)}\Gamma_{21}^{(1)} = |\Gamma_{12}^{(1)}|^2$ in Eq. (5.1.12) survives. In this case, $\gamma^{(2)}(\tau)$ takes the value $2 > \gamma^{(2)}(\tau) > 1$. From the view point of interference, $2 > \gamma^{(2)} > 1$ means the measured radiation fields measured at (r_1, t_1) and (r_2, t_2) are second-order partially coherent.

(3) When $\tau = 0$, the cross-interference term $\Gamma_{12}^{(1)}\Gamma_{21}^{(1)} = |\Gamma_{12}^{(1)}|^2$ in Eq. (5.1.12) reaches its maximum value of one. In this case, $\gamma^{(2)}(\tau)$ takes the value $\gamma^{(2)}(\tau) = 2$. From the view point of interference, $\gamma^{(2)}(\tau) = 2$ means the radiation fields measured at (r_1, t_1) and (r_2, t_2) are second-order correlated coherent.

The trivial second-order coherence is thus define with

$$\gamma^{(2)}(\mathbf{r}_1, t_1; \mathbf{r}_2, t_2) = 1. \tag{5.1.14}$$

The nontrivial second-order constructive interference is specified with

$$\gamma^{(2)}(\mathbf{r}_1, t_1; \mathbf{r}_2, t_2) > 1 \tag{5.1.15}$$

and the second-order destructive interference is specified with

$$\gamma^{(2)}(\mathbf{r}_1, t_1; \mathbf{r}_2, t_2) < 1. \tag{5.1.16}$$

We thus have enough reason to name $\Gamma^{(2)}(\mathbf{r}_1, t_1; \mathbf{r}_2, t_2)$ the second-order coherence function.

From the above analysis, we can see that the nontrivial second-order correlation of intensity fluctuations is the cross-interference term of the second-order interference:

$$\langle \Delta I(\mathbf{r}_1, t_1 \Delta I(\mathbf{r}_2, t_2) \rangle = \Gamma_{12}^{(1)} \Gamma_{21}^{(1)}$$

$$= \left[\int d\omega\, E^*(\omega) E(\omega)\, e^{i\omega(\tau_1 - \tau_2)} \right] \left[\int d\omega'\, E^*(\omega') E(\omega')\, e^{-i\omega'(\tau_1 - \tau_2)} \right].$$

$$\tag{5.1.17}$$

The intensity fluctuation correlation observed in the joint-photodetection of D_1 and D_2 is indeed the result of an interference in which a pair of Maxwell plane waves interfering with the pair itself. We name this interference second-order interference. We conclude that $\Gamma^{(2)}(r_1, t_1; r_2, t_2)$ is a measure of the second-order coherence of radiation fields, and it is often said that the nontrivial second-order correlation is measured from the intensity fluctuations, however, the intensity fluctuation correlation is caused by the second-order interference.

5.2 Second-Order Coherence: Einstein's Granularity Picture

$\Gamma^{(2)}(\mathbf{r}_1, t_1; \mathbf{r}_2, t_2)$ is often called the "classical second-order coherence function", although it is not limited to the classical Maxwell radiation continuum picture. Einstein's light granularity picture is also widely used to calculate $\Gamma^{(2)}(\mathbf{r}_1, t_1; \mathbf{r}_2, t_2)$. Perhaps it is not appropriate to name it "classical" in the context of Einstein's granularity picture.

In the following, we evaluate the second-order correlation or coherence function $\Gamma^{(2)}(\mathbf{r}_1, t_1; \mathbf{r}_2, t_2)$ of thermal radiation in Einstein's granularity picture. Assuming a light source of finite size that contains a large number of independent and randomly radiating point-like sub-sources, such as millions of spontaneous atomic transitions, the radiation fields $E(\mathbf{r}_1, t_1)$ and $E(\mathbf{r}_2, t_2)$, measured at photodetectors D_1 and D_2, respectively, are the results of superposition among a large number of subfields, labeled by $E_m(\mathbf{r}_{0m}, t_{0m})$, originated from each of these independent sub-sources. The second-order correlation or coherence function $\Gamma^{(2)}(\mathbf{r}_1, t_1; \mathbf{r}_2, t_2)$ measured by photodetectors D_1 and D_2, jointly, is therefore

$$
\begin{aligned}
\Gamma^{(2)}&(\mathbf{r}_1, t_1; \mathbf{r}_2, t_2) \\
&= \left\langle E^*(\mathbf{r}_1, t_1)E(\mathbf{r}_1, t_1)E^*(\mathbf{r}_2, t_2)E(\mathbf{r}_2, t_2) \right\rangle \\
&= \left\langle \sum_{m,n,p,q} E_m^*(\mathbf{r}_1, t_1)\, E_n(\mathbf{r}_1, t_1)E_p^*(\mathbf{r}_2, t_2)\, E_q(\mathbf{r}_2, t_2) \right\rangle \\
&= \sum_m E_m^*(\mathbf{r}_1, t_1)\, E_m(\mathbf{r}_1, t_1) \sum_n E_n^*(\mathbf{r}_2, t_2)\, E_n(\mathbf{r}_2, t_2) \\
&\quad + \sum_{m \neq n} E_m^*(\mathbf{r}_1, t_1)\, E_n(\mathbf{r}_1, t_1)\, E_n^*(\mathbf{r}_2, t_2)\, E_m(\mathbf{r}_2, t_2) \\
&= \sum_{m,n} \left| \frac{1}{\sqrt{2}} \left[E_m(\mathbf{r}_1, t_1)E_n(\mathbf{r}, t_2) + E_n(\mathbf{r}_1, t_1)E_m(\mathbf{r}_2, t_2) \right] \right|^2 ,
\end{aligned}
\tag{5.2.1}
$$

where m, n, p, q label the randomly created and randomly distributed subfields and their corresponding point-like sub-sources. Considering the random relative phases of the subfields, as the result of an interference cancelation when taking into account all possible random phases of the subfields, the only surviving terms from the ensemble sum are the following: (1) $m = q$ and $n = p$, (2) $m = p$ and $n = q$. Interestingly, in Einstein's granularity picture, the second-order correlation function $\Gamma^{(2)}(\mathbf{r}_1, t_1; \mathbf{r}_2, t_2)$ of thermal field is the sum of the following set of superpositions or interferences:

$$
\Gamma_{mn}^{(2)}(\mathbf{r}_1, t_1; \mathbf{r}_2, t_2) \equiv \left| \frac{1}{\sqrt{2}} \left[E_m(\mathbf{r}_1, t_1)E_n(\mathbf{r}, t_2) + E_n(\mathbf{r}_1, t_1)E_m(\mathbf{r}_2, t_2) \right] \right|^2
$$

$$
\tag{5.2.2}
$$

corresponding to two different yet indistinguishable alternatives for a pair of randomly created and randomly paired subfields, or photons, to produce a joint-photodetection event: (1) The mth subfield, or photon, produces a photodetection event at D_1, while the nth subfield, or photon, produces a photodetection event at D_2; (2) the nth subfield, or photon, produces a photodetection event at D_1, while the mth subfield, or photon, produces a photodetection event at D_2, namely, a two-photon interference — a randomly created and randomly paired subfields (photons) interfering with the pair itself.

On the other hand, we find

$$\langle I(\mathbf{r}_1, t_1)\rangle = \sum_m E_m^*(\mathbf{r}_1, t_1)\, E_m(\mathbf{r}_1, t_1),$$

$$\langle I(\mathbf{r}_2, t_2)\rangle = \sum_n E_n^*(\mathbf{r}_2, t_2)\, E_n(\mathbf{r}_2, t_2), \tag{5.2.3}$$

and

$$\langle \Delta I(\mathbf{r}_1, t_1)\Delta I(\mathbf{r}_2, t_2)\rangle$$

$$= \left\langle \sum_{m\neq n} E_m^*(\mathbf{r}_1, t_1)\, E_n(\mathbf{r}_1, t_1) \sum_{p\neq q} E_p^*(\mathbf{r}_2, t_2)\, E_q(\mathbf{r}_2, t_2) \right\rangle$$

$$= \sum_{m\neq n} E_m^*(\mathbf{r}_1, t_1)\, E_n(\mathbf{r}_1, t_1)\, E_n^*(\mathbf{r}_2, t_2)\, E_m(\mathbf{r}_2, t_2). \tag{5.2.4}$$

We thus write $\Gamma^{(2)}(\mathbf{r}_1, t_1; \mathbf{r}_2, t_2)$ as

$$\Gamma^{(2)}(\mathbf{r}_1, t_1; \mathbf{r}_2, t_2) = \langle I(\mathbf{r}_1, t_1)\rangle\langle I(\mathbf{r}_2, t_2)\rangle + \langle \Delta I(\mathbf{r}_1, t_1)\Delta I(\mathbf{r}_2, t_2)\rangle. \tag{5.2.5}$$

The first terms of Eq. (5.2.5) is a product of two mean intensities measured by D_1 and D_2, respectively. The second term of Eq. (5.2.5) is the intensity fluctuation correlation measured by D_1 and D_2, jointly. For thermal radiations, the first term usually contributes a constant to $\Gamma^{(2)}(\mathbf{r}_1, t_1; \mathbf{r}_2, t_2)$, and the second term contributes a nontrivial function of space–time coordinates $(\mathbf{r}_1, t_1)$ and $(\mathbf{r}_2, t_2)$ to $\Gamma^{(2)}(\mathbf{r}_1, t_1; \mathbf{r}_2, t_2)$, in terms of either temporal delay or spatial separation between the two measurements.

The nontrivial second-order coherence function of thermal field is observable from the correlation measurement of intensity fluctuations. What is the cause of this correlation of intensity fluctuations at distance? In Einstein's picture, the correlation of intensity fluctuations is caused by two-photon interference. Mathematically, $\langle \Delta I(\mathbf{r}_1, t_1)\Delta I(\mathbf{r}_2, t_2)\rangle$ corresponds to the cross-interference term of the two-photon superposition in Eqs. (5.2.1) and (5.2.2). The intensity fluctuation correlation is a coherent phenomenon. We thus name it two-photon interference induced intensity fluctuation correlation. In terms of the measurement, the correlation is observed from

$$\Delta I(\mathbf{r}_1, t_1) = \sum_{m\neq n} E_m^*(\mathbf{r}_1, t_1)E_n(\mathbf{r}_1, t_1),$$

which is measured by D_1 and

$$\Delta I(\mathbf{r}_2, t_2) = \sum_{p \neq q} E_p^*(\mathbf{r}_2, t_2) E_q(\mathbf{r}_2, t_2),$$

which is measured by D_2. However, only the $m = p$ and $n = q$ terms survive in the correlation measurement, due to nonlocal two-photon constructive-destructive interferences.

In an idealized measurement, assuming two idealized photodetectors and an idealized correlation measurement circuit, all intensity fluctuations measured by D_1 and D_2 contribute to the nontrivial correlation of $\Gamma^{(2)}(\mathbf{r}_1, t_1; \mathbf{r}_2, t_2)$. $\langle \Delta I(\mathbf{r}_1, t_1) \Delta I(\mathbf{r}_2, t_2) \rangle$ of thermal field is usually written in terms of the first-order coherence functions:

$$\langle \Delta I(\mathbf{r}_1, t_1) \Delta I(\mathbf{r}_2, t_2) \rangle$$
$$\simeq \sum_m E_m^*(\mathbf{r}_1, t_1) E_m(\mathbf{r}_2, t_2) \sum_n E_n^*(\mathbf{r}, t_2) E_n(\mathbf{r}_1, t_1)$$
$$= \Gamma^{(1)}(\mathbf{r}_1, t_1; \mathbf{r}_2, t_2) \Gamma^{(1)}(\mathbf{r}_2, t_2; \mathbf{r}_1, t_1). \tag{5.2.6}$$

$\Gamma^{(2)}(\mathbf{r}_1, t_1; \mathbf{r}_2, t_2)$ of thermal field is also formally written in terms of the first-order coherence functions:

$$\Gamma^{(2)}(\mathbf{r}_1, t_1; \mathbf{r}_2, t_2) = \Gamma_{11}^{(1)} \Gamma_{22}^{(1)} + \Gamma_{12}^{(1)} \Gamma_{21}^{(1)} = \Gamma_{11}^{(1)} \Gamma_{22}^{(1)} + |\Gamma_{12}^{(1)}|^2, \tag{5.2.7}$$

where we have written $\Gamma^{(1)}(\mathbf{r}_j, t_j; \mathbf{r}_k, t_k)$, $j = 1, 2; k = 1, 2$, with the short-hand notations $\Gamma_{jk}^{(1)}$:

$$\Gamma_{11}^{(1)} = \sum_m E_m^*(\mathbf{r}_1, t_1) E_m(\mathbf{r}_1, t_1), \quad \Gamma_{22}^{(1)} = \sum_n E_n^*(\mathbf{r}_2, t_2) E_n(\mathbf{r}_2, t_2),$$

$$\Gamma_{12}^{(1)} = \sum_m E_m^*(\mathbf{r}_1, t_1) E_m(\mathbf{r}_2, t_2), \quad \Gamma_{21}^{(1)} = \sum_n E_n^*(\mathbf{r}_2, t_2) E_n(\mathbf{r}_1, t_1). \tag{5.2.8}$$

All four $\Gamma_{ij}^{(1)}$s of thermal radiation have been calculated in Chapter 4. We recognize that $\Gamma_{12}^{(1)} \Gamma_{21}^{(1)} = |\Gamma_{12}^{(1)}|^2$ contributes a nontrivial function of $\Gamma^{(2)}(\tau)$ to $\Gamma^{(2)}(\mathbf{r}_1, t_1; \mathbf{r}_2, t_2)$, while $\Gamma_{11}^{(1)} \Gamma_{22}^{(1)}$ contributes a constant to $\Gamma^{(2)}(\mathbf{r}_1, t_1; \mathbf{r}_2, t_2)$. Under the condition of perfect correlation measurement, mathematically, the second-order coherence function of thermal field can be obtained from the first-order coherent functions.

It should be emphasized that, here, $\Gamma_{12}^{(1)} \Gamma_{21}^{(1)}$ is measured by two independent photodetectors D_1 and D_2 at different space–time coordinates

$(\mathbf{r}_1, t_1)$ and $(\mathbf{r}_2, t_2)$. Although Eq. (5.2.7) is helpful for calculations, we should keep in mind (1) the first-order coherence functions $\Gamma_{12}^{(1)}$ and $\Gamma_{21}^{(1)}$ are the results of first-order interference, i.e., a quantized wavepacket, or photon, interference with the quantized wavepacket, or photon, itself. The second-order coherence function, or correlation, is the result of two-photon interference, i.e., a pair of quantized wavepackets, or photons, interference with the pair itself. (2) $\Gamma_{12}^{(1)}$ and $\Gamma_{21}^{(1)}$ are measured differently in the first-order coherence function that was defined in Chapter 4, especially when taking into account nonidealized correlation measurement devices due to the "slow" responses of the electronics. We emphasize this point again in later discussions.

(I) Second-order temporal coherence of natural light

In the following, we calculate the second-order temporal coherence of natural light in Einstein's granularity picture. Consider a natural point-like radiation source, such as a distant star, which consists of a large number of randomly radiated and randomly distributed sub-sources, such as millions of randomly radiated spontaneous atomic transitions.

First, we treat each subfield, or quantized bundle of ray, as a single mode light quantum with energy $\hbar\omega$ similar to what Einstein did in 1905. This approximation is reasonable since the spectrum of each wavepacket, or subfield, is much narrower than the broad spectrum of the star radiation. In this simple approximation, the indexes of m and n can be replaced by ω and ω'. Similar to Eq. (5.2.1), the second-order coherence of natural light is calculated as follows:

$$
\Gamma^{(2)}(r_1, t_1; r_2, t_2)
$$

$$
= \left\langle \sum_{\omega} E^*(\omega)\, e^{i\omega\tau_1} \sum_{\omega'} E(\omega')\, e^{-i\omega'\tau_1} \sum_{\omega''} E^*(\omega'')\, e^{i\omega''\tau_2} \right.
$$

$$
\left. \times \sum_{\omega'''} E(\omega''')\, e^{-i\omega'''\tau_2} \right\rangle
$$

$$
= \sum_{\omega=\omega'} |E(\omega)|^2 \sum_{\omega''=\omega'''} |E(\omega'')|^2
$$

$$
+ \sum_{\omega=\omega'''} E^*(\omega)E(\omega)\, e^{i\omega(\tau_1-\tau_2)} \sum_{\omega'=\omega''} E^*(\omega')E(\omega')\, e^{-i\omega'(\tau_1-\tau_2)}
$$

$$
= \left\langle \sum_{\omega,\omega'} \left| \frac{1}{\sqrt{2}}\left[E(\omega)\, e^{i\omega\tau_1}\, E(\omega')\, e^{i\omega'\tau_2} + E(\omega)\, e^{i\omega\tau_2}\, E(\omega')\, e^{i\omega'\tau_1} \right] \right|^2 \right\rangle
$$

$$
\tag{5.2.9}
$$

corresponding to two different yet indistinguishable alternatives for a pair of randomly created and randomly paired subfields, or photons with energy $\hbar\omega$ and $\hbar\omega'$ to produce a joint-photodetection event: (1) the light quantum of $\hbar\omega$ produces a photodetection event at D_1, while the light quantum of $\hbar\omega'$ produces a photodetection event at D_2; (2) the light quantum of $\hbar\omega'$ produces a photodetection event at D_1, while the light quantum of $\hbar\omega$ produces a photodetection event at D_2, namely two-photon interference — a randomly created and randomly paired light quanta (photons) interfering with the pair itself. The above calculated second-order coherence function $\Gamma^{(2)}(r_1, t_1; r_2, t_2)$ is usually written in terms of the first-order coherence functions:

$$\Gamma^{(2)}(r_1, t_1; r_2, t_2)$$

$$= \sum_{\omega=\omega'} |E(\omega)|^2 \sum_{\omega''=\omega'''} |E(\omega'')|^2$$

$$+ \sum_{\omega=\omega'''} E^*(\omega)E(\omega)\, e^{i\omega(\tau_1-\tau_2)} \sum_{\omega'=\omega''} E^*(\omega')E(\omega')\, e^{-i\omega'(\tau_1-\tau_2)}$$

$$= \Gamma_{11}^{(1)}(0)\Gamma_{22}^{(1)}(0) + \Gamma_{12}^{(1)}(\tau)\Gamma_{21}^{(1)}(\tau), \tag{5.2.10}$$

where $\tau = \tau_1 - \tau_2$. In idealized correlation measurements,

$$\langle \Delta I(\tau_1)\Delta I(\tau_2)\rangle = \Gamma_{12}^{(1)}(\tau)\Gamma_{21}^{(1)}(\tau) = |\Gamma_{12}^{(1)}(\tau)|^2. \tag{5.2.11}$$

Next, we consider a more complicated model where the quantized bundle of rays or subfields are viewed as wavepackets of finite spectral width with carrier frequency ω_{0j}, similar to what we did in previous chapters. The wavepackets may have the same carrier frequency or different carrier frequencies. Let $A(\omega_{0j})$ denotes the probability amplitude for a wavepacket to have carrier frequency ω_{0j}, the second-order coherence function $\Gamma^{(2)}(\mathbf{r}_1, t_1; \mathbf{r}_2, t_2)$ of natural radiation is calculated as follows:

$$\Gamma^{(2)}(r_1, t_1; r_2, t_2)$$

$$= \left\langle \left[\sum_j \sum_m A^*(\omega_{0j})\mathcal{F}^*_{\tau_{1m}}\{a(\nu)\}e^{i\omega_{0j}\tau_{1m}}\right]\left[\sum_r \sum_q A(\omega_{0r})\mathcal{F}_{\tau_{1q}}\{a(\nu)\}e^{-i\omega_{0r}\tau_{1q}}\right]\right.$$

$$\times \left.\left[\sum_k \sum_n A^*(\omega_{0k})\mathcal{F}^*_{\tau_{2n}}\{a(\nu)\}e^{i\omega_{0k}\tau_{2n}}\right]\left[\sum_s \sum_p A(\omega_{0s})\mathcal{F}_{\tau_{2p}}\{a(\nu)\}e^{-i\omega_{0s}\tau_{2p}}\right]\right\rangle$$

$$
= \left[\sum_j \sum_m A^*(\omega_{0j}) \mathcal{F}^*_{\tau_{1m}}\{a(\nu)\} e^{i\omega_{0j}\tau_{1m}} \right]\left[\sum_j \sum_m A(\omega_{0k}) \mathcal{F}_{\tau_{1m}}\{a(\nu)\} e^{-i\omega_{0j}\tau_{1m}} \right]
$$

$$
\times \left[\sum_k \sum_n A^*(\omega_{0k}) \mathcal{F}^*_{\tau_{2n}}\{a(\nu)\} e^{i\omega_{0k}\tau_{2n}} \right]\left[\sum_k \sum_n A(\omega_{0k}) \mathcal{F}_{\tau_{2n}}\{a(\nu)\} e^{-i\omega_{0k}\tau_{2n}} \right]
$$

$$
+ \left[\sum_j \sum_m A(\omega_{0j}) \mathcal{F}_{\tau_{1m}}\{a(\nu)\} e^{i\omega_{0j}\tau_{1m}} \right]\left[\sum_j \sum_m A^*(\omega_{0j}) \mathcal{F}^*_{\tau_{2m}}\{a(\nu)\} e^{-i\omega_{0j}\tau_{2m}} \right]
$$

$$
\times \left[\sum_k \sum_n A(\omega_{0k}) \mathcal{F}_{\tau_{1n}}\{a(\nu)\} e^{-i\omega_{0k}\tau_{1n}} \right]\left[\sum_k \sum_n A^*(\omega_{0k}) \mathcal{F}^*_{\tau_{2n}}\{a(\nu)\} e^{i\omega_{0k}\tau_{2n}} \right]
$$

$$
= \sum_{j,k,m,n} \frac{1}{2} \left| \left[A(\omega_{0j})\, \mathcal{F}_{\tau_{1m}}\{a(\nu)\} e^{-i\omega_{0j}\tau_{1m}} \right]\left[A(\omega_{0k})\, \mathcal{F}_{\tau_{2n}}\{a(\nu)\} e^{-i\omega_{0k}\tau_{2n}} \right] \right.
$$

$$
\left. + \left[A(\omega_{0k})\, \mathcal{F}_{\tau_{1n}}\{a(\nu)\} e^{-i\omega_{0k}\tau_{1n}} \right]\left[A(\omega_{0j})\, \mathcal{F}^*_{\tau_{2m}}\{a(\nu)\} e^{-i\omega_{0j}\tau_{2m}} \right] \right|^2 \tag{5.2.12}
$$

corresponding to two different yet indistinguishable alternatives for a pair of randomly created and randomly paired photons, or subfields, i.e., the mth subfield or wavepacket with carrier frequencies ω_{0j} and the nth subfield or wavepacket with carrier frequency ω_{0k} to produce a joint-photodetection event of D_1 and D_2: (1) The mth subfield or wavepacket with carrier frequency ω_{0j} produces a photodetection event at D_1, while the nth subfield or wavepacket with carrier frequency ω_{0k} produces a photodetection event at D_2; (2) the nth subfield or wavepacket with carrier frequency ω_{0k} produces a photodetection event at D_1, while the mth subfield or wavepacket with carrier frequency ω_{0j} produces a photodetection event at D_2, namely two-photon interference — a randomly created and randomly paired photons interfering with the pair itself.

Case (1): Identical carrier frequency $\omega_{0j} = \omega_{0k} = \omega_0$

In this case, all the measured wavepackets have the same carrier frequency ω_0. The normalized degree of second-order coherence function $\gamma^{(2)}(z_1, t_1; z_2, t_2)$ is therefore

$$
\gamma^{(2)}(z_1, t_1; z_2, t_2) \cong 1 + \left| \mathcal{F}_\tau\{a^2(\nu)\} \right|^2. \tag{5.2.13}
$$

where, again, $\tau \equiv \tau_1 - \tau_2$. In idealized correlation measurements, the intensity fluctuation correlation is usually considered to be

$$
\langle \Delta I(\tau_1) \Delta I(\tau_2) \rangle = \left| \mathcal{F}_\tau\{a^2(\nu)\} \right|^2. \tag{5.2.14}
$$

Case (2): Large number of different carrier frequencies $\omega_{0j} \neq \omega_{0k}$

Calculating $\Gamma_{12}^{(1)}$ for a broad spectrum thermal field in which the bandwidth of the total field is much greater than that of the individual wavepacket, such as sunlight, an additional sum or integral on the carrier frequency ω_0 is necessary:

$$\simeq \left\langle \left[\sum_j \sum_m A(\omega_{0j}) \mathcal{F}_{\tau_{1m}}\{a(\nu)\} e^{i\omega_{0j}\tau_{1m}} \right] \right.$$

$$\times \left. \left[\sum_j \sum_m A^*(\omega_{0j}) \mathcal{F}^*_{\tau_{2m}}\{a(\nu)\} e^{-i\omega_{0j}\tau_{2m}} \right] \right\rangle$$

$$\simeq \sum_{j,m} |A(\omega_{0j})|^2 \, e^{i\omega_{0j}\tau} \, \mathcal{F}^*_{\tau_{1m}}\{a(\nu)\} \, \mathcal{F}_{\tau_{2m}}\{a(\nu)\}$$

$$\simeq \mathcal{F}_\tau\{|A(\nu_0)|^2\} \, \mathcal{F}_\tau\{a^2(\nu)\} \, e^{i\bar{\omega}_0\tau}, \qquad (5.2.15)$$

where $\bar{\omega}_0$ is the mean carrier frequency and $\nu_0 = \omega_0 - \bar{\omega}_0$. $\gamma^{(2)}(z_1, t_1; z_2, t_2)$ is thus approximated as

$$\gamma^{(2)}(z_1, t_1; z_2, t_2) = 1 + \left| \mathcal{F}_\tau\{A^2(\nu_0)\} \mathcal{F}_\tau\{a^2(\nu)\} \right|^2. \qquad (5.2.16)$$

In idealized correlation measurements, the intensity fluctuation correlation is thus

$$\langle \Delta I(\tau_1) \Delta I(\tau_2) \rangle = \left| \mathcal{F}_\tau\{A^2(\nu_0)\} \mathcal{F}_\tau\{a^2(\nu)\} \right|^2. \qquad (5.2.17)$$

The temporal width of the second-order coherence function is mainly determined by $\mathcal{F}_\tau\{A^2(\nu_0)\}$ due to its greater spectrum bandwidth.

It is obvious that the second-order temporal coherence as well as the degree of second-order temporal coherence of thermal field depend on the relative delay

$$\tau = \tau_1 - \tau_2 = [(t - \tau_1^e) - z_1/c] - [(t - \tau_2^e) - z_2)/c]$$

$$= (z_2 - z_1)/c + (\tau_2^e - \tau_1^e)$$

and it is symmetric with respect to τ

$$\gamma^{(2)}(\tau) = \gamma^{(2)}(-\tau). \qquad (5.2.18)$$

It should be emphasized that t_1 and t_2 are defined from the joint-detection time t, but not from the source, consistent with the Heisenberg picture. From the perspective of classical theory, apparently, the nontrivial second-order temporal correlation reveals a paradoxical behavior of thermal radiation: Although D_1 and D_2 are randomly triggered by the subfields from time to time at any time t_1 and t_2, if one of them is triggered at a certain time the other one has a twice greater chance of being triggered simultaneously at $\tau_1 = \tau_2$. It seems that the randomly radiated and randomly distributed subfields in thermal state must be nonrandomly "bunched" temporally in joint measurement between two distant photodetectors.

(II) **Second-order far-field spatial coherence of thermal field**

Assume a perfect degree of second-order temporal coherence by means of $\gamma^{(2)}(\tau) \simeq 2$, we evaluate the second-order far-field spatial coherence function of thermal field in terms of the transverse spatial separation of two measured fields. This calculation is under Einstein's granularity picture by assuming a disk-like radiation source with finite angular diameter, $\Delta\theta$, or transverse size of D that contains a large number of independent and randomly distribute point-like sub-sources (spontaneous atomic transitions) on the entire disk, such as the mth and nth sub-sources. To simplify the mathematics, we restrict our calculation to 1-D, as shown in Fig. 5.1.1. The angular size of the radiation source, defined as the angle subtended by the source at the detector, is $\Delta\theta$ ($\Delta\theta \sim D/z$, with $z_1 = z_2 = z$, D is the transverse size of the source).

We start the calculation of the second-order coherence function of thermal field from

$$\Gamma^{(2)}(x_1, t_1; x_2, t_2)$$

$$= \sum_{m,n} \left| \frac{1}{\sqrt{2}} \left[E_m(x_1, t_1) E_n(x_2, t_2) + E_n(x_1, t_1) E_m(x_2, t_2) \right] \right|^2 \qquad (5.2.19)$$

where we have simplified the problem in 1-D and assume all the mth and the nth sub-sources are randomly distributed along the x_0 axis and spatially distinguishable. The mth and the nth sub-sources are identified by their transverse coordinates of the x_0 axis. Equation (5.2.19) indicates a two-photon interference: two independent subfields, created by the mth sub-source at x_{0m} and the nth sub-sources at x_{0n}, excited a joint-photodetection event between D_1 and D_2 with two different yet indistinguishable alternatives — (1) the mth and the nth wavepackets are detected by D_1 and D_2, respectively, and (2) the mth and the nth wavepackets are detected by D_2 and D_1, respectively.[1]

Further simplifying the mathematics, we assume monochromatic radiation. Substituting the plane-wave approximation into Eq. (5.2.19), and assuming $t_1 \simeq t_2$ by arranging a symmetrical optical and electronic experimental setup for the photodetection events of D_1 and D_2, $\Gamma^{(2)}(x_1, x_2)$ is approximately

$$\Gamma^{(2)}(x_1, x_2)$$

$$\simeq \int dx_0 \int dx_0' \left| \frac{1}{\sqrt{2}} \left[a(x_0)e^{i\varphi(x_0)}e^{-ikr(x_0,x_1)}a(x_0')e^{i\varphi(x_0')}e^{-ikr(x_0',x_2)} \right. \right.$$

$$\left. \left. + a(x_0')e^{i\varphi(x_0')}e^{-ikr(x_0',x_1)}a(x_0)e^{i\varphi(x_0)}e^{-ikr(x_0,x_2)} \right] \right|^2$$

$$\simeq \int dx_0 \, a^2(x_0) \int dx_0' \, a^2(x_0') + \left| \int dx_0 \, a^2(x_0) \, e^{-ikx_0(x_1-x_2)/z} \right|^2$$

$$\simeq I_0^2 \left[1 + \mathrm{sinc}^2 \left(\frac{\pi \Delta\theta(x_1 - x_2)}{\lambda} \right) \right], \tag{5.2.20}$$

where we have applied the far-field approximation and treated $a(x_0) \simeq a$ as a constant. Comparing with our early calculation of the first-order spatial coherence function for the distant star, we see that $x_1 - x_2$ is equivalent to the spatial separation b between the two pinholes. Note, due to the chosen positive directions of x_1 and x_2, we have $x_1 - x_2 = b$.

The degree of second-order spatial coherence $\gamma^{(2)}$ is thus

$$\gamma^{(2)}(x_1, t_1; x_2, t_2) = 1 + \mathrm{sinc}^2 \left[\frac{\pi \Delta\theta(x_1 - x_2)}{\lambda} \right]. \tag{5.2.21}$$

If the angular size $\Delta\theta$ of the thermal source is not too small, for short wavelength radiations the sinc function in Eqs. (5.2.20) and (5.2.21) quickly drops from its maximum to minimum when $x_1 - x_2$ goes from zero to a value such that $\Delta\theta(x_1 - x_2)/\lambda = 1$. In this case, we effectively have a

---

[1] In the language of quantum mechanics, again, this superposition corresponds to a two-photon interference phenomenon, which involves a superposition between two different yet indistinguishable two-photon probability amplitudes: (1) Photon m and photon n are annihilated at D_1 and D_2, respectively, and (2) photon m and photon n are annihilated at D_2 and D_1, respectively.

"point"-to-"point" relationship between the x_1 plane and the x_2 plane. Note that Eqs. (5.2.20) and (5.2.21) are functions of $x_1 - x_2$, which is independent of the absolute values of either x_1 or x_2. This is a very important and useful property of thermal field. It signifies that whatever transverse coordinate x_1 we chooses for D_1, there is a unique position x_2 for D_2 where the maximum joint-detection between D_1 and D_2 is expected, i.e., maximum constructive interference between $E_m(x_1, t_1) E_n(x_2, t_2)$ and $E_n(x_1, t_1) E_m(x_2, t_2)$ in Eq. (5.2.19) is observable at that unique position.

(III) Second-order near-field spatial coherence of thermal field

In the following, we attempt a calculation starting from Eq. (5.2.1) for the Fresnel near-field second-order spatial coherence function of thermal field $\Gamma^{(2)}(\vec{\rho}_1, z_1; \vec{\rho}_2, z_2)$:

$$\Gamma^{(2)}(\vec{\rho}_1, z_1; \vec{\rho}_2, z_2)$$

$$= \sum_{m,n} \left| \frac{1}{\sqrt{2}} \left[E_m(\vec{\rho}_1, z_1) E_n(\vec{\rho}_2, z_2) + E_n(\vec{\rho}_1, z_1) E_m(\vec{\rho}_2, z_2) \right] \right|^2$$

$$= \sum_{m,n} \Gamma^{(2)}_{mn}(\vec{\rho}_1, z_1; \vec{\rho}_2, z_2), \tag{5.2.22}$$

where we have used the transverse and longitudinal coordinates of $\vec{\rho}$ and z to specify the spatial coordinates of the photodetectors. In Eq. (5.2.22), we have ignored the temporal variables by assuming a perfect degree of second-order temporal coherence as usual. In the near-field, we apply the Fresnel approximation to propagate the field from each sub-source to the photodetectors. Equation (5.2.22) can be written in terms of Green's functions:

$$\Gamma^{(2)}_{mn}(\vec{\rho}_1, z_1; \vec{\rho}_2, z_2)$$

$$= I^2_{mn} \left| \frac{1}{\sqrt{2}} \left[g_m(\vec{\rho}_1, z_1)\, g_n(\vec{\rho}_2, z_2) + g_m(\vec{\rho}_2, z_2)\, g_n(\vec{\rho}_1, z_1) \right] \right|^2 \tag{5.2.23}$$

with

$$g_m(\vec{\rho}_j, z_j) = \frac{c_0}{z_j} e^{i\frac{\omega z_j}{c}} e^{i\frac{\omega}{2cz_j}|\vec{\rho}_j - \vec{\rho}_{0m}|^2},$$

where c_0 is a normalization consistent. In Eq. (5.2.23), we have also assumed $|E_m|^2 = |E_n|^2 = I_{mn}$. Next, we approximate the sum of $\Gamma^{(2)}_{mn}(\vec{\rho}_1, z_1; \vec{\rho}_2, z_2)$ in terms of the mth and nth sub-source into the integrals of $\vec{\rho}_0$ and $\vec{\rho}_0'$ on the entire source plane:

$$\Gamma^{(2)}(\vec{\rho}_1, z_1; \vec{\rho}_2, z_2)$$

$$= I_0^2 \left[\left| \int d\vec{\rho}_0 \, g_{\vec{\rho}_0}(\vec{\rho}_1, z_1) \int d\vec{\rho}_0' \, g_{\vec{\rho}_0'}(\vec{\rho}_2, z_2) \right|^2 \right.$$

$$\left. + \int d\vec{\rho}_0 \, g_{\vec{\rho}_0}^*(\vec{\rho}_1, z_1) g_{\vec{\rho}_0}(\vec{\rho}_2, z_2) \int d\vec{\rho}_0' \, g_{\vec{\rho}_0'}^*(\vec{\rho}_2, z_2) g_{\vec{\rho}_0'}(\vec{\rho}_1, z_1) \right]$$

$$= \Gamma_{11}^{(1)} \Gamma_{22}^{(1)} + \Gamma_{12}^{(1)} \Gamma_{21}^{(1)}. \tag{5.2.24}$$

Substitute Green's functions into Eq. (5.2.24), we obtain

$$\Gamma_{11}^{(1)} \Gamma_{22}^{(1)} \sim \text{ constant}$$

and

$$\Gamma_{12}^{(1)}(\vec{\rho}_1, z_1; \vec{\rho}_2, z_2)$$

$$\propto \frac{1}{z_1 z_2} \int d\vec{\rho}_0 \, a^2(\vec{\rho}_0) \, e^{-i\frac{\omega}{c} z_1} \, e^{-i\frac{\omega}{2cz_1}|\vec{\rho}_1 - \vec{\rho}_0|^2} \, e^{i\frac{\omega}{c} z_2} \, e^{i\frac{\omega}{2cz_2}|\vec{\rho}_2 - \vec{\rho}_0|^2},$$

Assuming $a^2(\vec{\rho}_0) \sim$ constant, and taking $z_1 = z_2 = d$, we obtain

$$\Gamma_{12}^{(1)}(\vec{\rho}_1; \vec{\rho}_2) \propto \int d\vec{\rho}_0 \, e^{-i\frac{\omega}{2cd}|\vec{\rho}_1 - \vec{\rho}_0|^2} \, e^{i\frac{\omega}{2cd}|\vec{\rho}_2 - \vec{\rho}_0|^2}$$

$$\propto e^{-i\frac{\omega}{2cd}(|\vec{\rho}_1|^2 - |\vec{\rho}_2|^2)} \int d\vec{\rho}_0 \, e^{i\frac{\omega}{cd}(\vec{\rho}_1 - \vec{\rho}_2)\cdot\vec{\rho}_0}$$

$$\propto e^{-i\frac{\omega}{2cd}(|\vec{\rho}_1|^2 - |\vec{\rho}_2|^2)} \, \text{somb}\left[\frac{R}{d}\frac{\omega}{c}|\vec{\rho}_1 - \vec{\rho}_2| \right], \tag{5.2.25}$$

where we have assumed a disk-like light source with a finite radius of R, and again, $\text{somb}(x) = J_1(x)/x$, $J_1(x)$ is the first-order Bessel Function. In Eq. (5.2.25), we have absorbed all constants into the proportionality constant. The second-order spatial correlation function $\Gamma^{(2)}(\vec{\rho}_1; \vec{\rho}_2)$ is thus

$$\Gamma^{(2)}(\vec{\rho}_1, \vec{\rho}_2) = I_0^2 \left[1 + \text{somb}^2\left(\frac{R}{d}\frac{\omega}{c}|\vec{\rho}_1 - \vec{\rho}_2| \right) \right]. \tag{5.2.26}$$

Consequently, the degree of second-order spatial coherence is

$$\gamma^{(2)}(\vec{\rho}_1, \vec{\rho}_2) = 1 + \text{somb}^2\left(\frac{R}{d}\frac{\omega}{c}|\vec{\rho}_1 - \vec{\rho}_2| \right). \tag{5.2.27}$$

For a large value of $2R/d \sim \Delta\theta$, where $\Delta\theta$ is the angular size of the radiation source viewed at the photodetectors, the point-to-"spot" sombrero-like function can be approximated as a δ-function of $|\vec{\rho}_1 - \vec{\rho}_2|$. We thus effectively have a "point"-to-"point" correlation between the transverse planes of $z_1 = d$ and $z_2 = d$.

The above discussions on the second-order spatial coherence or correlation function of thermal field, either far-field or near-field, ended with an interesting result. Analogous to EPR's language, the photodetectors D_1 and D_2 have equal chance to be triggered at any position on the transverse planes of $z_1 = z_2$, however, if D_1 is triggered at a certain transverse position of $\vec{\rho}_1$, D_2 has twice chance to be trigged at a transverse position $\vec{\rho}_2 = \vec{\rho}_1$.

5.3 Second-Order Coherence: Quantum Theory

We now introduce the second-order coherence function, $G^{(2)}(\mathbf{r}_1, t_1; \mathbf{r}_2, t_2)$, which is formulated from the quantum theory of light. One may ask immediately the following: Why do we need quantum theory if Einstein's granularity picture or the concept of quantized subfield gives a successful explanation to the nontrivial second-order coherence function of thermal field? (1) Einstein's granularity picture does not work for a certain quantum states, such as entangled states; (2) although Einstein's picture gives a reasonable solution to the nontrivial second-order coherence function of thermal field, the concept of subfield itself is still under the framework of classical electromagnetic field theory. Recall the earlier story about the size of a single photon after one-year propagation in which Einstein trapped himself and his students into a self-contradictory situation. We encounter a more serious problem when dealing with second-order correlations of the thermal field. In addition to the collapse of one single-photon wavepacket, we are facing the collapse of two wavepackets, and the two collapses are correlated in space–time! Either in their temporal correlation or spatial correlation, the randomly created and randomly paired wavepackets, respectively, may collapse to any localized space–time points, however, if one wavepacket collapses to a certain space–time point, the other one has twice the chance to collapse at a unique localized space–time area. Einstein realized that his concept of quantized realistic electromagnetic subfield may imply an action-at-a-distance. Einstein may accept the two-photon interference picture of the second-order coherence of thermal field, however, Einstein can never accept action-at-a-distance. Einstein would ask the following: How long does it take for the nonlocal superposition of the two subfields to complete? We may have to accept the quantum mechanical concepts of probability and probability amplitude as Bohr suggested from the beginning of quantum theory.

Figure 5.3.1 illustrates a schematic setup of the measurement of $G^{(2)}(\mathbf{r}_1, t_1; \mathbf{r}_2, t_2)$, or $\langle \Delta n(\mathbf{r}_1, t_1) \Delta n(\mathbf{r}_2, t_2) \rangle$, for different quantum states. To be consistent with the classical analysis, we use the same setup as that of Fig. 5.1.1, except a single-photon joint-detection measurement circuit (including hardware and software). Assume a weak light source at single-photon level, and two point-like photodetectors D_1 and D_2 in photon counting mode are placed behind pinhole P_1 and P_2, respectively, for joint-photodetection measurement. The measured radiation is in a certain quantum state, for instance, in thermal state, in coherent state, or in entangled state. To simplify the discussion, we assume idealized point-like photodetectors, D_1 and D_2, and are both scannable together with P_1 and P_2 along the longitudinal axis and on the transverse plan to observe the temporal or spatial correlation.

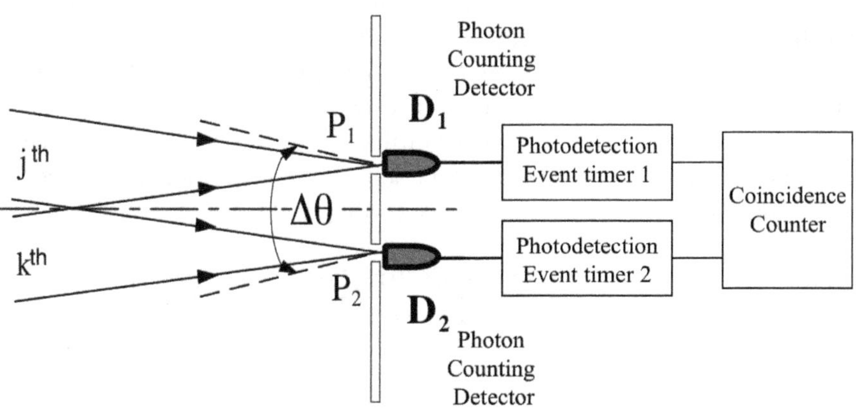

Fig. 5.3.1 A schematic measurement of second-order coherence function $G^{(2)}(\mathbf{r}_1, t_1; \mathbf{r}_2, t_2)$, which is defined as the fourth-order correlation of the field operators evaluated from different quantum states that are created from different radiation sources. In terms of the concept of photon, $G^{(2)}(\mathbf{r}_1, t_1; \mathbf{r}_2, t_2)$ is proportional to $\langle n(\mathbf{r}_1, t_1) n(\mathbf{r}_2, t_2) \rangle$. D_1 and D_2 are idealized point-like photon counting detectors. D_1 and D_2 are placed behind pinhole P_1 and P_2, respectively, for joint-photodetection measurement. The photodetection event timers record the number of photons counted by D_1 and D_2 within a chosen coincidence time window along time axis t_1 and t_2, respectively. The coincidence counter, a hardware and software circuit combination, evaluates $n(t_1) \times n(t_2)$ for all values of t_1 and t_2. Usually, the circuit is also able to provide a histogram, which evaluates the total measured $n(t_1) \times n(t_2)$ as a function of $t_1 - t_2$.

Following Glauber's theory, the probability of observing a pair of photons, either randomly paired photons by chance or coherently prepared photon pairs from a certain physical process, to excite a joint-detection event at $(\mathbf{r}_1, t_1)$ and $(\mathbf{r}_2, t_2)$ is

$$G^{(2)}(\mathbf{r}_1, t_1; \mathbf{r}_2, t_2)$$

$$\equiv \langle\, E^{(-)}(\mathbf{r}_1, t_1) E^{(-)}(\mathbf{r}_2, t_2) E^{(+)}(\mathbf{r}_2, t_2) E^{(+)}(\mathbf{r}_1, t_1)\,\rangle$$

$$= \langle\langle\, E^{(-)}(\mathbf{r}_1, t_1) E^{(-)}(\mathbf{r}_2, t_2) E^{(+)}(\mathbf{r}_2, t_2) E^{(+)}(\mathbf{r}_1, t_1)\,\rangle_{\mathrm{QM}}\rangle_{\mathrm{En}} \qquad (5.3.1)$$

which is recognized as the expectation value of the normal-ordered field operators at space–time coordinates $(\mathbf{r}_1, t_1)$ and $(\mathbf{r}_2, t_2)$. In the view of quantum theory of light, $G^{(2)}(\mathbf{r}_1, t_1; \mathbf{r}_2, t_2)$ is thus a measure of the probability for jointly observing a pair of photons at $(\mathbf{r}_1, t_1)$ and $(\mathbf{r}_2, t_2)$ from a radiation field. Differing from the first-order coherence function, the second-order coherence function is directly measurable by two photodetectors at space–time coordinates $(\mathbf{r}_1, t_1)$ and $(\mathbf{r}_2, t_2)$. It is very important to keep in mind that, here, t_1 and t_2 are the physical times of the two photodetection events, or the registration times of a pair of photons. In the first-order coherence function $G^{(1)}(\mathbf{r}_1, t_1; \mathbf{r}_2, t_2)$, however, t_1 and t_2 are the "referred early times" of a photodetection event at $(\mathbf{r}, t)$.

To be consistent with the theory of first-order coherence, a normalized second-order quantum coherence function, namely the degree of second-order quantum coherence, is defined accordingly:

$$g^{(2)}(\mathbf{r}_1, t_1; \mathbf{r}_2, t_2) \equiv \frac{\langle\, E^{(-)}(\mathbf{r}_1, t_1) E^{(-)}(\mathbf{r}_2, t_2) E^{(+)}(\mathbf{r}_2, t_2) E^{(+)}(\mathbf{r}_1, t_1)\,\rangle}{\langle\, E^{(-)}(\mathbf{r}_1, t_1) E^{(+)}(\mathbf{r}_1, t_1)\,\rangle \langle\, E^{(-)}(\mathbf{r}_2, t_2) E^{(+)}(\mathbf{r}_2, t_2)\,\rangle}.$$

$$(5.3.2)$$

The second-order quantum coherence function $G^{(2)}(\mathbf{r}_1, t_1; \mathbf{r}_2, t_2)$ as well as the degree of second-order quantum coherence $g^{(2)}(\mathbf{r}_1, t_1; \mathbf{r}_2, t_2)$ of a radiation is obviously determined by the state of the jointly measured photon pairs. It is not surprising that different two-photon states lead to different second-order coherences or correlations. In the following, we calculate the degree of second-order quantum coherence $g^{(2)}(\mathbf{r}_1, t_1; \mathbf{r}_2, t_2)$ of thermal field in the single-photon state representation and in the coherent state representation.

5.4 Second-Order Coherence of Thermal State

In this section, we evaluate the second-order coherence for thermal field from the view of quantum theory.

(I) $G^{(2)}(\mathbf{r}_1, t_1; \mathbf{r}_2, t_2)$ of thermal field in single-photon state representation

Suppose the light source is weak enough that no more than one pair of photons are observed within the chosen coincidence time window ΔT. Applying the same model for thermal field that has been discussed in Chapter 3, we may approximate the mixed state of the thermal field as

$$|\tilde{\Psi}\rangle = \prod_m \{|0\rangle + \epsilon\, c_m\, |\Psi_m\rangle\}$$

$$\simeq |0\rangle + \epsilon \left[\sum_m c_m\, |\Psi_m\rangle\right] + \epsilon^2 \left[\sum_{m<n} c_m\, c_n |\Psi_m\rangle |\Psi_n\rangle\right] + \cdots,$$

where c_m is the complex amplitude of the mth state. To simplify the mathematics, we assume equal chance of photon creation at each sub-source, $|c_m|^2 = $ constant, which is physically reasonable. However, c_m contains a random phase. Since $|\epsilon| \ll 1$, we only list the first-order and the second-order approximations on ϵ. The necessary lowest-order approximation that contributes to the second-order correlation measurement is

$$|\tilde{\Psi}\rangle = \sum_{m<n} c_m\, c_n |\Psi_m\rangle |\Psi_n\rangle \tag{5.4.1}$$

with

$$|\Psi_m\rangle = \int d\mathbf{k}\, f_m(\mathbf{k})\, \hat{a}_m^\dagger(\mathbf{k})|0\rangle.$$

The state of Eq. (5.4.1) is a mixed two-photon state which characterizes the state of a large number of jointly measured photon pairs, each created from a randomly radiated and randomly paired atomic transitions, as an inhomogeneous ensemble. The mixed state of Eq. (5.4.1) is conventionally written in the form of mixed ensemble:

$$\hat{\rho} = \sum_{m,n} P_{mn} |\Psi_m\rangle |\Psi_n\rangle \langle \Psi_n| \langle \Psi_m|, \tag{5.4.2}$$

where P_{mn} is the probability for observing the m-nth pair of photons.

The field operators are similar to that of Eq. (4.5.12), except approximated into a more general form of 3-D:

$$\hat{E}^{(+)}(\mathbf{r}, t) = \sum_m \int d\mathbf{k}\, \hat{a}_m(\mathbf{k})\, g_m(\mathbf{k}; \mathbf{r}, t),$$

$$\hat{E}^{(-)}(\mathbf{r}, t) = \sum_m \int d\mathbf{k}\, \hat{a}_m^\dagger(\mathbf{k})\, g_m^*(\mathbf{k}; \mathbf{r}, t), \tag{5.4.3}$$

where $g_m(\mathbf{k}; \mathbf{r}, t)$ is Green's function that propagates the $\mathbf{k}$-mode of the mth subfield from the mth sub-source to space-time coordinate $(\mathbf{r}, t)$. We have also ignored the constants. We thus have

$$G^{(2)}(\mathbf{r}_1, t_1; \mathbf{r}_2, t_2)$$

$$= \left\langle \langle \Psi | \hat{E}^{(-)}(\mathbf{r}_1, t_1) \hat{E}^{(-)}(\mathbf{r}_2, t_2) | 0 \rangle \langle 0 | \hat{E}^{(+)}(\mathbf{r}_2, t_2) \hat{E}^{(+)}(\mathbf{r}_1, t_1) | \Psi \rangle \right\rangle_{En}$$

$$= \left\langle \sum_m c_m^* \psi_m^*(\mathbf{r}_1, t_1) \sum_n c_n^* \psi_n^*(\mathbf{r}_2, t_2) \sum_p c_p \psi_p(\mathbf{r}_2, t_2) \sum_q c_q \psi_q(\mathbf{r}_1, t_1) \right\rangle_{En}$$

$$= \sum_m |c_m|^2 \psi_m^*(\mathbf{r}_1, t_1) \psi_m(\mathbf{r}_1, t_1) \sum_n |c_n|^2 \psi_n^*(\mathbf{r}_2, t_2) \psi_n(\mathbf{r}_2, t_2)$$

$$+ \sum_{m \neq n} |c_m|^2 |c_n|^2 \, \psi_m^*(\mathbf{r}_1, t_1) \, \psi_n(\mathbf{r}_1, t_1) \, \psi_n^*(\mathbf{r}_2, t_2) \, \psi_m(\mathbf{r}_2, t_2)$$

$$\propto \sum_{m,n} \left| \frac{1}{\sqrt{2}} [\psi_m(\mathbf{r}_1, t_1) \psi_n(\mathbf{r}_2, t_2) + \psi_m(\mathbf{r}_2, t_2) \psi_n(\mathbf{r}_1, t_1)] \right|^2, \tag{5.4.4}$$

where $\psi_m(\mathbf{r}_j, t_j)$, $j = 1, 2$, is defined as the effective wavefunction of the mth photon at space–time coordinate $(\mathbf{r}_j, t_j)$:

$$\psi_m(\mathbf{r}_j, t_j) \equiv \langle 0 | \hat{E}^{(+)}(\mathbf{r}_2, t_2) | \Psi_m \rangle = \int d\mathbf{k}\, f_m(\mathbf{k})\, g_m(\mathbf{k}; r_j, t_j). \tag{5.4.5}$$

We may find immediately that the second-order coherence function $G^{(2)}(\mathbf{r}_1, t_1; \mathbf{r}_2, t_2)$ of thermal state, which is formulated from quantum theory of light in single-photon state representation, is the same as the second-order coherence function $\Gamma^{(2)}(\mathbf{r}_1, t_1; \mathbf{r}_2, t_2)$ that we have calculated previously in Einstein's picture, except the subfield is replaced by the effective wavefunction of a photon. We also find that the second-order coherence function $G^{(2)}(\mathbf{r}_1, t_1; \mathbf{r}_2, t_2)$ of thermal state is the result of two-photon interference which involves the superposition of two different yet indistinguishable two-photon amplitudes: (1) The mth photon is annihilated at $(\mathbf{r}_1, t_1)$, while the nth photon is annihilated at $(\mathbf{r}_2, t_2)$, and (2) the mth photon is annihilated at $(\mathbf{r}_2, t_2)$, while the nth photon is annihilated at $(\mathbf{r}_1, t_1)$. We name this superposition two-photon interference: a pair of randomly created and randomly distributed photons in thermal state interfering with the pair itself.

Similar to Eq. (5.2.3) in Einstein's picture, we find the constant part of $G^{(2)}(\mathbf{r}_1, t_1; \mathbf{r}_2, t_2)$ is a product of the following two mean number of photons measured by D_1 and D_2, respectively,

$$\langle n(\mathbf{r}_1, t_1) \rangle \propto \sum_m \psi_m^*(\mathbf{r}_1, t_1)\, \psi_m(\mathbf{r}_1, t_1),$$

$$\langle n(\mathbf{r}_2, t_2) \rangle \propto \sum_n \psi_n^*(\mathbf{r}_2, t_2)\, \psi_n(\mathbf{r}_2, t_2), \tag{5.4.6}$$

except the subfield is replaced by the effective wavefunction of a photon. The cross-interference term of Eq. (5.4.4), which is the nontrivial term of the second-order coherence function $G^{(2)}(\mathbf{r}_1, t_1; \mathbf{r}_2, t_2)$, corresponding to the photon number fluctuation correlation,

$$\langle \Delta n(\mathbf{r}_1, t_1) \Delta n(\mathbf{r}_2, t_2) \rangle \propto \sum_{m \neq n} \psi_m^*(\mathbf{r}_1, t_1)\psi_n(\mathbf{r}_1, t_1)\psi_n^*(\mathbf{r}_2, t_2)\psi_m(\mathbf{r}_2, t_2),$$

$$\tag{5.4.7}$$

is also the same as the intensity fluctuation correlation we have calculated previously in Einstein's picture, except the subfield is replaced by the effective wavefunction of a photon.

In the case of a perfect measurement system, including idealized photodetectors and idealized correlation measurement electronics, the second-order coherence function, $G^{(2)}(\mathbf{r}_1, t_1; \mathbf{r}_2, t_2)$, of thermal state can be formally written in terms of its first-order quantum coherence functions:

$$G^{(2)}(\mathbf{r}_1, t_1; \mathbf{r}_2, t_2) = G_{11}^{(1)} G_{22}^{(1)} + G_{12}^{(1)} G_{21}^{(1)} = G_{11}^{(1)} G_{22}^{(1)} + |G_{12}^{(1)}|^2, \tag{5.4.8}$$

where

$$G_{11}^{(1)} = \sum_m \psi_m^*(\mathbf{r}_1, t_1)\psi_m(\mathbf{r}_1, t_1) \quad G_{22}^{(1)} = \sum_n \psi_n^*(\mathbf{r}_2, t_2)\psi_n(\mathbf{r}_2, t_2),$$

$$G_{12}^{(1)} = \sum_m \psi_m^*(\mathbf{r}_1, t_1)\psi_m(\mathbf{r}_2, t_2) \quad G_{21}^{(1)} = \sum_n \psi_n^*(\mathbf{r}_2, t_2)\psi_n(\mathbf{r}_1, t_1). \tag{5.4.9}$$

It should be emphasized that, here, $G_{12}^{(1)}$ and $G_{21}^{(1)}$ are physically measured by two independent photodetectors D_1 and D_2 at different space-time coordinates $(\mathbf{r}_1, t_1)$ and $(\mathbf{r}_2, t_2)$, which is very different from the measurement of the first-order coherence function. Although Eq. (5.4.8) is helpful for mathematical calculations, we must always keep this in mind, especially when applying nonidealized measurement system. The realistic finite response time of the photodetectors and the correlation measurement circuit will complicate the calculation, especially when dealing with thermal field of broadband spectrum.

(II) Two-photon effective wavefunction of thermal state in single-photon state representation

The effective wavefunction of the m-nth photon pair in thermal state can be defined directly from the following product state:

$$|\Psi_{mn}\rangle = |\Psi_m\rangle|\Psi_n\rangle, \tag{5.4.10}$$

with

$$|\Psi_m\rangle = \int d\omega\, f_m(\omega)\, \hat{a}^\dagger(\omega)|0\rangle, \quad |\Psi_n\rangle = \int d\omega\, f_n(\omega)\, \hat{b}^\dagger(\omega)|0\rangle,$$

where the complex amplitudes $f_m(\omega)$ and $f_n(\omega)$ will be treated with identical spectrum distribution but random phases that are determined by the initial condition of the mth and nth atomic transitions. The second-order coherence function of the m-nth photon pair is thus

$$
\begin{aligned}
G^{(2)}_{mn}&(\mathbf{r}_1, t_1; \mathbf{r}_2, t_2)\\
&= \langle\Psi_{mn}|\, E^{(-)}(\mathbf{r}_1, t_1)E^{(-)}(\mathbf{r}_2, t_2)E^{(+)}(\mathbf{r}_2, t_2)E^{(+)}(\mathbf{r}_1, t_1)\,|\Psi_{mn}\rangle\\
&= \sum_l \langle\Psi_{mn}|\, E^{(-)}(\mathbf{r}_1, t_1)E^{(-)}(\mathbf{r}_2, t_2)\,|l\rangle\,\langle l|\, E^{(+)}(\mathbf{r}_2, t_2)E^{(+)}(\mathbf{r}_1, t_1)\,|\Psi_{mn}\rangle\\
&= \langle\Psi_{mn}|E^{(-)}(\mathbf{r}_1, t_1)E^{(-)}(\mathbf{r}_2, t_2)|0\rangle\langle 0|E^{(+)}(\mathbf{r}_2, t_2)E^{(+)}(\mathbf{r}_1, t_1)|\Psi_{mn}\rangle\\
&= \big|\psi_{mn}(\mathbf{r}_1, t_1; \mathbf{r}_2, t_2)\big|^2, \tag{5.4.11}
\end{aligned}
$$

where, in the third line, we have used the completeness relation

$$\sum_l |l\rangle\langle l| = 1.$$

The effective wavefunction of the m–nth photon pair is defined as

$$\psi_{mn}(\mathbf{r}_1, t_1; \mathbf{r}_2, t_2) \equiv \langle 0|E^{(+)}(\mathbf{r}_2, t_2)E^{(+)}(\mathbf{r}_1, t_1)|\Psi_{mn}\rangle. \tag{5.4.12}$$

The second-order quantum coherence function $G^{(2)}(\mathbf{r}_1, t_1; \mathbf{r}_2, t_2)$ of thermal field can be also calculated from the density operator of Eq. (5.4.2), which characterizes an ensemble of mixed two-photon product states:

$$G^{(2)}(\mathbf{r}_1, t_1; \mathbf{r}_2, t_2)$$

$$= \langle\langle E^{(-)}(\mathbf{r}_1, t_1) E^{(-)}(\mathbf{r}_2, t_2) E^{(+)}(\mathbf{r}_2, t_2) E^{(+)}(\mathbf{r}_1, t_1) \rangle_{\mathrm{QM}}\rangle_{\mathrm{En}}$$

$$= tr\left[\hat{\rho}\, E^{(-)}(\mathbf{r}_1, t_1) E^{(-)}(\mathbf{r}_2, t_2)\, E^{(+)}(\mathbf{r}_2, t_2) E^{(+)}(\mathbf{r}_1, t_1)\right]$$

$$= \sum_{m,n} P_{mn} \langle \Psi_{mn}| E^{(-)}(\mathbf{r}_1, t_1) E^{(-)}(\mathbf{r}_2, t_2) E^{(+)}(\mathbf{r}_2, t_2) E^{(+)}(\mathbf{r}_1, t_1) |\Psi_{mn}\rangle$$

$$= \sum_{m,n} P_{mn}\, G^{(2)}_{mn}(\mathbf{r}_1, t_1; \mathbf{r}_2, t_2) = \sum_{m,n} P_{mn}\, |\psi_{mn}(\mathbf{r}_1, t_1; \mathbf{r}_2, t_2)|^2, \qquad (5.4.13)$$

where P_{mn} is the probability to find the radiation in the two-photon product state $|\Psi_{mn}\rangle$. Obviously, $P_{mn} = \text{constant}$ for thermal field, and thus we have

$$G^{(2)}(\mathbf{r}_1, t_1; \mathbf{r}_2, t_2) = \sum_{m,n} |\psi_{mn}(\mathbf{r}_1, t_1; \mathbf{r}_2, t_2)|^2. \qquad (5.4.14)$$

(III) $G^{(2)}(\mathbf{r}_1, t_1; \mathbf{r}_2, t_2)$ **of thermal state in coherent state representation**

Consider the mixed thermal state in coherent state representation that has been introduced in Eq. (3.5.7):

$$|\tilde{\Psi}\rangle = \prod_m |\Psi_m\rangle = \prod_{m,\mathbf{k}} |\alpha_m(\mathbf{k})\rangle,$$

where we have explicitly separated index m, indicating the mth photon or the mth group of indistinguishable photons, from $\mathbf{k}$, indicating the $\mathbf{k}$th mode. Coherent state may represent a group of identical photons with $\bar{n} = |\alpha| \gg 1$. Note that $\hat{a}_m(\mathbf{k})|\alpha_m(\mathbf{k})\rangle = \alpha_m(\mathbf{k})|\alpha_m(\mathbf{k})\rangle$ is an eigen state of the annihilation operator $\hat{a}_m(\mathbf{k})$. In the mixed thermal state, the complex eigen value $\alpha_m(\mathbf{k})$ contains a random phase. Coherent state representation has not only simplified the calculation of second-order coherence function significantly but also made it possible to calculate the coherence function of indistinguishable photon groups. The field operators are the same, as shown in Eq. (5.4.3). We thus have

$$\langle \Psi| \hat{E}^{(-)}(\mathbf{r}, t) |\Psi\rangle = \langle \Psi| \left[\sum_m \int d\mathbf{k}\, \hat{a}^\dagger_m(\mathbf{k})\, g^*_m(\mathbf{k}; \mathbf{r}, t)\right] |\Psi\rangle$$

$$= \langle \Psi| \left[\sum_m \int d\mathbf{k}\, \alpha^*_m(\mathbf{k})\, g^*_m(\mathbf{k}; \mathbf{r}, t)\right] |\Psi\rangle$$

$$= \sum_m \psi^*_m(\mathbf{r}, t),$$

$$\langle\Psi|\hat{E}^{(+)}(\mathbf{r},t)|\Psi\rangle = \langle\Psi|\left[\sum_n \int d\mathbf{k}\,\hat{a}_n(\mathbf{k})\,g_n(\mathbf{k};\mathbf{r},t)\right]|\Psi\rangle$$

$$= \langle\Psi|\left[\sum_n \int d\mathbf{k}\,\alpha_n(\mathbf{k})\,g_n(\mathbf{k};\mathbf{r},t)\right]|\Psi\rangle$$

$$= \sum_n \psi_n(\mathbf{r},t). \tag{5.4.15}$$

Therefore, we have

$$G^{(2)}(\mathbf{r}_1,t_1;\mathbf{r}_2,t_2)$$

$$= \left\langle \sum_m \psi_m^*(\mathbf{r}_1,t_1) \sum_n \psi_n^*(\mathbf{r}_2,t_2) \sum_p \psi_p(\mathbf{r}_2,t_2) \sum_q \psi_q(\mathbf{r}_1,t_1) \right\rangle_{\mathrm{En}}$$

$$= \sum_m \psi_m^*(\mathbf{r}_1,t_1)\psi_m(\mathbf{r}_1,t_1) \sum_n \psi_n^*(\mathbf{r}_2,t_2)\psi_n(\mathbf{r}_2,t_2)$$

$$+ \sum_{m\neq n} \psi_m^*(\mathbf{r}_1,t_1)\psi_n(\mathbf{r}_1,t_1)\psi_n^*(\mathbf{r}_2,t_2)\psi_m(\mathbf{r}_2,t_2)$$

$$= \sum_{m,n} \left| \frac{1}{\sqrt{2}}[\psi_m(\mathbf{r}_1,t_1)\psi_n(\mathbf{r}_2,t_2) + \psi_n(\mathbf{r}_1,t_1)\psi_m(\mathbf{r}_2,t_2)] \right|^2, \tag{5.4.16}$$

where $\psi_m(\mathbf{r}_j,t_j)$, $j=1,2$, is defined as the effective wavefunction of the mth photon or the mth group identical photons at space–time coordinate $(\mathbf{r}_j,t_j)$ of the jth photodetector D_j:

$$\psi_m(\mathbf{r}_j,t_j) = \int d\mathbf{k}\,\alpha_m(\mathbf{k})\,g_m(\mathbf{k};\mathbf{r}_j,t_j). \tag{5.4.17}$$

The second-order coherence function $G^{(2)}(\mathbf{r}_1,t_1;\mathbf{r}_2,t_2)$ of thermal field, which is formulated from quantum theory of light in coherent state representation, is the result of two-photon interference which involves the superposition of two different yet indistinguishable two-photon amplitudes: (1) The mth photon, or the mth group indistinguishable photons, is detected at $(\mathbf{r}_1,t_1)$, while the nth photon, or the nth group indistinguishable photons, is detected at $(\mathbf{r}_2,t_2)$, and (2) the mth photon, or the mth group indistinguishable photons, is detected at $(\mathbf{r}_2,t_2)$, while the nth photon, or the mth group indistinguishable photons, is detected at $(\mathbf{r}_1,t_1)$. We name this superposition two-photon interference: a pair of randomly created and randomly distributed photons, or a pair of indistinguishable photon groups, in thermal state interfering with the pair itself.

Similar to Eq. (5.4.6), we find the constant part of $G^{(2)}(\mathbf{r}_1, t_1; \mathbf{r}_2, t_2)$ is a product of the following two mean number of photons measured by D_1 and D_2, respectively,

$$\langle n(\mathbf{r}_1, t_1) \rangle \propto \sum_m \psi_m^*(\mathbf{r}_1, t_1) \, \psi_m(\mathbf{r}_1, t_1),$$

$$\langle n(\mathbf{r}_2, t_2) \rangle \propto \sum_n \psi_n^*(\mathbf{r}_2, t_2) \, \psi_n(\mathbf{r}_2, t_2),$$

while the cross-interference term, which is the nontrivial term of the second-order coherence function $G^{(2)}(\mathbf{r}_1, t_1; \mathbf{r}_2, t_2)$, corresponding to the photon number fluctuation correlation measured by D_1 and D_2, jointly,

$$\langle \Delta n(\mathbf{r}_1, t_1) \Delta n(\mathbf{r}_2, t_2) \rangle \propto \sum_{m \neq n} \psi_m^*(\mathbf{r}_1, t_1) \psi_n(\mathbf{r}_1, t_1) \psi_n^*(\mathbf{r}_2, t_2) \psi_m(\mathbf{r}_2, t_2),$$

is also the same as the photon number fluctuation correlation we have calculated in the single-photon state representation.

We find, again, the second-order quantum coherence function $G^{(2)}(\mathbf{r}_1, t_1; \mathbf{r}_2, t_2)$ calculated from the quantum coherent state representation is the same as the second-order coherence function $\Gamma^{(2)}(\mathbf{r}_1, t_1; \mathbf{r}_2, t_2)$ that we have calculated previously in Einstein's picture, except the subfield is replaced by the effective wavefunction of a photon or a group of identical photons.

In the case of a perfect measurement system, including idealized photodetectors and idealized correlation measurement electronics, $G^{(2)}(\mathbf{r}_1, t_1; \mathbf{r}_2, t_2)$ can be formally written in terms of the first-order quantum coherence functions:

$$G^{(2)}(\mathbf{r}_1, t_1; \mathbf{r}_2, t_2) = G_{11}^{(1)} G_{22}^{(1)} + G_{12}^{(1)} G_{21}^{(1)} = G_{11}^{(1)} G_{22}^{(1)} + |G_{12}^{(1)}|^2. \quad (5.4.18)$$

It should be emphasized that, again, here, $G_{12}^{(1)}$ is measured by two independent photodetectors D_1 and D_2 at different space–time coordinates $(\mathbf{r}_1, t_1)$ and $(\mathbf{r}_2, t_2)$, which is very different from the measurement of the first-order coherence function. The realistic finite response time of the photodetectors and correlation measurement circuit will complicate the calculation, especially when dealing with thermal state of broadband spectrum.

(IV) Two-photon effective wavefunction of thermal state in coherent state representation

Similar to the single-photon state representation, we may define a two-photon effective wavefunction for the m-nth pair of photons, or the m-nth pair of groups of identical photons, in the coherent state representation:

$$\psi_{mn}(\mathbf{r}_1, t_1; \mathbf{r}_2, t_2) = \psi_m(\mathbf{r}_1, t_1)\psi_n(\mathbf{r}_2, t_2).$$

The second-order coherence function $G^{(2)}(\mathbf{r}_1, t_1; \mathbf{r}_2, t_2)$ is thus written as the superposition between two-photon effective wavefunctions:

$$G^{(2)}(\mathbf{r}_1, t_1; \mathbf{r}_2, t_2) = \sum_{m,n} \left| \frac{1}{\sqrt{2}} [\psi_{mn}(\mathbf{r}_1, t_1; \mathbf{r}_2, t_2) + \psi_{nm}(\mathbf{r}_1, t_1; \mathbf{r}_2, t_2)] \right|^2.$$

In fact, the effective wavefunction of the m-nth photon pair, or the m-nth pair of groups of indistinguishable photons, can be defined directly from a product state

$$|\Psi_{mn}\rangle = |\Psi_m\rangle|\Psi_n\rangle,$$

with

$$|\Psi_m\rangle = \prod_{\mathbf{k}} |\alpha_m(\mathbf{k})\rangle, \quad |\Psi_n\rangle = \prod_{\mathbf{k}'} |\alpha_n(\mathbf{k}')\rangle.$$

The second-order coherence function of the m-nth photon pair is thus

$$
\begin{aligned}
G^{(2)}_{mn}&(\mathbf{r}_1, t_1; \mathbf{r}_2, t_2) \\
&= \langle \Psi_{mn}| \, E^{(-)}(\mathbf{r}_1, t_1)E^{(-)}(\mathbf{r}_2, t_2)E^{(+)}(\mathbf{r}_2, t_2)E^{(+)}(\mathbf{r}_1, t_1) \, |\Psi_{mn}\rangle \\
&= \langle \Psi_{mn}| \, E^{(-)}(\mathbf{r}_1, t_1)E^{(-)}(\mathbf{r}_2, t_2) \, |\Psi_{mn}\rangle \\
&\quad \times \langle \Psi_{mn}| \, E^{(+)}(\mathbf{r}_2, t_2)E^{(+)}(\mathbf{r}_1, t_1) \, |\Psi_{mn}\rangle \\
&= |\psi_{mn}(\mathbf{r}_1, t_1; \mathbf{r}_2, t_2)|^2.
\end{aligned}
$$

The effective wavefunction of the m-nth photon pair, which is characterized by a product state, is defined as

$$\psi_{mn}(\mathbf{r}_1, t_1; \mathbf{r}_2, t_2) = \langle \Psi_{mn}|E^{(+)}(\mathbf{r}_2, t_2)E^{(+)}(\mathbf{r}_1, t_1)|\Psi_{mn}\rangle. \qquad (5.4.19)$$

Thermal state concerns an ensemble of mixed two-photon product states; the second-order quantum coherence function can be calculated from the density operator:

$$G^{(2)}(\mathbf{r}_1, t_1; \mathbf{r}_2, t_2)$$
$$= \langle\langle E^{(-)}(\mathbf{r}_1, t_1)E^{(-)}(\mathbf{r}_2, t_2)E^{(+)}(\mathbf{r}_2, t_2)E^{(+)}(\mathbf{r}_1, t_1)\rangle_{\mathrm{QM}}\rangle_{\mathrm{En}}$$
$$= tr\left[\hat{\rho}\, E^{(-)}(\mathbf{r}_1, t_1)E^{(-)}(\mathbf{r}_2, t_2)\, E^{(+)}(\mathbf{r}_2, t_2)E^{(+)}(\mathbf{r}_1, t_1)\right]$$
$$= \sum_{m,n}\langle\Psi_{mn}|\, E^{(-)}(\mathbf{r}_1, t_1)E^{(-)}(\mathbf{r}_2, t_2)E^{(+)}(\mathbf{r}_2, t_2)E^{(+)}(\mathbf{r}_1, t_1)\,|\Psi_{mn}\rangle$$
$$= \sum_{m,n} G^{(2)}_{mn}(\mathbf{r}_1, t_1; \mathbf{r}_2, t_2) = \sum_{m,n}|\psi_{mn}(\mathbf{r}_1, t_1; \mathbf{r}_2, t_2)|^2, \qquad (5.4.20)$$

where we have considered $P_{mn} =$ constant, again, for thermal state.

(V) Second-order quantum temporal coherence function of thermal state

Assuming a far-field joint photon counting measurement illustrated in Fig. 5.3.1, the mth and the nth photons both have 50%-50% chances to be detected by D_1 and D_2 in the following two different yet indistinguishable alternative ways: (1) The mth photon is annihilated at D_1, while the nth photon is annihilated at D_2; (2) the mth photon is annihilated at D_2, while the nth photon is annihilated at D_1. The 1-D effective two-photon wavefunction of the m–nth photon pair is calculated from the field operators and the product state:

$$\psi_{mn}(z_1, t_1; z_2, t_2) = \frac{1}{\sqrt{2}}\left[\psi_{mn}(z_1, t_1; z_2, t_2) + \psi_{nm}(z_1, t_1; z_2, t_2)\right],$$

where

$$\psi_{mn}(z_1, t_1; z_2, t_2) = \langle\Psi_{mn}|\hat{E}_n^{(+)}(z_2, t_2)\hat{E}_m^{(+)}(z_1, t_1)|\Psi_{mn}\rangle$$
$$= \langle\Psi_m|\hat{E}_m^{(+)}(z_1, t_1)|\Psi_m\rangle\langle\Psi_n|\hat{E}_n^{(+)}(z_2, t_2)|\Psi_n\rangle$$
$$= \psi_m(z_1, t_1)\psi_n(z_2, t_2),$$
$$\psi_{nm}(z_1, t_1; z_2, t_2) = \langle\Psi_{mn}|\hat{E}_n^{(+)}(z_1, t_1)\hat{E}_m^{(+)}(z_2, t_2)|\Psi_{mn}\rangle$$
$$= \langle\Psi_n|\hat{E}_n^{(+)}(z_1, t_1)|\Psi_n\rangle\langle\psi_m|\hat{E}_m^{(+)}(z_2, t_2)|\Psi_m\rangle$$
$$= \psi_n(z_1, t_1)\psi_m(z_2, t_2).$$

The two effective wavefunctions in the above equation represent two different yet indistinguishable two-photon amplitudes: (1) The mth and nth photon or group identical photons are measured at D_1 and D_2, respectively; (2) the mth and nth photon or group identical photons are measured at D_2 and D_1, respectively. The superposition of the above two-photon amplitudes yields a nontrivial second-order correlation:

$$G^{(2)}(z_1, t_1; z_2, t_2) \propto \sum_{m,n} \frac{1}{2} \left| \psi_{mn}(z_1, t_1; z_2, t_2) + \psi_{nm}(z_1, t_1; z_2, t_2) \right|^2$$

$$= G^{(1)}_{11} G^{(1)}_{22} + G^{(1)}_{12} G^{(1)}_{21}$$

with

$$G^{(1)}_{11} G^{(1)}_{22} = \sum_m \left| \psi_m(z_1, t_1) \right|^2 \sum_n \left| \psi_n(z_2, t_2) \right|^2$$

$$= \sum_m \left| \mathcal{F}_{\tau_1 - t_{0m}} \{ f(\nu) \} \right|^2 \sum_n \left| \mathcal{F}_{\tau_2 - t_{0n}} \{ f(\nu) \} \right|^2$$

$$\simeq \int dt_{0m} \left| \mathcal{F}_{\tau_1 - t_{0m}} \{ f(\nu) \} \right|^2 \int dt_{0n} \left| \mathcal{F}_{\tau_1 - t_{0n}} \{ f(\nu) \} \right|^2$$

$$= G_0^2,$$

$$G^{(1)}_{12} G^{(1)}_{21} = \sum_{m,n} \psi_m^*(z_1, t_1) \psi_m(z_2, t_2) \psi_n^*(z_2, t_2) \psi_n(z_1, t_1)$$

$$= \sum_m \left[\mathcal{F}^*_{\tau_1 - t_{0m}} \{ f(\nu) \} \mathcal{F}_{\tau_2 - t_{0m}} \{ f(\nu) \} \right]$$

$$\times \sum_n \left[\mathcal{F}^*_{\tau_2 - t_{0n}} \{ f(\nu) \} \mathcal{F}_{\tau_1 - t_{0n}} \{ f(\nu) \} \right]$$

$$\simeq \int dt_{0m} \left[\mathcal{F}^*_{\tau_1 - t_{0m}} \{ f(\nu) \} \mathcal{F}_{\tau_2 - t_{0m}} \{ f(\nu) \} \right]$$

$$\times \int dt_{0n} \left[\mathcal{F}^*_{\tau_1 - t_{0n}} \{ f(\nu) \} \mathcal{F}_{\tau_2 - t_{0n}} \{ f(\nu) \} \right]$$

$$= G_0^2 \left| \mathcal{F}_\tau \{ f^2(\nu) \} \right|^2,$$

where we assume a randomly radiated thermal light source with equal probability of creating a photon at time t_{om}, and a perfect joint measurement system, including the photon counting detectors and the joint measurement circuit, and have taken the wavepackets in the form of their Fourier transforms. It is clearly shown in the above exercise that this unusual behavior is determined by the product nature of the two-photon state.

The second-order quantum temporal coherence function of thermal state is thus

$$G^{(2)}(z_1, t_1; z_2, t_2) = G_0 \left[1 + \left| \mathcal{F}_\tau \{ f^2(\nu) \} \right|^2 \right] = G^{(2)}(\tau) \qquad (5.4.21)$$

which is a function of the temporal delay τ, allowing for accumulative measurement.

Based on the above analysis, either in terms of the single-photon state representation or in terms of the coherent state representation, we may conclude that the calculation of the second-order quantum coherence function $G^{(2)}(\mathbf{r}_1, t_1; \mathbf{r}_2, t_2)$ of thermal field, either temporal or spatial, is the same as that of our previously calculated second-order coherence $\Gamma^{(2)}(\mathbf{r}_1, t_1; \mathbf{r}_2, t_2)$ by simply replacing Einstein's subfield with the effective wavefunction of a photon or a group of indistinguishable photons. Consequently, repeating the previous calculations in Einstein's picture, we may conclude the same second-order temporal-spatial coherence function as well as the degree of second-order temporal-spatial coherence of thermal field. Physically, we have not been able to find any differences between the second-order temporal and spatial coherence functions measured in joint photon-counting mode and measured in continuous analog mode.

Figure 5.4.1 shows a typical measurement of the degree of second-order temporal coherence of a pseudo-thermal field, $g^{(2)}(t_1 - t_2)$, in photon-counting regime. The experimental results clearly show that the photons have twice greater chance of been jointly detected at $t_1 = t_2$ by photodetectors D_1 and D_2 located at $z_1 = z_2$. These experimental effects were considered "photon bunching" in the history. The observation of "photon bunching" is quite a surprise, which apparently contradicts with the random nature of thermal light. Remember, the two photons in thermal state are created randomly with equal probabilities at any time and any sub-source, t_{0m} and t_{0n}. There is no physical mechanism to create photons in "bunching" from stochastic radiation processes. In the view of quantum optics, "photon bunching" is nothing but a *two-photon interference* phenomenon: Although the photons are all created independently and randomly from the thermal light source, the two-photon amplitudes are superposed constructively when $\tau_1 - \tau_2 = 0$. It should be emphasized that *two-photon interference* is not the interference between two photons. The interference takes place between two different yet indistinguishable alternatives, namely the two-photon amplitudes that contributed

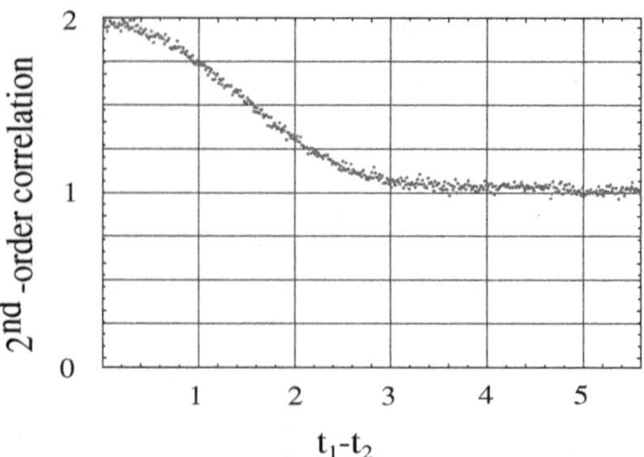

Fig. 5.4.1　Experimental data for the measurement of second-order temporal coherence function $G^{(2)}(\tau)$ of a pseudo-thermal field by joint photon-counting. In this measurement, D_1 and D_2 were fixed at $z_1 = z_2$. The measurement is on the second-order correlation function $G^{(2)}(t_1 - t_2)$. Two histograms of the photodetection events, n_j vs t_j, of D_1 and D_2 in the jth time window, were recorded by an "event timer". The second-order time correlation was calculated from the two histograms. The horizontal axis $t_1 - t_2$, in the unit of 10^{-8} s, labels the registration time differences of the photon pair measured by photon-counting detectors D_1 and D_2 at $z_1 = z_2$.

to a joint-photodetection event. Although the quantum interference picture for the phenomenon is quite straightforward, one may feel uncomfortable, perhaps, due to the nonlocal nature of two-photon interference. The two-photon superposition is "nonlocal" which involves two separated space-time coordinates through the measurement of two independent photodetectors. Thermal light is considered "classical" radiation in some theories. However, we should not forget quantum theory begins with the study of thermal light: Blackbody is a typical thermal source for generating thermal radiation, and the Planck theory which leads to the quantization of electromagnetic fields is all about blackbody radiation.

(VI) Second-order quantum near-field spatial coherence function of thermal state

We have concluded that the quantum second-order coherence function $G^{(2)}(\mathbf{r}_1, t_1; \mathbf{r}_2, t_2)$ can be calculated from Einstein's picture by simply replacing the quantized subfields with the effective wavefunctions. The following is an exercise to derive the second-order quantum near-field spatial coherence function $G^{(2)}(\vec{\rho}_1, z_1; \vec{\rho}_2, z_2)$ from our previously calculated second-order near-field coherence function $\Gamma^{(2)}(\vec{\rho}_1, z_1; \vec{\rho}_2, z_2)$:

$$G^{(2)}(\vec{\rho}_1, z_1; \vec{\rho}_2, z_2)$$

$$= \sum_{m,n} \left| \frac{1}{\sqrt{2}} \left[\psi_{mn}(\vec{\rho}_1, z_1; \vec{\rho}_2, z_2) + \psi_{nm}(\vec{\rho}_1, z_1; \vec{\rho}_2, z_2) \right] \right|^2$$

$$= \sum_{m,n} \left| \frac{1}{\sqrt{2}} \left[\psi_m(\vec{\rho}_1, z_1)\psi_n(\vec{\rho}_2, z_2) + \psi_n(\vec{\rho}_1, z_1)\psi_m(\vec{\rho}_2, z_2) \right] \right|^2, \quad (5.4.22)$$

by replacing Einstein's subfield with quantum effective wavefunction. In the above calculation, again, we have ignored the temporal variables by assuming a perfect second-order temporal correlation similar to our early calculations for second-order spatial correlation. Substitute Green's function or the near-field Fresnel propagator into the effective wavefunctions, either derived from single-photon state representation or from coherence state representation, we have

$$G^{(2)}_{mn}(\vec{\rho}_1, z_1; \vec{\rho}_2, z_2)$$

$$= \left| |\alpha_{mn}|^2 \frac{1}{\sqrt{2}} \left[g_m(\vec{\rho}_1, z_1)\, g_n(\vec{\rho}_2, z_2) + g_n(\vec{\rho}_1, z_1)\, g_m(\vec{\rho}_2, z_2) \right] \right|^2 \quad (5.4.23)$$

with

$$g_m(\vec{\rho}_j, z_j) = \frac{c_0}{z_j} e^{i\frac{\omega z_j}{c}} e^{i\frac{\omega}{2cz_j}|\vec{\rho}_j - \vec{\rho}_m|^2},$$

where c_0 is a normalization consistent. Now, consider the sum of $G^{(2)}_{mn}(\vec{\rho}_1, z_1; \vec{\rho}_2, z_2)$ by turning the sum of m and n into the integrals of $\vec{\rho}_0$ and $\vec{\rho}_0'$ on the entire source plane:

$$G^{(2)}(\vec{\rho}_1, z_1; \vec{\rho}_2, z_2)$$

$$= G_0^2 \left[\left| \int d\vec{\rho}_0 g_{\vec{\rho}_0}(\vec{\rho}_1, z_1) \int d\vec{\rho}_0' g_{\vec{\rho}_0'}(\vec{\rho}_2, z_2) \right|^2 \right.$$

$$\left. + \int d\vec{\rho}_0 g^*_{\vec{\rho}_0}(\vec{\rho}_1, z_1) g_{\vec{\rho}_0}(\vec{\rho}_2, z_2) \int d\vec{\rho}_0' g^*_{\vec{\rho}_0'}(\vec{\rho}_2, z_2) g_{\vec{\rho}_0'}(\vec{\rho}_1, z_1) \right]$$

$$= G^{(1)}_{11} G^{(1)}_{22} + G^{(1)}_{12} G^{(1)}_{21}. \quad (5.4.24)$$

We then follow the same mathematical procedure as that in the calculation of $\Gamma^{(2)}(\vec{\rho}_1, \vec{\rho}_2)$; the second-order near-field spatial quantum correlation function is calculated to be

$$G^{(2)}(\vec{\rho}_1, \vec{\rho}_2) = G_0^2 \left[1 + \mathrm{somb}^2\left(\frac{R}{d} \frac{\omega}{c} |\vec{\rho}_1 - \vec{\rho}_2| \right) \right], \tag{5.4.25}$$

which is the same as $\Gamma^{(2)}(\vec{\rho}_1, \vec{\rho}_2)$. Consequently, the degree of second-order spatial coherence is

$$g^{(2)}(\vec{\rho}_1, \vec{\rho}_2) = 1 + \mathrm{somb}^2\left(\frac{R}{d} \frac{\omega}{c} |\vec{\rho}_1 - \vec{\rho}_2| \right), \tag{5.4.26}$$

which is also the same as $\gamma^{(2)}(\vec{\rho}_1, \vec{\rho}_2)$ calculated from Einstein's picture. This result has been experimentally confirmed.

5.5   Second-Order Coherence of Entangled State and Number State

Quantum entanglement is discussed in Chapter 6. In this section, we simply introduce an entangled state and a number state of $n = 2$ and calculate their second-order quantum coherence function $G^{(2)}(\mathbf{r}_1, t_1; \mathbf{r}_2, t_2)$. The incoherent or coherent radiation sources in the experimental setup of Fig. 5.3.1 are replaced by the entangled two-photon state or the number state of $n = 2$.

(I) Entangled biphoton state of SPDC

In the following, an entangled two-photon state generated from a nonlinear optical process of spontaneous parametric down-conversion (SPDC) is applied for the calculation of the second-order coherence function $G^{(2)}(\mathbf{r}_1, t_1; \mathbf{r}_2, t_2)$. Roughly speaking, an entangled two-photon state is a nonfactorizable state of two-photons: The state cannot be written as a product of that of two photons. In other words, the two photons cannot be considered as independent. The two-photon state of SPDC is a very special entangled two-particle state. The two subsystems are "perfectly correlated", i.e., by measuring the state of one, the state of the other one is determined with certainty, although none of the subsystems is in a defined state before the measurement. The entangled signal–idler photon pair of SPDC was named biphoton by Klyshko. To simplify the mathematics, the following calculation is approximated in 1-D by considering the following nonfactorizable biphoton state with total photon number $n = 2$:

$$|\Psi\rangle = \int d\omega_s d\omega_i \, \delta(\omega_s + \omega_i - \omega_p) \, f(\omega_s, \omega_i) \, \hat{a}^\dagger(\omega_s) \, \hat{a}^\dagger(\omega_i) \, |0\rangle$$

$$\simeq \int_{-\infty}^{\infty} d\nu \, f(\nu) \, a^\dagger(\omega_s^0 + \nu) \, a^\dagger(\omega_i^0 - \nu)|0\rangle \tag{5.5.1}$$

with

$$\omega_s = \omega_s^0 + \nu, \quad \omega_i = \omega_i^0 - \nu, \quad \omega_s + \omega_i = \omega_p = \text{constant},$$

where $\delta(\omega_p - \omega_s - \omega_i)$ has been applied and ω_s^0 and ω_i^0 are the center frequencies for the signal mode ω_s and the idler mode ω_i, respectively. We have generalized the integral to infinity with the help of the normalized spectral density function $f(\nu)$. The biphoton state of Eq. (5.5.1) is a pure two-photon state by means: (1) The state of a signal–idler pair is a state vector in the form of a coherent superposition of all possible Fock states of $n = 2$, $|\ldots, 0, 1_{\omega_s}, \ldots, 1_{\omega_i}, 0, \ldots\rangle$; (2) a large number of jointly measured signal–idler photon pairs are in the same state, which can be treated as a homogenous ensemble.

Assume a simple far-field 1-D measurement in which a point-like photon counting detector D_1 registers a photodetection event at (z_1, t_1) and another point-like photon counting detector D_2 observes another photodetection event at (z_2, t_2), the second-order temporal coherence function is thus calculated as

$$
\begin{aligned}
G^{(2)}&(z_1, t_1; z_2, t_2) \\
&= \langle \Psi | \hat{E}^{(-)}(z_1, t_1) \hat{E}^{(-)}(z_2, t_2) \hat{E}^{(+)}(z_2, t_2) \hat{E}^{(+)}(z_1, t_1) | \Psi \rangle \\
&= \sum_n \langle \Psi | \hat{E}^{(-)}(z_1, t_1) \hat{E}^{(-)}(z_2, t_2) |n\rangle \langle n| \hat{E}^{(+)}(z_2, t_2) \hat{E}^{(+)}(z_1, t_1) | \Psi \rangle \\
&= | \langle 0 | \hat{E}^{(+)}(z_2, t_2) \hat{E}^{(+)}(z_1, t_1) | \Psi \rangle |^2 \\
&= | \Psi(z_1, t_1; z_2, t_2) |^2,
\end{aligned}
\tag{5.5.2}
$$

where, in the third line, we have used the completeness relation

$$\sum_n |n\rangle \langle n| = 1.$$

The effective wavefunction of a pair of entangled signal–idler photons, characterized by the state of Eq. (5.5.1), is thus defined in Eq. (5.5.2):

$$\Psi(z_1, t_1; z_2, t_2) = \langle 0 | \hat{E}^{(+)}(z_2, t_2) \hat{E}^{(+)}(z_1, t_1) | \Psi \rangle. \tag{5.5.3}$$

The two-photon effective wavefunction is then calculated as

$$
\begin{aligned}
\Psi&(z_1, t_1; z_2, t_2) \\
&= \langle 0 | \left[\int d\omega\, \hat{a}(\omega)\, e^{-i(\omega t_2 - k z_2)} \right] \left[\int d\omega'\, \hat{a}(\omega')\, e^{-i(\omega' t_1 - k' z_1)} \right] \\
&\quad \times \left[\int_{-\infty}^{\infty} d\nu\, f(\nu)\, a^\dagger(\omega_s^0 + \nu)\, a^\dagger(\omega_i^0 - \nu)|0\rangle \right]
\end{aligned}
$$

$$\cong \int d\omega \int d\omega' \int_{-\infty}^{\infty} d\nu \, f(\nu) \, e^{-i\omega'\tau_2} \, e^{-i\omega\tau_1}$$

$$\times \, \langle \, 0 \, | \, \hat{a}(\omega) \, \hat{a}(\omega') \, a^{\dagger}(\omega_s^0 + \nu) \, a^{\dagger}(\omega_i^0 - \nu) | \, 0 \, \rangle. \tag{5.5.4}$$

We consider a simplified joint measurement in which D_1 that is located at z_1 only receives the signal and D_2 that is located at z_2 only receives the idler. The signal and idler can be distinguished experimentally either by spectral filters when $\omega_s^0 \neq \omega_i^0$, or by polarization analyzers when the signal and idler are orthogonal polarized. In this case, the only surviving term in Eq. (5.5.4) is

$$\Psi(z_1, t_1; z_2, t_2) \cong \int_{-\infty}^{\infty} d\nu \, f(\nu) \, e^{-i(\omega_i^0 - \nu)\tau_2} \, e^{-i(\omega_s^0 + \nu)\tau_1}$$

$$= \Psi_0 e^{-i(\omega_s^0 \tau_1 + \omega_i^0 \tau_2)} \, \mathcal{F}_{\tau_1 - \tau_2} \left\{ f(\nu) \right\}$$

$$= \Psi_0 e^{-i\frac{\omega_p}{2}(\tau_1 + \tau_2)} \, \mathcal{F}_{\tau_1 - \tau_2} \left\{ f(\nu) \right\} e^{-i\omega_d(\tau_1 - \tau_2)}, \tag{5.5.5}$$

where Ψ_0 is a normalization constant and $\omega_d = (\omega_s^0 - \omega_i^0)/2$. It is clear that the two-photon effective wavefunction of the biphoton system cannot be written as a product of two effective wavefunctions of signal photon and idler photon. Different from the randomly created and randomly paired photons in thermal state, the entangled signal–idler photon pair is characterized by a nonlocal wavepacket instead of two wavepackets. We discuss this in detail in the following chapter. The second-order coherence function is thus

$$G^{(2)}(z_1, t_1; z_2, t_2) = \left| \Psi(z_1, t_1; z_2, t_2) \right|^2 = |\Psi_0|^2 \left| \mathcal{F}_{\tau} \left\{ f(\nu) \right\} \right|^2, \tag{5.5.6}$$

indicating a temporal correlation function of $\tau = \tau_1 - \tau_2$. It is easy to see from Eqs. (5.5.5) and (5.5.6) that this correlation is the result of a two-photon interference: an entangled signal–idler photon pair interference with the pair itself by means of a superposition among infinite number of two-photon amplitudes $f(\nu) \, e^{-i(\omega_i^0 - \nu)\tau_2} \, e^{-i(\omega_s^0 + \nu)\tau_1}$ with $-\infty \leq \nu \leq \infty$.

A schematic picture of a biphoton wavepacket described by Eq. (5.5.5), is illustrated in Fig. 5.5.1. It shows clearly that $\Psi(z_1, t_1; z_2, t_2)$ of the entangled two-photon system is represented by a nonfactorizable 2-dimension wavepacket instead of a product of two wavepackets. The nonfactorizable wavepacket, again, from a different point of view, illustrates the entangled nature of the two-photon state. One should not be surprised by the fact that the system does not have two individual wavepackets associated with

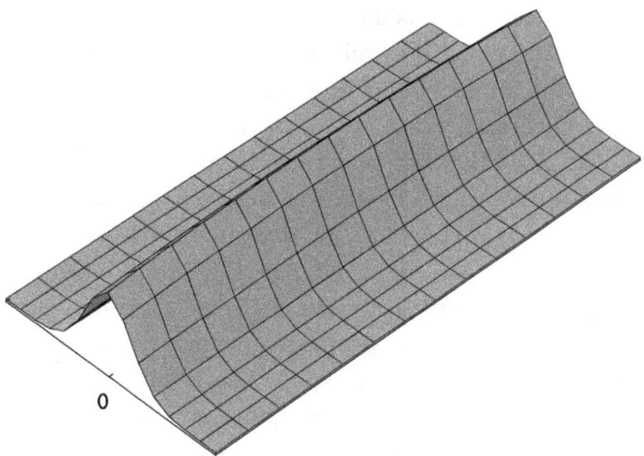

Fig. 5.5.1 A schematic envelope of a biphoton wavepacket with a Gaussian shape along $\tau_1 - \tau_2$. The wavepacket is uniformly distributed along $\tau_1 + \tau_2$ due to the assumption of $\omega_p = $ constant.

the signal photon and the idler photon. The biphoton state is formed from a coherent superposition of Fock states of $n = 2$ instead of a product of two individual single-photon wavepackets. The coherent superposition of these Fock states results in the nonfactorizable effective wavefunction or the biphoton wavepacket. This is very different from the classical case. It may be easier to see the connection between the picture of a single-photon wavepacket and the concept of a photon. It is definitely not easy to accept the fact that the connection between the biphoton wavepacket and the concept of biphoton is the same as that of the single-photon. The biphoton wavepacket describes the dynamic property of the entangled photon pair as one system in space–time. We will return to the discussion of these important physics again later.

Phenomenologically, unlike the single-photon wavepacket, (1) the biphoton wavepacket is dynamically propagating along the axis z_1 and z_2 in different directions; however, (2) the propagation is never independent. The joint photoelectron events happen in such a way that whenever a photoelectron event occurs at space–time $(\mathbf{r}_1, t_1)$, the other one can happen only in the neighborhood of space–time point $(\mathbf{r}_2, t_2)$, which is related to $(\mathbf{r}_1, t_1)$ by $(t_1 - t_2) - (z_1 - z_2)/c = 0$. The statement of point (1) is formulated by $e^{-i(\omega_0 t_2 - k_0 z_2)} e^{-i(\omega_0 t_1 - k_0 z_1)}$ or $e^{-i\omega_0(\tau_1 + \tau_2)}$ along the $\tau_1 + \tau_2$ axis. The statement of point (2) is formulated by $\mathcal{F}_{\tau_1 - \tau_2}\{f(\nu)\}$ along the

$\tau_1 - \tau_2$ axis. This phenomenon is interesting. On one hand, the signal photon and the idler photon are both in mixed single-photon state with randomly distributed probability of producing a photoelectron event along both z_1 and z_2 axes in space and along t_1 and t_2 in time. It means that neither the signal photon nor the idler photon is localized in space-time. On the other hand, the space–time correlation between the joint photoelectron events, $(t_1 - t_2) - (z_1 - z_2)/c = 0$, determines the location and time of finding the idler (signal) photon after the annihilation of the signal (idler) photon. How could the detection of the signal (idler) photon localize the idler (signal) photon? The problem gets more interesting when the two photodetection events become space-like separated events: The two detectors could be separated in light-years but the determination, however, is made instantaneously. It seems that we may have to give up the concept of two photons when thinking about an EPR two-photon state. The signal–idler system might be better treated as a whole, as described by the nonfactorizable biphoton wavepacket. If one insists that the signal photon and the idler photon are two individual systems, the EPR paradox may be unavoidable. This point is emphasized again in the following chapter.

(II) Number state of $n_\omega = 2$

In Chapter 2, we have given an example of number state of $n_\omega = 2$:

$$|\Psi\rangle = \sum_\omega f(\omega)[\hat{a}^\dagger(\omega)]^2|0\rangle \simeq \int_{-\infty}^\infty d\nu f(\nu)[a^\dagger(\omega_0 + \nu)]^2|0\rangle. \tag{5.5.7}$$

The two-photon effective wavefunction is calculated as

$$\Psi(z_1, t_1; z_2, t_2)$$
$$= \langle 0| \left[\int d\omega\, a(\omega)\, e^{-i(\omega t_2 - k z_2)}\right] \left[\int d\omega'\, a(\omega')\, e^{-i(\omega' t_1 - k' z_1)}\right]$$
$$\times \left[\int_{-\infty}^\infty d\nu\, f(\nu)\, a^\dagger(\omega_0 + \nu)\, a^\dagger(\omega_0 + \nu)|0\rangle\right]$$
$$\cong \int d\omega \int d\omega' \int_{-\infty}^\infty d\nu\, f(\nu)\, e^{-i\omega' \tau_2}\, e^{-i\omega \tau_1}$$
$$\times \langle 0| a(\omega) a(\omega')\, a^\dagger(\omega_0 + \nu)\, a^\dagger(\omega_0 + \nu)|0\rangle. \tag{5.5.8}$$

The photodetectors D_1 that is located at z_1 and D_2 that is located at z_2 cannot distinguish the two modes of $\omega = \omega_0 + \nu$. In this case, the surviving terms are these when $\omega = \omega' = \omega_0 + \nu$, thus

$$\Psi(z_1, t_1; z_2, t_2) \cong \int_{-\infty}^{\infty} d\nu \, f(\nu) \left[e^{-i(\omega_0+\nu)(\tau_2+\tau_1)} \right]$$

$$= \Psi_0 \, \mathcal{F}_{\tau_2+\tau_1} \{ f(\nu) \} e^{-i\omega_0(\tau_2+\tau_1)}, \qquad (5.5.9)$$

where Ψ_0 is a normalization constant. Again, $\Psi(z_1, t_1; z_2, t_2)$ cannot be written as the product of $\Psi(z_1, t_1)$ and $\Psi(z_2, t_2)$. The number state of $n_\omega = 2$ is also characterized by a nonfactorizable 2-dimensional function. However, this 2-D function is very different from either Eq. (5.5.5). The second-order coherence is thus

$$G^{(2)}(z_1, t_1; z_2, t_2) = |\Psi_0|^2 \left| \mathcal{F}_{\tau_2+\tau_1} \{ f(\nu) \} \right|^2. \qquad (5.5.10)$$

It is interesting to see that $G^{(2)}(z_1, t_1; z_2, t_2)$ will be a delta-function like function of $\tau_2 + \tau_1$ if $\Delta\nu \sim \infty$. In this case, $(t_1 + t_2) \simeq (z_1 + z_2)/c$. The physics behind is very different from that of the entangled states too.

5.6 Second-Order Coherence of Laser Beam

Figure 5.6.1 is a schematic experimental setup for measuring the second-order temporal and spatial coherence of laser beams. A well-collimated laser beam is divided by a 50%-50% beamsplitter. Point-like photodetectors D_1 and D_2 are scannable longitudinally and transversely in the far-field zones of the transmitted and the reflected arms for the joint-detection of the radiation. The joint-detection circuit contains a current-current linear multiplier which has been described earlier. The output of the

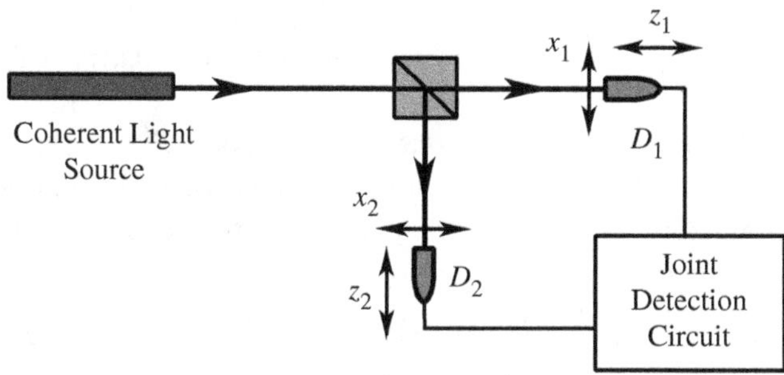

Fig. 5.6.1 A schematic measurement of the second-order temporal coherence function $\Gamma^{(2)}(z_1, t_1; z_2, t_2)$ or the spatial coherence function $\Gamma^{(2)}(x_1, t_1; x_2, t_2)$ of a coherent radiation source.

current-current linear multiplier, which is proportional to the temporal coherence function $\Gamma^{(2)}(z_1, t_1; z_2, t_2)$, or the spatial coherence function $\Gamma^{(2)}(x_1, t_1; x_2, t_2)$ is recorded as a function of the relative delay or spatial separation of the measured fields, either by scanning the electronic delays or the longitudinal optical delays; or by scanning D_1 and D_2 transversely, as shown in the figure.

Light is conventionally classified into coherent radiation and incoherent radiation under the framework of classical electromagnetic theory of light. The radiation produced by a laser, either CW or pulse, is usually considered coherent light. As we pointed out in Chapter 1, this "convention" is very confusing. In Einstein's picture, a single cavity-mode CW laser beam is the result of a coherent superposition of subfields and Fourier modes; but the superposition of cavity modes of a multi-cavity-mode CW laser is incoherent, since the relative phases between cavity modes are random. Furthermore, the second-order correlation of CW laser beams, either single cavity-mode or multi-cavity-mode, are usually considered to be trivial, since the intensity fluctuations of CW laser beams are negligible, $\bar{I} \gg |\Delta I|$. As we will see from the following discussions, these conventional views are improper.

(I) Second-order temporal coherence of single cavity-mode CW laser beam

In Einstein's picture, a single cavity-mode laser beam is the result of coherent superposition of subfields and Fourier modes. The electromagnetic field measured by the jth photodetector is the following coherent superposition:

$$E(z_j, t_j) = \int d\omega\, E(\omega)\, e^{-i\omega\tau_j} = \mathcal{F}_{\tau_j}\{E(\nu)\} e^{-i\omega_0\tau_j} \tag{5.6.1}$$

where we have taken into account the finite spectral bandwidth of the cavity mode with a spectral distribution function $E(\omega)$, and ω_0 is the central frequency of the mode; $E(\nu) = \sum_m a_m(\nu)$ is the total amplitude contributed by all possible stimulatory emitted subfields; $\tau_j \equiv t_j - r_j/c$ as usual.

In general, the second-order coherence function $\Gamma^{(2)}(z_1, t_1; z_2, t_2)$ of a single cavity mode laser beam is simply a factorizeable function of two expected intensities measured by D_1 and D_2 at coordinates (z_1, t_1) and (z_2, t_2), respectively,

$$\begin{aligned}
\Gamma^{(2)}(z_1, t_1; z_2, t_2) &= \langle E^*(z_1, t_1)E(z_1, t_1)E^*(z_2, t_2)E(z_2, t_2)\rangle \\
&= \langle \mathcal{F}^*_{\tau_1}\{E(\nu)\} e^{i\omega_0\tau_1}\, \mathcal{F}_{\tau_1}\{E(\nu)\} e^{-i\omega_0\tau_1} \\
&\quad \times \mathcal{F}^*_{\tau_2}\{E(\nu)\} e^{i\omega_0\tau_2}\mathcal{F}_{\tau_2}\{E(\nu)\} e^{-i\omega_0\tau_2}\rangle \\
&= \Gamma^{(1)}(z_1, t_1; z_1, t_1)\,\Gamma^{(1)}(z_2, t_2; z_2, t_2), \tag{5.6.2}
\end{aligned}$$

and consequently the normalized degree of second-order coherence of coherent radiation turns to be

$$\gamma^{(2)}(z_1, t_1; z_2, t_2) = 1. \tag{5.6.3}$$

This type of second-order correlation is named a "trivial" correlation. Statistically, this trivial correlation means no correlation between the two temporally separated radiations measured at (z_1, t_1) and (z_2, t_2).

The above result is consistent with that of quantum theory. Consider the state of radiation created by a single cavity-mode CW laser as a coherent state

$$|\Psi\rangle = \prod_\omega |\alpha(\omega)\rangle \tag{5.6.4}$$

representing the state of a group of indistinguishable photons. Note, the above state is in single-mode, but not in single-frequency. When the finite spectral bandwidth of the cavity-mode is taken into account, for a far-field measurement, the field operators at $(\mathbf{r}_1, t_1)$ and $(\mathbf{r}_2, t_2)$ are:

$$\hat{E}^{(+)}(z_j, t_j) = \int d\omega' \, \hat{a}(\omega') \, g_0(\omega; z_j, t_j)$$

$$\hat{E}^{(-)}(z_j, t_j) = \int d\omega \, \hat{a}^\dagger(\omega) \, g_0^*(\omega; z_j, t_j). \tag{5.6.5}$$

Since coherent state is an eigen state of the annihilation operator, the calculation of $G^{(2)}(z_1, t_1; z_2, t_2)$ is straightforward,

$$G^{(2)}(z_1, t_1; z_2, t_2)$$

$$= \prod_\omega \langle \alpha(\omega)| \, \hat{E}^{(-)}(z_1, t_1)\hat{E}^{(-)}(z_2, t_2)\hat{E}^{(+)}(z_2, t_2)\hat{E}^{(+)}(z_1, t_1) \prod_\omega |\alpha(\omega')\rangle$$

$$= \int d\omega \, \alpha^*(\omega) \, e^{i\omega\tau_1} \int d\omega'' \, \alpha^*(\omega'') \, e^{i\omega''\tau_2} \int d\omega''' \, \alpha(\omega''') \, e^{-i\omega'''\tau_2}$$

$$\times \int d\omega' \, \alpha(\omega') \, e^{-i\omega'\tau_1}$$

$$= \Psi^*(z_1, t_1)\Psi^*(z_2, t_2)\Psi(z_2, t_2)\Psi(z_1, t_1)$$

$$= G^{(1)}(z_1, t_1; z_1, t_1) \, G^{(1)}(z_2, t_2; z_2, t_2) \tag{5.6.6}$$

where $\Psi(z_j, t_j)$, $j = 1, 2$, is the effective wavefunction defined in previous Chapters. It is easy to find from Eq. (5.6.6) that

$$g^{(2)}(\mathbf{r}_1, t_1; \mathbf{r}_2, t_2) = 1. \tag{5.6.7}$$

(II) Second-order temporal coherence of multi-cavity-mode CW laser beam

Consider a CW laser beam consisting of N incoherently superposed cavity modes. For laser cavities that are several kilometers long, which has been achieved in fiber lasers, the number of cavity modes N may approach a million. The incoherent superposition of these cavity modes cannot be ignored when measure the second-order coherence functions of the multi-cavity-mode CW laser beams.

We start our calculation with the superposed electromagnetic fields measured by the photodetector D_j,

$$E(z_j, t_j) = \sum_{m=1}^{N} E_m(z_j, t_j) = \sum_{m=1}^{N} e^{i\varphi_m} \mathcal{F}_{\tau_j}\{E(\nu)\} e^{-i\omega_m \tau_j} \tag{5.6.8}$$

where m labels the mth cavity mode, again, we have taken into account the finite spectral bandwidth of each cavity mode with a spectral distribution function $E(\omega)$. To emphasize the incoherent superposition of cavity modes, the random phase φ_m is explicitly written out. The cavity produces equal spacing ω_b between cavity modes, such that the carrier frequency of the mth mode can be written as $\omega_m = \omega_0 + m\omega_b$.

The second-order temporal coherence function of the multi-cavity-mode CW laser beam is calculated as follows:

$$\Gamma^{(2)}(\tau_1, \tau_2) = \left\langle \sum_{m=1}^{N} E_m^*(\tau_1) \sum_{n=1}^{N} E_n(\tau_1) \sum_{p=1}^{N} E_p^*(\tau_2) \sum_{q=1}^{N} E_q(\tau_2) \right\rangle$$

$$= \sum_{m=1}^{N} E_m^*(\tau_1) E_m(\tau_1) \sum_{n=1}^{N} E_n^*(\tau_2) E_n(\tau_2)$$

$$+ \sum_{m \neq n}^{N} E_m^*(\tau_1) E_n(\tau_1) E_n^*(\tau_2) E_m(\tau_2)$$

$$= \sum_{m,n} \left| \frac{1}{\sqrt{2}} [E_m(\tau_1) E_n(\tau_2) + E_n(\tau_1) E_m(\tau_2)] \right|^2$$

$$= \langle I(\tau_1)\rangle \langle I(\tau_2)\rangle + \langle \Delta I(\tau_1) \Delta I(\tau_2)\rangle. \tag{5.6.9}$$

Due to the random relative phases between the modes, the only surviving terms of the ensemble average are the terms where $m = n$ and $q = p$, and the terms where $m = q$ and $n = p$. Eq. (5.6.9) specifies two different yet indistinguishable ways for a pair of cavity modes to produce a joint-detection of D_1 and D_2: (1) one or more indistinguishable photons from the mth longitudinal mode triggers D_1 at (z_1, t_1) while one or more indistinguishable photons from the nth longitudinal mode triggers D_2 at (z_2, t_2); and (2) one or more indistinguishable photons from the mth cavity mode triggers D_2 at (z_2, t_2) while one or more indistinguishable photons from the nth cavity mode triggers D_1 at (z_1, t_1). The second-order temporal coherence function, or correlation function, of a multi-mode CW laser beam is thus the result of a nonlocal interference: a pair of randomly created and randomly paired groups of indistinguishable photons interferes with the pair itself. Indeed, the cross interference term of Eq. (5.6.9) corresponds to the intensity fluctuation correlation:

$$\langle \Delta I(\tau_1) \Delta I(\tau_2) \rangle = \sum_{m \neq n}^{N} E_m^*(\tau_1) E_n(\tau_1) E_n^*(\tau_2) E_m(\tau_2). \tag{5.6.10}$$

Similar to thermal light, the measured intensity fluctuations of D_1 and D_2 are

$$\Delta I(z_1, t_1) = \sum_{m \neq n}^{N} E_m^*(\tau_1) E_n(\tau_1)$$

$$\Delta I(z_2, t_2) = \sum_{p \neq q}^{N} E_p^*(\tau_2) E_q(\tau_2), \tag{5.6.11}$$

indicating that $\Delta I(z_1, t_1)$ and $\Delta I(z_2, t_2)$, measured by D_1 and D_2, respectively, are independent each other without any correlation; however, only the $m = q$ and $n = p$ terms survive in the joint-detection of D_1 and D_2 due to the nonlocal constructive-destructive two-mode interferences. The only contribution to the intensity fluctuation correlation are these survive terms.

The nontrivial contribution is calculated as follows:

$$\langle \Delta I(\tau_1) \Delta I(\tau_2) \rangle = \sum_{m \neq n}^{N} E_m^*(\tau_1) E_n(\tau_1) E_n^*(\tau_2) E_m(\tau_2)$$

$$= \sum_{m \neq n}^{N} \mathcal{F}^*_{\tau_{1m}} \{ E(\nu) \} e^{i\omega_m \tau_{1m}} \mathcal{F}_{\tau_{1n}} \{ E(\nu) \} e^{-i\omega_n \tau_{1n}}$$

$$\times \mathcal{F}^*_{\tau_{2n}} \{ E(\nu) \} e^{i\omega_n \tau_{2n}} \mathcal{F}_{\tau_{2m}} \{ E(\nu) \} e^{-i\omega_m \tau_{2m}}$$

$$= \sum_{m}^{N} \left| \mathcal{F}^*_{\tau_1 m}\{E(\nu)\} \mathcal{F}_{\tau_2 m}\{E(\nu)\} e^{i(\omega_0 + m\omega_b)\tau} \right|^2$$

$$= \left| \mathcal{F}_{\tau}\{E(\nu)\} \right|^2 \frac{\sin^2(N\omega_b \tau/2)}{\sin^2(\omega_b \tau/2)} \tag{5.6.12}$$

The nontrivial part of the second-order temporal coherence function is a set of comb-like correlation peaks with an equal spacing of $\Delta t \simeq 1/\nu_b N$, which has been called Ghost Frequency Comb (GFC).

The above result is consistent with that of quantum theory. Consider the state of a multi-cavity-mode CW laser beam as a mixed coherent state

$$|\tilde{\Psi}\rangle = \prod_{m} |\Psi_m\rangle = \prod_{m} |\alpha_m(\omega)\rangle, \tag{5.6.13}$$

where m labels the mth cavity mode, or the mth group of indistinguishable photons; $|\alpha_m(\omega)\rangle$ is an eigenstate of the annihilation operator $\hat{a}_m(\omega)$ with a complex eigenvalue $\alpha_m(\omega) = a_m(\omega) e^{i\varphi_m(\omega)}$, which has a real and positive amplitude and a phase,

$$\hat{a}_m(\omega)|\alpha_m(\omega)\rangle = \alpha_m(\omega)|\alpha(\omega)\rangle. \tag{5.6.14}$$

The field operators for the 1-D measurement are

$$\hat{E}^{(+)}(z_j, t_j) = \sum_{m} \hat{a}_m(\omega)\, g_m(\omega; z_j, t_j)$$

$$\hat{E}^{(-)}(z_j, t_j) = \sum_{m} \hat{a}^\dagger_m(\omega)\, g^*_m(\omega; z_j, t_j), \tag{5.6.15}$$

where $g_m(\omega; r_j, t_j)$ is the Green's function that propagates the mth mode from the source to space-time coordinates (r_j, t_j) of the jth photodetector D_j, $j = 1, 2$. We have also ignored the constants.

The second order coherence function, or correlation function, is calculated from

$$G^{(2)}(z_1, t_1; z_2, t_2)$$

$$= \langle \hat{E}^{(-)}(z_1, t_1)\hat{E}^{(-)}(z_2, t_2)\hat{E}^{(+)}(z_2, t_2)\hat{E}^{(+)}(z_1, t_1) \rangle$$

$$= \left\langle \sum_{m} \psi^*_m(z_1, t_1) \sum_{n} \psi^*_n(z_2, t_2) \sum_{p} \psi_p(z_2, t_2) \sum_{q} \psi_q(z_1, t_1) \right\rangle, \tag{5.6.16}$$

where we define $\psi_m(z_j, t_j)$ the effective wavefunction of the mth "subfield" measured by D_j,

$$\langle \Psi | \hat{E}^{(+)}(z_j, t_j) | \Psi \rangle = \langle \Psi | \sum_m \hat{a}_m(\omega)\, g_m(\omega; z_j, t_j) | \Psi \rangle$$

$$= \sum_m \alpha_m(\omega)\, g_m(\omega; z_j, t_j)$$

$$= \sum_m \psi_m(z_j, t_j). \tag{5.6.17}$$

Note, each mth "subfield", or mth cavity-mode, consists of a large number of indistinguishable photons.

Due to the random relative phases between the subfields, the only surviving terms in Eq. 5.6.16 are the $m = q$ and $n = p$ terms as well as the $m = p$ and $n = q$ terms. So we are left with

$$G^{(2)}(z_1, t_1; z_2, t_2)$$

$$= \sum_m \psi_m^*(z_1, t_1)\psi_m(z_1, t_1) \sum_n \psi_n^*(z_2, t_2)\psi_n(z_2, t_2)$$

$$+ \sum_{m \neq n} \psi_m^*(z_1, t_1)\psi_n(z_1, t_1)\psi_n^*(z_2, t_2)\psi_m(z_2, t_2)$$

$$= \sum_{m,n} \left| \frac{1}{\sqrt{2}} \left[\psi_m(z_1, t_1)\psi_n(z_2, t_2) + \psi_n(z_1, t_1)\psi_m(z_2, t_2) \right] \right|^2. \tag{5.6.18}$$

indicating the superposition of two different yet indistinguishable quantum probability amplitudes: (1) one or more indistinguishable photons from the mth longitudinal mode triggers D_1 at (z_1, t_1) while one or more indistinguishable photons from the nth longitudinal mode triggers D_2 at (z_2, t_2); and (2) one or more indistinguishable photons from the mth longitudinal mode triggers D_2 at (z_2, t_2) while one or more indistinguishable photons from the nth longitudinal mode triggers D_1 at (z_1, t_1). The second-order temporal coherence function, or correlation function, of a multi-mode CW laser beam is thus the result of a nonlocal interference: a pair of randomly created and randomly paired groups of indistinguishable photons interferes with the pair itself.

In Eq. 5.6.18, the cross-interference term is the "nontrivial" contribution to the second-order coherence function $G^{(2)}(z_1, t_1; z_2, t_2)$, corresponding to the intensity fluctuation correlation of the CW laser beam measured jointly by D_1 and D_2,

$$\langle \Delta I(z_1, t_1) \Delta I(z_2, t_2) \rangle = \sum_{m \neq n} \psi_m^*(z_1, t_1)\, \psi_n(z_1, t_1)\, \psi_n^*(z_2, t_2)\, \psi_m(z_2, t_2)$$

$$(5.6.19)$$

although the intensity fluctuations measured by D_1 and D_2, respectively, are completely independent without any correlation

$$\Delta I(z_1, t_1) = \sum_{m \neq n} \psi_m^*(z_1, t_1)\psi_n(z_1, t_1)$$

$$\Delta I(z_1, t_1) = \sum_{p \neq q} \psi_p^*(r_2, t_2)\psi_q(r_2, t_2). \qquad (5.6.20)$$

The product of the mean intensities measured by each photodetector is

$$\langle I(z_1, t_1) \rangle \langle I(z_2, t_2) \rangle = \sum_m \psi_m^*(z_1, t_1)\psi_m(z_1, t_1) \sum_n \psi_n^*(z_2, t_2)\psi_n(z_2, t_2).$$

$$(5.6.21)$$

The normalized $G^{(2)}(z_1, t_1; z_2, t_2)$ is thus related to the intensity fluctuation correlation as

$$g^{(2)}(z_1, t_1; z_2, t_2) \propto 1 + \frac{\langle \Delta I(z_1, t_1) \Delta I(z_2, t_2) \rangle}{\langle I(z_1, t_1) \rangle \langle I(z_2, t_2) \rangle}. \qquad (5.6.22)$$

Taking into account the finite spectral bandwidth of a cavity-mode, the effective wavefunction of the mth group of indistinguishable photons is calculated as:

$$\psi_m(z_j, t_j) = \int_{\Delta\omega} d\omega\, \alpha_m(\omega)\, g_m(\omega; z_j, t_j) = \mathcal{F}_{\tau_{jm}}\{\alpha(\nu)\}\, e^{-i\omega_m \tau_{jm}},$$

$$(5.6.23)$$

where $\alpha(\nu)$ is the spectral distribution function of the cavity-mode. Here, each cavity-mode has been modeled as a wave packet with carrier frequency ω_m and an "envelope", which is the Fourier transform of the spectral distribution function of the mth mode. We have also introduced the notations $\tau_j \equiv t_j - z_j/c$ and $\tau_{jm} = \tau_j - t_{0m}$, where t_{0m} is the creation time of the mth cavity-mode. The intensity fluctuation correlation is calculated as follows:

$$\langle \Delta I(\tau_1)\Delta I(\tau_2)\rangle = \sum_{m,n} \mathcal{F}^*_{\tau_{1m}}\{\alpha(\nu)\}\, e^{i\omega_m \tau_{1m}}\, \mathcal{F}_{\tau_{1n}}\{\alpha(\nu)\}\, e^{-i\omega_n \tau_{1m}}$$

$$\times\, \mathcal{F}^*_{\tau_{2n}}\{\alpha(\nu)\}\, e^{i\omega_n \tau_{2m}}\, \mathcal{F}_{\tau_{2m}}\{\alpha(\nu)\}\, e^{-i\omega_m \tau_{2m}}$$

$$= \left[\sum_m \mathcal{F}^*_{\tau_{1m}}\{\alpha(\nu)\}\mathcal{F}_{\tau_{2m}}\{\alpha(\nu)\}\, e^{i\omega_m \tau}\right]$$

$$\times \left[\sum_n \mathcal{F}_{\tau_{1n}}\{\alpha(\nu)\}\mathcal{F}^*_{\tau_{2n}}\{\alpha(\nu)\}\, e^{-i\omega_n \tau}\right] \tag{5.6.24}$$

where $\tau \equiv \tau_1 - \tau_2 = (t_1 - t_2) - (z_1 - z_2)/c$. Calculating one summation at a time, we find

$$\sum_m \mathcal{F}^*_{\tau_{1m}}\{a(\nu)\}\mathcal{F}_{\tau_{2m}}\{a(\nu)\}\, e^{i\omega_m \tau}$$

$$\simeq \int dt_{0m}\, \mathcal{F}^*_{\tau_{1m}}\{a(\nu)\}\mathcal{F}_{\tau_{2m}}\{a(\nu)\}$$

$$\times \left[e^{i\omega_1 \tau}\left[1 + e^{i\omega_b \tau} + e^{i2\omega_b \tau} + \cdots + e^{i(N-1)\omega_b \tau}\right]\right]$$

$$\simeq \mathcal{F}_\tau\{a^2(\nu)\}\left[e^{i(N-1)\omega_b \tau/2}\frac{\sin(N\omega_b \tau/2)}{\sin(\omega_b \tau/2)}\right]. \tag{5.6.25}$$

Here ω_b is the *beat* frequency between neighboring cavity-modes and N is the total number of cavity-modes. The intensity fluctuation correlation, corresponding to the cross-interference term of Eq. 5.6.18, then reduces to

$$\langle \Delta I(\tau_1)\Delta I(\tau_2)\rangle \propto \left|\mathcal{F}_\tau\{a^2(\nu)\}\right|^2 \left[\frac{\sin^2(N\omega_b \tau/2)}{\sin^2(\omega_b \tau/2)}\right]. \tag{5.6.26}$$

The comb-like multipole periodic correlation peaks at $\omega_b \tau/2 = n\pi$, for $n = 0, \pm 1, \pm 2, \ldots, \pm(N-1)$,

$$(t_1 - t_2)_n = \frac{1}{\nu_b}n + (r_1 - r_2)/c \tag{5.6.27}$$

where we have used $\omega_b = 2\pi\nu_b$ and defined $(t_1 - t_2)_n$ as the measured value of $t_1 - t_2$ at the nth comb peak. The temporal width of each correlation peak, measured between neighboring zeros of the correlation function, is approximately

$$\Delta t \simeq \frac{1}{\nu_b N}. \tag{5.6.28}$$

Fig. 5.6.2 illustrates a calculated $\langle \Delta I(\tau_1) \Delta I(\tau_2) \rangle$ for the second-order correlation measurement of a multi-cavity-mode laser beam. Fig. 5.6.3 gives a typical experimentally observed ghost frequency comb (GFC).

Examining the comb-like correlation function, it appears that the intensity fluctuation of the laser beam is correlated with $\gamma > 1$ within these periodic, precise, and narrow time windows. However, as long as the relative delay $\tau_1 - \tau_2$ falls into the region between the periodic sharp correlation peaks, the intensity fluctuation of the laser beam becomes uncorrelated with $\gamma = 1$. Do we have any CW laser beam with such statistical nature of intensity fluctuations? Of course not! A question naturally arises: Can the observed GFC be considered as statistical correlation of intensity fluctuations of the CW laser beam? Perhaps, the nonlocal interference picture is more acceptable. The GFC in Eq. (5.6.24) is indeed calculated from the cross-interference term of Eq. (5.6.9). As we have pointed out, this interference occurs at distant space-time coordinates (z_1, t_1) and (z_2, t_2),

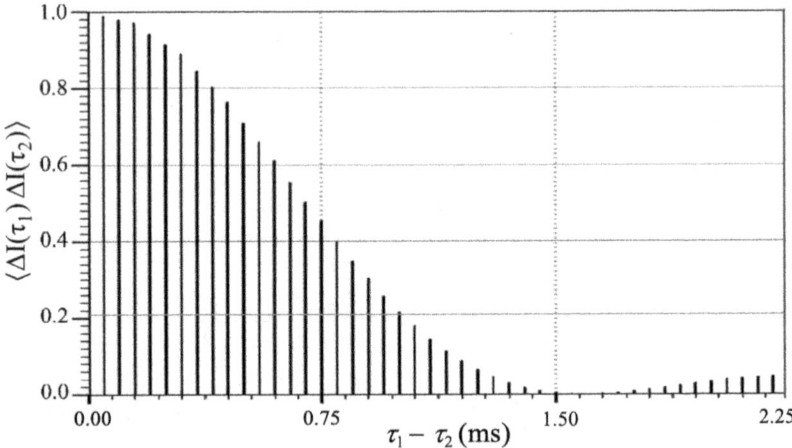

Fig. 5.6.2　Numerical simulation of a GFC: a comb-function with sharp correlation peaks is observable in the intensity fluctuation correlation measurement of D_1 and D_2, although both D_1 and D_2 measure constant intensities as expected from CW laser beams. In this simulation, we assumed 100,000 cavity modes with mode separation $\omega_b/2\pi = 20$ kHz. We also assumed finite spectral bandwidth of 200 Hz for each cavity mode which results in a sinc-function envelope of 1.50 ms temporal width on top of the comb-function. The temporal width of each comb peak is 500 ps.

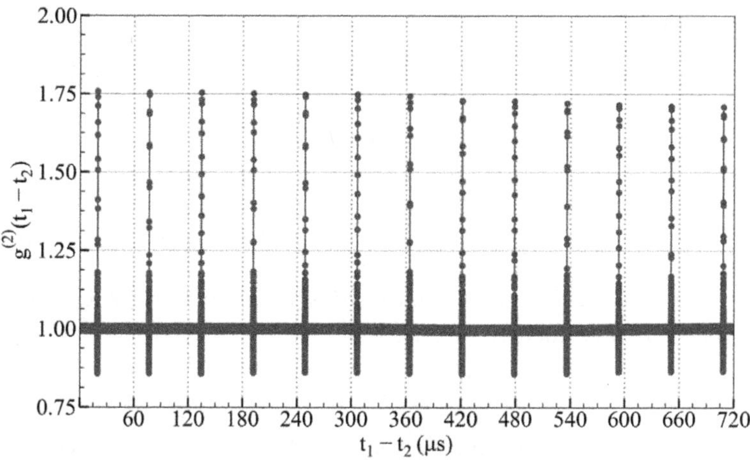

Fig. 5.6.3 A typical experimentally observed GFC. In this measurement, a fiber laser of 11.7 km ring cavity generated approximately 500,000 evenly spaced cavity-modes all in the TEM_{00} transverse profile at $1562 - 1563$ nm wavelength. The measured mode separation was $17,425.8336 \pm 0.0003$ Hz. Note the first GFC peak was at $\tau = 19,569.025 \pm 0.004$ ns, corresponding to $3,999.1195 \pm 0.0008$ m optical distance between D_1 and D_2, agreeing with the optical length of the fiber delay line placed between D_1 and D_2.

and the two photodetection events can be easily managed to achieve space-like separation. Apparently, the observation of GFC is strong evidence of nonlocal interference.

(III) Second-order temporal correlation of laser pulses

In early discussions, we have given four simplified models of radiation to clarify four different types of superposition among a large number of sub-fields and Fourier-modes. A Q-switched laser pulse can be approximately as a radiation with large number of coherently radiated subfields and coherently excited Fourier-modes. Similar to our early analysis, we assume a Gaussian spectrum in each of the sub-radiation. The coherent superposition of indistinguishable subfields and the coherent superposition of Fourier-modes of Gaussian distribution together produce a Gaussian wavepackets as given in previous chapters,

$$E(z,t) = E_0 \, e^{-\tau^2/4\sigma^2} \, e^{-i[\omega_0 t - k(\omega_0)z - \varphi_0]}.$$

It is easy to find that the second-order temporal coherence function $\Gamma^{(2)}(z_1, t_1; z_2, t_2)$ becomes a factorizeable function of $\Gamma^{(1)}(z_1, t_1; z_1, t_1)$ and

$$\Gamma^{(1)}(z_2, t_2; z_2, t_2)$$

$$\Gamma^{(2)}(z_1, t_1; z_2, t_2) = \left\langle E^*(z_1, t_1) E(z_1, t_1) E^*(z_2, t_2) E(z_2, t_2) \right\rangle$$

$$= I_0 e^{-\tau_1^2/2\sigma^2} \left| e^{-i[\omega_0 t_1 - k(\omega_0) z_1 - \varphi_0]} \right|^2$$

$$\times I_0 e^{-\tau_2^2/2\sigma^2} \left| e^{-i[\omega_0 t_2 - k(\omega_0) z_2 - \varphi_0]} \right|^2$$

$$= \Gamma^{(1)}(z_1, t_1; z_1, t_1) \, \Gamma^{(1)}(z_2, t_2; z_2, t_2) \qquad (5.6.29)$$

which is the product of two mean intensities or two first-order self-coherence functions, and consequently,

$$\gamma^{(2)}(z_1, t_1; z_2, t_2) = 1.$$

This type second-order correlation is usually called "trivial" correlation. Statistically, this trivial correlation means no correlation between the two temporally separated radiations measured at (z_1, t_1) and (z_2, t_2), although the radiation may hold a well-defined temporal profile in a deterministic manner.

(IV) Second-order spatial coherence of TEM$_{00}$ laser beams.

The calculation of second-order spatial coherence of a TEM$_{00}$ laser beam is similar to that of second-order temporal correlation of single-mode CW laser beam. In 1-D measurements, the trivial correlation is a produce of the first-order self-coherence functions $\Gamma^{(1)}(x_1, x_1)$ and $\Gamma(x_2.x_2)$

$$\Gamma^{(2)}(x_1, x_2) = \Gamma_{01} \operatorname{sinc}^2 \frac{\pi D x_1}{\lambda z_1} \, \Gamma_{02} \operatorname{sinc}^2 \frac{\pi D x_2}{\lambda z_2}$$

$$= \Gamma^{(1)}(x_1, x_1) \, \Gamma^{(1)}(x_2, x_2) \qquad (5.6.30)$$

where D is the diameter of the laser beam (transverse size of the source) and λ is the wavelength of the radiation. In 1-D measurements, we thus have

$$\gamma^{(2)}(x_1, x_2) = 1. \qquad (5.6.31)$$

The above results are consistent with that of the quantum theory. In the electromagnetic field theory of light and the quantum theory of light, we both conclude the second-order spatial coherence function of TEM$_{00}$ laser beam is factorizeable into a product of the first-order self-coherence functions, and the degree of second-order coherence of is a trivial constant of one.

Why the second-order temporal correlation of a single-longitudianal-mode CW laser beam and the second-order spatial correlation of a single-transverse-mode TEM_{00} laser beam are trivial? Why do the statistical properties of intensity fluctuations in a single-mode laser beam differ from those in a multi-mode laser beam? These questions are difficult to answer from the perspective of classical statistics. We provide the following answer in terms of "two-photon" interference. The state of a single-mode CW laser beam and the state of a single-mode TEM_{00} laser beam can be approximated as a pure coherent state, representing the state of a group *indistinguishable* photons. The only interference we can observe from this state is the self-interference of the group with the group itself. A multi-mode laser beam, either longitudinal or transverse, can be approximated as mixed coherent states, representing a mixed ensemble of distinguishable groups of indistinguishable photons. A pair of distinguishable groups of indistinguishable photons is able to interfere with the pair itself, resulting in a nontrivial second-order coherence function.

5.7 Entangled Coherent State

Since Einstein–Podolsky–Rosen (EPR) suggested a *gedankenexperiment* and introduced an entangled two-particle system in 1935, we have experimentally verified the nonlocal EPR correlation of entangled photon pairs created in positronium annihilation, in atomic cascade decay, and in spontaneous parametric down-conversion (SPDC). In light of new technology, it is now possible to generate a large number of stimulated longitudinal-cavity-modes in an optical fiber laser. Similarly, a large number of paired longitudinal-cavity-modes can be generated from an optical parametric oscillator (OPO) such as that shown in Fig. 5.7.1. When the OPO operates with a pump power well above its threshold, each cavity mode can be described by a coherent state. The state of the output field of a multi-cavity-mode OPO is thus approximated as:

$$|\Psi\rangle = \prod_{s,i} \delta\left(\omega_s + \omega_i - \omega_p\right) \delta\left(\mathbf{k}_s + \mathbf{k}_i - \mathbf{k}_p\right) |\alpha_s(\mathbf{k}_s)\rangle|\alpha_i(\mathbf{k}_i)\rangle \qquad (5.7.1)$$

where $|\alpha_s(\mathbf{k}_s)\rangle$ and $|\alpha_i(\mathbf{k}_i)\rangle$, $|\alpha| \gg 1$, are the state vectors of the signal-idler modes of the OPO in the coherent state representation; and ω_j and $\mathbf{k}_j$, for $j = s, i, p$, are the frequency and wavevector of the signal (s) modes, idler (i) modes, and the pump (p) laser beam. Consider each cavity mode a group of indistinguishable photons, this state indicates: (1) each group of

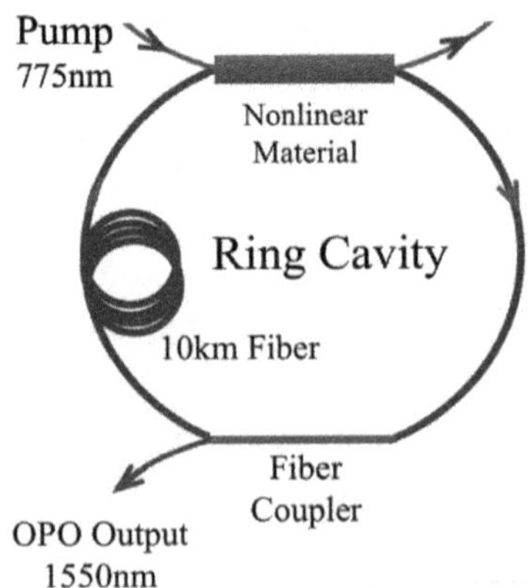

Fig. 5.7.1   Schematic of a fiber Optical Parametric Oscillator (OPO). A $\sim$10 km fiber cavity is able to produce $\sim$500,000 entangled pairs of longitudinal-cavity-mode within $\sim$10 GHz bandwidth of phase matching. Half-million mode pairs added constructively in an entangled and coherent manner form the state of Eq. (5.7.1).

indistinguishable photons represented by $|\alpha_s(\mathbf{k}_s)\rangle$ is entangled with another unique group of indistinguishable photons represented by $|\alpha_i(\mathbf{k}_i)\rangle$, by means of energy conservation $\hbar\omega_s + \hbar\omega_i = \hbar\omega_p$ and momentum conservation $\hbar\mathbf{k}_s + \hbar\mathbf{k}_i = \hbar\mathbf{k}_p$; and (2) a large number of such signal-idler pairs are created and superposed coherently to form a pair of entangled signal-idler laser beams.

Based on the entangled coherent states of Eq. (5.7.1), we now calculate the second-order coherence function $G^{(2)}(\mathbf{r}_1, t_1; \mathbf{r}_2, t_2)$. Assume a simple measurement in which the signal beam and the idler beam, respectively, are directed to two point-like photodetectors D_1 and D_2. These detectors measure the intensities of the signal beam and the idler beam, respectively, and are used to measure the intensity correlation between the signal beam and the idler beam, jointly. It is no surprise that both D_1 and D_2 observe constant intensities, respectively, as expected from a CW laser beam. In the joint measurement of intensity correlation of two CW laser beams, do we expect any surprise?

In this 1-D measurement, it is reasonable to concentrate our calculation to the temporal correlation $G^{(2)}(r_1, t_1; r_2, t_2)$. We therefore simplify the state described by Eq. (5.7.1) to 1-D version:

$$|\Psi\rangle = \prod_{s,i} \delta\left(\omega_s + \omega_i - \omega_p\right) |\alpha_s(\omega_s)\rangle|\alpha_i(\omega_i)\rangle. \tag{5.7.2}$$

The second-order temporal correlation function of the entangled bi-mode coherent state of Eq.(5.7.2) is thus

$$G^{(2)}(r_1, t_1; r_2, t_2) = \langle \Psi | \hat{E}^{(-)}(r_1, t_1) \hat{E}^{(-)}(r_2, t_2) \hat{E}^{(+)}(r_2, t_2) \hat{E}^{(+)}(r_1, t_1) | \Psi \rangle$$

$$= | \langle \Psi | \hat{E}^{(+)}(r_2, t_2) \hat{E}^{(+)}(r_1, t_1) | \Psi \rangle |^2$$

$$\equiv | \Psi(r_1, t_1; r_2, t_2) |^2 \tag{5.7.3}$$

where, again, we assume D_1 is triggered by one or more indistinguishable photons from the signal beam at (r_1, t_1) while D_2 is triggered by one or more indistinguishable photons from the idler beam at (r_2, t_2). $\Psi(r_1, t_1; r_2, t_2)$ is the probability amplitude to observe a joint photodetection event at space-time (r_1, t_1) and (r_2, t_2). We name it the effective wavefunction of the entangled signal-idler coherent states or entangled signal-idler beams. To evaluate $\Psi(r_1, t_1; r_2, t_2)$ and $G^{(2)}(r_1, t_1; r_2, t_2)$, we need to propagate the field operators from the source to space-time coordinates (r_1, t_1) and (r_2, t_2).

In general, the field operator $\hat{E}^{(+)}(r, t)$ at space-time point (r, t) can be written in terms of the Green's function which propagates each Fourier mode from space-time point (r_0, t_0) to (r, t):

$$\hat{E}^{(+)}(r, t) = \sum_{\omega} g_0(\omega; r, t) \, \hat{E}^{(+)}(\omega) \tag{5.7.4}$$

To simplify the notation, we have assumed one polarization. Under the 1-D assumption, the field operators become

$$\hat{E}^{(+)}(r_j, t_j) \cong \sum_{\omega} e^{-i\omega\tau_j} \hat{a}(\omega) \tag{5.7.5}$$

where $\tau_j \equiv t_j - r_j/c$, r_j is the longitudinal coordinate of the *j*th photodetector. We have chosen $r_0 = 0$ at the output plane of the entangled bright light source. Since $|\alpha(\omega)\rangle$ is an eigenstate of the annihilation operator with an eigenvalue $\alpha(\omega)$

$$\hat{a}(\omega)|\alpha(\omega)\rangle = \alpha(\omega)|\alpha(\omega)\rangle, \quad \hat{a}(\omega)|\Psi\rangle = \alpha(\omega)|\Psi\rangle,$$

the effective wavefunction $\Psi(\tau_1, \tau_2)$ of the signal-idler beams is calculated as

$$\Psi(\tau_1, \tau_2) = \langle \Psi | \sum_{\omega, \omega'} e^{-i\omega\tau_2} \hat{a}(\omega) \, e^{-i\omega'\tau_1} \hat{a}(\omega') | \Psi \rangle \tag{5.7.6}$$

$$= \Psi_0 \sum_{\omega_s, \omega_i} \delta(\omega_s + \omega_i - \omega_p) \alpha(\omega_s) e^{-i\omega_s\tau_1} \alpha(\omega_i) e^{-i\omega_i\tau_2}$$

indicating the coherent superposition of a large number of entangled signal-idler pairs: the probability of observing a joint photodetection event at (r_1, t_1) and (r_2, t_2) is the result of a coherent superposition of a large number of probability amplitudes of the signal-idler pairs.

The superposition can be calculated easily with the help of the δ-function,

$$\Psi(\tau_1, \tau_2) = \Psi_0 \sum_{\omega_s, \omega_i} \delta(\omega_s + \omega_i - \omega_p)\, \alpha(\omega_s)\alpha(\omega_i)\, e^{-i\frac{1}{2}(\omega_s+\omega_i)(\tau_1+\tau_2)}$$

$$\times\, e^{-i\frac{1}{2}(\omega_s-\omega_i)(\tau_1-\tau_2)}$$

$$= \Psi_0\, e^{-i\omega_p(\tau_1+\tau_2)/2} \sum_{\nu} f(\nu)\, e^{-i\nu(\tau_1-\tau_2)} \tag{5.7.7}$$

where we have assumed degenerate OPO, i.e., $\omega_s^0 = \omega_i^0$ with ω_s^0 and ω_i^0 the central frequencies of the signal and the idler, ν is the detuning frequency $\nu = \omega_s - \omega_s^0$.

Furthermore, we can assume an OPO with a large number N of cavity modes, $\nu \equiv n\omega_b$, $n = 0, 1, \ldots, N-1$, and a constant mode distribution function $f(\nu)$,

$$\Psi(\tau_1, \tau_2) \simeq \Psi_0\, e^{-i\omega_p(\tau_1+\tau_2)/2} \sum_{n=0}^{N-1} e^{-in\omega_b(\tau_1-\tau_2)}$$

$$= \Psi_0\, e^{-i\omega_p(\tau_1+\tau_2)/2} \left[e^{i(N-1)\omega_b\tau/2}\, \frac{\sin N\omega_b\tau/2}{\sin \omega_b\tau/2} \right] \tag{5.7.8}$$

where ω_b is the beating frequency between neighboring cavity modes, and $\tau \equiv \tau_1 - \tau_2$. In the above calculation, we have assumed each cavity mode contains a single-frequency. Realistically, we may have to take into account the finite spectral bandwidth of each cavity mode,

$$\Psi(\tau_1, \tau_2) = \langle\Psi| \sum_{\omega', \omega''} e^{-i\omega'\tau_2} \hat{a}(\omega') e^{-i\omega''\tau_1} \hat{a}(\omega'') |\Psi\rangle$$

$$= \Psi_0 \sum_{\omega_s, \omega_i} \delta(\omega_s + \omega_i - \omega_p)\, \mathcal{F}_{\tau_1}\{\alpha(\omega_s)\} e^{-i\omega_s\tau_1} \mathcal{F}_{\tau_2}\{\alpha(\omega_i)\} e^{-i\omega_i\tau_2}$$

$$= \Psi_0 \sum_{\omega_s, \omega_i} \delta(\omega_s + \omega_i - \omega_p)\, \mathcal{F}_{\tau_1}\{\alpha(\omega_s)\} \mathcal{F}_{\tau_2}\{\alpha(\omega_i)\}$$

$$\times\, e^{-i\frac{1}{2}(\omega_s+\omega_i)(\tau_1+\tau_2)}\, e^{-i\frac{1}{2}(\omega_s-\omega_i)(\tau_1-\tau_2)}$$

$$= \Psi_0 e^{-i\omega_p(\tau_1+\tau_2)/2} \sum_{n=0}^{N-1} \mathcal{F}_{\tau_1}\{\alpha(\omega_s^0 + n\omega_b)\}$$

$$\times \mathcal{F}_{\tau_2}\{\alpha(\omega_i^0 - n\omega_b)\} \left[e^{i(N-1)\omega_b\tau/2} \frac{\sin N\omega_b\tau/2}{\sin \omega_b\tau/2} \right] \tag{5.7.9}$$

where $\mathcal{F}_{\tau_1}\{\alpha(\omega_s^0 + n\omega_b)\}$ and $\mathcal{F}_{\tau_2}\{\alpha(\omega_i^0 - n\omega_b)\}$ are the Fourier transforms of the spectral functions of the signal and idler cavity-modes, respectively. It is clear, the amplitude of the effective wavefunction is determined by two functions: (1) the value of $\sin Nx / \sin x$; and (2) the degree of overlap between the Fourier transforms $\mathcal{F}_{\tau_1}\{\alpha(\omega_s^0 + n\omega_b)\}$ and $\mathcal{F}_{\tau_2}\{\alpha(\omega_i^0 - n\omega_b)\}$. For a large number of cavity modes, the summation can be approximated by a convolution integral

$$\sum_{n=0}^{N-1} \mathcal{F}_{\tau_1}\{\alpha(\omega_s^0 + n\omega_b)\}\mathcal{F}_{\tau_2}\{\alpha(\omega_i^0 - n\omega_b)\}$$

$$\simeq \int dt_{0n}\, \mathcal{F}_{\tau_{1n}}\{\alpha(\omega_s)\}\mathcal{F}_{\tau_{2n}}\{\alpha(\omega_i)\}$$

$$\simeq \mathcal{F}_{\tau}\{\alpha^2(\omega)\} \tag{5.7.10}$$

where $\alpha(\omega)$ specifies the spectral distribution of the cavity modes, t_{0n} is the creation time of the nth OPO cavity mode for which we have assumed a random distribution of t_{0n}. We thus have

$$G^{(2)}(\tau) \propto \left| \mathcal{F}_{\tau}\{\alpha^2(\omega)\} \right|^2 \left[\frac{\sin^2 N\omega_b\tau/2}{\sin^2 \omega_b\tau/2} \right]. \tag{5.7.11}$$

If we assume a constant distribution function $a(\omega)$ for each cavity mode within its spectral bandwidth $\Delta\omega$, and for all cavity modes,

$$G^{(2)}(\tau) \propto \left[\text{sinc}^2 \frac{\Delta\omega\tau}{2} \right] \left[\frac{\sin^2 N\omega_b\tau/2}{\sin^2 \omega_b\tau/2} \right] \tag{5.7.12}$$

where $\Delta\omega$ is the spectral bandwidth, or line-width, of the cavity mode, again, $\tau \equiv \tau_1 - \tau_2 = (t_1 - t_2) - (r_1 - r_2)/c$. We name this comb-like function Quantum Ghost Frequency Comb (QGFC), due to its 100% contrast. The QGFC has multipole periodic correlation peaks at $\omega_b\tau/2 = n\pi$, for $n = 0, \pm1, \pm2, \ldots, \pm(N-1)$;

$$(t_1 - t_2)_n = \frac{1}{\nu_b} n + (r_1 - r_2)/c \tag{5.7.13}$$

where we have used $\omega_b = 2\pi\nu_b$ and defined $(t_1 - t_2)_n$ as the measured value of $t_1 - t_2$ at the nth comb peak. The temporal width of each QGFC peak, measured between neighboring zeros of the correlation function, is approximately

$$\Delta t \simeq \frac{1}{\nu_b N}. \qquad (5.7.14)$$

Figure 5.7.2 illustrates an example of a QGFC, which is calculated by assuming a kilometers-long OPO cavity with half-million pairs of cavity-modes.

Indeed it is surprising that a 100% contrast comb-function is observable from the joint-detection of the entangled signal-idler laser beams at distance, although both detectors, D_1 and D_2, measure constant light intensities, consistent with the expectation of CW laser beams. We are facing the following serious questions: (1) What is the cause of the periodic multi-correlation-peaks? (2) Can we trust zero-coincidence observations from a direct measurement of CW laser beams? These questions will be addressed in Chapter 6.

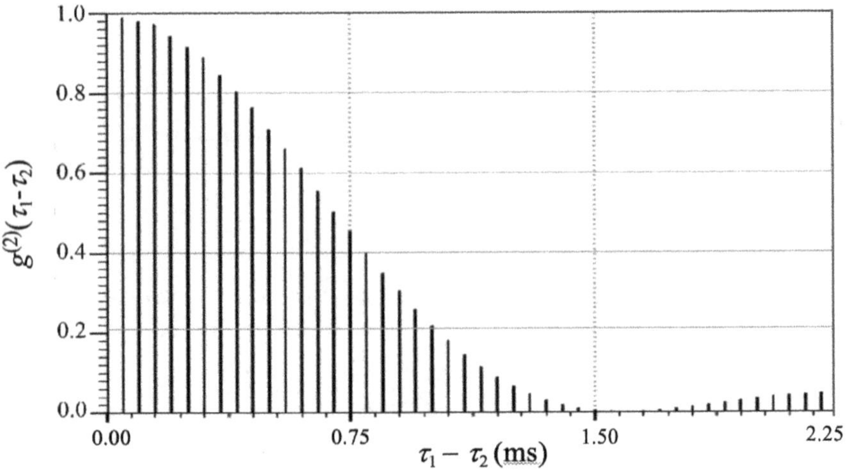

Fig. 5.7.2 Numerical simulation of QGFC: 100% contrast comb-function with sinc-function envelope observable in the joint photodetection of the entangled signal-idler beams, although both D_1 and D_2 measure constant intensities as expected from CW laser beams. In this simulation, we assumed 100,000 cavity modes with mode separation $\omega_b/2\pi = 20$ kHz. We also assumed finite spectral bandwidth of 200 Hz for each cavity mode which results in a sinc-function envelope of 1.50 ms temporal width on top of the comb-function. The temporal width of each comb peak is 500 ps.

5.8 Measurement of Second-Order Coherence

We have discussed two types of setups for the measurement of second-order coherence function in Figs. 5.1.1 and 5.3.1. Figure 5.1.1 uses two analog photodetectors and an analog joint-detection circuit for the measurement of $\langle i_1(t)i_2(t)\rangle \propto \langle I(\mathbf{r}_1,t_1)I(\mathbf{r}_2,t_2)\rangle \propto \langle n(\mathbf{r}_1,t_1)n(\mathbf{r}_2,t_2)\rangle$, while Fig. 5.3.1 uses two photon counting detectors and a digital joint-detection circuit for the measurement of $\langle n(\mathbf{r}_1,t_1)n(\mathbf{r}_2,t_2)\rangle \propto \langle I(\mathbf{r}_1,t_1)I(\mathbf{r}_2,t_2)\rangle$. Although the setup in Fig. 5.1.1 is usually applied to strong radiation, and the setup in Fig. 5.3.1 is usually applied at weak light condition, both can be used for the measurement of $\Gamma^{(2)}(\mathbf{r}_1,t_1;\mathbf{r}_2,t_2)$ and $G^{(2)}(\mathbf{r}_1,t_1;\mathbf{r}_2,t_2)$. The calculations of $\Gamma^{(2)}(\mathbf{r}_1,t_1;\mathbf{r}_2,t_2)$ and $G^{(2)}(\mathbf{r}_1,t_1;\mathbf{r}_2,t_2)$ are different; their predicted results, however, are the same for the same type of radiation. The second-order coherence calculation for thermal radiation is a good example. In Einstein's granulation picture of light, we conclude that the second-order coherence function $\Gamma^{(2)}(\mathbf{r}_1,t_1;\mathbf{r}_2,t_2)$ of thermal field, which is calculated from the electromagnetic theory, is the same as $G^{(2)}(\mathbf{r}_1,t_1;\mathbf{r}_2,t_2)$ of thermal state, which is calculated from the quantum theory. Interestingly, both calculations show that the nontrivial correlation of thermal light is the result of two-photon interference: A randomly created pair of subfields, or photons, interferes with the pair itself. The second-order coherence function $G^{(2)}(\mathbf{r}_1,t_1;\mathbf{r}_2,t_2)$ of entangled states can only be calculated by quantum theory, because entangled states cannot be defined in classical electromagnetic theory.

In the following discussion, we focus on $G^{(1)}(\mathbf{r}_1,t_1;\mathbf{r}_2,t_2)$.

Unlike the first-order coherence function $G^{(1)}(\mathbf{r}_1,t_1;\mathbf{r}_2,t_2)$, the second-order coherence function $G^{(2)}(\mathbf{r}_1,t_1;\mathbf{r}_2,t_2)$ is a directly measurable function of space–time variables $f(\mathbf{r}_1,t_1;\mathbf{r}_2,t_2)$. In the view of quantum mechanics, $G^{(2)}(\mathbf{r}_1,t_1;\mathbf{r}_2,t_2)$ indicates the probability of jointly observe two photons at space–time coordinates $(\mathbf{r}_1,t_1;\mathbf{r}_2,t_2)$. To emphasize the "two-photon" interference nature, we have written the second-order coherent function $G^{(2)}(\mathbf{r}_1,t_1;\mathbf{r}_2,t_2)$ as a function of the space-time coordinates of the photodetection events, $(\mathbf{r}_1,t_1)$ and $(\mathbf{r}_2,t_2)$, of D_1 and D_2. However, this does not prevent $G^{(2)}(\mathbf{r}_1,t_1;\mathbf{r}_2,t_2)$ from being a correlation function of $(t_1 - t_2)$ or $(\mathbf{r}_1 - \mathbf{r}_2)$, such as $G^{(2)}(\tau_1 - \tau_2)$ or $G^{(2)}(\vec{\rho}_1 - \vec{\rho}_2)$. In fact, if $G^{(2)}(\mathbf{r}_1,t_1;\mathbf{r}_2,t_2)$ is a nontrivial function of either space variable or time variable, we consider the radiation is second-order spatially correlated or temporally correlated. The second-order temporal correlation function is usually written as $G^{(2)}(\tau)$ with $\tau \equiv t_1 - t_2$, if there exist such a correlation, when the coordinates of $\mathbf{r}_1$ and $\mathbf{r}_2$ are taken fixed values.

If $G^{(2)}(\mathbf{r}_1, t_1; \mathbf{r}_2, t_2)$ is a correlation function of $(t_1 - t_2)$, the measurement of the second-order coherence function, or the second-order correlation function, can be performed accumulatively by means of statistical histograms of a large number of joint-photodetection events as a function of $t_1 - t_2$. Accumulative measurement is helpful for an observation under weak light, especially in single-photon level. The photon registration events of the photodetectors D_1 and D_2 can be recorded in the accuracy of 10^{-12} s in modern technologies. $G^{(2)}$ thus can be measured as a function of $t_1 - t_2$ in picosecond resolution by recording the registration time for each individual photon pairs. In the following, we show two examples of $G^{(2)}$ measurements in photon counting regime.

(I) Measurement of $G^{(2)}(t_1 - t_2)$ and $g^{(2)}(t_1 - t_2)$

In these type of measurements, the spatial coordinates of the photodetectors, $\mathbf{r}_1$ and $\mathbf{r}_2$, are chosen to be constants during the measurement. The measurement electronics is able to read and record the registration times t_1 and t_2 or the registration time difference $t_1 - t_2$ for each of the measured photon pairs. A statistical histogram, showing the number of joint-photodetection events against $t_1 - t_2$, is made after counting a large number of joint-photodetection events. This distribution corresponds to the second-order correlation function $G^{(2)}(t_1 - t_2)$. In previous section, we have shown a typical measured degree of second-order coherence function $g^{(2)}(t_1 - t_2)$, i.e., the normalized second-order coherence function, for thermal (pseudo-thermal) field with narrow spectrum in Fig. 5.4.1.

Figure 5.8.1 is an measured $G^{(2)}(t_1 - t_2)$ function for entangled photon pairs of SPDC. The horizontal axis $t_1 - t_2$ labels the registration time differences of the photon pairs, corresponding to $\tau_1 - \tau_2$ with constant spatial coordinates $\mathbf{r}_1$ and $\mathbf{r}_2$. The vertical axis indicates the number of photon pairs that are observed with the value of $t_1 - t_2$. We may draw two conclusions from Fig. 5.8.1 directly: (1) The entangled photon pair is likely to be generated simultaneously; and (2) the measured $t_1 - t_2$ are independent of the creation time of the pair.

The above two examples are typical measurements of $G^{(2)}(t_1 - t_2)$ and $g^{(2)}(t_1 - t_2)$ in photon counting. The measurement electronics records the joint-photodetection time difference $t_1 - t_2$ for each photon pairs. The observed function is based on the statistics of a large number of photon pairs. The distribution function or the histogram that illustrates number

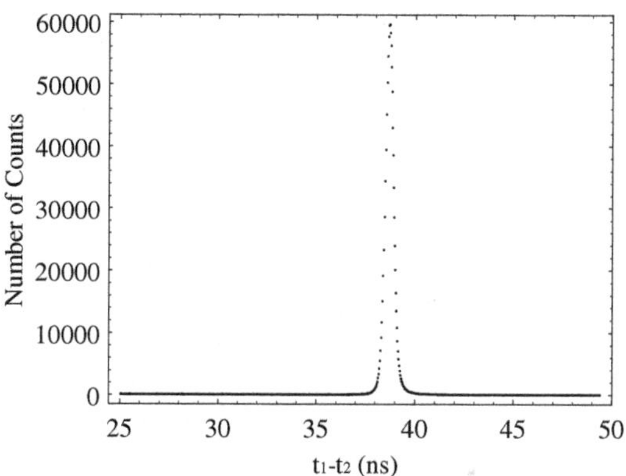

Fig. 5.8.1 Second-order correlation function for entangled photon pairs. The horizontal axis $t_1 - t_2$, which labels the measured registration time difference of the photon pair, corresponds to $\tau_1 - \tau_2$ with fixed values of z_1 and z_2. The calculated temporal with of $G^{(2)}(t_1 - t_2)$ is a few femtoseconds; the measured width, however, is about 700 ps, due to the relatively slow photodetectors.

of counts against $t_1 - t_2$, statistically corresponds to the second-order correlation function $G^{(2)}(t_1 - t_2)$ that is calculated based on the knowledge of the state of the radiation.

Due to the limited ability of the photodetectors in determining the registration time of a photoelectron, obviously, the responds time of the photodetectors and the associated electronics will affect the measurement of $G^{(2)}(t_1 - t_2)$. For instance, the calculated width of $G^{(2)}(t_1 - t_2)$ for entangled signal–idler pairs of SPDC is typically in the order of a few femtosecond to a few hundred femtosecond. The observed width in Fig. 5.8.1, however, is about 700 ps. The broadening is due to the slow response of the photodetectors. In the observation of a photoelectron event, accompanied with the annihilation of a photon at time t, the photodetector exports an pulse of electric current to an electronic circuit which is able to analyze the pulse and to determine the electronic registration time $\tilde{t}$ within a certain uncertainty. The jitter of the leading edge of the pulse as well as the fluctuations of the pulse height, all contribute to the uncertainty of the electronic registration times, or the measured times, of $\tilde{t}_1$ and $\tilde{t}_2$. In Chapter 1, we have characterized this uncertainty as a response function of the photodetector, $D(\tilde{t}-t)$, where t is the photon annihilation time and $\tilde{t}$ is the electronic registration time, and have treated the joint photodetection measurement of D_1 and D_2 as a convolution between the response functions and the second-order coherence function $G^{(2)}(t_1 - t_2)$:

$$G^{(2)}(\tilde{t}_1 - \tilde{t}_2) = \frac{1}{t_c^2} \int dt_1 \int dt_2 \, G^{(2)}(t_1 - t_2) \, D(\tilde{t}_1 - t_1) \, D(\tilde{t}_2 - t_2),$$

(5.8.1)

where we have chosen

$$\int dt \, D(\tilde{t} - t) = t_c,$$

(5.8.2)

where t_c is defined as the response time of the photodetector. When the width of the response functions are much narrower then that of $G^{(2)}(t_1 - t_2)$, the response function can be treated as delta functions, $D(\tilde{t} - t) \sim t_c \delta(\tilde{t} - t)$. In this extreme case, the measured second-order correlation function will reveal the theoretical expectation of $G^{(2)}(t_1 - t_2)$:

$$G^{(2)}(\tilde{t}_1 - \tilde{t}_2) = \int dt_1 \int dt_2 \, G^{(2)}(t_1 - t_2) \, \delta(\tilde{t}_1 - t_1) \, \delta(\tilde{t}_2 - t_2)$$

$$= G^{(2)}(t_1 - t_2).$$

(5.8.3)

However, when the response time is in the same order or even greater then the width of the correlation function, the situation is completely different. When the width of the response functions are much wider then that of $G^{(2)}(t_1 - t_2)$, the second-order correlation function itself can then be treated as a delta function. In this extreme case, Eq. (5.8.1) turns into the following convolution between the response functions of the two photon counting detectors:

$$G^{(2)}(\tilde{t}_1 - \tilde{t}_2) = G^{(2)}(0) \left(\frac{\tau_0}{t_c^2} \right) \int d\tau \, D(\tilde{t}_1 - \tilde{t}_2 - \tau) \, D(\tau).$$

(5.8.4)

Where we have chosen

$$\int dt \, G^{(2)}(t) = G^{(2)}(0)\tau_0 \simeq \int dt \, G^{(2)}(0)\tau_0 \, \delta(t),$$

(5.8.5)

where $\tau_0 \sim \tau_c$ is a constant and is named as the second-order correlation time. Equation (5.8.4) indicates the following: (1) The width of the observed $G^{(2)}(\tilde{t}_1 - \tilde{t}_2)$ is now determined by the responds function of the photodetectors, which could be significantly broadened. This is what has happened in the measurement of Fig. 5.8.1. (2) The relative slow response time of the photodetectors may reduce the magnitude of the measured second-order correlation. For Gaussian-type respond functions, the convolution yields a Gaussian function of $G^{(2)}(\tilde{t}_1 - \tilde{t}_2)$ with a reduced central value of $G^{(2)}(0)$. The reduction factor is roughly τ_0/t_c.

The above reduction may not be a problem in the measurement of entangled states, however, it may affect the measurement of $G^{(2)}$ function for thermal or pseudo-thermal light significantly. The $G^{(2)}$ function of thermal or pseudo-thermal field has two terms corresponding to (1) the product of two measured mean photon numbers and (2) the photon number fluctuation correlation:

$$G^{(2)}(\tilde{t}_1 - \tilde{t}_2) \propto \frac{1}{t_c^2} \int dt_1 \int dt_2 \, \langle n_1 n_2 \rangle D(\tilde{t}_1 - t_1) \, D(\tilde{t}_2 - t_2)$$

$$= \frac{1}{t_c^2} \int dt_1 \int dt_2 \, [\langle n_1 \rangle \langle n_2 \rangle + \langle \Delta n_1 \Delta n_2 \rangle] D(\tilde{t}_1 - t_1) \, D(\tilde{t}_2 - t_2).$$

$$(5.8.6)$$

The time average over t_c may reduce the magnitude of the second term many orders smaller than that of the first term. For example, the time integral does not change the value of the first term of Eq. (5.8.6) for a CW thermal or pseudo-thermal field with broad spectrum:

$$\frac{1}{t_c^2} \int dt_1 \int dt_2 \, [\langle n_1 \rangle \langle n_2 \rangle D(\tilde{t}_1 - t_1) \, D(\tilde{t}_2 - t_2)$$

$$= \frac{1}{t_c^2} \left[\bar{n}_1 \bar{n}_2 \int dt_1 \, D(\tilde{t}_1 - t_1) \int dt_2 \, D(\tilde{t}_2 - t_2) \right]$$

$$= \bar{n}_1 \bar{n}_2. \qquad (5.8.7)$$

However, it reduces the value of the second term significantly when $\tau_0 \ll t_c$

$$\frac{1}{t_c^2} \int dt_1 \int dt_2 \, \langle \Delta n_1 \Delta n_2 \rangle D(\tilde{t}_1 - t_1) D(\tilde{t}_2 - t_2)$$

$$\simeq \frac{\bar{n}_1 \bar{n}_2}{t_c^2} \int dt_1 \int dt_2 \, \tau_0 \delta(t_1 - t_2) D(\tilde{t}_1 - t_1) D(\tilde{t}_2 - t_2)$$

$$\simeq \frac{\bar{n}_1 \bar{n}_2}{t_c^2} \tau_0 \int d\tau \, D(\tilde{t}_1 - \tilde{t}_2 - \tau) \, D(\tau). \qquad (5.8.8)$$

For Gaussian-type respond functions, the magnitude of the second term is thus roughly τ_0 / t_c times smaller than that of the first term. The coherence time of sunlight is in the order of 10^{-15} s. A nanosecond photodetector yields $\tau_0 / t_c \sim 10^{-6}$. It is definitely uneasy to distinguish the nontrivial photon number fluctuation correlation, which is one part of 10^6 relative to the trivial constant $\bar{n}_1 \bar{n}_2$, from the measurement of $G^{(2)}(\tilde{t}_1 - \tilde{t}_2)$. This has been the major difficulty for HBT measurement of sunlight and other type natural radiations.

How to identify the nontrivial second-order coherence of thermal field with short coherence time τ_0? The obvious approach is to use fast photodetectors with $t_c \sim \tau_0$. The state-of-the-art technology, however, has not been able to achieve that fast yet. One realistic approach, perhaps, is to use short pulsed thermal or pseudo-thermal light. If the temporal width of $\langle n_j \rangle$, $j = 1, 2$, is in the same order of τ_0, the time integral in Eq. (5.8.6) shall have similar effects on the trivial correlation part of $G^{(2)}$:

$$\frac{1}{t_c^2} \int dt_1 \int dt_2 \langle n(t_1) \rangle \langle n(t_2) \rangle D(\tilde{t}_1 - t_1) D(\tilde{t}_2 - t_2)$$

$$\simeq \frac{\bar{n}_1 \bar{n}_2}{t_c^2} \int dt_1 \int dt_2 \, \tau_0 \delta(t_1 - t_2) D(\tilde{t}_1 - t_1) D(\tilde{t}_2 - t_2)$$

$$\simeq \frac{\bar{n}_1 \bar{n}_2}{t_c^2} \tau_0 \int d\tau \, D(\tilde{t}_1 - \tilde{t}_2 - \tau) \, D(\tau), \tag{5.8.9}$$

which is the same as Eq. (5.8.8). In Eq. (5.8.9), we have approximated the short pulsed $\langle n(t_1) \rangle \langle n(t_2) \rangle$ a δ-function, $\langle n(t_1) \rangle \langle n(t_2) \rangle \sim \bar{n}_1 \bar{n}_2 \tau_0 \delta(t_1 - t_2)$. This mechanism may be suitable for short pulsed HBT type applications, such as synchrotron X-ray ghost imaging-microscope, although the spectrum of an X-ray source is much wider than that of sunlight. Regarding sunlight HBT measurements, of course, it is impossible to have a short pulsed sun, however, we can definitely "gate" our photodetectors to "expose" D_1 and D_2 within a short time window. In addition to a specially designed measurement circuit to separate $\langle \Delta n_1 \Delta n_2 \rangle$ from $\langle n_1 \rangle \langle n_2 \rangle$, an observable nontrivial correlation of $\langle \Delta n_1 \Delta n_2 \rangle$ is achievable for sunlight HBT measurements.

(II) About "coincidence" measurement

A coincidence measurement does not measure $G^{(2)}(t_1 - t_2)$ but rather cumulatively counts the joint detection events that fall into a certain coincidence time window Δt_c around a chosen value of $t_1 - t_2 = \tau$, where τ is a time constant which is determined by a particular experimental arrangement. Mathematically, this is equivalent to have a time integral on the second-order coherence function $G^{(2)}(t_1 - t_2)$:

$$R_c = \int_{\Delta T} dt_1 \, dt_2 \, G^{(2)}(t_1 - t_2) \, S(t_1 - t_2), \tag{5.8.10}$$

where R_c is the coincidence counting rate and $S(t_1 - t_2)$ represents the time window of the coincidence circuit

$$S(t_1 - t_2) = \begin{cases} 1 & \tau - \Delta t_c/2 \leq t_1 - t_2 \leq \tau + \Delta t_c/2 \\ 0 & \text{otherwise.} \end{cases}$$

(III) Measurement of $G^{(2)}(z_1 - z_2)$ or $G^{(2)}(\vec{\rho}_1 - \vec{\rho}_2)$

The longitudinal and transverse spatial correlation measurements have been discussed earlier. A coincidence counter or a linear multiplier and associated electronics can be used to measure the correlation by scanning the point-like photodetector longitudinally or transversely. In the photon counting measurements, the coincidence counter counts and records all the the joint-detection events fall into the coincidence time window of $\tau - \Delta t_c/2 \leq t_1 - t_2 \leq \tau + \Delta t_c/2$ as a function of $z_1 - z_2$ or $\vec{\rho}_1 - \vec{\rho}_2$. In the current–current correlation measurements, the linear multiplier and associated electronics integrates and records the output reading of the linear multiplier that is proportional to $i_1(t)i_2(t) \propto I(t_1)I(t_2)$. The delays of the electronic cables between the linear multiplier and the photodetectors D_1 and D_2 define the early times $t_1 = t - \tau_1^e$, $t_2 = t - \tau_2^e$ of the measured photodetection events,[2] as well as $t_1 - t_2 = \tau_1^e - \tau_2^e$. $G^{(2)}(z_1 - z_2)$ is measured by means of reading and recording the coincidence counting rate or the integrated output of the linear multiplier as a function of $z_1 - z_2$ or $\vec{\rho}_1 - \vec{\rho}_2$. The variation of $z_1 - z_2$ or $\vec{\rho}_1 - \vec{\rho}_2$ can be achieved by scanning one of the photodetectors along its longitudinal axis or its transverse plane.[3]

It should be emphasized that the temporal and the transverse correlation may not be treated as independent. For instance, in the measurement of the second-order transverse spatial coherence function $G^{(2)}(\vec{\rho}_1 - \vec{\rho}_2)$, the value of $\tau_1 - \tau_2$ must be carefully chosen to achieve a nonzero constant value

---

[2]The linear multiplier measures the current–current correlation of the photodetectors D_1 and D_2 at time t: $V_{12}(t) \propto i_1(t)i_2(t) \propto I(t_1)I(t_2)$, where $V_{12}(t)$ is the output reading of the linear multiplier at time t and $i_1(t)$ and $i_2(t)$ are the output currents of the photodetectors at time t. The delays of the electronic cables between the linear multiplier and the photodetectors D_1 and D_2 define the early times $t_1 = t - \tau_1^e$, $t_2 = t - \tau_2^e$ of the measured photodetection events, as well as $t_1 - t_2 = \tau_1^e - \tau_2^e$.

[3]The linear multiplier can also be used for the measurement of $G^{(2)}(t_1 - t_2)$ of stationary light by varying the cable length of the delay-line or by other means of the electronic delays with fixed values of $\mathbf{r}_1$ and $\mathbf{r}_2$.

(usually the maximum value) of $G^{(2)}(z_1, t_1; z_2, t_2)$ during the scanning of the transverse coordinates of the point-like photodetectors.

5.9   The Hanbury Brown and Twiss Interferometer

In 1956, Hanbury Brown and Twiss published two interferometers that measure the second-order temporal coherence function and the second-order spatial coherence function of thermal field, namely the so-called temporal intensity interferometer and spatial intensity interferometer.

(I) Temporal HBT interferometer

The temporal intensity interferometer of HBT is schematically shown in Fig. 5.9.1. In this interferometer, radiation emitted from a far-field thermal radiation source (such as a distant star) is divided at a beamsplitter into two beams of equal intensities. The two independent light beams

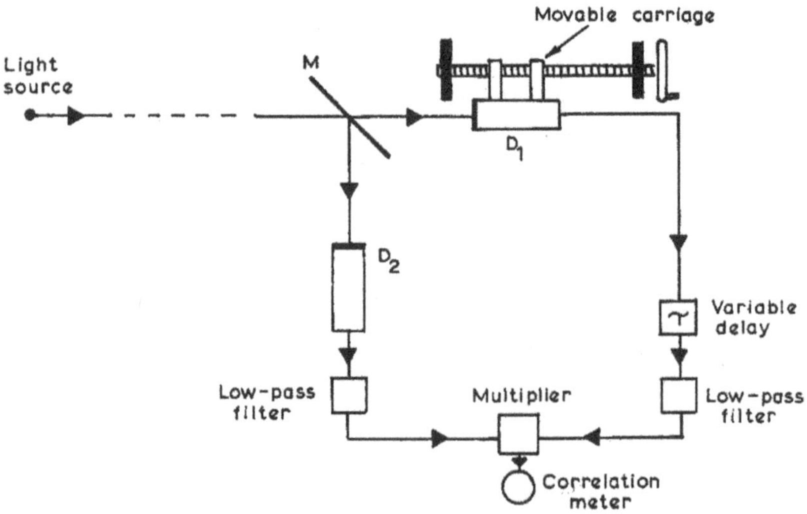

Fig. 5.9.1   Schematic of the historical Hanbury Brown and Twiss temporal intensity interferometer. This HBT interferometer measures the second-order temporal coherence of thermal radiation in the joint-detection of two spatially separated photodetectors D_1 and D_2. In the time of HBT, the response time of the best correlation multiplier may achieve 10^{-6} s. The coherent time of a natural white light is in the order of 10^{-14} s. HBT coupled a "low-pass filter" with each of their photomultipliers to increase the contrast of the measured $\Gamma^{(2)}(z_1, t_1; z_2, t_2)$ to an observable level.

are then meet with two photodetectors D_1 and D_2, one of which can be scanned longitudinally along the optical path. The output photocurrents of D_1 and D_2 are recorded, respectively, and multiplied electronically by a linear multiplier at a later time, $t = t_1 + \tau_1^e = t_2 + t_1^e$. HBT found that although both $\langle I_1 \rangle$ and $\langle I_2 \rangle$ kept constant values during the scanning, the statistical correlation of $\langle I_1 I_2 \rangle$, however, turned to be a nontrivial function of $\tau = \tau_1 - \tau_2 = (z_2 - z_1)/c + (\tau_2^e - \tau_1^e)$:

$$\langle I(\tau_1)I(\tau_2) \rangle \propto 1 + \alpha \left| \mathcal{F}_\tau \{ a^2(\nu) \} \right|^2, \qquad (5.9.1)$$

where, again, $\tau_j = (t - \tau_j^e) - z_j/c = t_j - z_j/c$, $j = 1, 2$, $0 < \alpha < 1$ is a positive real number that determines the contrast of the measured second-order temporal coherence function $\gamma^{(2)}(z_1, t_1; z_2, t_2)$. When $\alpha = 1$, Eq. (5.9.1) is the same as our theoretical expectation of Eq. (5.2.13). The experimentally observed contrasts of $\gamma^{(2)}(z_1, t_1; z_2, t_2)$ from historical HBT interferometers, are quite low with $\alpha \ll 1$.

As we have discussed in previous section, the low contrast $\alpha \ll 1$ was the result of (1) relatively broadband spectrum thermal field and (2) relatively slower photodetectors and associated correlation measurement electronics used in the historical HBT interferometer. Recall that Eq. (5.2.13) was derived by assuming a "perfect" correlation measurement with "perfect" photodetectors and "perfect" electronics. Due to the limited ability of the photodetectors in determining the registration time of a photoelectron, the relative slow responds time of the photodetectors and the associated electronics will affect the measurement of $G^{(2)}(t_1 - t_2)$.

The widths of the response functions of the photodetectors used in the historical HBT interferometers are much wider than that of the $G^{(2)}(t_1 - t_2)$ of natural light in their astrophysics observations. The second-order correlation function can be treated as a delta function in the convolution between the response functions of the two photon counting detectors:

$$G^{(2)}(\tilde{t}_1 - \tilde{t}_2)$$

$$= \frac{1}{t_c^2} \int dt_1 \int dt_2\, G^{(2)}(t_1 - t_2)\, D(\tilde{t}_1 - t_1)\, D(\tilde{t}_2 - t_2)$$

$$\propto \frac{1}{t_c^2} \int dt_1 \int dt_2\, [\langle n_1 \rangle \langle n_2 \rangle + \langle \Delta n_1 \Delta n_2 \rangle]\, D(\tilde{t}_1 - t_1)\, D(\tilde{t}_2 - t_2). \quad (5.9.2)$$

The time integral does not change the value of the first term of Eq. (5.9.2) simply because it is a constant:

$$\frac{1}{t_c^2} \int dt_1 \int dt_2 \left[\langle n_1 \rangle \langle n_2 \rangle D(\tilde{t}_1 - t_1)\, D(\tilde{t}_2 - t_2) \right.$$

$$= \frac{1}{t_c^2} \left[\bar{n}_1 \bar{n}_2 \int dt_1\, D(\tilde{t}_1 - t_1) \int dt_2\, D(\tilde{t}_2 - t_2) \right]$$

$$= \bar{n}_1 \bar{n}_2.$$

However, it reduces the value of the second term significantly when $\tau_0 \ll t_c$:

$$\frac{1}{t_c^2} \int dt_1 \int dt_2\, \langle \Delta n_1 \Delta n_2 \rangle D(\tilde{t}_1 - t_1) D(\tilde{t}_2 - t_2)$$

$$\simeq \frac{\bar{n}_1 \bar{n}_2}{t_c^2} \int dt_1 \int dt_2\, \tau_0 \delta(t_1 - t_2) D(\tilde{t}_1 - t_1) D(\tilde{t}_2 - t_2)$$

$$\simeq \frac{\bar{n}_1 \bar{n}_2}{t_c^2} \tau_0 \int d\tau\, D(\tilde{t}_1 - \tilde{t}_2 - \tau)\, D(\tau).$$

For Gaussian-type respond functions, the magnitude of the second term is thus roughly τ_0/t_c times smaller than that of the first term. The coherence time of sunlight is in the order of 10^{-15} s. A nanosecond photodetector yields $\tau_0/t_c \sim 10^{-6}$. It is definitely uneasy to distinguish the nontrivial photon number fluctuation correlation, which is one part of 10^6 relative to the trivial constant $\bar{n}_1 \bar{n}_2$, from the measurement of $G^{(2)}(\tilde{t}_1 - \tilde{t}_2)$. This has been the major difficulty for HBT measurement of sunlight and other type natural radiations.

In the beginning of the 1970s, right after the invention of laser, the concept of pseudo-thermal light was introduced into the measurement of the second-order coherence of thermal field to achieve $\alpha \sim 1$. A pseudo-thermal light source usually contains a laser beam and a fast rotating defusing ground glass. The transversely expanded laser beam is scattered by millions of tiny scattering elements of the fast rotating ground glass. These tiny scattering elements play the role of point-like sub-sources. Millions of randomly scattered subfields, each may contain a number of identical photons, are scattered from each tiny scattering elements of the ground glass with *random relative phases*. The spectral bandwidth of a laser beam, especially operated in single-mode or single-frequency, is usually narrow enough to satisfy $\tau_0 \sim t_c$. Figure 5.9.2 is a collection of historically measured temporal degree of second-order coherence function of thermal or pseudo-thermal field by using similar experimental setups of the HBT

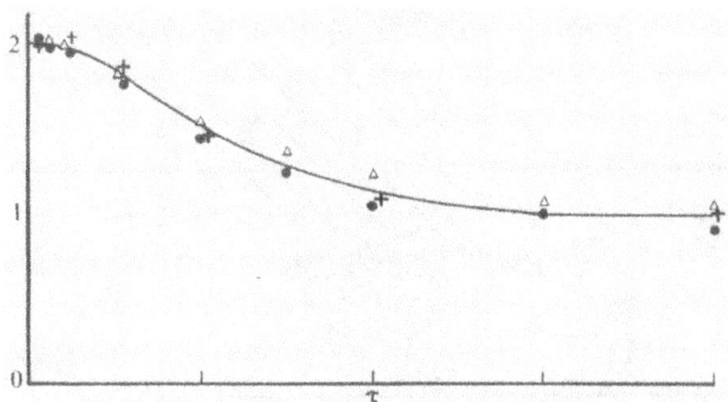

Fig. 5.9.2 Historical measurements of the second-order temporal coherence function of thermal field. Thermal radiation has almost twice the chance of been captured by two individual photodetectors at $\tau = \tau_1 - \tau_2 \sim 0$, although the thermal radiation is incoherently, randomly, and stochastically radiated from the source. Note that most of these measurements use "pseudo-thermal" light sources with relatively narrower spectrum.

interferometer. In these measurements, the two photodetectors D_1 and D_2 have almost twice chance to be excited at $\tau = \tau_1 - \tau_2 \sim 0$ then that of $\tau = \tau_1 - \tau_2 > \tau_c$, where τ_c is the coherence time of the thermal field.

(II) Far-field spatial HBT interferometer

Hanbury Brown and Twiss published their spatial intensity interferometer intensity interferometer in the same year as that of their temporal intensity interferometer. In fact, in 1956, Hanbury Brown and Twiss had successfully utilized the second-order far-field spatial coherence of thermal field

$$\gamma^{(2)}(x_1, t_1; x_2, t_2) = 1 + \text{sinc}^2 \frac{\pi \Delta \theta}{\lambda}(x_1 - x_2) \tag{5.9.3}$$

in astrophysics applications for measuring the angular diameter of distant stars or the angular separation of distant double-stars. A typical spatial HBT spatial intensity interferometer for such applications is schematically shown in Fig. 5.9.3. The spatial HBT interferometer is similar to the Michelson stellar interferometer, except that the observation is the second-order spatial correlation measured by two independent photodetectors, instead of the visibility of the first-order interference pattern measured by one photodetector. The long-base-line HBT intensity interferometer has been widely used in modern astronomical observations. The HBT interferometer is especially useful for measuring celestial bodies of smaller angular size and their angular separation. When the separation between D_1 and D_2 is "scanned" from $x_1 - x_2 = 0$ to $x_1 - x_2 = b$, see Fig. 5.9.3,

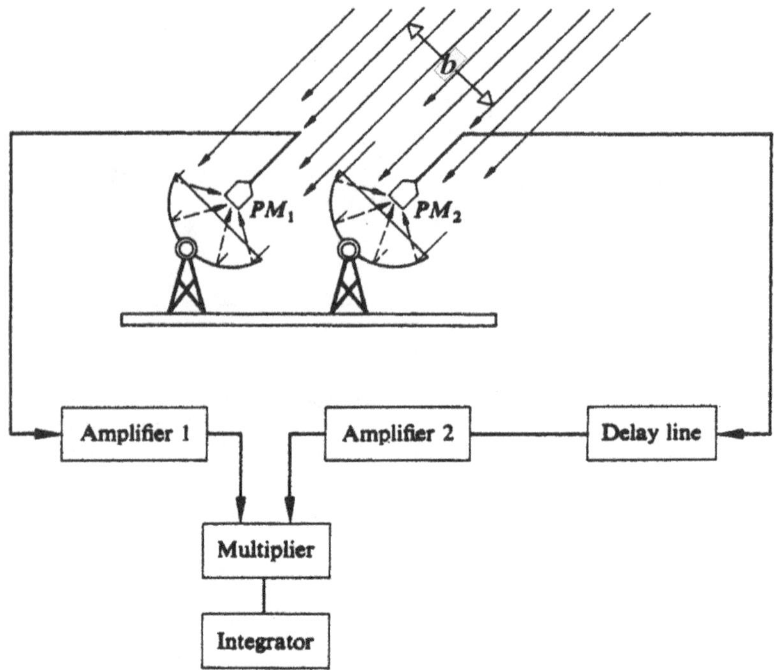

Fig. 5.9.3 Schematic of a spatial HBT interferometer for astrophysics applications. The interferometer measures the angular size of a distant thermal radiation source by means of its second-order spatial coherence, or spatial correlation. When the separation between D_1 and D_2 is "scanned" from $x_1 - x_2 = 0$ to $x_1 - x_2 = b$, the measured intensity fluctuation correlation drops from its maximum value of "one" to its minimum value of "zero". The greater the value of b, the smaller the $\Delta\theta$ that is measurable. In modern applications, b, which is usually called the "base line", can be as long as kilometers or more.

the measured intensity fluctuation correlation drops from its maximum value of "one" to its minimum value of "zero". The greater the value of b, the smaller the $\Delta\theta$ that is measurable. In modern applications, b, which is usually called "base line", can be as long as kilometers or more.

(III) Near-field spatial HBT interferometer

In the early days, spatial HBT intensity interferometer was considered working in far-field only. Until the advent of the HBT interferometer for half a century, we proved that the spatial HBT interferometer can also work in the Fresnel near-field too. The second-order spatial coherence function of thermal field $G^{(2)}(\mathbf{r}_1, t_1; \mathbf{r}_2, t_2)$ in near-field was experimentally demonstrated by Scarcelli *et al.* in the years 2005–2006. Their experimental setup, schematically illustrated in Fig. 5.3.1, is similar to that of the historical HBT experiment, except the far-field distant star is replaced by a Fresnel near-field pseudo-thermal radiation source. An unfolded version of the schematic is shown in Fig. 5.9.4, which might be easier for analyzing the physics. Scarcelli *et al.* found that although the single detector counting rates or the output currents of D_1 and D_2 were monitored, respectively, to be constants during the measurement, nontrivial second-order correlations were observable with almost 50% contrast while the two point-like-photodetectors, D_1 and D_2, are aligned symmetrically on the transverse planes of x_1 and x_2, as indicated in the upper and the lower cases of Fig. 5.9.4. In the upper measurement, D_1 and D_2 are aligned symmetrically on the optical axis. A sinc-like function of $\langle \Delta I_1 \Delta I_2 \rangle$ is observed by scanning either D_1 or D_2 transversely in the neighborhood of the optical axis. When D_1 is moved a few millimeters up (or down) from its symmetrical position, as shown in the middle, the nontrivial correlation disappears with $\langle \Delta I_1 \Delta I_2 \rangle \sim 0$ when scanning either D_1 or D_2 in the

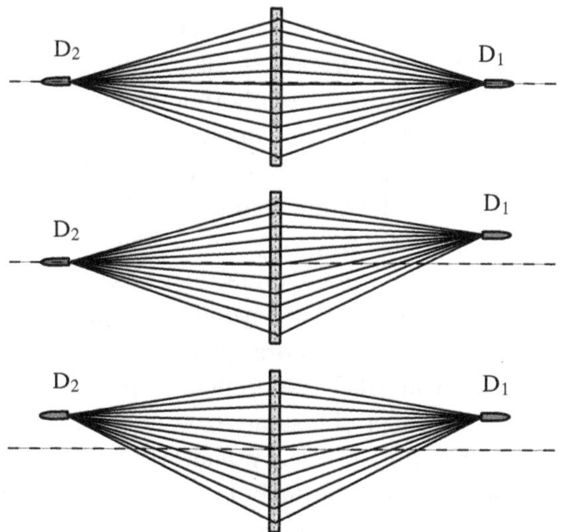

Fig. 5.9.4 Second-order near-field spatial correlation measurement by Scarcelli *et al.* Upper: D_1 and D_2 are placed at equal distances from the source and aligned symmetrically on the optical axis, a nontrivial $g^{(2)}(x_1 - x_2)$ is observed with maximum value of ~ 2. Middle: D_1 is moved up to a nonsymmetrical position, $g^{(2)}(x_1 - x_2)$ becomes a constant of 1. Lower: D_2 is moved up to the symmetrical position with D_1, the nontrivial $g^{(2)}(x_1 - x_2)$ is observable with a maximum value of ~ 2, again.

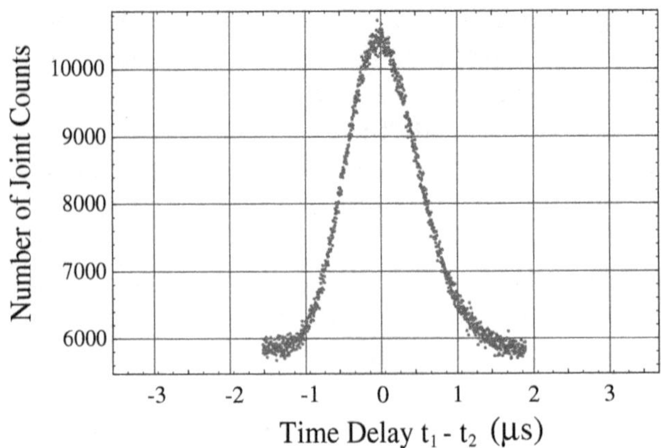

Fig. 5.9.5   Second-order temporal coherence of the pseudo-thermal radiation used in the near-field HBT experiment. The coherent time is found in the order of μs, corresponding to a longitudinal optical delay of $\sim$300 m.

neighborhood of that unsymmetrical position. In the lower measurement, D_2 is moved up (or down) to a symmetrical position, again with respect to D_1. A similar sinc-like function of $\langle \Delta I_1 \Delta I_2 \rangle$ is observed by scanning either D_1 or D_2 in the neighborhood of their new symmetrical position. Note that *equal distance* between the photodetectors and the light source is required for the observation of the sinc-function like correlation. For a large source of transverse dimension, a few millimeter difference may cause a complete disappearing of the nontrivial correlation.

What is the reason for $\langle \Delta I_1 \Delta I_2 \rangle \sim 0$ when D_1 is moved a few millimeters up (or down) from its symmetrical position? Is it possible the spatial move caused a temporal delay beyond the temporal coherence of the thermal field? A second-order temporal coherence of the pseudo-thermal field used in the near-field HBT experiment was measured in the same experimental setup and under the same experimental condition. Scarcelli *et al.* found the coherent time of their pseudo-thermal field is in the order of μs, corresponding to an optical delay of $\sim$300 m. A few millimeters up (or down) of the photodetectors in Fresnel near-field can never result in such a delay.

The near-field second-order coherence function has been calculated in last section

$$g^{(2)}(\vec{\rho}_1, \vec{\rho}_2) = 1 + \mathrm{somb}^2 \left(\frac{\pi \Delta \theta}{\lambda} |\vec{\rho}_1 - \vec{\rho}_2| \right). \tag{5.9.4}$$

The source-angular-size dependent of the point-to-spot correlation has been experimentally confirmed in 1-D measurements. Figure 5.9.6 shows the experimental results of Zhou *et al.* with different angular sized sources. The fitting curves are calculated from Eq. (5.9.4). The calculated $\gamma^{(2)}(x_1 - x_2)$ functions agree with the experimental results within the experimental error. It is interesting to find that a point-to-"point" HBT correlation is achievable with a large angular sized thermal field source.

The above experimental observations are quite surprise. First, it tells us that our 50 years belief about the far-field condition of HBT correlation is never true; the nontrivial second-order spatial correlation of thermal field is observable in the Fresnel near-field. Second, apparently, we have observed an EPR-type nonlocal phenomenon: Although D_1 and D_2 both show constant counting rates when scanning their transverse positions, if D_1 observes a photon or wavepacket at a transverse position x_1, D_2 has twice chance to observe its randomly paired photon or wavepacket at a certain transverse coordinate x_2. This nontrivial second-order near-field spatial correlation is not only interesting in fundamental concerns but also useful in practical applications. The point-to-point correlation between two near-field planes has turned into a useful new imaging technology: lensless ghost imaging, which is especially useful for these radiations for which no effective imaging lenses can be used to produce the classic point-to-point

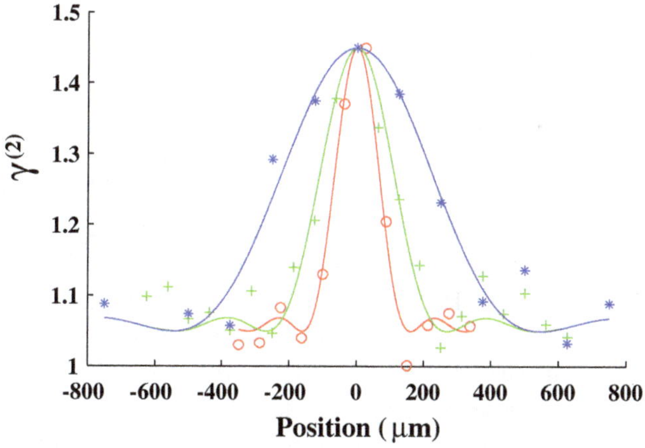

Fig. 5.9.6 Measured point-to-"spot" spatial correlation of a pseudo-thermal radiation with different source sizes (diameters) 1 mm (blue star), 2 mm (green cross), and 4 mm (red circle), respectively. The fitting curves are calculated from Eq. (5.9.4).

image forming function. We have a detailed discussion on the lensless ghost imaging in Chapter 10.

(IV) **The physical cause of the HBT correlation**

What is the physical cause of the HBT correlation? What is the reason for two randomly created and randomly distributed photons to have twice chance of being jointly captured within a temporal delay that equals the coherence time of the thermal field and within a transverse area that equals the coherence area of the thermal radiation? From the view point of quantum optics, either in terms of Einstein's granularity picture of light or in terms of the concepts of quantum mechanics, HBT correlation is a two-photon interference phenomenon: a pair of randomly created and randomly paired subfields or photons interfering with the pair itself. A large number of constructively destructively added two-photon interferences produced the nontrivial temporal and spatial intensity fluctuation correlation of HBT.

Interestingly, the physical cause of HBT correlation remains yet the subject of today's debate. A naive argument of rejecting two-photon interference considers thermal light "classical": "How could quantum theory apply to classical light?" Perhaps this argument forgot that blackbody radiation is a typical thermal radiation and Planck's quantization is indeed for thermal field. We cannot forget that the starting point of quantum theory is the classically incomprehensible behavior of the thermal field! A profound reservation about two-photon interference might be its nonlocal nature. The concept of nonlocal interference is indeed uneasy to accept.

Historically, a widely accepted theory suggests that the temporal and spatial HBT correlations of thermal field are pre-prepared in the radiation source as a statistical property of thermal light. Thermal light was named "correlated light" accordingly. It is true that, statistically, the intensity of light, or the number of photons, created from a thermal source fluctuates temporally and spatially. However, we cannot find any physical mechanism to "correlate" these pre-prepared fluctuations. For example, the HBT interferometers on earth receive radiations from all possible locations of the sun, including its upper edge and lower edge. The intensity of light, or the number of photons, created from the upper edge of the sun and from the lower edge of the sun may both fluctuate statistically. Do we have any physical mechanism that is able to prepare radiations from the upper edge of the sun and from the lower edge of the sun with identical intensity fluctuations or photon number fluctuations? If there is any, this mechanism is indeed nonlocal. The creation process of sunlight is stochastic. Either in the view of classical electromagnetic theory of light or in the view of quantum theory of light, sunlight should be created randomly and distributed randomly.

It was HBT that suggested a reasonable interpretation to the HBT correlation. In a HBT intensity interferometer, see Fig. 5.9.3, the measurement is in the far-field of the thermal radiation source, which is equivalent to the Fourier transform plane. When D_1 and D_2 are moved side by side, the two detectors measure the same spatial mode of the radiation field. HBT believe that the measured intensities should have the same fluctuations while the two photodetectors receive the same spatial mode and thus yield a maximum value of $\langle \Delta I(\mathbf{k})\Delta I(\mathbf{k})\rangle$ and gives $\gamma^{(2)} \sim 2$. When the two photodetectors move apart to a certain distance, D_1 and D_2 start to measure different spatial modes of the radiation field. In this case, the measured intensities have different fluctuations. The measurement yields $\langle \Delta I(\mathbf{k})\Delta I(\mathbf{k}')\rangle = 0$ and gives $\gamma^{(2)} \sim 1$. Figure 5.9.7 illustrates the above two different situations.

For 50 years, this theory has convinced us to believe that the observation of the nontrivial second-order correlation of thermal field only occurs in the Fraunhofer far-field. What will happen if we move the two HBT

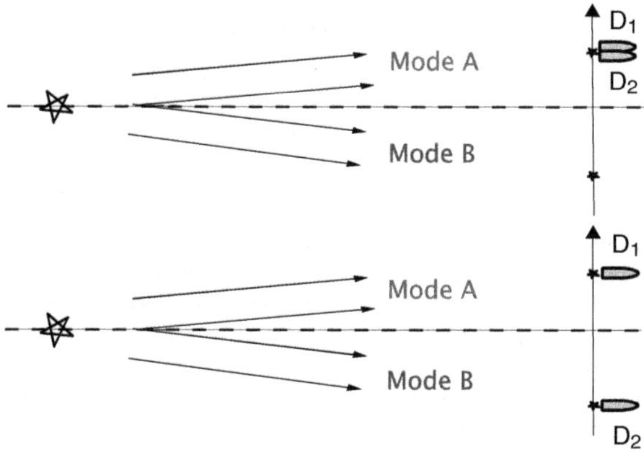

Fig. 5.9.7 A phenomenological interpretation of the historical HBT experiment. Upper: The two photodetectors receive identical modes of the far-field radiation and thus experience identical intensity fluctuations. The joint measurement of D_1 and D_2 gives a maximum value of $\langle \Delta I_1 \Delta I_2\rangle$. Lower: The two photodetectors receive different modes of the far-field radiation. In this case, the joint measurement gives $\langle \Delta I_1 \Delta I_2\rangle = 0$.

photodetectors to the Fresnel "near-field"[4] as shown in the unfolded schematic of Fig. 5.9.8? Is the nontrivial second-order correlation still observable in the Fresnel near-field? In fact, the second-order correlation of thermal field in the Fresnel near-field has been observed by Scarcelli *et al.* during the years 2005–2006.

Examining Fig. 5.9.8, it is easy to fine that in the Fresnel near-field, (1) each photodetector, D_1 and D_2, is able to receive radiations from a large number of sub-sources, which are spatially distributed on the entire source, and spatial modes, which are angularly distributed within the entire opening angle of the source; (2) in a joint-detection between D_1 and D_2, the two photodetectors have more chances to be triggered by radiations coming from different sub-sources and different spatial modes. The ratio between the joint-detections triggered by radiations coming from the same sub-source (mode) and these triggered by different sub-sources (modes) is roughly $N/N^2 = 1/N$, and (3) despite the scanning position of D_1 (or D_2), the ratio of $1/N$ does not change. If we believe the HBT correlation of thermal field comes from the measurement of identical spatial mode $\langle \Delta I(\mathbf{k})\Delta I(\mathbf{k})\rangle$ and the measurement of different spatial modes results in $\langle \Delta I(\mathbf{k})\Delta I(\mathbf{k}')\rangle = 0$, we would easily conclude a constant second-order spatial coherence in the Fresnel near-field. i.e., $\langle I_1 I_2\rangle \sim 1$ despite the scanning position of D_1 (or D_2), which is in false.

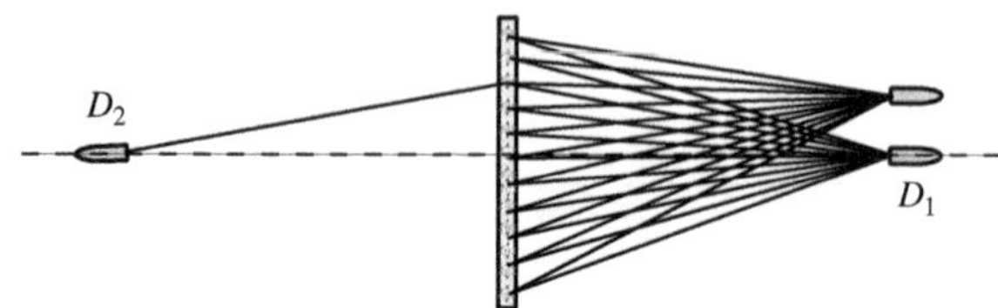

Fig. 5.9.8 (1) Each photodetector is able to receive radiations from a large number of sub-sources or spatial modes; (2) in a joint-detection between D_1 and D_2, the two photodetectors have more chances to be triggered by radiations coming from different sub-sources and different spatial modes. The ratio between the joint-detections triggered by radiations coming from the same sub-source (mode) and these triggered by different sub-sources (modes) is roughly $N/N^2 = 1/N$, and (3) despite the scanning position of D_1 (or D_2), the ratio of $1/N$ does not change.

---

[4]The concept of "near-field" was defined by Fresnel to be distinct from the Fraunhofer far-field. The Fresnel near-field is defined for a light source with angular size satisfying $\Delta\theta > \lambda/D$, where D is the diameter of the source. The Fresnel near-field is different from the "near-surface-field". The "near-surface-field" considers a distance of a few wavelengths from a surface.

We know that the observed intensity fluctuation of thermal light may have different causes. Statistically, intensity of light, or number of photons, created from a thermal source may fluctuate. The index variations of the medium of propagation may also induce additional intensity fluctuations. However, these random fluctuations can never produce a nontrivial $\langle \Delta I_1(\mathbf{r}_1, t_1) \Delta I_2(\mathbf{r}_2, t_2) \rangle$ that depends on the temporal delay and spatial separation of the photodetection events of D_1 and D_2. Furthermore, the nontrivial $\langle \Delta I_1(\mathbf{r}_1, t_1) \Delta I_2(\mathbf{r}_2, t_2) \rangle$ is invariant under these additional "turbulence"-induced fluctuations, i.e., turbulence resistant. We may say the following: The HBT correlation is *observed* from the intensity fluctuations, or the photon number fluctuations, of the thermal field, however, it is neither *caused* by the statistical fluctuations originated from the stochastic photon creation processes nor *caused* by the additional fluctuations induced by its propagation medium.

5.10 Nth-Order Coherence of Light

In this section, we introduce the concept of higher-order coherence or correlation of light. Formulated from electromagnetic field, the Nth-order ($N > 2$) coherence of light is defined as

$$
\begin{aligned}
\Gamma^{(N)} &(\mathbf{r}_1, t_1; \mathbf{r}_2, t_2; \ldots; \mathbf{r}_N, t_N) \\
&= \langle I(\mathbf{r}_1, t_1)\, I(\mathbf{r}_2, t_2) \ldots I(\mathbf{r}_N, t_N) \rangle \\
&= \langle E^*(\mathbf{r}_1, t_1) E(\mathbf{r}_1, t_1)\, E^*(\mathbf{r}_2, t_2) E(\mathbf{r}_2, t_2) \ldots E^*(\mathbf{r}_N, t_N) E(\mathbf{r}_N, t_N) \rangle
\end{aligned}
\tag{5.10.1}
$$

and the degree of Nth-order coherence is defined as

$$
\begin{aligned}
\gamma^{(N)} &(\mathbf{r}_1, t_1; \mathbf{r}_2, t_2; \ldots; \mathbf{r}_N, t_N) \\
&= \frac{\Gamma^{(N)}(\mathbf{r}_1, t_1; \mathbf{r}_2, t_2; \ldots; \mathbf{r}_N, t_N)}{\Gamma^{(1)}(\mathbf{r}_1, t_1; \mathbf{r}_1, t_1) \Gamma^{(1)}(\mathbf{r}_2, t_2; \mathbf{r}_2, t_2) \ldots \Gamma^{(1)}(\mathbf{r}_N, t_N; \mathbf{r}_N, t_N)},
\end{aligned}
\tag{5.10.2}
$$

where the ensemble average, $\langle \ldots \rangle$ denotes, again, *taking into account all possible realizations of the field.* In classical theory, $\gamma^{(N)}(\mathbf{r}_1, t_1; \mathbf{r}_2, t_2; \ldots; \mathbf{r}_N, t_N)$ is the normalized correlation of N intensities measured at space-time coordinates $(\mathbf{r}_1, t_1)$, $(\mathbf{r}_2, t_2), \ldots$, and $(\mathbf{r}_N, t_N)$.

Formulated from quantum theory of light, the Nth-order $(N > 2)$ coherence of light is defined as

$$G^{(N)}(\mathbf{r}_1, t_1; \mathbf{r}_2, t_2; \ldots; \mathbf{r}_N, t_N)$$

$$= \big\langle \langle \hat{E}^{(-)}(\mathbf{r}_1, t_1) \hat{E}^{(-)}(\mathbf{r}_2, t_2) \ldots \hat{E}^{(-)}(\mathbf{r}_N, t_N)$$

$$\times \hat{E}^{(+)}(\mathbf{r}_N, t_N) \ldots \hat{E}^{(+)}(\mathbf{r}_2, t_2) \hat{E}^{(+)}(\mathbf{r}_1, t_1) \rangle_{\text{QM}} \big\rangle_{\text{En}} \qquad (5.10.3)$$

and the degree of Nth-order coherence is defined as

$$g^{(N)}(\mathbf{r}_1, t_1; \mathbf{r}_2, t_2; \ldots; \mathbf{r}_N, t_N)$$

$$= \frac{G^{(N)}(\mathbf{r}_1, t_1; \mathbf{r}_2, t_2; \ldots; \mathbf{r}_N, t_N)}{G^{(1)}(\mathbf{r}_1, t_1; \mathbf{r}_1, t_1) G^{(1)}(\mathbf{r}_2, t_2; \mathbf{r}_2, t_2) \ldots G^{(1)}(\mathbf{r}_N, t_N; \mathbf{r}_N, t_N)},$$

$$(5.10.4)$$

where $\langle \ldots \rangle_{\text{En}}$ denotes, again, an ensemble average. In the view of quantum mechanics, $g^{(N)}(\mathbf{r}_1, t_1; \mathbf{r}_2, t_2; \ldots; \mathbf{r}_N, t_N)$ is the probability of jointly observing N photons at space-time coordinates $(\mathbf{r}_1, t_1)$, $(\mathbf{r}_2, t_2)$, $\ldots$, and $(\mathbf{r}_N, t_N)$.

There should be no surprise that the Nth-order coherence function of thermal field is a nontrivial function of space-time coordinates $(\mathbf{r}_1, t_1; \mathbf{r}_2, t_2 \ldots; \mathbf{r}_N, t_N)$ of the N-fold joint photodetection events, caused by N-photon interference: the randomly created and randomly grouped N photons in thermal state interfering with the group itself. To show this, we start from the discussion of third-order coherence of thermal light in terms of Einstein's granularity picture and the quantum theory of light, we than generalize it to the Nth-order coherence function of thermal radiation.

(I) Third-order coherence of thermal light

We first calculate the third-order coherence of thermal field in Einstein's picture. Assuming a simple experimental setup, in which a thermal light source is facing three photodetectors, D_1, D_2, and D_3, either in analog modes or in photon counting modes. D_1, D_2, and D_3 as well as the correlation circuit are prepared for a three-folding joint measurement or a three-folding joint photon counting. In Einstein's picture, the thermal radiation source constants of a large number of randomly distributed and randomly radiating independent point-like sub-sources, such as trillions of independent and randomly radiating atomic transitions, contribute to the measurement of each of the three detectors. Each point-like sub-source contributes an independent wavepacket as a subfield of complex amplitude $E_m = a_m e^{i\varphi_m}$, where a_m is the real and positive amplitude of the mth subfield and φ_m is a *random* phase associated with the mth sub-field. Basically, we have the following pictures for the source: (I) a large number of independent point-like sub-sources distributed randomly in space (counted spatially); (II) each point-source contains a large number of independently and randomly radiating atoms (counted temporally); (III) a large number of sub-sources, either counted spatially or temporally, may contribute to each of the independent radiation mode $(\vec{\kappa}, \omega)$ at each of the individual point photodetectors (counted by mode).

It is easy to find that the expectation value of the first-order jth self-correlation function, or the jth intensity measured by the jth photodetector, D_j, $j = 1, 2, 3$, is a constant:

$$\Gamma^{(1)}(\mathbf{r}_j, t_j; \mathbf{r}_j, t_j) = \langle E^*(\mathbf{r}_j, t_j) E(\mathbf{r}_j, t_j) \rangle = \text{constant},$$

where the expectation operation has taken into account all possible values of the phases of the subfields. Although each and all the measured intensities are constants, it does not prevent a nontrivial third-order coherence in the joint measurement of three independent photodetectors:

$$\Gamma^{(3)}(\mathbf{r_1}, t_1; \mathbf{r_2}, t_2; \mathbf{r_3}, t_3)$$

$$= \langle E^*(\mathbf{r_1}, t_1) E(\mathbf{r_1}, t_1) E^*(\mathbf{r_2}, t_2) E(\mathbf{r_2}, t_2) E^*(\mathbf{r_3}, t_3) E(\mathbf{r_3}, t_3) \rangle$$

$$= \left\langle \sum_a E^*_{a1} \sum_b E_{b1} \sum_c E^*_{c2} \sum_d E_{d2} \sum_e E^*_{e3} \sum_f E_{f3} \right\rangle$$

$$= \sum_{a,b,c} \left| \frac{1}{\sqrt{6}} \left[E_{a1} E_{b2} E_{c3} + E_{a1} E_{b3} E_{c2} + E_{a2} E_{b1} E_{c3} \right. \right.$$

$$\left. \left. + E_{a2} E_{b3} E_{c1} + E_{a3} E_{b1} E_{c2} + E_{a3} E_{b2} E_{c1} \right] \right|^2, \tag{5.10.5}$$

where E_{aj}, $a = 1, 2, \ldots, \infty$, $j = 1, 2, 3$, is short for $E_a(\mathbf{r}_j, t_j)$, indicating the field at coordinate $(\mathbf{r}_j, t_j)$ that is originated from the ath sub-source. Similar to the calculation of the second-order coherence function of thermal light, a partial ensemble average has been taken in Eq. (5.10.5) by means of *taking into account all possible random phase values of the subfields.*

Equation (5.10.5) indicates that $\Gamma^{(3)}(\mathbf{r_1}, t_1; \mathbf{r_2}, t_2; \mathbf{r_3}, t_3)$ is the sum of a large number of interferences, and each interference corresponds to the following superposition:

$$\Gamma^{(3)}_{abc} = \left| \frac{1}{\sqrt{6}} \left[E_{a1} E_{b2} E_{c3} + E_{a1} E_{b3} E_{c2} + E_{a2} E_{b1} E_{c3} \right. \right.$$

$$\left. \left. + E_{a2} E_{b3} E_{c1} + E_{a3} E_{b1} E_{c2} + E_{a3} E_{b2} E_{c1} \right] \right|^2 . \tag{5.10.6}$$

The superposed six terms in Eq. (5.10.6) indicate six different yet indistinguishable alternatives for three independent quantized subfields, or photons, to trigger a three-fold joint-detection event of D_1, D_2, and D_3. These six alternatives are schematically illustrated in Fig. 5.10.1.

Similar to the second-order coherence function of thermal light, in an idealized correlation measurement by assuming three perfect photodetectors and a perfect three-fold correlation circuit, in general, we may write $\Gamma^{(3)}(\mathbf{r_1}, t_1; \mathbf{r_2}, t_2; \mathbf{r_3}, t_3)$ in terms of the first-order coherence functions

$$\Gamma^{(3)}(\mathbf{r_1}, t_1; \mathbf{r_2}, t_2; \mathbf{r_3}, t_3)$$

$$= \Gamma^{(1)}_{11} \Gamma^{(1)}_{22} \Gamma^{(1)}_{33} + \left| \Gamma^{(1)}_{12} \right|^2 \Gamma^{(1)}_{33} + \left| \Gamma^{(1)}_{23} \right|^2 \Gamma^{(1)}_{11}$$

$$+ \left| \Gamma^{(1)}_{31} \right|^2 \Gamma^{(1)}_{22} + \Gamma^{(1)}_{12} \Gamma^{(1)}_{23} \Gamma^{(1)}_{31} + \Gamma^{(1)}_{21} \Gamma^{(1)}_{32} \Gamma^{(1)}_{13} . \tag{5.10.7}$$

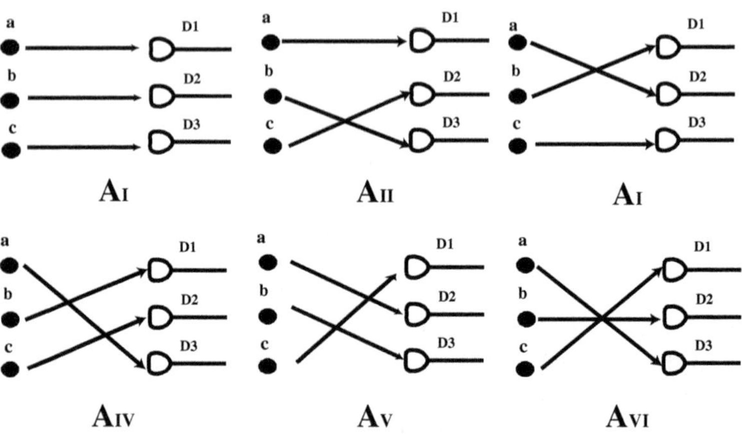

Fig. 5.10.1   Three independent quantized subfields, or photons, a, b, c have six alternatives of triggering a joint-detection event between D_1, D_2, and D_3. The sum of a large number of these interferences yield a nontrivial third-order coherence function of thermal field $\Gamma^{(3)}(\mathbf{r_1}, t_1; \mathbf{r_2}, t_2; \mathbf{r_3}, t_3)$.

Equation (5.10.7) can be normalized as

$$\gamma^{(3)}(\mathbf{r}_1, t_1; \mathbf{r}_2, t_2; \mathbf{r}_3, t_3)$$

$$= 1 + |\gamma_{12}^{(1)}|^2 + |\gamma_{13}^{(1)}|^2 + |\gamma_{23}^{(1)}|^2 + \gamma_{12}^{(1)}\gamma_{23}^{(1)}\gamma_{31}^{(1)} + \gamma_{21}^{(1)}\gamma_{32}^{(1)}\gamma_{13}^{(1)}. \quad (5.10.8)$$

The third-order coherence formulated from quantum theory can be easily calculated. Similar to the first-order and second-order coherence of thermal field, we may model thermal radiation either in single-photon state representation or in coherent state representation. In either cases, the second-order quantum coherence of thermal field $G^{(N)}(\mathbf{r}_1, t_1; \mathbf{r}_2, t_2; \mathbf{r}_3, t_3)$ can be easily calculated:

$$G^{(3)}(\mathbf{r}_1, t_1; \mathbf{r}_2, t_2; \mathbf{r}_3, t_3)$$

$$= \langle\langle \hat{E}^{(-)}(\mathbf{r}_1, t_1)\hat{E}^{(-)}(\mathbf{r}_2, t_2)\hat{E}^{(-)}(\mathbf{r}_3, t_3)$$

$$\times \hat{E}^{(+)}(\mathbf{r}_3, t_3)\hat{E}^{(+)}(\mathbf{r}_2, t_2)\hat{E}^{(+)}(\mathbf{r}_1, t_1)\rangle_{\mathrm{QM}}\rangle_{\mathrm{En}}$$

$$= \sum_{a,b,c} \left| \frac{1}{\sqrt{6}} \left[\psi_{a1}\psi_{b2}\psi_{c3} + \psi_{a1}\psi_{b3}\psi_{c2} + \psi_{a2}\psi_{b1}\psi_{c3} \right.\right.$$

$$\left.\left. + \psi_{a2}\psi_{b3}\psi_{c1} + \psi_{a3}\psi_{b1}\psi_{c2} + \psi_{a3}\psi_{b2}\psi_{c1} \right] \right|^2, \quad (5.10.9)$$

where ψ_{aj} (ψ_{bj}) (ψ_{cj}), $j = 1, 2, 3$, is the effective wavefunction of the ath $(b$th$)$ $(c$th$)$ photon that is created from the ath $(b$th$)$ $(c$th$)$ sub-source and is annihilated at the jth photodetector D_j. Equation (5.10.9) indicates that $G^{(3)}(\mathbf{r}_1, t_1; \mathbf{r}_2, t_2; \mathbf{r}_3, t_3)$ is the sum of a large number of interferences, and each interference corresponds to the following superposition of three-photon amplitudes:

$$G_{abc}^{(3)} = \left| \frac{1}{\sqrt{6}} \left[\psi_{a1}\psi_{b2}\psi_{c3} + \psi_{a1}\psi_{b3}\psi_{c2} + \psi_{a2}\psi_{b1}\psi_{c3} \right.\right.$$

$$\left.\left. + \psi_{a2}\psi_{b3}\psi_{c1} + \psi_{a3}\psi_{b1}\psi_{c2} + \psi_{a3}\psi_{b2}\psi_{c1} \right] \right|^2. \quad (5.10.10)$$

The superposed six terms in Eq. (5.10.10) indicate six different yet indistinguishable alternatives for three randomly created photons to trigger a three-fold joint-detection event of D_1, D_2, and D_3. These six three-photon amplitudes are also schematically illustrated in Fig. 5.10.1.

In an idealized correlation measurement by assuming three perfect photodetectors and a perfect three-fold correlation measurement circuit, we may write $G^{(3)}(\mathbf{r}_1, t_1; \mathbf{r}_2, t_2; \mathbf{r}_3, t_3)$ in terms of the first-order quantum coherence functions:

$$G^{(3)}(\mathbf{r}_1, t_1; \mathbf{r}_2, t_2; \mathbf{r}_3, t_3)$$

$$= G_{11}^{(1)} G_{22}^{(1)} G_{33}^{(1)} + \left|G_{12}^{(1)}\right|^2 G_{33}^{(1)} + \left|G_{23}^{(1)}\right|^2 G_{11}^{(1)}$$

$$+ \left|G_{31}^{(1)}\right|^2 G_{22}^{(1)} + G_{12}^{(1)} G_{23}^{(1)} G_{31}^{(1)} + G_{21}^{(1)} G_{32}^{(1)} G_{13}^{(1)}. \tag{5.10.11}$$

Equation (5.10.11) can be normalized as

$$g^{(3)}(\mathbf{r}_1, t_1; \mathbf{r}_2, t_2; \mathbf{r}_3, t_3)$$

$$= 1 + \left|g_{12}^{(1)}\right|^2 + \left|g_{13}^{(1)}\right|^2 + \left|g_{23}^{(1)}\right|^2 + g_{12}^{(1)} g_{23}^{(1)} g_{31}^{(1)} + g_{21}^{(1)} g_{32}^{(1)} g_{13}^{(1)}. \tag{5.10.12}$$

No surprise, we found that $G^{(3)}(\mathbf{r}_1, t_1; \mathbf{r}_2, t_2; \mathbf{r}_3, t_3)$ has the same form as $\Gamma^{(3)}(\mathbf{r}_1, t_1; \mathbf{r}_2, t_2; \mathbf{r}_3, t_3)$, except replacing Einstein's sub-fields with the effective wavefunctions.

The third-order temporal and spatial coherence function of pseudo-thermal field have been experimentally measured by Zhou *et al.* around the year 2010. The experimental setup of Zhou *et al.* is similar to that of the HBT interferometer, except the use of three photodetectors and a 3-fold joint-detection coincidence counter.

(1) Third-order temporal coherence of thermal light

In the third-order temporal correlation measurement of Zhou *et al.*, their three point-like photodetectors were set with equal distances, $z_1 = z_2 = z_3$, from their point-like pseudo-thermal source (simulated by a pinhole).

Based on the experimentally setup of Zhou *et al.*, the calculation of the third-order temporal coherence of $\gamma^{(3)}(t_1, t_2, t_3)$ or $g^{(3)}(t_1, t_2, t_3)$ is straightforward:

$$g^{(3)}(t_1, t_2, t_3)$$

$$= 1 + \mathrm{sinc}^2\left[\frac{\Delta\omega(t_1 - t_2)}{2\pi}\right]$$

$$+ \mathrm{sinc}^2\left[\frac{\Delta\omega(t_2 - t_3)}{2\pi}\right] + \mathrm{sinc}^2\left[\frac{\Delta\omega(t_3 - t_1)}{2\pi}\right]$$

$$+ 2\,\mathrm{sinc}\left[\frac{\Delta\omega(t_1 - t_2)}{2\pi}\right] \mathrm{sinc}\left[\frac{\Delta\omega(t_2 - t_3)}{2\pi}\right] \mathrm{sinc}\left[\frac{\Delta\omega(t_3 - t_1)}{2\pi}\right]. \tag{5.10.13}$$

In Eq. (5.10.13), we have assumed a constant spectrum distribution $f(\omega)$ within the bandwidth of the radiation field $\Delta\omega$ to simplify the mathematics. It is easy to see that when $t_1 = t_2 = t_3$, $g^{(3)}(t_1, t_2, t_3) = 6$. The third-order correlation function of thermal field achieves a maximum contrast of 6 to 1 ($\sim 71\%$ visibility).

The measured $g^{(3)}(t_1, t_2, t_3)$ is shown in Fig. 5.10.2. The experimental observation of Zhou *et al.* and the simulation are in agreement within the experimental error.

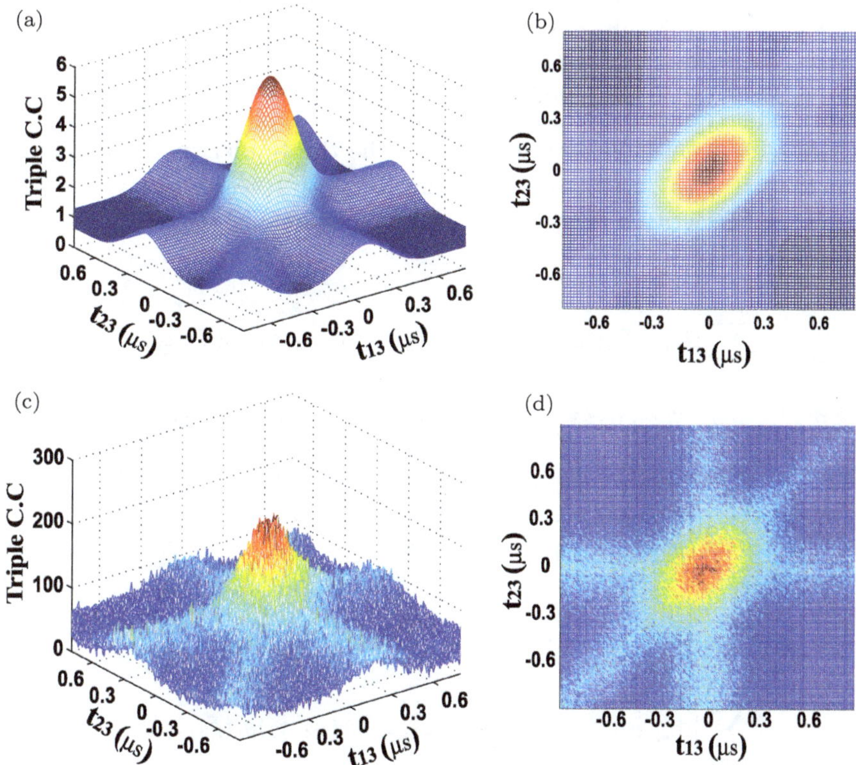

Fig. 5.10.2 Calculated (upper, a and b) and measured (lower, c and d) third-order temporal correlation of thermal light. The 3-D three-photon joint detection histogram is plotted as a function of $t_{13} \equiv t_1 - t_3$ and $t_{23} \equiv t_2 - t_3$. The simulation function is calculated from Eq. (5.10.13). In addition, the single detector counting rates for D_1, D_2, and D_3 are all monitored to be constants. The experimental data and the simulation are in agreement within the experimental error.

(2) **Third-order near-field spatial coherence of thermal light**

In the following, we briefly discuss the third-order near-field spatial coherence of thermal light. Now, considering a measurement similar to that of the second-order near-field spatial coherence measurement, except the scanning of three point-like photodetectors in the transverse planes of $\vec{\rho}_1$, $\vec{\rho}_2$ and $\vec{\rho}_3$, respectively. We assume a perfect joint-measurement system, including three perfect photodetectors and a perfect 3-fold joint-measurement coincidence circuit, and a maximum temporal correlation at single wavelength thermal field to simplify the mathematics.

Similar to the calculation of near-field second-order spatial coherence, we applying the near-field Fresnel Green's function to propagate each subfield from each sub-source to the photodetectors:

$$g_a(\vec{\rho}_j, z_j) = \frac{c_o}{z_j} e^{i\frac{\omega z_j}{c}} e^{i\frac{\omega}{2cz_j}|\vec{\rho}_j - \vec{\rho}_a|^2},$$

where c_0 is a normalization consistent. The third-order spatial coherence functions $\Gamma^{(3)}_{abc}(\vec{\rho}_1, z_1; \vec{\rho}_2, z_2; \vec{\rho}_3, z_3)$ and $G^{(3)}_{abc}(\vec{\rho}_1, z_1; \vec{\rho}_2, z_3; \vec{\rho}_3, z_3)$, respectively, can be written in terms of Green's functions:

$$\Gamma^{(3)}_{abc}(\vec{\rho}_1, z_1; \vec{\rho}_2, z_2; \vec{\rho}_3, z_3)$$

$$= \sum_{a,b,c} |E_a|^2 |E_b|^2 |E_c|^2 \left| \frac{1}{\sqrt{6}} \left[g_a(\vec{\rho}_1, z_1) g_b(\vec{\rho}_2, z_2) g_c(\vec{\rho}_3, z_3) \right. \right.$$

$$+ g_a(\vec{\rho}_1, z_1) g_b(\vec{\rho}_3, z_3) g_c(\vec{\rho}_2, z_2) + g_a(\vec{\rho}_2, z_2) g_b(\vec{\rho}_1, z_1) g_c(\vec{\rho}_3, z_3)$$

$$+ g_a(\vec{\rho}_2, z_2) g_b(\vec{\rho}_3, z_3) g_c(\vec{\rho}_1, z_1) + g_a(\vec{\rho}_3, z_2) g_b(\vec{\rho}_1, z_1) g_c(\vec{\rho}_2, z_2)$$

$$\left. \left. + g_a(\vec{\rho}_3, z_3) g_b(\vec{\rho}_2, z_2) g_c(\vec{\rho}_1, z_1) \right] \right|^2$$

and

$$G^{(3)}_{abc}(\vec{\rho}_1, z_1; \vec{\rho}_2, z_2; \vec{\rho}_3, z_3)$$

$$= \sum_{a,b,c} \left| |\alpha_a|^2 |\alpha_b|^2 |\alpha_c|^2 \right| \frac{1}{\sqrt{6}} \left[g_a(\vec{\rho}_1, z_1) g_b(\vec{\rho}_2, z_2) g_c(\vec{\rho}_3, z_3) \right.$$

$$+ g_a(\vec{\rho}_1, z_1) g_b(\vec{\rho}_3, z_3) g_c(\vec{\rho}_2, z_2) + g_a(\vec{\rho}_2, z_2) g_b(\vec{\rho}_1, z_1) g_c(\vec{\rho}_3, z_3)$$

$$+ g_a(\vec{\rho}_2, z_2) g_b(\vec{\rho}_3, z_3) g_c(\vec{\rho}_1, z_1) + g_a(\vec{\rho}_3, z_2) g_b(\vec{\rho}_1, z_1) g_c(\vec{\rho}_2, z_2)$$

$$\left. + g_a(\vec{\rho}_3, z_3) g_b(\vec{\rho}_2, z_2) g_c(\vec{\rho}_1, z_1) \right]^2 .$$

Now, consider the sum of $\Gamma^{(3)}_{abc}(\vec{\rho}_1, z_1; \vec{\rho}_2, z_2; \vec{\rho}_3, z_3)$ and $G^{(3)}_{abd}(\vec{\rho}_1, z_1; \vec{\rho}_2, z_2; \vec{\rho}_3, z_3)$ by turning the sum of a, b, and c into the integrals of $\vec{\rho}_0$, $\vec{\rho}_0'$, and $\vec{\rho}_0''$ on the entire source plane:

$$
\begin{aligned}
\Gamma^{(3)}(\vec{\rho}_1, \vec{\rho}_2, \vec{\rho}_3) &= \sum_{a,b,c} \Gamma^{(3)}_{abc}(\vec{\rho}_1, \vec{\rho}_2, \vec{\rho}_3) \\
&= \Gamma^{(1)}(\vec{\rho}_1, \vec{\rho}_1)\Gamma^{(1)}(\vec{\rho}_2, \vec{\rho}_2)\Gamma^{(1)}(\vec{\rho}_3, \vec{\rho}_3) \\
&\quad + \left|\Gamma^{(1)}(\vec{\rho}_1, \vec{\rho}_2)\right|^2 \Gamma^{(1)}(\vec{\rho}_3, \vec{\rho}_3) \\
&\quad + \left|\Gamma^{(1)}(\vec{\rho}_2, \vec{\rho}_3)\right|^2 \Gamma^{(1)}(\vec{\rho}_1, \vec{\rho}_1) \\
&\quad + \left|\Gamma^{(1)}(\vec{\rho}_3, \vec{\rho}_1)\right|^2 \Gamma^{(1)}(\vec{\rho}_2, \vec{\rho}_2) \\
&\quad + \Gamma^{(1)}(\vec{\rho}_1, \vec{\rho}_2)\Gamma^{(1)}(\vec{\rho}_2, \vec{\rho}_3)\Gamma^{(1)}(\vec{\rho}_3, \vec{\rho}_1) \\
&\quad + \Gamma^{(1)}(\vec{\rho}_2, \vec{\rho}_1)\Gamma^{(1)}(\vec{\rho}_3, \vec{\rho}_2)\Gamma^{(1)}(\vec{\rho}_1, \vec{\rho}_3),
\end{aligned}
\tag{5.10.14}
$$

where

$$
\Gamma^{(1)}_{jk} = I_0 \int d\vec{\rho}_0 \, g^*_{\vec{\rho}_0}(\vec{\rho}_j, z_j) g_{\vec{\rho}_0}(\vec{\rho}_k, z_k)\big],
$$

where we have absorbed all constants into constant I_0:

$$
\begin{aligned}
G^{(3)}(\vec{\rho}_1, \vec{\rho}_2, \vec{\rho}_3) &= \sum_{a,b,c} G^{(3)}_{abc}(\vec{\rho}_1, \vec{\rho}_2, \vec{\rho}_3) \\
&= G^{(1)}(\vec{\rho}_1, \vec{\rho}_1)G^{(1)}(\vec{\rho}_2, \vec{\rho}_2)G^{(1)}(\vec{\rho}_3, \vec{\rho}_3) \\
&\quad + \left|G^{(1)}(\vec{\rho}_1, \vec{\rho}_2)\right|^2 G^{(1)}(\vec{\rho}_3, \vec{\rho}_3) \\
&\quad + \left|G^{(1)}(\vec{\rho}_2, \vec{\rho}_3)\right|^2 G^{(1)}(\vec{\rho}_1, \vec{\rho}_1) \\
&\quad + \left|G^{(1)}(\vec{\rho}_3, \vec{\rho}_1)\right|^2 G^{(1)}(\vec{\rho}_2, \vec{\rho}_2) \\
&\quad + G^{(1)}(\vec{\rho}_1, \vec{\rho}_2)G^{(1)}(\vec{\rho}_2, \vec{\rho}_3)G^{(1)}(\vec{\rho}_3, \vec{\rho}_1) \\
&\quad + G^{(1)}(\vec{\rho}_2, \vec{\rho}_1)G^{(1)}(\vec{\rho}_3, \vec{\rho}_2)G^{(1)}(\vec{\rho}_1, \vec{\rho}_3),
\end{aligned}
\tag{5.10.15}
$$

where

$$
G^{(1)}_{jk} = |\alpha_0|^2 \int d\vec{\rho}_0 \, g^*_{\vec{\rho}_0}(\vec{\rho}_j, z_j) g_{\vec{\rho}_0}(\vec{\rho}_k, z_k)\big],
$$

where we have absorbed all constants into constant $|\alpha_0|^2$.

Completing the integral on $\vec{\rho}_0$ for each $\Gamma^{(1)}_{jk}$ and $G^{(1)}_{jk}$ and taking $z_j = z_k = d$, similar to the near-field second-order spatial coherence, the normalized third-order coherence function $\gamma^{(3)}(\vec{\rho}_1, \vec{\rho}_2, \vec{\rho}_3)$ and $g^{(3)}(\vec{\rho}_1, \vec{\rho}_2, \vec{\rho}_3)$, respectively, is then approximated as

$$g^{(3)}(\vec{\rho}_1, \vec{\rho}_2, \vec{\rho}_3)$$

$$\propto 1 + \mathrm{somb}^2 \left[\frac{D}{d} \frac{\pi}{\lambda} |\vec{\rho}_1 - \vec{\rho}_2| \right]$$

$$+ \mathrm{somb}^2 \left[\frac{D}{d} \frac{\pi}{\lambda} |\vec{\rho}_2 - \vec{\rho}_3| \right] + \mathrm{somb}^2 \left[\frac{D}{d} \frac{\pi}{\lambda} |\vec{\rho}_3 - \vec{\rho}_1| \right]$$

$$+ 2\,\mathrm{somb} \left[\frac{D}{d} \frac{\pi}{\lambda} |\vec{\rho}_1 - \vec{\rho}_2| \right] \mathrm{somb} \left[\frac{D}{d} \frac{\pi}{\lambda} |\vec{\rho}_2 - \vec{\rho}_3| \right] \mathrm{somb} \left[\frac{D}{d} \frac{\pi}{\lambda} |\vec{\rho}_3 - \vec{\rho}_1| \right],$$

$$(5.10.16)$$

where $D/d = \Delta\theta$ is the angular size of the source from the view points of the photodetectors.

In 1-D, the normalized third-order coherence function $\gamma^{(3)}(x_1, x_2, x_3)$ or $g^{(3)}(x_1, x_2, x_3)$ is approximated to be

$$g^{(3)}(x_1, x_2, x_3)$$

$$\propto 1 + \mathrm{sinc}^2 \left[\frac{D}{d} \frac{\pi}{\lambda} (x_1 - x_2) \right]$$

$$+ \mathrm{sinc}^2 \left[\frac{D}{d} \frac{\pi}{\lambda} (x_2 - x_3) \right] + \mathrm{somb}^2 \left[\frac{D}{d} \frac{\pi}{\lambda} (x_3 - x_1) \right]$$

$$+ 2\,\mathrm{sinc} \left[\frac{D}{d} \frac{\pi}{\lambda} (x_1 - x_2) \right] \mathrm{sinc} \left[\frac{D}{d} \frac{\pi}{\lambda} (x_2 - x_3) \right] \mathrm{sinc} \left[\frac{D}{d} \frac{\pi}{\lambda} (x_3 - x_1) \right].$$

$$(5.10.17)$$

For a large angular-sized thermal source, the sinc-like functions effectively turn into δ-functions of $(x_j - x_k)$, $j = 1, 2, 3$, $k = 1, 2, 3$. The point-to-point-to-point nontrivial correlation between three transverse planes encourages three-photon lensless ghost imaging of thermal light with enhanced spatial resolution. The three-photon ghost imaging of thermal light exhibits a number of unusual interesting features that may be useful for certain applications.

(II) Nth-order coherence of thermal light

The concept and calculation on thermal light third-order near-field spatial coherence are easily generalizable to higher orders. The physics and mathematics may all start from Eq. (5.10.18) or Eq. (5.10.19), which is considered the key equation to understand the N-photon interference nature of the nontrivial Nth-order coherence of thermal field.

The Nth-order ($N \geq 2$) coherence of thermal light can be extended from $\Gamma^{(3)}(\mathbf{r}_1, t_1; \mathbf{r}_2, t_2; \mathbf{r}_3, t_3)$ to $\Gamma^{(N)}(\mathbf{r}_1, t_1; \ldots \mathbf{r}_N, t_N)$:

$$\Gamma^{(N)}(\mathbf{r}_1, t_1; \ldots; \mathbf{r}_N, t_N)$$

$$\equiv \langle E^*(\mathbf{r}_1, t_1) E(\mathbf{r}_1, t_1) \ldots E^*(\mathbf{r}_N, t_N) E(\mathbf{r}_N, t_N) \rangle$$

$$\simeq \sum_{a,b,c\ldots} \left| \frac{1}{\sqrt{N!}} \left[\sum_{N!} E_{a1} E_{b2} E_{c3} \ldots \right] \right|^2, \tag{5.10.18}$$

where $\sum_{N!} E_{a1} E_{b2} E_{c3} \ldots$ means the sum of $N!$ terms of N-fold subfields, permuting in the orders of $1, 2, 3 \ldots, N$, $1, 3, 2 \ldots, N$, $2, 1, 3 \ldots, N$, etc., corresponding to $N!$ alternative ways for the N independent subfields (photons), created from a large number of independent sub-sources that are labeled by $a, b, c \ldots$, to produce a N-fold joint-photodetection event between N independent photodetectors.

Replacing the subfields by the effective wavefunctions, the quantum Nth-order ($N \geq 2$) coherence of thermal light can be easily extend from $G^{(3)}(\mathbf{r}_1, t_1; \mathbf{r}_2, t_2; \mathbf{r}_3, t_3)$ to $G^{(N)}(\mathbf{r}_1, t_1; \ldots \mathbf{r}_N, t_N)$:

$$G^{(N)}(\mathbf{r}_1, t_1; \ldots; \mathbf{r}_N, t_N)$$

$$\equiv \langle\langle \hat{E}^{(-)}(\mathbf{r}_1, t_1) \ldots \hat{E}^{(-)}(\mathbf{r}_N, t_N) \hat{E}^{(+)}(\mathbf{r}_N, t_N) \ldots \hat{E}^{(+)}(\mathbf{r}_1, t_1) \rangle_{\text{QM}} \rangle_{\text{En}}$$

$$\simeq \sum_{a,b,c\ldots} \left| \frac{1}{\sqrt{N!}} \left[\sum_{N!} \psi_{a1} \psi_{b2} \psi_{c3} \ldots \right] \right|^2. \tag{5.10.19}$$

(III) Nth-order coherence of coherent light

Similar to the second-order coherence of coherent state, there should be no surprise that the degree of the Nth-order coherence light in coherent state is one. In fact, it is easy to show that the Nth-order coherence function of coherent light is factorizable into N independent first-order self-correlation functions (mean intensities), and consequently the normalized degree of Nth-order coherence of coherent radiation is

$$\gamma_{\text{coh}}^{(N)}(\mathbf{r}_1, t_1; \mathbf{r}_2, t_2 \ldots; \mathbf{r}_N, t_N) = 1,$$

$$g_{\text{coh}}^{(N)}(\mathbf{r}_1, t_1; \mathbf{r}_2, t_2 \ldots; \mathbf{r}_N, t_N) = 1. \tag{5.10.20}$$

Bibliography

Fano U., *Am. J. Phys.* **29**, 539 (1961).

Glauber R.J., *Quantum Optics*, in S.M. Kay and A. Maitland (Eds.), Academic Press, New York, 1970.

Hanbury-Brown R., and Twiss R.Q., *Nature* **177**, 27 (1956); **178**, 1046 (1956); **178**, 1447 (1956).

Hanbury-Brown R., *Intensity Interferometer*, Taylor and Francis Ltd, London, 1974.

Joshi B., Smith T.A., and Shih Y.H., *Phys. Rev. A* 110, L031702 (2024).

Liu J. and Shih Y.H., *Phys. Rev. A* **79**, 203819 (2009).

Loudon R., *The Quantum Theory of Light*, Oxford Science, 2000.

Martienssen W. and Spiller E., *Am. J. Phys.* **32**, 919 (1964).

Scarcelli G., Berardi V., and Shih Y.H., *Phys. Rev. Lett.* **96**, 063602 (2006).

Scully M.O. and Zubairy M.S., *Quantum Optics*, Cambridge, 1997.

Shih Y.H., *Phys. Rev. A*, to be published (2024).

Valencia A., Scarcelli G., D'Angelo M., and Shih Y.H., *Phys. Rev. Lett.* **94**, 063601 (2005).

Zhou Y., Simon J., Liu J., and Shih Y.H., *Phys. Rev. A* (2010).

Chapter 6

Quantum Entanglement

In quantum theory, *a particle* is allowed to be in a state of coherent superposition among a set of orthogonal states. A vivid picture of this concept is the "Schrödinger cat": *a cat* is alive and dead simultaneously. This would be no surprise if one means an ensemble of cats with 50% alive and 50% dead. But, "Schrödinger cat" is *a cat*. We are talking about *a cat* being alive and dead simultaneously. In mathematics, the best word to characterize the relationship between states "alive" and "dead" is perhaps "orthogonal". In quantum mechanics, the coherent superposition of orthogonal states is used to describe the state of a quantum object, or a particle. The superposition principle is indeed a mystery in terms of our everyday life experience.

In this chapter, we turn to another surprising consequence of quantum mechanics, namely, quantum entanglement. Quantum entanglement involves a multi-particle system and coherent superposition of orthogonal multi-particle states. The Schrödinger cat is perhaps still the best example to picture quantum entanglement. Now, we are talking about a pair of Schrödinger cats "propagating" to distant locations. The two cats are nonclassical by means of the following two features: (1) each of the cats has a 50%–50% chance to be alive and dead; (2) the two has to be observed both alive or both dead whenever we look at them, despite the distance between the two. There would be probably no surprise if our observation is based on a large number of twin cats that are prepared to be half alive-alive and another half dead-dead. In this case, obviously, we have 50% chance to observe an alive-alive pair and 50% chance to observe a dead-dead pair for each joint-observation. However, we are talking about *a pair* of cats, i.e., each pair of cats, to be in the state of alive-alive and dead-dead simultaneously, and, in addition, each of the cats in the pair must be alive and dead simultaneously. It seems impossible to have such cats in classical reality.

Beyond the coherent superposition of orthogonal single-particle states, the coherent superposition of orthogonal multi-particle states represents an even more troubling concepts. It is not only because the superposition of multi-particle states has no counterpart in classical theory, but also because it represents nonlocal behavior of particles which may never be explained classically.

6.1 EPR Experiment and EPR State

The concept of quantum entanglement started in 1935. Einstein, Podolsky and Rosen (EPR) suggested a *gedankenexperiment* and introduced an entangled two-particle system based on the coherent superposition of continuously distributed infinite number of two-particle wavefunctions. The EPR system composes two distant interaction-free particles which are characterized by the following nonfactorizable wavefunction:

$$\Psi(x_1, x_2) = \frac{1}{2\pi\hbar} \int dp_1 dp_2 \, \delta(p_1 + p_2) \, e^{ip_1(x_1-x_0)/\hbar} e^{ip_2 x_2/\hbar}$$

$$= \delta(x_1 - x_2 - x_0), \tag{6.1.1}$$

where $e^{ip_1(x_1-x_0)/\hbar}$ and $e^{ip_2 x_2/\hbar}$ are the eigenfunctions, with eigenvalues $p_1 = p$ and $p_2 = -p$, respectively, of the momentum operators $\hat{p}_1$ and $\hat{p}_2$ associated with particles 1 and 2; x_1 and x_2 are the coordinate variables to describe the positions of particles 1 and 2, respectively; and x_0 is a constant. The EPR state is very peculiar. Although there is no interaction between the two distant particles, the two-particle superposition cannot be factorized into a product of two individual superposition of two particles. Quantum theory does not prevent such states.

What can we learn from the EPR state of Eq. (6.1.1)?

(1) In coordinate representation, the wavefunction is a delta-function: $\delta(x_1 - x_2 - x_0)$. The two particles are always separated in space with a constant value of $x_1 - x_2 = x_0$, although the coordinates x_1 and x_2 of the two particles are both unspecified.

(2) The delta-wavefunction $\delta(x_1 - x_2 - x_0)$ is the result of a two-particle superposition, which contains infinite number of continuously distributed plane-wavefunction pairs of free particle one, $e^{ip_1(x_1-x_0)/\hbar}$, and free particle two, $e^{ip_2 x_2/\hbar}$. The momentum of the two-particle system is conserved to a constant value of zero. It is the momentum conservation $\delta(p_1 + p_2)$ that

made the superposition special: although the momentum of particle one and particle two may take on any values, the delta function restricts the superposition with only these terms in which the total momentum of the system takes a constant value of zero.

Now, we transfer the wavefunction from coordinate representation to momentum representation:

$$\Psi(p_1, p_2) = \frac{1}{2\pi\hbar} \int dx_1 dx_2 \, \delta(x_1 - x_2 - x_0) \, e^{-ip_1(x_1-x_0)/\hbar} e^{-ip_2 x_2/\hbar}$$

$$= \delta(p_1 + p_2). \tag{6.1.2}$$

What can we learn from the EPR state of Eq. (6.1.2)?

(1) In momentum representation, the wavefunction is a delta-function: $\delta(p_1 + p_2)$. The total momentum of the two-particle system takes a constant value of $p_1 + p_2 = 0$, although the momenta p_1 and p_2 are both unspecified.

(2) The delta-wavefunction $\delta(p_1 + p_2)$ is the result of a two-particle superposition, which contains infinite number of continuously distributed plane-wavefunction pairs of free particle one, $e^{-ip_1(x_1-x_0)/\hbar}$, and free particle two, $e^{-ip_2 x_2/\hbar}$, with a particular relationship in their positions $\delta(x_1 - x_2 - x_0)$. It is $\delta(x_1 - x_2 - x_0)$ that made the superposition special: although the coordinates of particle one and particle two may take on any values, the delta function restricts the superposition with only these terms in which $x_1 - x_2$ is a constant value of x_0.

In an EPR system, the value of the momentum (position) for neither single subsystem is determined. However, if one of the subsystems is measured to be at a certain momentum (position), the other one is determined with a unique corresponding value, despite the distance between them. An idealized EPR state of a two-particle system is therefore characterized by $\Delta(p_1 + p_2) = 0$ and $\Delta(x_1 - x_2) = 0$ simultaneously, even if the momentum and position of each individual free particle are completely undefined, i.e., $\Delta p_j \sim \infty$ and $\Delta x_j \sim \infty$, $j = 1, 2$. In other words, each of the subsystems may have completely random values or all possible values of momentum and position in the course of their motion, but the correlations of the two subsystems are determined with certainty whenever a joint measurement is performed.

The EPR states of Eqs. (6.1.1) and (6.1.2) are simply the results of the quantum mechanical *nonlocal superposition of infinite number of continuously distributed two-particle states*. The physics behind EPR states is far beyond the acceptable limit of Einstein.

Does a free particle have a defined momentum and position in the state of Eqs. (6.1.1) and (6.1.2), regardless of whether we measure it or not? On one hand, the momentum and position of neither individual particle is specified and the superposition is taken over all possible values of the momentum and position. We may have to believe that the particles do not have any defined momentum and position, or have all possible values of momentum and position within the superposition, during the course of their motion. On the other hand, if the measured momentum or position of one particle uniquely determines the momentum or position of the other distant particle, it would be hard for anyone who believes no action-at-a-distance to imagine that the momentum and position of the two particles are not predetermined with defined values before the measurement. EPR thus put us into a paradoxical situation. It seems reasonable for us to ask the same question that EPR had asked in 1935: "Can quantum-mechanical description of physical reality be considered complete?"

In their 1935 article, Einstein, Podolsky and Rosen argued that the existence of the entangled two-particle state of Eqs. (6.1.1) and (6.1.2), a straightforward quantum mechanical superposition of a large number of two-particle states, led to the violation of the uncertainty principle of quantum theory. To draw their conclusion, EPR started from the following criteria:

Locality: There is no action-at-a-distance;

Reality: "If, without in any way disturbing a system, we can predict with certainty the value of a physical quantity, then there exist an element of physical reality corresponding to this quantity." According to the delta wavefunctions, we can predict with certainty the outcome result of measuring the momentum (position) of particle 1 by measuring the momentum (position) of particle 2, and the measurement of particle 2 cannot cause any disturbance to particle 1, if the measurements are space-like separated events. Thus, the momentum and position of particle 1 must both be elements of physical reality regardless of whether we measure it or not. This, however, is not allowed by quantum theory. Now consider:

Completeness: "Every element of the physical reality must have a counterpart in the complete theory." This led to the question as the title of their 1935 article: "Can Quantum-Mechanical Description of Physical Reality Be Considered Complete?"

EPR's arguments were never appreciated by Copenhagen. One objection was regarding EPR's criterion of physical reality: "it is too narrow", was a criticism from Bohr. It is perhaps not easy to find a wider criterion.

A memorable quote from Wheeler, "No elementary quantum phenomenon is a phenomenon until it is a recorded phenomenon", summarizes what Copenhagen has been trying to tell us. By 1927, most physicists accepted the Copenhagen interpretation as the standard view of quantum formalism. Einstein, however, refused to compromise. As Pais recalled in his book, during a walk around 1950, Einstein suddenly stopped and "asked me if I really believed that the moon (pion) exists only if I look at it."

There has been arguments considering $\Delta(p_1 + p_2)\Delta(x_1 - x_2) = 0$ a violation of the uncertainty principle. This argument is false. It is easy to find that $p_1 + p_2$ and $x_1 - x_2$ are not conjugate variables. As we know, nonconjugate variables correspond to commuting operators in quantum mechanics, if the corresponding operators exist.[1] To have $\Delta(p_1 + p_2) = 0$ and $\Delta(x_1 - x_2) = 0$ simultaneously, or to have $\Delta(p_1 + p_2)\Delta(x_1 - x_2) = 0$, is not a violation of the uncertainty principle. This point can be easily seen from the following two dimensional Fourier transform:

$$\Psi(x_1, x_2) = \frac{1}{2\pi\hbar} \int dp_1 \, dp_2 \, \delta(p_1 + p_2) \, e^{ip_1(x_1-x_0)/\hbar} \, e^{ip_2 x_2/\hbar}$$

$$= \frac{1}{2\pi\hbar} \int d(p_1 + p_2) \, \delta(p_1 + p_2) \, e^{i(p_1+p_2)(x_1'+x_2)/2\hbar}$$

$$\times \int d(p_1 - p_2)/2 \, e^{i(p_1-p_2)(x_1'-x_2)/2\hbar}$$

$$= 1 \times \delta(x_1 - x_2 - x_0),$$

where $x' = x_1 - x_0$;

$$\Psi(p_1, p_2) = \frac{1}{2\pi\hbar} \int dx_1 \, dx_2 \, \delta(x_1 - x_2 - x_0) \, e^{-ip_1(x_1-x_0)/\hbar} \, e^{-ip_2 x_2/\hbar}$$

$$= \frac{1}{2\pi\hbar} \int d(x_1' + x_2) \, e^{-i(p_1+p_2)(x_1'+x_2)/2\hbar}$$

$$\times \int d(x_1' - x_2)/2 \, \delta(x_1' - x_2) \, e^{-i(p_1-p_2)(x_1'-x_2)/2\hbar}$$

$$= \delta(p_1 + p_2) \times 1.$$

---

[1] It is possible that a measurable physical variable has no quantum mechanical operator associated with, such as time t. From this perspective, an uncertainty relation based on variables rather than operators is more general. We should also remember that there is no position operator defined for a photon, however, this does not prevent us to discuss the uncertainty of observing a corresponding photodetection event in terms of its spatial coordinate or position Δx, and the uncertain relation of $\Delta x \Delta p \geq h$.

The Fourier conjugate variables are $(x_1 + x_2) \Leftrightarrow (p_1 + p_2)$ and $(x_1 - x_2) \Leftrightarrow (p_1 - p_2)$. Although it is possible to have $\Delta(x_1 - x_2) \sim 0$ and $\Delta(p_1 + p_2) \sim 0$ simultaneously, the uncertainty relations must hold for the Fourier conjugates $\Delta(x_1 + x_2)\Delta(p_1 + p_2) \geq \hbar$, and $\Delta(x_1 - x_2)\Delta(p_1 - p_2) \geq \hbar$; with $\Delta(p_1 - p_2) \sim \infty$ and $\Delta(x_1 + x_2) \sim \infty$.

In fact, in their 1935 paper, Einstein–Podolsky–Rosen never questioned $\Delta(x_1 - x_2)\Delta(p_1 + p_2) = 0$ as a violation of the uncertainty principle. The violation of the uncertainty principle was probably not Einstein's concern at all, although their 1935 paradox was based on the argument of the uncertainty principle. What really bothered Einstein so much? In all his life, Einstein, a true believer of realism, never accepted that a particle does not have a defined momentum and position during its motion, but rather is specified by probability amplitude of certain momentum and position. "God does not play dice" was the most vivid criticism from Einstein to refuse the Schrödinger cat. The entangled two-particle system was used as an example to clarify and to reinforce Einstein's realistic opinion. To Einstein, the acceptance of Schrödinger cat probably means action-at-a-distance or an inconsistency between quantum mechanics and the theory of relativity, when dealing with the entangled EPR two-particle system.

In the last few years of his life, Einstein posed his students a question: suppose a photon of energy $\hbar\omega$ is created from a point source, such as an atomic transition; how big is the photon after propagating one year? This question seems easy to answer for his students. Since the photon is created from a point source, it would propagate in the form of a spherical wave and its wavefront must be a big sphere with a diameter of two lightyears after one year propagating. We may find a "consensus" or "common sense" behind the above statement: the total energy of the photon $\hbar\omega$ must be carried by the wavefront of the photon and thus uniformly distributed on the big sphere. Einstein then asked again: suppose that photon is annihilated by a point-like photon counting detector located on the surface of the big sphere, how long does it take for the energy on the other side of the big sphere to arrive at the detector? This question is not easy to answer. Two years? Bohr provided a famous answer to this question: The "wavefunction collapses" instantaneously! Why does the wavefunction need to "collapse"? Bohr did not explain. Nevertheless, Bohr conveyed an important message to us: The quantum mechanical picture of the photon is very different, and at least its wavefunction, if any, is *nonlocal*.

The EPR paper of 1935 posted a similar question to us. Now, assume an entangled pair of photons is created from a light source, such as an atomic cascade decay. In the creation process of the photon pair, the energy and momentum of the system must be conserved, i.e., (1) although the energy of each created photon may have considerable uncertainty, the total energy of the two photons must be $\hbar\omega_1 + \hbar\omega_2 = E_2 - E_1 \simeq$ constant. (2) Although each of the photons may propagate to all possible 4π directions, the pair must be observed simultaneously at opposite directions. After one-year propagation, EPR ask the following: If photon-1 is measured by a point-like photon counting detector located on one-side of the surface of the big sphere with energy $\hbar\omega_1$, what is the energy of photon-2 on the other side of the big sphere? According to the two-photon state, photon-2 must measured with energy $[(E_2 - E_1) - \hbar\omega_1)]$, although photon-2 may take all possible energy values during the course of its propagation. How long does it take for photon-2, which is measured on the other side of the big sphere, to learn the energy value of photon-1?

Since 1935, scientists have attempted to provide answers to EPR. A "simple" answer to Einstein's question follows from a statistical model. In this model, instead of a pure state of entangled two photons, which is represented by a coherent superposition of a large number of two-photon states, the state of the photon pair is treted as a statistical mixture: each pair of photons is created with defined energies, $\hbar\omega_1$, $\hbar\omega_2$, and with defined momenta, $\mathbf{p}_1$, $\mathbf{p}_2$. The defined values of the energy and momentum may differ from pair to pair, but the energy and momentum must be conserved in each pair with $\hbar\omega_1 + \hbar\omega_2 = E_2 - E_1$ and $\mathbf{p}_1 + \mathbf{p}_2 = 0$. In this model, the photon number are equally distributed on the big sphere after the measurement of an ensemble of photon pairs. However, in the measurement of each pair, if photon-1 is measured at a certain point on the big sphere with energy $\hbar\omega_1$, photon-2 must be observed at a unique point on the opposite side of the big sphere with energy $[(E_2 - E_1) - \hbar\omega_1)]$. Basically, this "simple" model abandoned the coherent superposition of two-photon amplitudes.

EPR may question this model immediately, since the state of the photon pair in this model is completely different from theirs. The EPR state is an entangled two-photon state, represented by a coherent superposition of a large number of two-photon states. The EPR correlation is the result of two-photon interference: a pair of entangled photons interfering with the pair itself. From another perspective, we can think that this model is questioning the existence of EPR state in nature. In fact, there have been quite a few doubts in history: Does EPR state really exist? The answer is positive. Over the past 50 years, we have experimentally proven the existence of EPR states! The entangled two-photon state, represented by a coherent superposition of a large number of two-photon states, does exist in nature and can be generated by a laser beam from nonlinear optical interaction. Furthermore, two-photon interference is not only the phenomenon of quantum entanglement, a randomly created and randomly paired photons in thermal state is also able to interfere with the pair itself at distance.

Is it possible to have a "better" theory which provides correct predictions of the behavior of a particle similar to quantum theory and, at the same time, respects its description of physical reality by EPR as "complete"? The followers of Einstein's realism have tried in many different ways to formulate a realistic theory of quantum mechanics. It was Bohm who first attempted a version of a so called "hidden variable theory", which seemed to satisfy these requirements. The hidden variable theory was successfully applied to many different quantum phenomena until 1964, when Bell proved a theorem to show that an inequality, which is violated by certain quantum mechanical statistical predictions, can be used to distinguish local hidden variable theory from quantum mechanics. Since then, the testing of Bell's inequalities became a standard instrument for the study of fundamental problems of quantum theory. The experimental testing of Bell inequality started from the early 1970s. Most of the historical experiments concluded the violation of Bell inequality and thus disapproved the local hidden variable theory.

6.2 Entangled State vs Product State and Classically Correlated State

(1) Product state

In general, a product state describes the behavior of a system composing of two or more independent subsystems. Usually, the measurement of a product state involves the join-detection of two or more independent particles, such as the coincidence measures between two or more particle counting detectors within a certain time window. The experimental setup is similar to that of the EPR correlation measurement, however, the measurement only deals with independent particles. The independent particle system can be characterized as a product state, i.e., the state of the system is factorizable into a product of states of two or more subsystems.

For example,

$$\hat{\rho} = \hat{\rho}_1 \times \hat{\rho}_2. \tag{6.2.1}$$

The density matrix $\hat{\rho}$ in Eq. (6.2.1) characterizes a quantum system composed of two independent subsystems of $\hat{\rho}_1$ and $\hat{\rho}_2$. In Eq. (6.2.1), each independent subsystem can be in any state. The simplest case is a product of two pure states:

$$|\Psi\rangle = \sum_{a,b} f(a)g(b)\,|a\rangle\,|b\rangle = \sum_a f(a)\,|a\rangle \cdot \sum_b g(b)\,|b\rangle, \tag{6.2.2}$$

where $\{|a\rangle\}$ and $\{|b\rangle\}$ are two sets of orthogonal vectors for subsystems 1 and 2, respectively, and

$$\hat{\rho} = |\Psi\rangle\langle\Psi|, \quad \hat{\rho}_1 = \sum_{a,a'} f(a)f^*(a')|a\rangle\langle a'|, \quad \hat{\rho}_2 = \sum_{b,b'} f(b)f^*(b')|b\rangle\langle b'|.$$

(2) Entangled state

Differing from product states, entangled states describe the behavior of entangled quantum systems. The entangled two-particle state was mathematically formulated by Schrödinger. Consider a pure state for a system composed of two spatially separated subsystems:

$$\hat{\rho} = |\Psi\rangle\langle\Psi|, \quad |\Psi\rangle = \sum_{a,b} c(a,b)\,|a\rangle\,|b\rangle, \tag{6.2.3}$$

where, again, $\{|a\rangle\}$ and $\{|b\rangle\}$ are two sets of orthogonal vectors for subsystems 1 and 2, respectively, and $\hat{\rho}$ is the density matrix. If $c(a,b)$ does not factor into a product of the form $f(a) \times g(b)$, then it follows that the state does not factor into a product state for subsystems 1 and 2:

$$c(a,b) \neq f(a) \times g(b), \quad \hat{\rho} \neq \hat{\rho}_1 \times \hat{\rho}_2. \tag{6.2.4}$$

The state was defined by Schrödinger as the entangled state.

Following this notation, the first classic example of a two-particle entangled state, the EPR state of Eqs. (6.1.1) and (6.1.2), is thus,

$$|\Psi\rangle_{\mathrm{EPR}} = \sum_{p_1,p_2} \delta(p_1 + p_2)\,|p_1\rangle|p_2\rangle$$

$$= \sum_{x_1,x_2} \delta(x_1 - x_2 + x_0)\,|x_1\rangle|x_2\rangle, \tag{6.2.5}$$

where we have described the EPR entangled system as the coherent super-position of the two-particle momentum eigenstates as well as the coherent superposition of the two-particle position eigenstates. One clear property of the EPR state is its independence of vector bases. The two δ-functions in Eq. (6.2.5) represent, respectively and simultaneously, the perfect position–position correlation and momentum–momentum correlation. Although the two distant particles are interaction-free, the superposition selects only the eigenstates which are specified by the δ-function. We may use the following statement to summarize the surprising feature of the EPR state: *the values of the momentum and the position for neither interaction-free single subsystem is determinate. However, if one of the subsystems is measured to be at a certain value of momentum and/or position, the momentum and/or position of the other one is 100% determined, despite the distance between them.*

It is necessary to emphasize that Eq. (6.2.5) is true, simultaneously, in the conjugate momentum and position space.

(3) Classically correlated state

There exist many different types of classically correlated states. The following are two typical classically correlated states that have been used to simulate the EPR state of Eq. (6.2.5):

$$\hat{\rho} = \sum_{p_1,p_2} \delta(p_1 + p_2) \, | \, p_1 \rangle | \, p_2 \rangle \langle \, p_2 \, | \langle \, p_1 \, | \qquad (6.2.6)$$

or

$$\hat{\rho} = \sum_{x_1,x_2} \delta(x_1 - x_2 + x_0) \, | \, x_1 \rangle | \, x_2 \rangle \langle \, x_2 \, | \langle \, x_1 \, |. \qquad (6.2.7)$$

Differing from entangled states, Eqs. (6.2.6) and (6.2.7) cannot be simultaneously true in one system as we have discussed earlier.

(4) Entangled states in spin variables

A simplified classic example of entangled two-particle system was suggested by Bohm. Instead of using continuous space–time variables, Bohm simplified the entangled two-particle state to discrete spin variables. EPR–Bohm state is a singlet state of two spin 1/2 particles:

$$|\Psi\rangle = \frac{1}{\sqrt{2}} \left[\, |\uparrow\rangle_1 \, |\downarrow\rangle_2 - |\downarrow\rangle_1 \, |\uparrow\rangle_2 \, \right], \qquad (6.2.8)$$

where the kets $|\uparrow\rangle$ and $|\downarrow\rangle$ represent the orthogonal states of spin "up" and spin "down", respectively, along an *arbitrary* direction. Again, for this state, *the spin of neither particle is determined; however, if one particle is measured to be spin up along a certain direction, the other one must be spin down along that direction, despite the distance between the two spin 1/2 particles.* Unlike the original EPR state, Eq. (6.2.8) contains only the coherent superposition of two terms of the two-particle state, rather than the integral of infinite number of continuously distributed two-particle states. Similar to the original EPR state, Eq. (6.2.8) is independent of the choice of the spin directions and the eigenstates of the associated noncommuting spin operators. It is easy to show that classically correlated states are coordinate dependent. The state does not hold the same form if choosing the other orthogonal spin directions. Entangled state is very different from classically correlated state, in this regard.

The most widely used entangled two-particle states might have been the "Bell states" (or EPR–Bohm–Bell states). Bell states are a set of polarization states for a pair of entangled photons. The four Bell states which form a complete orthonormal basis of two-photon state are usually represented as

$$\left| \Phi_{12}^{(\pm)} \right\rangle = \frac{1}{\sqrt{2}} \left[\, |\, 0_1 \, 0_2 \,\rangle \pm |\, 1_1 \, 1_2 \,\rangle \, \right],$$

$$\left| \Psi_{12}^{(\pm)} \right\rangle = \frac{1}{\sqrt{2}} \left[\, |\, 0_1 \, 1_2 \,\rangle \pm |\, 1_1 \, 0_2 \,\rangle \, \right], \tag{6.2.9}$$

where $|0\rangle$ and $|1\rangle$ represent two arbitrary orthogonal polarization bases, for example, $|0\rangle = |H\rangle$ (horizontal) and $|1\rangle = |V\rangle$ (vertical) linear polarization, respectively. We have a detailed discussion on Bell states later.

6.3 Entangled Biphoton State

The state of a signal–idler photon pair created in the nonlinear optical process of spontaneous parametric down conversion (SPDC) is a typical EPR state. Roughly speaking, the process of SPDC involves sending a pump laser beam into a nonlinear material, such as a noncentrosymmetric crystal. Occasionally, the nonlinear interaction leads to the annihilation of a high frequency pump photon and the simultaneous creation of a pair of lower frequency signal–idler photons into an entangled two-photon state:

$$|\Psi\rangle = \Psi_0 \sum_{s,i} \delta \left(\omega_s + \omega_i - \omega_p \right) \delta \left(\mathbf{k}_s + \mathbf{k}_i - \mathbf{k}_p \right) a_s^\dagger(\mathbf{k}_s) \, a_i^\dagger(\mathbf{k}_i) \, | \, 0 \rangle, \tag{6.3.1}$$

where ω_j, $\mathbf{k}_j$ (j = s, i, p) are the frequency and wavevector of the signal (s), idler (i), and pump (p), $a_s^\dagger$ and $a_i^\dagger$ are creation operators for the signal and the idler photon, respectively, and Ψ_0 is a normalization constant. We have assumed a CW monochromatic laser pump, i.e., ω_p and $\mathbf{k}_p$ are considered as constants. The two delta functions in Eq. (6.3.1) are technically named as phase matching condition:

$$\omega_p = \omega_s + \omega_i, \quad \mathbf{k}_p = \mathbf{k}_s + \mathbf{k}_i. \tag{6.3.2}$$

The names *signal* and *idler* are historical leftovers. The names probably came about due to the fact that in the early days of SPDC, most of the experiments were done with non-degenerate processes. One radiation was in the visible range (and thus easily detected, the signal), and the other was in IR range (usually not detected, the idler). We see in the following discussions that the role of the idler is not any less than that of the signal. The SPDC process is referred to as type-I if the signal and idler photons have identical polarizations, and type-II if they have orthogonal polarizations. The process is said to be *degenerate* if the SPDC photon pair have the same free space wavelength (e.g., $\lambda_i = \lambda_s = 2\lambda_p$), and *nondegenerate* otherwise. In general, the pair exit the crystal *noncollinearly*, that is, propagate to different directions defined by the second equation in Eq. (6.3.2) and the Snell's law. Of course, the pair may also exit *collinearly*, in the same direction, together with the pump.

The state of the signal–idler pair can be derived, quantum mechanically, by the first order perturbation theory with the help of the nonlinear interaction Hamiltonian. The SPDC interaction arises in a nonlinear crystal driven by a pump laser beam. The polarization, i.e., the dipole moment per unit volume, is given by

$$P_i = \chi_{i,j}^{(1)} E_j + \chi_{i,j,k}^{(2)} E_j E_k + \chi_{i,j,k,l}^{(3)} E_j E_k E_l + \cdots, \tag{6.3.3}$$

where $\chi^{(m)}$ is the mth order electrical susceptibility tensor. In SPDC, it is the second order nonlinear susceptibility $\chi^{(2)}$ that plays the role. The second order nonlinear interaction Hamiltonian can be written as

$$H = \epsilon_0 \int_V d\mathbf{r} \, \chi_{ijk}^{(2)} \, E_i E_j E_k, \tag{6.3.4}$$

where the integral is taken over the interaction volume V.

It is convenient to use the Fourier representation for the electrical fields in Eq. (6.3.4):

$$\mathbf{E}(\mathbf{r},\,t) = \int d\mathbf{k} \,[\, \mathbf{E}^{(-)}(\mathbf{k})e^{-i(\omega(\mathbf{k})t-\mathbf{k}\cdot\mathbf{r})} + \mathbf{E}^{(+)}(\mathbf{k})e^{i(\omega(\mathbf{k})t-\mathbf{k}\cdot\mathbf{r})} \,]. \qquad (6.3.5)$$

Substituting Eq. (6.3.5) into Eq. (6.3.4) and keeping only the terms of interest, we obtain the SPDC Hamiltonian in the interaction representation:

$$\hat{H}_{\text{int}}(t) = \epsilon_0 \int_V d\mathbf{r} \int d\mathbf{k}_s\, d\mathbf{k}_i\, \chi^{(2)}_{lmn}\, \hat{E}^{(+)}_{p\,l}\, e^{i(\omega_p t - \mathbf{k}_p\cdot\mathbf{r})}$$

$$\times\, \hat{E}^{(-)}_{s\,m}\, e^{-i(\omega_s(\mathbf{k}_s)t-\mathbf{k}_s\cdot\mathbf{r})}\, \hat{E}^{(-)}_{i\,n}\, e^{-i(\omega_i(\mathbf{k}_i)t-\mathbf{k}_i\cdot\mathbf{r})} + \text{h.c.}, \qquad (6.3.6)$$

where h.c. stands for Hermitian conjugate. To simplify the calculation, we have also assumed the pump field to be plane and monochromatic with wave vector $\mathbf{k}_p$ and frequency ω_p.

It is easily notable that in Eq. (6.3.6), the volume integration can be done for some simplified cases. At this point, we assume that V is infinitely large. Later, we see that the finite size of V in longitudinal and/or transversal directions may have to be taken into account. For an infinite volume V, the interaction Hamiltonian Eq. (6.3.6) is written as

$$\hat{H}_{\text{int}}(t) = \epsilon_0 \int d\mathbf{k}_s\, d\mathbf{k}_i\, \chi^{(2)}_{lmn}\, \hat{E}^{(+)}_{p\,l}\, \hat{E}^{(-)}_{s\,m}\, \hat{E}^{(-)}_{i\,n}$$

$$\times\, \delta(\mathbf{k}_p - \mathbf{k}_s - \mathbf{k}_i)e^{i(\omega_p - \omega_s(\mathbf{k}_s) - \omega_i(\mathbf{k}_i))t} + h.c. \qquad (6.3.7)$$

It is reasonable to consider the pump field classical, which is usually a laser beam, and quantize the signal and idler fields, which are both in single-photon level:

$$\hat{E}^{(-)}(\mathbf{k}) = -i\sqrt{\frac{2\pi\hbar\omega}{V}}\,a^{\dagger}(\mathbf{k})$$

$$\hat{E}^{(+)}(\mathbf{k}) = i\sqrt{\frac{2\pi\hbar\omega}{V}}\,a(\mathbf{k}), \qquad (6.3.8)$$

where $a^{\dagger}(\mathbf{k})$ and $a(\mathbf{k})$ are photon creation and annihilation operators, respectively. The state of the emitted photon pair can be calculated by applying perturbation to the first order,

$$|\Psi\rangle \simeq -\frac{i}{\hbar}\int dt\, H_{\text{int}}(t)\,|0\rangle. \qquad (6.3.9)$$

By using vacuum $|0\rangle$ for the initial state in Eq. (6.3.9), we assume that there is no input radiation in any signal and idler modes, that is, we have a spontaneous parametric down conversion (SPDC) process.

Further assuming an infinite interaction time, evaluating the time integral in Eq. (6.3.9) and omitting altogether the constants and slow (square root) functions of ω, we obtain the *entangled* two-photon state of Eq. (6.3.1) in the form of integral:

$$|\Psi\rangle \simeq \Psi_0 \int d\mathbf{k}_s d\mathbf{k}_i \, \delta[\omega_p - \omega_s(\mathbf{k}_s) - \omega_i(\mathbf{k}_i)]\delta(\mathbf{k}_p - \mathbf{k}_s - \mathbf{k}_i)a_s^\dagger(\mathbf{k}_s)a_i^\dagger(\mathbf{k}_i)|0\rangle,$$

$$(6.3.10)$$

where Ψ_0 is a normalization constant which has absorbed all omitted constants. Equation (6.3.10) has been used in Chapter 5 for the calculation of second-order correlation function.

The way of realizing the delta functions in Eq. (6.3.10), i.e., the way of achieving "phase matching", basically determines how the signal–idler pair "looks". For example, in a negative uniaxial crystal, one can use a linearly polarized pump laser beam as an extraordinary ray of the crystal to generate a signal–idler pair both polarized as the ordinary rays of the crystal, which is defined as type-I phase matching. One can alternatively generate a signal–idler pair with one ordinary polarized and another extraordinary polarized, which is defined as type II phase matching. Figure 6.3.1 shows three examples of SPDC two-photon source. All three schemes have been widely used for different experimental purposes. Technical details can be found from text books and research references in nonlinear optics.

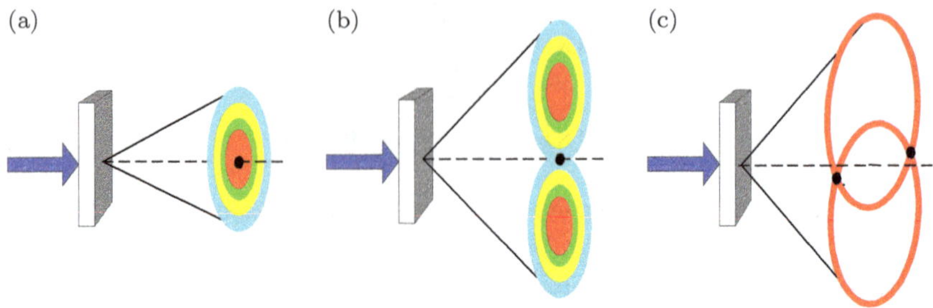

Fig. 6.3.1   Three widely used SPDC setups. (a) Type-I SPDC. (b) Collinear degenerate type-II SPDC. Two rings overlap at one region. (c) Noncollinear degenerate type-II SPDC. For clarity, only two degenerate rings, one for e-polarization and the other for o-polarization, are shown.

The two-photon state in the forms of Eq. (6.3.10) is a pure state which contains a coherent superposition of infinite number of continuously distributed two-photon states. The entangled state of Eq. (6.3.1) and Eq. (6.3.10) describe the behavior of a signal–idler photon pair mathematically. Does the signal or the idler photon in the EPR state of Eq. (6.3.1) or Eq. (6.3.10) have a defined energy and momentum regardless of whether we measure it or not? Quantum mechanics answers: No! However, if one of the subsystems is measured with a certain energy and momentum, the other one is determined with certainty, despite the distance between them.

In the above calculation of the two-photon state we have approximated an infinite large volume of nonlinear interaction. For a finite volume of nonlinear interaction, we may write the state of the signal–idler photon pair in a more general form:

$$|\Psi\rangle = \int d\mathbf{k_s}\, d\mathbf{k_i}\, F(\mathbf{k}_s, \mathbf{k}_i)\, a_i^\dagger(\mathbf{k}_s)\, a_s^\dagger(\mathbf{k}_i)|\, 0\,\rangle, \qquad (6.3.11)$$

where

$$F(\mathbf{k}_s, \mathbf{k}_i) = \epsilon\, \delta(\omega_p - \omega_s - \omega_i)\, f(\Delta_z L)\, h_{tr}(\vec{\kappa}_1 + \vec{\kappa}_2)$$

$$f(\Delta_z L) = \int_L dz\, e^{-i(k_p - k_{sz} - k_{iz})z}$$

$$h_{tr}(\vec{\kappa}_1 + \vec{\kappa}_2) = \int_A d\vec{\rho}\, \tilde{h}_{tr}(\vec{\rho})\, e^{-i(\vec{\kappa}_s + \vec{\kappa}_i)\cdot\vec{\rho}}$$

$$\Delta_z = k_p - k_{sz} - k_{iz}, \qquad (6.3.12)$$

where ϵ is named as parametric gain, ϵ is proportional to the second order electric susceptibility $\chi^{(2)}$ and is usually treated as a constant; L is the length of the nonlinear interaction; the integral in $\vec{\kappa}$ is evaluated over the cross-section A of the nonlinear material illuminated by the pump, $\vec{\rho}$ is the transverse coordinate vector, $\vec{\kappa}_j$ (with $j = s, i$) is the transverse wavevector of the signal and idler, and $f(|\,\vec{\rho}\,|)$ is the transverse profile of the pump, which can be treated as a Gaussian in most of the experimental conditions. The functions $f(\Delta_z L)$ and $h_{tr}(\vec{\kappa}_1 + \vec{\kappa}_2)$ can be approximated as δ-functions for an infinitely long ($L \sim \infty$) and wide ($A \sim \infty$) nonlinear interaction region. The reason we have chosen the form of Eq. (6.3.12) is to separate the "longitudinal" and the "transverse" correlations. We show that $\delta(\omega_p - \omega_s - \omega_i)$ and $f(\Delta_z L)$ together can be rewritten as a function of $\omega_s - \omega_i$. To simplify the mathematics, we assume near co-linearly SPDC. In this situation, $|\,\vec{\kappa}_{s,i}\,| \ll |\,\mathbf{k}_{s,i}\,|$.

Basically, function $f(\Delta_z L)$ determines the "longitudinal" space–time correlation. Finding the solution of the integral is straightforward:

$$f(\Delta_z L) = \int_0^L dz\, e^{-i(k_p - k_{sz} - k_{iz})z}$$

$$= e^{-i\Delta_z L/2} \operatorname{sinc}(\Delta_z L/2). \tag{6.3.13}$$

Now, we consider $f(\Delta_z L)$ with $\delta(\omega_p - \omega_s - \omega_i)$ together, and taking advantage of the δ-function in frequencies by introducing a detuning frequency ν to evaluate function $f(\Delta_z L)$:

$$\omega_s = \omega_s^0 + \nu,$$

$$\omega_i = \omega_i^0 - \nu, \tag{6.3.14}$$

$$\omega_p = \omega_s + \omega_i = \omega_s^0 + \omega_i^0.$$

The dispersion relation $k(\omega)$ allows us to express the wave numbers through the detuning frequency ν:

$$k_s \approx k(\omega_s^0) + \nu \left.\frac{dk}{d\omega}\right|_{\omega_s^0} = k(\omega_s^0) + \frac{\nu}{u_s},$$

$$k_i \approx k(\omega_i^0) - \nu \left.\frac{dk}{d\omega}\right|_{\omega_i^0} = k(\omega_i^0) - \frac{\nu}{u_i}, \tag{6.3.15}$$

where u_s and u_i are group velocities for the signal and the idler, respectively. Now, we connect Δ_z with the detuning frequency ν:

$$\Delta_z = k_p - k_{sz} - k_{iz}$$

$$= k_p - \sqrt{(k_s)^2 - (\vec{\kappa}_s)^2} - \sqrt{(k_i)^2 - (\vec{\kappa}_i)^2}$$

$$\cong k_p - k_s - k_i + \frac{(\vec{\kappa}_s)^2}{2k_s} + \frac{(\vec{\kappa}_i)^2}{2k_i}$$

$$\cong k_p - k(\omega_s^0) - k(\omega_i^0) + \frac{\nu}{u_s} - \frac{\nu}{u_i} + \frac{(\vec{\kappa}_s)^2}{2k_s} + \frac{(\vec{\kappa}_i)^2}{2k_i}$$

$$\cong D\nu, \tag{6.3.16}$$

where $D \equiv 1/u_s - 1/u_i$. We have also applied $k_p - k(\omega_s^0) - k(\omega_i^0) = 0$ and $|\vec{\kappa}_{s,i}| \ll |\mathbf{k}_{s,i}|$. The "longitudinal" wavevector correlation function is rewritten as a function of the detuning frequency ν: $f(\Delta_z L) \cong f(\nu D L)$. In addition to the above approximations, we have inexplicitly assumed the angular independence of the wavevector $k = n(\theta)\omega/c$. For type II SPDC, the refraction index of the extraordinary ray depends on the angle between the wavevector and the optical axis and an additional term appears in the expansion. Making the approximation valid, we have restricted our calculation to near-collinear process. Thus, for a good approximation, in the near-collinear experimental setup:

$$\Delta_z L \cong \nu DL = (\omega_s - \omega_i)DL/2. \tag{6.3.17}$$

Type-I degenerate SPDC is a special case. Due to the fact that $u_s = u_i$, and hence, $D = 0$, the expansion of $k(\omega)$ should be carried out up to the second order. Instead of (6.3.17), we have

$$\Delta_z L \cong -\nu^2 D' L = -(\omega_s - \omega_i)^2 D'L/4, \tag{6.3.18}$$

where

$$D' \equiv \frac{d}{d\omega}\left(\frac{1}{u}\right)\bigg|_{\omega^0}.$$

The two-photon state of the signal–idler pair is then approximated as

$$|\Psi\rangle = \int d\nu\, d\vec{\kappa}_s\, d\vec{\kappa}_i\, f(\nu)\, h_{tr}(\vec{\kappa}_s + \vec{\kappa}_i)\, a_s^\dagger(\omega_s^0 + \nu, \vec{\kappa}_s)\, a_i^\dagger(\omega_i^0 - \nu, \vec{\kappa}_i)|0\rangle, \tag{6.3.19}$$

where the normalization constant has been absorbed into $f(\nu)$.

6.4 EPR Correlation of Entangled Biphoton System

The EPR state generated from SPDC is a pure state that characterizes the behavior of an entangled biphoton system. The state is in the form of coherent superposition of a set of specially selected two-photon states. The EPR correlation, or the nontrivial second-order coherence function $G^{(2)}(\mathbf{r}_1, t_1; \mathbf{r}_2, t_2)$, is the result of a nonlocal interference among this peculiar set of biphoton amplitudes, which specifies the probability for jointly detecting a signal–idler photon pair at coordinates $(\mathbf{r}_1, t_1)$ and $(\mathbf{r}_2, t_2)$. In principle, one joint detection event of an EPR pair involves the superposition of all biphoton amplitudes that are specified by the EPR state. One measurement of an EPR pair is able to provide all information about the interference. A question naturally comes in this regard: Can we then observe the EPR correlation from the measurement of one EPR pair? Generally speaking, quantum mechanics teaches us that we may never learn any meaningful physics from the measurement of one pair of particles. On one hand, in photon counting, the outcome of a measurement is just a *yes* (a count or a "click") or *no* (no count).

In a joint measurement of two photon counting detectors, the outcome of *yes* means a *yes-yes* or a "click-click" joint detection event. A joint detection event of *yes* may occur at any pair of space–time coordinates $(\mathbf{r}_1, t_1; \mathbf{r}_2, t_2)$ for a measurement, but definitely cannot occur at all points of an interference pattern. On the other hand, quantum mechanics does not predict a precise coordinates $(\mathbf{r}_1, t_1)$ and $(\mathbf{r}_2, t_2)$ for a photon pair to appear. Rather, quantum mechanics predict the probability for a joint detection event to occur at a space–time point $(\mathbf{r}_1, t_1$ and $\mathbf{r}_2, t_2)$. To learn the EPR correlation, an ensemble measurement on a large number of *identical* pairs are necessary, where *identical* means that all pairs which are involved in the measurement must be prepared in the same EPR state. The measurement of quantum coherence or correlation is typically statistical. Statistically, if the outcome of a joint measurement has its maximum probability for *yes* to occur at a certain set of values of physical observable or under a certain relationship between physical variables and 100% *no* otherwise, the measured quantum system is considered to have an EPR correlation on that observable. As a good example, EPR's *gedankenexperiment* suggested to us a system of particle pairs with perfect correlation $\delta(x_1 - x_2 + x_0)$ in position. To confirm the EPR correlation, we need to observe *yes* only when the positions of the two distant detectors satisfy $x_1 - x_2 = x_0$, and 100% *no* otherwise when $x_1 - x_2 \neq x_0$. To observe this peculiar EPR correlation, a realistic approach is to measure the joint detection counting rate at each pair of chosen coordinates x_1 and x_2 by scanning all possible values of $x_1 - x_2$. In quantum optics, this means the measurement of the second-order correlation function of $G^{(2)}(\tau_1 - \tau_2)$ (longitudinal), and/or $G^{(2)}(\vec{\rho}_1 - \vec{\rho}_2)$ (transverse), where $\tau_j = t_j - z_j/c$, $j = 1, 2$, and $\vec{\rho}_j$ is the transverse coordinate of the jth point-like photon counting detector.

Now, we exam the second-order coherence function of the entangled signal–idler photon pair of SPDC

$$G^{(2)}(\mathbf{r}_1, t_1; \mathbf{r}_2, t_2) = \langle \hat{E}^{(-)}(\mathbf{r}_1, t_1)\hat{E}^{(-)}(\mathbf{r}_2, t_2)\hat{E}^{(+)}(\mathbf{r}_2, t_2)\hat{E}^{(+)}(\mathbf{r}_1, t_1)\rangle,$$

$$(6.4.1)$$

where $\hat{E}^{(-)}$ and $\hat{E}^{(+)}$ are the negative-frequency and the positive-frequency field operators of the detection events at space–time points $(\mathbf{r}_1, t_1)$ and $(\mathbf{r}_2, t_2)$. For the entangled biphoton state of SPDC,

$$G^{(2)}(\mathbf{r}_1, t_1; \mathbf{r}_2, t_2) = |\langle 0 |\hat{E}^{(+)}(\mathbf{r}_2, t_2)\hat{E}^{(+)}(\mathbf{r}_1, t_1)|\Psi\rangle|^2$$

$$= |\Psi(\mathbf{r}_1, t_1; \mathbf{r}_2, t_2)|^2, \qquad (6.4.2)$$

where $|\Psi\rangle$ is the biphoton state, and $\Psi(\mathbf{r}_1, t_1; \mathbf{r}_2, t_2)$ is the biphoton effective wavefunction that we have discussed in previous chapter. To evaluate $G^{(2)}(\mathbf{r}_1, t_1; \mathbf{r}_2, t_2)$ and $\Psi(\mathbf{r}_1, t_1; \mathbf{r}_2, t_2)$, we need to propagate the field operators from the source to the space–time coordinates $(\mathbf{r}_1, t_1)$ and $(\mathbf{r}_2, t_2)$.

In general, the field operator $\hat{E}^{(+)}(\mathbf{r}, t)$ at space–time point $(\mathbf{r}, t)$ can be written in terms of Green's function which propagates each Fourier mode from space–time point $(\mathbf{r}_0, t_0)$ to $(\mathbf{r}, t)$:

$$\hat{E}^{(+)}(\mathbf{r}, t) = \sum_{\mathbf{k}} \hat{E}^{(+)}(\mathbf{k}, \mathbf{r}_0, t_0)\, g(\mathbf{k}, \mathbf{r} - \mathbf{r}_0, t - t_0). \tag{6.4.3}$$

To simplify the notation, we have assumed one polarization in Eq. (6.4.3). Green's function $g(\mathbf{k}, \mathbf{r} - \mathbf{r}_0, t - t_0)$ is also named optical transfer function. For different experimental setup, $g(\mathbf{k}, \mathbf{r} - \mathbf{r}_0, t - t_0)$ can be quite different.

Considering an idealized simple experimental setup shown in Fig. 6.4.1, in which the signal–idler pair is received by two point-like photon counting detectors D_1 (signal) and D_2 (idler), respectively and jointly, for the measurements of the second-order temporal (longitudinal) coherence function $G^{(2)}(\tau_1 - \tau_2)$ and spatial (transverse) coherence function $G^{(2)}(\vec{\rho}_1 - \vec{\rho}_2)$. To simplify the mathematics, we further assume paraxial experimental condition. In the discussion of longitudinal and transverse correlation measurements, it is convenient to write the field $\hat{E}^{(+)}(\mathbf{r}_j, t_j)$ in terms of its

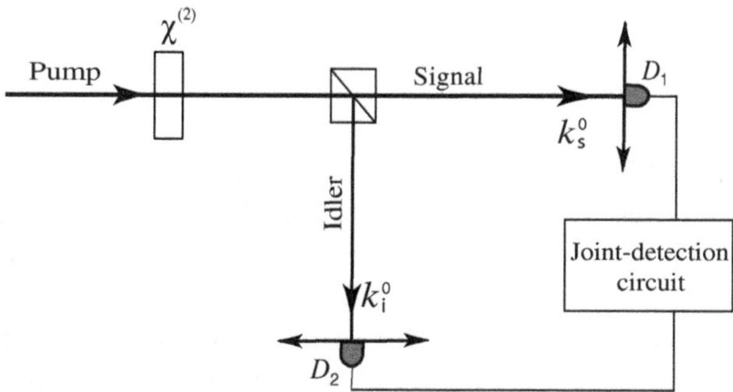

Fig. 6.4.1 Simplified measurement of second-order temporal and spatial coherence or correlation for entangled biphoton state: The signal–idler photon pair are received by two point-like photodetectors D_1 (signal) and D_2 (idler), respectively and jointly. Assuming paraxial propagation of the signal–idler, the z_1 and z_2 are chosen along the central wavevector $\mathbf{k}_s^0$ and $\mathbf{k}_i^0$.

longitudinal and transverse space–time variables under the Fresnel paraxial approximation:

$$\hat{E}^{(+)}(\vec{\rho}_j, z_j, t_j) \cong \int d\omega\, d\vec{\kappa}\; g(\vec{\kappa}, \omega; \vec{\rho}_j, z_j)\, e^{-i\omega t_j}\, \hat{a}(\omega, \vec{\kappa})$$

$$\cong \int d\omega\, d\vec{\kappa}\; \gamma(\vec{\kappa}, \omega; \vec{\rho}_j, z_j)\, e^{-i\omega \tau_j}\, \hat{a}(\omega, \vec{\kappa}), \qquad (6.4.4)$$

where $g(\vec{\kappa}, \omega; \vec{\rho}_j, z_j) = \gamma(\vec{\kappa}, \omega; \vec{\rho}_j, z_j)\, e^{i\omega z_j/c}$ is the spatial part of Green's function, $\vec{\rho}_j$ and z_j are the transverse and longitudinal coordinates of the jth photodetector and $\vec{\kappa}$ is the transverse wavevector. We have chosen $z_0 = 0$ and $t_0 = 0$ at the output plane of the SPDC. For convenience, all constants associated with the field are absorbed into $g(\vec{\kappa}, \omega; \vec{\rho}_j, z_j)$.

The two-photon effective wavefunction $\Psi(\vec{\rho}_1, z_1, t_1; \vec{\rho}_2, z_2, t_2)$ is thus calculated as

$$\Psi(\vec{\rho}_1, z_1, t_1; \vec{\rho}_2, z_2, t_2)$$

$$= \langle 0 | \int d\omega'\, d\vec{\kappa}'\; g(\vec{\kappa}', \omega'; \vec{\rho}_2, z_2)\, e^{-i\omega' t_2}\, \hat{a}(\omega', \vec{\kappa}')$$

$$\times \int d\omega''\, d\vec{\kappa}''\; g(\vec{\kappa}'', \omega''; \vec{\rho}_1, z_1)\, e^{-i\omega'' t_1}\, \hat{a}(\omega'', \vec{\kappa}'')$$

$$\times \int d\nu\, d\vec{\kappa}_s\, d\vec{\kappa}_i\, f(\nu)\, h_{tr}(\vec{\kappa}_s + \vec{\kappa}_i)\, a_s^\dagger(\omega_s^0 + \nu, \vec{\kappa}_s)\, a_i^\dagger(\omega_i^0 - \nu, \vec{\kappa}_i) | 0 \rangle$$

$$= e^{-i(\omega_s^0 \tau_1 + \omega_i^0 \tau_2)} \int d\nu\, d\vec{\kappa}_s\, d\vec{\kappa}_i\, f(\nu)\, h_{tr}(\vec{\kappa}_s + \vec{\kappa}_i)$$

$$\times e^{-i\nu(\tau_1 - \tau_2)}\, \gamma(\vec{\kappa}_s, \nu; \vec{\rho}_1, z_1)\, \gamma(\vec{\kappa}_i, -\nu; \vec{\rho}_2, z_2). \qquad (6.4.5)$$

Although Eq. (6.4.5) cannot be factorized into a product of longitudinal and transverse integrals, it is not difficult to calculate the temporal correlation and the transverse spatial correlation separately by choosing suitable experimental conditions.

Typical experiment may be designed for measuring either temporal (longitudinal) or spatial (transverse) correlation only. Thus, based on different experimental setups, we may simplify the calculation to either temporal (longitudinal) correlation:

$$\Psi(\tau_1; \tau_2) = \Psi_0\, e^{-i(\omega_s^0 \tau_1 + \omega_i^0 \tau_2)} \int d\nu\, f(\nu)\, e^{-i\nu(\tau_1 - \tau_2)} \qquad (6.4.6)$$

or spatial (transverse) correlation:

$$\Psi(\vec{\rho}_1, z_1; \vec{\rho}_2, z_2)$$

$$= \Psi_0 \int d\vec{\kappa}_s \, d\vec{\kappa}_i \, h_{tr}(\vec{\kappa}_s + \vec{\kappa}_i) \, g(\vec{\kappa}_s, \omega_s; \vec{\rho}_1, z_1) \, g(\vec{\kappa}_i, \omega_i; \vec{\rho}_2, z_2). \qquad (6.4.7)$$

In certain experimental conditions, we may approximate $h_{tr}(\vec{\kappa}_s + \vec{\kappa}_i) \sim \delta(\vec{\kappa}_s + \vec{\kappa}_i)$.

(I) Biphoton temporal correlation:

To measure the biphoton temporal correlation of SPDC, we select a pair of transverse wavevector $\vec{\kappa}_s = -\vec{\kappa}_i$ in Eq. (6.4.5) by using appropriate optical apertures. The effective two-photon wavefunction is thus simplified to that of Eq. (6.4.6):

$$\Psi(\tau_1; \tau_2) \cong \Psi_0 \, e^{-i(\omega_s^0 \tau_1 + \omega_i^0 \tau_2)} \int d\nu \, f(\nu) \, e^{-i\nu(\tau_1 - \tau_2)}$$

$$= \left[\Psi_0 \, e^{-\frac{i}{2}(\omega_s^0 + \omega_i^0)(\tau_1 + \tau_2)} \right] \left[\mathcal{F}_{\tau_1 - \tau_2}\left\{ f(\nu) \right\} e^{-\frac{i}{2}(\omega_s^0 - \omega_i^0)(\tau_1 - \tau_2)} \right],$$
$$(6.4.8)$$

where, again, $\mathcal{F}_{\tau_1 - \tau_2}\left\{ f(\nu) \right\}$ is the Fourier transform of the spectrum amplitude function $f(\nu)$. Equation (6.4.8) indicates a 2-D wavepacket: a δ-function like narrow envelope along axis $\tau_1 - \tau_2$ with constant amplitude in axis $\tau_1 + \tau_2$. For fixed positions of the photodetectors D_1 and D_2, the 2-D wavepacket means the following: the signal–idler pair may be jointly detected at any time; however, if the signal is registered at a certain time t_1, the idler must be registered at a unique time of $t_2 \sim t_1 - (z_1 - z_2)/c$. In other words, although the joint detection of the pair may happen at any times of t_1 and t_2 with equal probability ($\Delta(t_1 + t_2) \sim \infty$), the registration time difference of the pair must happen within $\Delta(t_1 - t_2) \sim 0$. A schematic of the wavepacket has been shown earlier.

The longitudinal correlation function $G^{(2)}(\tau_1 - \tau_2)$ is thus

$$G^{(2)}(\tau_1 - \tau_2) \propto |\, \mathcal{F}_{\tau_1 - \tau_2}\left\{ f(\nu) \right\} |^2,$$

which is a δ-function-like function in the case of SPDC. Thus, we have shown the entangled signal–idler photon pair of SPDC hold a typical EPR correlation in energy and time:

$$\Delta(\omega_s + \omega_i) \sim 0 \quad \& \quad \Delta(t_1 - t_2) \sim 0$$

$$\text{with} \qquad \Delta\omega_s \sim \infty, \quad \Delta\omega_i \sim \infty, \quad \Delta t_1 \sim \infty, \quad \Delta t_2 \sim \infty.$$

(II) **Biphoton spatial correlation**:

In the following, we study the spatial correlation $G^{(2)}(\vec{\rho}_1, z_1; \vec{\rho}_2, z_2)$ of entangled biphoton state in three different measurements: (1) in the output plane of the source; (2) in the Fraunhofer far-field, and (3) in the Fresnel near-field. Similar to that of the biphoton temporal correlation, we calculate the effective biphoton wavefunction of the signal–idler pair under the above three conditions. To simplify the mathematics, we concentrate our calculation on the spatial function only by choosing a pair of monochromatic conjugate frequencies ω_s and ω_i. In this case, the effective two-photon wavefunction of Eq. (6.4.5) is simplified to that of Eq. (6.4.7):

$$\Psi(\vec{\rho}_1, z_1; \vec{\rho}_2, z_2) = \Psi_0 \int d\vec{\kappa}_s \, d\vec{\kappa}_i \, \delta(\vec{\kappa}_s + \vec{\kappa}_i) \, g(\vec{\kappa}_s, \omega_s, \vec{\rho}_1, z_1) \, g(\vec{\kappa}_i, \omega_i, \vec{\rho}_2, z_2),$$

where we have assumed $h_{tr}(\vec{\kappa}_s + \vec{\kappa}_i) \sim \delta(\vec{\kappa}_s + \vec{\kappa}_i)$, which is reasonable for a large transverse sized SPDC.

(1) $\Psi(\vec{\rho}_1, z_1; \vec{\rho}_2, z_2)$ **in the output plane of SPDC**

The integral of $d\vec{\kappa}_s$ and $d\vec{\kappa}_i$ can be evaluated easily with the help of the EPR type transverse wavevector correlation $\delta(\vec{\kappa}_s + \vec{\kappa}_i)$:

$$\int d\vec{\kappa}_s \, d\vec{\kappa}_i \, \delta(\vec{\kappa}_s + \vec{\kappa}_i) \, e^{-i\vec{\kappa}_s \cdot \vec{\rho}_s} \, e^{-i\vec{\kappa}_i \cdot \vec{\rho}_i}$$

$$= \int d(\vec{\kappa}_s + \vec{\kappa}_i) \, \delta(\vec{\kappa}_s + \vec{\kappa}_i) \, e^{-i(\vec{\kappa}_s + \vec{\kappa}_i) \cdot (\vec{\rho}_s + \vec{\rho}_i)/2}$$

$$\times \int d(\vec{\kappa}_s - \vec{\kappa}_i) \, e^{-i(\vec{\kappa}_s - \vec{\kappa}_i) \cdot (\vec{\rho}_s - \vec{\rho}_i)/2}$$

$$\simeq 1 \times \delta(\vec{\rho}_s - \vec{\rho}_i), \tag{6.4.9}$$

where we have assumed a uniformly distributed signal–idler fields with incident (or output) angle $\theta_s \sim |\vec{\kappa}_s|/|\mathbf{k}_s|$ and $\theta_i \sim |\vec{\kappa}_i|/|\mathbf{k}_i|$ relative to the output plane: $A(\vec{\rho}_s) \sim e^{-i\vec{\kappa}_s \cdot \vec{\rho}_s}$ and $A(\vec{\rho}_i) \sim e^{-i\vec{\kappa}_i \cdot \vec{\rho}_i}$. Equation (6.4.9) indicates an idealized 2-D wavepacket in terms of its transverse spatial coordinates: a normalized constant distribution along the axis of $(\vec{\rho}_s + \vec{\rho}_i)$, which is the Fourier transform of $\delta(\vec{\kappa}_s + \vec{\kappa}_i)$, and a δ-function along the axis of $(\vec{\rho}_s - \vec{\rho}_i)$, which is the Fourier transform of a normalized constant.

Thus, we have shown that the entangled signal–idler photon pair of SPDC holds a typical EPR correlation in transverse momentum and position, which is very close to the original proposal of EPR:

$$\Delta(\vec{\kappa}_s + \vec{\kappa}_i) \sim 0 \ \ \& \ \Delta(\vec{\rho}_s - \vec{\rho}_i) \sim 0$$

$$\text{with} \ \ \Delta\vec{\kappa}_s \sim \infty, \ \ \Delta\vec{\kappa}_i \sim \infty, \ \ \Delta\vec{\rho}_s \sim \infty, \ \ \Delta\vec{\rho}_i \sim \infty.$$

In EPR's language, we may never know where to locate the signal photon and the idler photon on the output plane of the source, however, if the signal (idler) is found at a certain position, the idler (signal) must be observed at a corresponding unique position. Simultaneously, the signal photon and the idler may have any transverse momentum, however, if the signal (idler) is measured with a transverse momentum, the transverse momentum of the idler (signal) is uniquely determined with certainty. In the *collinear* SPDC: The signal–idler pair is always emitted from the same point in the output plane of the two-photon source, $\vec{\rho}_s = \vec{\rho}_i$, and if one of them propagates slightly off from the collinear axes, the other one must propagate to the opposite direction with $\vec{\kappa}_s = -\vec{\kappa}_i$. This interesting behavior has been experimentally utilized in quantum imaging.

One may not have too much problem to face the above statement. The interaction of spontaneous parametric down-conversion is nevertheless a local phenomenon occurs inside the nonlinear material. The nonlinear interaction coherently creates mode in pairs that satisfy the phase match conditions of Eq. (6.3.2) which is also named as the energy conservation and momentum conservation. The signal–idler photon pair can be excited to any of these coupled modes or in all of these coupled modes simultaneously, resulting in a particular two-photon superposition. It is this "selected" two-photon superposition made the signal–idler pair comes out from the same point of the source and propagates to opposite directions with $\vec{\kappa}_s = -\vec{\kappa}_i$.[2]

(2) $\Psi(\vec{\rho}_1, z_1; \vec{\rho}_2, z_2)$ in the Fraunhofer far-field

The two-photon superposition is getting more interesting when the signal–idler is propagated to a large distance, either by free propagation or

---

[2]Mathematically, we have approved $\vec{\kappa}_s \simeq -\vec{\kappa}_i$ and $\vec{\rho}_s \simeq \vec{\rho}_i$ in SPDC. Physically, this peculiar behavior distinguishes the entangled biphoton radiation from classical light. Applying $\vec{\kappa}_s = -\vec{\kappa}_i$ directly to the calculation, the integral of Eq. (6.4.9) can be simplified as

$$\int d\vec{\kappa}_s \, d\vec{\kappa}_i \, \delta(\vec{\kappa}_s + \vec{\kappa}_i) \, e^{-i\vec{\kappa}_s \cdot \vec{\rho}_s} \, e^{-i\vec{\kappa}_i \cdot \vec{\rho}_i} \simeq \int d\vec{\kappa}_s \, e^{-i\vec{\kappa}_s \cdot (\vec{\rho}_s - \vec{\rho}_i)} \simeq \delta(\vec{\rho}_s - \vec{\rho}_i)$$

and effectively gives $\vec{\rho}_s \simeq \vec{\rho}_i$. Physically and mathematically, based on $\vec{\kappa}_s \simeq -\vec{\kappa}_i$ and $\vec{\rho}_s \simeq \vec{\rho}_i$, we may find a better way to view the physics behind the observation and to simplify the calculation significantly.

guided by optical components such as lenses. In classical opinion, the signal photon and the idler photon are considered independent whenever they are released from the source because there is no interaction anymore between the pair in free space. Therefore, the signal photon and the idler photon should have independent and random distributions in terms of their transverse position $\vec{\rho}_1$ and $\vec{\rho}_2$. This classical picture, however, is incorrect. It is found that the signal–idler biphoton system would not lose its entangled nature in transverse position. This interesting behavior has been experimentally observed in quantum imaging, indicating an EPR type correlation of $\delta(\vec{\rho}_1 - \vec{\rho}_2)$. The subdiffraction limit spatial resolution observed in the "quantum lithography" experiment and the nonlocal correlation observed in the "ghost imaging" experiment are both the results of two-photon superposition. Two-photon superposition does happen to a distant joint detection event of a signal–idler photon pair. There is no surprise one would have difficulties to face this effect. The two-photon superposition is a nonlocal concept in this case. The biphoton amplitudes in the two-photon superposition correspond to different yet indistinguishable alternative ways of triggering a joint-photodetection event at distance. There is no counterpart of such concept in classical theory and may never be understood classically.

Now we consider propagating the signal–idler pair away from the source to the far-field observation points $(\vec{\rho}_1, z_1)$ and $(\vec{\rho}_2, z_2)$, respectively. To simplify the discussion, we place D_1 and D_2 on the plane of $z_1 = z_2$. In the far-field, either achieved by moving D_1 and D_2 to distances or by the use of a Fourier transform lens, we take a first-order approximation in the phase delay, i.e., $r \simeq r_0 - \vec{\kappa} \cdot \vec{\rho}_0$, where r is the distance from point $\vec{\rho}_0$ to D_1 or D_2, r_0 is the distance from the origin point of the coordinate system which is defined as the center point in the output plane of the SPDC, to D_1 or D_2. In this case, the effective biphoton wavefunction is approximated as

$$\Psi(\vec{\rho}_1, z_1; \vec{\rho}_2, z_2) \simeq \langle 0 | \frac{e^{-i(\omega t_1 - k_1 r_{01})}}{r_{01}} \int d\vec{\rho}_0 \, A(\vec{\rho}_0) \, e^{-i\vec{\kappa}_1 \cdot \vec{\rho}_0} \, \hat{a}(\vec{\kappa}_1)$$

$$\times \frac{e^{-i(\omega t_2 - k_2 r_{02})}}{r_{02}} \int d\vec{\rho'}_0 \, A(\vec{\rho'}_0) \, e^{-i\vec{\kappa}_2 \cdot \vec{\rho'}_0} \, \hat{a}(\vec{\kappa}_2)$$

$$\times \int d\vec{\kappa}_s \, d\vec{\kappa}_i \, \delta(\vec{\kappa}_s + \vec{\kappa}_i) \, \hat{a}^\dagger(\vec{\kappa}_s) \, \hat{a}^\dagger(\vec{\kappa}_i) \, | 0 \rangle$$

$$\propto \int d\vec{\rho}_0 \, d\vec{\rho'}_0 \, \delta(\vec{\rho}_0 - \vec{\rho'}_0) e^{-i\vec{\kappa}_1 \cdot \vec{\rho}_0} \, e^{-\vec{\kappa}_2 \cdot \vec{\rho'}_0}$$

$$\propto 1 \times \delta(\vec{\rho}_1 + \vec{\rho}_2). \tag{6.4.10}$$

The δ-function of $\delta(\vec{\rho}_0 - \vec{\rho}'_0)$ in Eq. (6.4.10) is obtained from the double integral of $d\vec{\kappa}_s$ and $d\vec{\kappa}_i$ with $A(\vec{\rho}_0) \sim A_0 e^{-i\vec{\kappa}_s \cdot \vec{\rho}_0}$ and $A(\vec{\rho}'_0) \sim A_0 e^{-i\vec{\kappa}_i \cdot \vec{\rho}'_0}$ where A_0 is a constant. The physics is very clear in the above calculation. (1) The far-field plane is the Fourier transform plane of the SPDC. Each point on the Fourier transform plane plane corresponds to a $\vec{\kappa}_s$ or $\vec{\kappa}_i$. The δ-function $\delta(\vec{\rho}_1 + \vec{\rho}_2)$ on the Fourier transform plane confirms $\delta(\vec{\kappa}_s + \vec{\kappa}_i)$. (2) The 2-D wavepacket in transverse spatial coordinates is the result of a superposition among a large number of biphoton amplitudes. Each biphoton amplitude starts from a point on the output plane of SPDC (at $\vec{\rho}_0 = \vec{\rho}'_0$) and ends at point photodetectors D_1 and D_2. The 2-D wavepacket indicates a typical EPR correlation: we may observe the signal photon and the idler photon at any point on a far-field plane of the entangled biphoton source, however, if one of them is observed at a certain point the other one can only be observed at a unique point.

In the the above calculation, we have treated all integrals to infinity. This approximation yields δ-function for constant distributions. In reality, the finite size of the biphoton source or the applied lenses in the experimental setup may have to be taken into account. Let us assume a finite integral on a circular area from $|\vec{\rho}_0| = 0$ to $|\vec{\rho}_0| = R$ for degenerate SPDC ($\omega_s = \omega_i$), the integral of Eq. (6.4.10) turns

$$\Psi(\vec{\rho}_1, z_1; \vec{\rho}_2, z_2) \propto \int_A d\vec{\rho}_0 \, e^{-i(\vec{\kappa}_1 + \vec{\kappa}_2) \cdot \vec{\rho}_0} \simeq \frac{2J_1\left[R\,|\vec{\kappa}_1 + \vec{\kappa}_2|\right]}{R\,|\vec{\kappa}_1 + \vec{\kappa}_2|}$$

$$= \text{somb}\left[\frac{R}{z}\frac{\omega}{c}(\vec{\rho}_1 + \vec{\rho}_2)\right], \tag{6.4.11}$$

indicating a Fraunhofer diffraction pattern in the joint detection of D_1 and D_2, which is a somb-function of $\vec{\rho}_1 + \vec{\rho}_2$. For large sized SPDC, nevertheless, δ-functions are good approximations.

(3) $\Psi(\vec{\rho}_1, z_1; \vec{\rho}_2, z_2)$ in the Fresnel near-field

We now consider the Fresnel near-field measurement by moving D_1 and D_2 to the near-field of the biphoton source. We further assume the two-photon source has a finite but large transverse dimension. Under this simple experimental setup, Green's function, or the optical transfer function describing arm-j, $j = 1, 2$, in which the signal and the idler freely propagate to photodetector D_1 and D_2, respectively. Substitute the $g_j(\omega, \vec{\kappa}; z_j, \vec{\rho}_j)$ of free propagation, $j = 1, 2$, into Eq. (6.4.7), the effective wavefunction is thus

$$\Psi(\vec{\rho}_1, z_1; \vec{\rho}_2, z_2)$$

$$= \Psi_0 \int d\vec{\kappa}_s \, d\vec{\kappa}_i \, \delta(\vec{\kappa}_s + \vec{\kappa}_i) \, e^{-i\vec{\kappa}_s \cdot \vec{\rho}_0} \, e^{-i\vec{\kappa}_i \cdot \vec{\rho}'_0}$$

$$\times \int_A d\vec{\rho}_0 \, \frac{-i\omega_s}{2\pi c z_1} \, e^{i\frac{\omega_s}{c} z_1} \, G\left(|\vec{\rho}_1 - \vec{\rho}_s|, \frac{\omega_s}{c z_1}\right)$$

$$\times \int_A d\vec{\rho}'_0 \, \frac{-i\omega_i}{2\pi c z_2} \, e^{i\frac{\omega_i}{c} z_2} \, G\left(|\vec{\rho}_2 - \vec{\rho}'_0|, \frac{\omega_i}{c z_2}\right)$$

$$\simeq \frac{-\omega_s \omega_i}{2\pi^2 c^2 z_1 z_2} \, e^{i\left(\frac{\omega_s}{c} z_1 + \frac{\omega_i}{c} z_2\right)} \int_A d\vec{\rho}_0 \, e^{i\left[\frac{\omega_s}{2 c z_1}|\vec{\rho}_1 - \vec{\rho}_0|^2 + \frac{\omega_i}{c z_2}|\vec{\rho}_2 - \vec{\rho}_0|^2\right]}, \quad (6.4.12)$$

where $\vec{\rho}_0$ $(\vec{\kappa}_s)$ and $\vec{\rho}'_0$ $(\vec{\kappa}_i)$ are the transverse coordinates (wavevectors) for the signal and the idler fields, respectively, defined on the output plane of the biphoton source. The superposition of the above large number of biphoton amplitudes will produce a biphoton Fresnel diffraction pattern as a function of $\vec{\rho}_1$ and $\vec{\rho}_2$. Mathematically, it may not be easy to find an analytical solution for arbitrary $\vec{\rho}_1$ and $\vec{\rho}_2$, numerical solutions are always helpful for comparing with the experimental observation.

6.5   Subsystem in an Entangled Biphoton State

The entangled EPR two-particle state is a pure state. The precise correlation of the subsystems is completely described by the state. The measurement, however, is not necessarily always for the two-particle system. It is an experimental choice to study only a subsystem and to ignore the other. What can we learn about a subsystem from these kinds of measurements? Mathematically, it is easy to show that by taking a partial trace of a two-particle pure state, the state of the subsystems are both in mixed states with entropy greater than zero. One can only learn statistical properties of the subsystems in this kind of measurement. In the following, again, we use the signal–idler pair of SPDC as an example to explore the physics.

(I) **The state of the signal (or idler):**

The biphoton state of SPDC is a pure state that satisfies

$$\hat{\rho}^2 = \hat{\rho}, \quad \hat{\rho} \equiv |\Psi\rangle \langle\Psi|, \quad (6.5.1)$$

where $\hat{\rho}$ is the density matrix operator corresponding to the biphoton state of SPDC. The single photon state of the signal and idler,

$$\hat{\rho}_s = tr_i \left| \Psi \right\rangle \left\langle \Psi \right|, \quad \hat{\rho}_i = tr_s \left| \Psi \right\rangle \left\langle \Psi \right|, \tag{6.5.2}$$

are not. To calculate the signal (idler) state from the biphoton state, we have to take a partial trace, as usual, summing over the idler (signal) modes.

We assume a type II SPDC. The orthogonally polarized signal and idler are degenerate in frequency around $\omega^0 \cong \omega_p/2$. To simplify the discussion, by assuming appropriate experimental conditions, we trivialize the transverse part of the state and write the biphoton state in the following simplified form:

$$\left| \Psi \right\rangle = \Psi_0 \int d\nu \; \Phi(\mathrm{DL}\nu) \; a_s^\dagger(\omega^0 + \nu) \; a_i^\dagger(\omega^0 - \nu) \left| 0 \right\rangle,$$

where $\Phi(\mathrm{DL}\nu)$ is a *sinc*-like function:

$$\Phi(\mathrm{DL}\nu) = \frac{1 - e^{-i\mathrm{DL}\nu}}{i\mathrm{DL}\nu},$$

which is a function of the crystal length L, and the difference of inverse group velocities of the signal (ordinary) and the idler (extraordinary), $\mathrm{D} \equiv 1/u_o - 1/u_e$. The constant Ψ_0 is calculated from the normalization $tr\,\hat{\rho} = \left\langle \Psi \mid \Psi \right\rangle = 1$. It is easy to calculate and to find $\hat{\rho}^2 = \hat{\rho}$ for the biphoton state of the signal–idler pair.

Summing over the idler modes, the density matrix of signal is given by

$$\hat{\rho}_s = \Psi_0^2 \int d\nu \; |\Phi(\nu)|^2 \; a_s^\dagger(\omega^0 + \nu) \left| 0 \right\rangle \left\langle 0 \right| a_s(\omega^0 + \nu), \tag{6.5.3}$$

with

$$|\Phi(\nu)|^2 = \mathrm{sinc}^2 \frac{\mathrm{DL}\nu}{2}, \tag{6.5.4}$$

where all constants coming from the integral have been absorbed into Ψ_0. First, we find immediately that $\hat{\rho}_s^2 \neq \hat{\rho}_s$. It means the state of the signal is a mixed state (as is the idler). Second, it is very interesting to find that the spectrum of the signal dependents on the group velocity of the idler. This, however, should not come as a surprise, because the state of the signal photon is calculated from the biphoton state by summing over the idler modes.

The spectrum of the signal and idler has been experimentally verified by Strekalov *et al* by using a Michelson interferometer in a standard

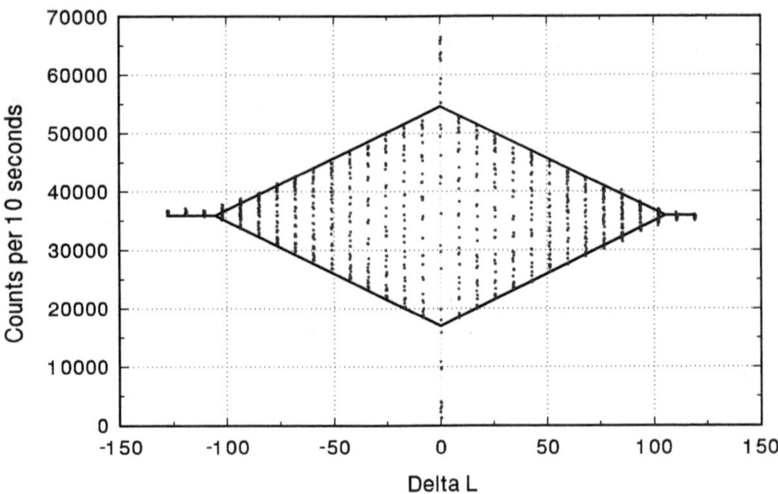

Fig. 6.5.1 Experimental data indicated a "double notch" envelope. Each of the doted single vertical line contains many cycles of sinusoidal modulation.

Fourier spectroscopy type measurement. The measured interference pattern is shown in Fig. 6.5.1. The envelope of the sinusoidal modulations (in segments) is fitted very well by two "notch" functions (upper and lower part of the envelope). The experimental data agrees with the theoretical analysis of the experiment.

The following is a simple calculation to explain the observed "notch" function. We first define the field operators:

$$\hat{E}^{(+)}(t, z_d) = \hat{E}_1^{(+)}\left(t - \frac{z_1}{c}, z_0\right) + \hat{E}_2^{(+)}\left(t - \frac{z_2}{c}, z_0\right),$$

where z_d is the position of the photodetector, z_0 is the input point of the interferometer, $t_1 = t - \frac{z_1}{c}$ and $t_2 = t - \frac{z_2}{c}$, respectively, are the early times before propagated to the photodetector at time t with time delays of z_1/c and z_2/c, where z_1 and z_2 are the optical paths in arm 1 and arm 2 of the interferometer. We have defined a very general field operator which is the superposition of two early fields propagated individually through arm 1 and arm 2 of any type of interferometer. The counting rate of the photon counting detector is thus

$$R_d = tr\left[\hat{\rho}_s \hat{E}^{(-)}(t, z_d)\hat{E}^{(+)}(t, z_d)\right]$$

$$= \Psi_0^2 \int d\nu \, |\Phi(\nu)|^2 \left|\langle 0| \left[\hat{E}_1^{(+)}\left(t - \frac{z_1}{c}, z_0\right)\right.\right.$$

$$\left. + \hat{E}_2^{(+)}\left(t - \frac{z_1}{c}, z_0\right)\right] a_s^\dagger(\omega^0 + \nu)\,|0\rangle\Big|^2$$

$$\propto 1 + Re\left\{e^{-i\omega^0\tau}\int d\nu \,\text{sinc}^2\frac{\text{DL}\nu}{2}\,e^{-i\nu\tau}\right\}, \tag{6.5.5}$$

where $\tau = (z_1 - z_2)/c$. The Fourier transform of $\text{sinc}^2(\text{DL}\nu/2)$ has a "notch" shape. It is noted that the base of the "notch" function is determined by parameter DL of the SPDC, which is easily confirmed from the experiment.

(II) **The entropy problem**:

Now we turn to another interesting aspect of physics, namely the physics of entropy. Entropy is an important concept in the information theory. The concept, named as Von Neumann entropy, is given by the state

$$S = -tr\left(\hat{\rho}\log\hat{\rho}\right). \tag{6.5.6}$$

It is easy to find that the entropy of the entangled two-photon pure state is zero. The entropy of its subsystems, however, are both greater than zero. The value of the Von Neumann entropy can be numerically evaluated from the measured spectrum. Note that the density operator of the subsystem is diagonal. Taking its trace is simply performing an integral over the frequency spectrum with the measured spectrum function. It is straightforward to find the entropy of the subsystems $S_s > 0$. This is an expected result due to the nature of the mixed state of the subsystems. Considering that the entropy of the two-photon system is zero and the entropy of the subsystems are both greater than zero, does this mean that negative entropy is present somewhere in the entangled two-photon system? According to classical information theory, for the entangled two-photon system, $S_s + S_{s|i} = 0$, where $S_{s|i}$ is the conditional entropy. It is this conditional entropy that must be negative, which means that *given the result of a measurement over one particle, the result of measurement over the other must yield negative information*. This paradoxical statement is similar and, in fact, closely related to the EPR "paradox". It comes from the same philosophy as that of the EPR.

6.6 Biphoton in Dispersive Media

In this section we study the propagation of entangled photons in optical dispersive medium. Dispersion contains rich physics in general. It is not our goal to discuss in detail about biphoton dispersion, here, we give a simple analysis on the propagation of the entangled singnal-idler pair of a CW-pumped SPDC in group dispersive optical fibers. As we have learned, although a biphoton wavepacket is defined with the two-photon state of SPDC, there is no wavepacket defined with the subsystems of the signal and the idler. After propagating the signal and the idler a certain distance in optical fibers, do we expect a "broadened" signal or idler photon? It is interesting to find that there is no wavepacket either broadened or unbroadened associated with the propagation of either signal or idler individually. It is the biphoton wavefunction $\psi(t_1, r_1; t_2, r_2)$ or the second-order correlation function $G^{(2)}(t_1, r_1; t_2, r_2)$ that is broadened in the dispersive medium. The biphoton wavepacket in dispersive medium propagates like a classical optical pulse.

Let us consider a simple experiment as shown in Fig. 6.6.1. We use a biphoton source of a CW laser pumped SPDC and two point-like photon counting detectors with a photon counting coincidence circuit to setup a standard second-order temporal correlation measurement. The photodetectors D_1 and D_2 are placed in the far-field zone of z_1 and z_2, respectively, for the detection of the signal and the idler photons. The photon-counting circuit records the registration time difference, $t_1 - t_2$, for each signal (registered at t_1) — idler (registered at t_2) pair. After measuring a large number of pairs, the circuit reports a histogram that summarizes how many total number of pairs (vertical axis) are measured with each value of $t_1 - t_2$ (horizontal axis). The measured histogram, number of joint detection counting rate against $t_1 - t_2$, corresponds to the seconder-order temporal correlation function $G^{(2)}(t_1 - t_2)$.

If we set up our experiment in the vacuum, or realistically in the air in which the dispersion can be ignored, and observed a certain function

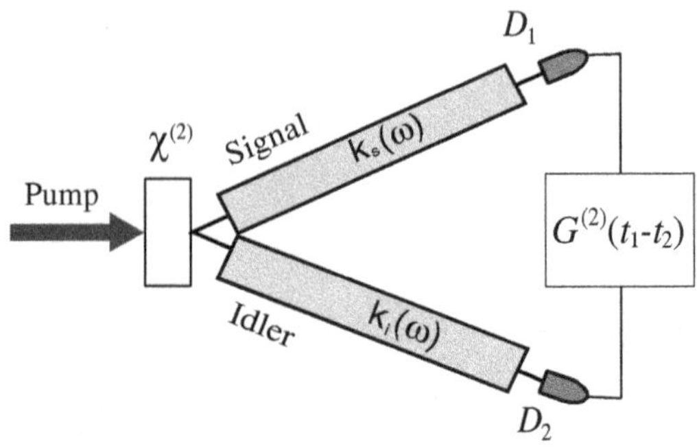

Fig. 6.6.1 Broadening of $G^{(2)}(t_1 - t_2)$ due to the propagation of the signal and idler photon pair in group dispersive media.

$G^{(2)}(t_1 - t_2)$ in the temporal correlation measurement. As we have learned earlier, the observed function $G^{(2)}(t_1-t_2)$ is mainly determined by the phase matching condition of the SPDC. There is no contribution from the group broadening of either signal or idler propagation. Now, we put dispersive medium, such as optical fibers, between the two-photon source and the photodetectors D_1 and D_2. The dispersive medium has unignorable first-order and second-order dispersion:

$$\left. \frac{dk(\omega)}{d\omega} \right|_{\omega_s^0} = k' \neq 0 \quad \text{and} \quad \left. \frac{d^2 k(\omega)}{d\omega^2} \right|_{\omega_i^0} = k'' \neq 0$$

at the neighborhood of the wavelengths ω_s^0 and ω_i^0, where ω_s^0 and ω_i^0 are the central frequencies of the signal–idler radiation. The following calculation answers the questions about what will happen to the $G^{(2)}(t_1 - t_2)$ function: Is the function $G^{(2)}(t_1 - t_2)$ broadened by the dispersive propagation of the signal and the idler?

The state of the signal–idler biphoton pair of SPDC, generated by a CW laser beam at frequency ω_p, can be written in the following simplified form as we have derived earlier:

$$|\Psi\rangle = \int_{-\infty}^{\infty} d\nu \, f(\nu) \, a_s^\dagger(\omega_s^0 + \nu) a_i^\dagger(\omega_i^0 - \nu) \, |0\rangle,$$

where, again, ω_s^0 and ω_i^0 are the central frequencies of the signal–idler; ν is the returning frequency. The spectral amplitude, $f(\nu)$, provides all the information about the correlation properties of the two-photon light. $f(\nu)$ is basically determined by the longitudinal part of the phase matching in SPDC. Based on the simplified biphoton state and the field operators,

$$\hat{E}_1^{(+)} = \int d\omega_s \, a(\omega_s) \, e^{-i[\omega_s t_1 - k(\omega_s) z_1]},$$

$$\hat{E}_2^{(+)} = \int d\omega_i \, a(\omega_i) \, e^{-i[\omega_i t_2 - k(\omega_i) z_2]},$$

the biphoton wavepacket is calculated to be:

$$\psi(t_1, r_1; t_2, r_2) = \int d\omega_s f(\omega_s) e^{-i[\omega_s t_1 - k(\omega_s) z_1]} e^{-i[\omega_i t_2 - k(\omega_i) z_2]}. \tag{6.6.1}$$

Considering the first-order and the second-order dispersion of the propagation along z_1 and z_2, The biphoton wavepacket turns out to be

$$\psi(t_1, z_1; t_2, z_2) = \left\{ e^{-i\{(\omega_s^0 t_1 + \omega_i^0 t_2) - [k(\omega_s^0)z_1 + k(\omega_i^0)z_2]\}} \right\}$$

$$\times \int_{-\infty}^{\infty} d\nu \, f(\nu) \, e^{i(k_s'' z_1 + k_i'' z_2)(\nu^2/2)} \, e^{-i\nu\left[(t_1 - t_2) - \left(\frac{z_1}{u_s} - \frac{z_2}{u_i}\right)\right]}$$

$$= \left\{ e^{-i\{(\omega_s^0 t_1 + \omega_i^0 t_2) - [k(\omega_s^0)z_1 + k(\omega_i^0)z_2]\}} \right\}$$

$$\times \, \mathcal{F}_\tau \left\{ f(\nu) e^{i(k_s'' z_1 + k_i'' z_2)(\nu^2/2)} \right\}, \tag{6.6.2}$$

where we have defined

$$\tau \equiv (t_1 - t_2) - [z_1/u_s - z_2/u_i]$$

$$= [t_1 - z_1/u_s] - [t_2 - z_2/u_i]$$

$$\equiv \tau_1 - \tau_2.$$

The second-order temporal coherence function $G^{(2)}(\tau)$ is thus

$$G^{(2)}(\tau) \propto \left| \mathcal{F}_\tau \left\{ f(\nu) e^{i(k_s'' z_1 + k_i'' z_2)(\nu^2/2)} \right\} \right|^2. \tag{6.6.3}$$

In Eqs. (6.6.2) and (6.6.3), $\mathcal{F}_\tau\{f(\nu)e^{i(k_s'' z_1 + k_i'' z_2)(\nu^2/2)}\}$ denotes as the Fourier transform of function $f(\nu)e^{i(k_s'' z_1 + k_i'' z_2)(\nu^2/2)}$. It is easy to find mathematically that the Gaussian function $e^{i(k_s'' z_1 + k_i'' z_2)\nu^2/2}$, corresponding to the second-order dispersion in both optical paths of the signal and the idler, contributes to the broadening of the biphoton wavepacket $\psi(t_1, z_1; t_2, z_2)$ as well as the second-order correlation function $G^{(2)}(\tau)$.

In summary, we have the following conclusions regarding to the propagation of biphoton wavepacket in dispersion media:

(1) The concept of broadening of a biphoton wavepacket is fundamentally different from the broadening of two classical wavepackets. The broadening of the second-order correlation function in classical case, which can be measured by an optical auto-correlator, is due to the broadening of each measured individual pulses. In the measurement of entangled biphoton system, however, there is no wavepacket either broadened or unbroadened associated with the propagation of either signal or idler individually. It is the $G^{(2)}(\tau)$ function been broadened by the group dispersion media in 2-D space of $\tau = \tau_1 - \tau_2$. Comparing with a classical pulse propagating in group dispersion medium, we find that it is $G^{(2)}(\tau)$ in Eq. (6.6.3) plays the role

of a pulse, except propagating in the 2-D space of $\tau = \tau_1 - \tau_2$. The second-order correlation function $G^{(2)}(t_1, 0; t_2, 0)$ observed at the output plane of SPDC has been calculated earlier

$$G^{(2)}(t_1 - t_2) \propto \left| \mathcal{F}_{t_1 - t_2}\{f(\nu)\} \right|^2 \sim \delta(t_1 - t_2)$$

for a large sized $\chi^{(2)}$ material. The δ-function like second-order correlation function will be broadened by the dispersive media after a certain distance propagation, which is given in Eq. (6.6.3). The broadening process is similar to that of a laser pulse propagating in the same dispersive medium.[3]

(2) Nonlocal dispersion cancellation: one may have realized already an interesting aspect of the physics from Eq. (6.6.2). It is easy to see that any positive (negative) group-dispersion of the signal can be cancelled nonlocally by applying a negative (positive) dispersive medium in the path of the idler by achieving

$$k_s'' z_1 + k_i'' z_2 = 0. \tag{6.6.4}$$

This interesting behavior has been named "nonlocal two-photon dispersion cancellation". No classical interpretation seems to be possible to explain this effect. Suppose we have two weak light pulses propagating through the positive and negative group-dispersion media channels, respectively, to reach D_1 and D_2 for second-order correlation measurement. In the view of classical physics, each pulse will be broadened locally and independently by the dispersion medium, and thus causing a statistically broadened temporal

---

[3]Behaving like classical pulse in optical fiber, when the length of a group-dispersive medium is greater than its "characteristic dispersion length", $z_{\text{dis}} = \Delta\tau_0^2/2\pi k''$, where $\Delta\tau_0$ is the unbroadened initial width of $G^{(2)}$, the second-order correlation function $G^{(2)}(\tau)$ is found to be

$$G^{(2)}(\tau) \sim \left| f\left(\frac{\tau}{k_s'' z_1 + k_i'' z_2}\right) \right|^2,$$

i.e., achieving the same function as the spectrum of the SPDC. Furthermore, the width of $G^{(2)}(\tau)$, $\Delta\tau$, is proportional to the propagation distance along the dispersive medium

$$\Delta\tau = \Delta\tau_0 \frac{z}{z_{\text{dis}}}.$$

In the far-field-zone, i.e., $z > z_{\text{dis}}$, $\Psi(t_1, z_1; t_2, z_2)$ acquires a stable form which takes the same "shape" as the spectrum function, $f(\nu)$, and the width of $G^{(2)}(\tau)$ is linearly proportional to the length of the known dispersive medium. This effect has been observed experimentally by Valencia *et al.*

width of the second-order correlation function:

$$\sigma_{12} \geq \sqrt{\sigma_1^2 + \sigma_2^2}, \tag{6.6.5}$$

where σ_{12}, σ_1, and σ_2 are the broadened widths of the correlation function and the two propagating pulses.

Two-photon dispersion cancellation is useful in certain applications, such as nonlocal timing and positioning and distant clock synchronization, in which delta-function-like EPR correlation in $t_1 - t_2$ or in $z_1 - z_2$ is expected.

6.7   Entangled Coherent State

In Chapter 5, we have introduced an entangled coherent state that is generated by a multi-cavity-mode optical parametric oscillator (OPO). The state of the output field of the multi-cavity-mode OPO can be approximated as:

$$|\Psi\rangle = \prod_{s,i} \delta\left(\omega_s + \omega_i - \omega_p\right) \delta\left(\mathbf{k}_s + \mathbf{k}_i - \mathbf{k}_p\right) |\alpha_s(\mathbf{k}_s)\rangle |\alpha_i(\mathbf{k}_i)\rangle \tag{6.7.1}$$

where $|\alpha_s(\mathbf{k}_s)\rangle$ and $|\alpha_i(\mathbf{k}_i)\rangle$, $|\alpha| \gg 1$, are the state vectors of the signal-idler modes of the OPO in the coherent state representation; and ω_j and $\mathbf{k}_j$, for $j = s, i, p$, are the frequency and wavevector of the signal (s) modes, idler (i) modes, and the pump (p) laser beam. Considering each cavity mode a group of indistinguishable photons, this state indicates: (1) each group of indistinguishable photons represented by $|\alpha_s(\mathbf{k}_s)\rangle$ is entangled with another unique group of identical photons represented by $|\alpha_i(\mathbf{k}_i\rangle$ by means of energy conservation $\hbar\omega_s + \hbar\omega_i = \hbar\omega_p$ and momentum conservation $\hbar\mathbf{k}_s + \hbar\mathbf{k}_i = \hbar\mathbf{k}_p$; and (2) a large number of such signal-idler pairs are created and superposed coherently to form an entangled pair of signal-idler laser beams.

In the following, we derive Eq. (6.7.1). Similar to that of SPDC, the state of the signal-idler cavity-mode pair can be calculated, quantum mechanically, by the perturbation theory with the help of the nonlinear interaction Hamiltonian. The nonlinear interaction arises in a nonlinear material driven by a pump laser beam. The polarization, i.e., the dipole moment per unit volume, is given by

$$P_i = \chi_{i,j}^{(1)} E_j + \chi_{i,j,k}^{(2)} E_j E_k + \chi_{i,j,k,l}^{(3)} E_j E_k E_l + \cdots \tag{6.7.2}$$

where $\chi^{(m)}$ is the mth order electrical susceptibility tensor. In the case of bi-mode state, it is the second order nonlinear susceptibility $\chi^{(2)}$ that plays the role. The second-order nonlinear interaction Hamiltonian can be written as

$$H = \epsilon_0 \int_V d\mathbf{r} \ \chi_{ijk}^{(2)} \ E_i E_j E_k \tag{6.7.3}$$

where the integral is taken over the interaction volume, V.

It is convenient to use the Fourier representation for the electrical fields in Eq. (6.7.3):

$$\mathbf{E}(\mathbf{r}, t) = \int d\mathbf{k} \ \mathbf{E}(\mathbf{k}) e^{-i(\omega(\mathbf{k})t - \mathbf{k} \cdot \mathbf{r})} + h.c. \tag{6.7.4}$$

Substituting Eq. (6.7.4) into Eq. (6.7.3) and keeping only the terms of interest, we obtain the three-wave mixing Hamiltonian in the interaction representation:

$$\hat{H}_{\mathrm{int}}(t) = \epsilon_0 \int_V d\mathbf{r} \int d\mathbf{k}_s \, d\mathbf{k}_i \, \chi_{lmn}^{(2)} \hat{E}_{pl}^{(+)} e^{i(\omega_p t - \mathbf{k}_p \cdot \mathbf{r})}$$

$$\hat{E}_{sm}^{(-)} e^{-i(\omega_s(\mathbf{k}_s)t - \mathbf{k}_s \cdot \mathbf{r})} \hat{E}_{in}^{(-)} e^{-i(\omega_i(\mathbf{k}_i)t - \mathbf{k}_i \cdot \mathbf{r})} + h.c., \tag{6.7.5}$$

To simplify the calculation, we have also assumed the pump field to be plane and monochromatic with single wave vector $\mathbf{k}_p$ and frequency ω_p.

It is easily noticeable that the volume integration in Eq. (6.7.5) can be done for some simplified cases. At this point, to simplify the calculation, we assume that V is infinitely large. The interaction Hamiltonian in Eq. (6.7.5) is written as

$$\hat{H}_{\mathrm{int}}(t) = \epsilon_0 \int d\mathbf{k}_s \, d\mathbf{k}_i \, \chi_{lmn}^{(2)} \ \hat{E}_{pl}^{(+)} \hat{E}_{sm}^{(-)} \hat{E}_{in}^{(-)}$$

$$\times \, \delta(\mathbf{k}_p - \mathbf{k}_s - \mathbf{k}_i) e^{i(\omega_p - \omega_s(\mathbf{k}_s) - \omega_i(\mathbf{k}_i))t} + h.c. \tag{6.7.6}$$

It is reasonable to consider the pump field a classical wave (a laser beam), and quantize the signal and idler fields,

$$\hat{E}^{(-)}(\mathbf{k}) = i\sqrt{\frac{2\pi\hbar\omega}{V}} \hat{a}^\dagger(\mathbf{k}), \quad \hat{E}^{(+)}(\mathbf{k}) = -i\sqrt{\frac{2\pi\hbar\omega}{V}} \hat{a}(\mathbf{k})$$

where $a^\dagger(\mathbf{k})$ [$\hat{a}(\mathbf{k})$] is photon creation [annihilation] operator. The state of the created signal-idler mode pair from the first nonlinear interaction, i.e., SPDC, can be obtained by applying perturbation to the first order,

$$|\Psi\rangle \simeq -\frac{i}{\hbar} \int dt \, H_{\mathrm{int}}(t) \, |0\rangle. \tag{6.7.7}$$

By using vacuum $|0\rangle$ as the initial state, we assume that there is no input radiation in any signal and idler modes, which is reasonable for the first nonlinear interaction. Further assuming an infinite interaction time, evaluating the time integral and omitting altogether the constants and slow (square root) functions of ω, the state of a pair of signal-idler mode from the first nonlinear interaction, namely SPDC, is calculated to be:

$$|\Psi\rangle = \left[|0\rangle + a_s^\dagger(\omega_s)\, a_i^\dagger(\omega_i)\, |0\rangle\right]_{\omega_s+\omega_i=\omega_p} \tag{6.7.8}$$

where we assumed single transverse and longitudinal modes of signal, ω_s, and idler, ω_i. Not to introduce any confusion, we mark the phase matching condition with each mode pair as a parameter index.

Below, we simplify the calculation of the state generated in the OPO into three steps as an approximation:

(1) Consider the first nonlinear interaction, i.e., SPDC, creates a set of entangled mode pairs simultaneously,

$$\left[|0\rangle + a_s^\dagger(\omega_s)\, a_i^\dagger(\omega_i)\, |0\rangle\right]_{\omega_s+\omega_i=\omega_p}.$$

In the case of SPDC, since all signal-idler mode pairs must be phase matched with the monochromatic single mode pump, all created mode pairs of SPDC added constructively in an entangled and coherent manner form the state

$$|\Psi\rangle = \prod_{s,i} \left[|0\rangle + a_s^\dagger(\omega_s)\, a_i^\dagger(\omega_i)\, |0\rangle\right]_{\omega_s+\omega_i=\omega_p}. \tag{6.7.9}$$

Taking the first-order approximation, we have

$$|\Psi\rangle = \sum_{s,i} \delta\left(\omega_s + \omega_i - \omega_p\right) a_s^\dagger(\omega_s)\, a_i^\dagger(\omega_i)\, |0\rangle \tag{6.7.10}$$

which is a 1-D version of the SPDC biphoton state.

(2) Consider the second, third, ... nth nonlinear interactions, in the case of OPO, the *stimulated* parametric amplification add the *stimulatory* creations constructively. The state of each cavity-mode can be approximated as

$$|\Psi\rangle \simeq \left[|0\rangle + a_s^\dagger(\omega_s)\, a_i^\dagger(\omega_i)\, |0\rangle\right]^n_{\omega_s+\omega_i=\omega_p}$$

$$\simeq |0\rangle + \frac{n}{1!}\, a_s^\dagger(\omega_s)\, a_i^\dagger(\omega_i)|0\rangle$$

$$+ \frac{n(n-1)}{2!}\, [a_s^\dagger(\omega_s)\, a_i^\dagger(\omega_i))]^2|0\rangle$$

$$+ \frac{n(n-1)(n-2)...(n-m+1)}{m!}\, [a_s^\dagger(\omega_s)\, a_i^\dagger(\omega_i)]^m|0\rangle$$

$$+ \cdots$$

$$\simeq \left[|\alpha_s(\omega_s)\rangle|\alpha_i(\omega_i)\rangle\right]_{\omega_s+\omega_i=\omega_p}. \tag{6.7.11}$$

(3) Since all signal-idler mode pairs must be phase matched with the monochromatic single mode pump, all the *stimulatory created* mode pairs added constructively in an entangled and coherent manner form the state of Eq. (6.7.1) in 1-D

$$|\Psi\rangle = \prod_{s,i} \delta\left(\omega_s + \omega_i - \omega_p\right) |\alpha_s(\omega_s)\rangle|\alpha_i(\omega_i)\rangle, \qquad (6.7.12)$$

indicating a vector in the space of coherent state.

Based on the entangled coherent states of Eq. (6.7.1), we have calculated the second-order coherence function $G^{(2)}(\mathbf{r}_1, t_1; \mathbf{r}_2, t_2)$ in Chapter 5 by assuming a simple measurement in which the signal beam and the idler beam, respectively, are directed to two point-like photodetectors D_1 and D_2. These detectors measure the intensities of the signal beam and the idler beam, respectively, and are used to measure the intensity correlation between the signal beam and the idler beam, jointly. It is no surprise that both D_1 and D_2 observe constant intensities, respectively, as expected from a CW laser beam. However, surprisingly, the calculation yielded a 100% contrast comb-like function of $\tau = \tau_1 - \tau_2 = (t_1 - t_2) - (r_1 - r_2)/c$,

$$G^{(2)}(\tau) = \left|\mathcal{F}_\tau\{\alpha^2(\omega)\}\right|^2 \left[\frac{\sin^2 N\omega_b\tau/2}{\sin^2 \omega_b\tau/2}\right]. \qquad (6.7.13)$$

If we assume that each cavity mode has a constant distribution function $a(\omega)$ within its spectral bandwidth $\Delta\omega$, and for all cavity-modes,

$$G^{(2)}(\tau) = \left[\text{sinc}^2\frac{\Delta\omega\tau}{2}\right] \left[\frac{\sin^2 N\omega_b\tau/2}{\sin^2 \omega_b\tau/2}\right] \qquad (6.7.14)$$

where $\Delta\omega$ is the spectral bandwidth, or line-width, of the cavity-mode. We name this comb-function quantum ghost frequency comb (QGFC) due tom its 100% contrast. A numerical simulated QGFC has been illustrated in Fig. 5.7.2 of Chapter 5. Similar GFCs of 50% contrast have been experimentally demonstrated from intensity fluctuation correlation measurement of a fiber laser beam with half a million cavity-modes.

Similar to the 50% contrast GFC observed from the correlation measurement of multi-cavity-mode laser beams, this 100% contrast QGFC also has multipole periodic correlation peaks at $\omega_b\tau/2 = n\pi$, for $n = 0, \pm1, \pm2, \cdots, \pm(N-1)$,

$$(t_1 - t_2)_n = \frac{1}{\nu_b}n + (r_1 - r_2)/c \qquad (6.7.15)$$

where we have used $\omega_b = 2\pi\nu_b$ and defined $(t_1 - t_2)_n$ as the measured value of $t_1 - t_2$ at the nth comb peak. The temporal width of each QGFC peak, measured between neighboring zeros of the correlation function, is approximately

$$\Delta t \simeq \frac{1}{\nu_b N}. \tag{6.7.16}$$

Indeed it is surprising that a 100% contrast comb-like function is observable from the joint-detection of the entangled signal-idler laser beams, although both detectors, D_1 and D_2, measure constant light intensities, consistent with the expectation of CW laser beams. We are facing two serious questions: (1) What is the cause of these sharp periodic correlation peaks? (2) Can we trust zero-coincidence observations from a direct measurement of CW laser beams?

We know that most of the observed correlation functions of entangled photon pairs, either created from atomic cascade decay or generated from SPDC, are Gaussian-like functions with a single peak at $\tau_1 - \tau_2 = 0$. A widely accepted idea is that this is due to the simultaneous generation of entangled photon pairs at the source. Since the entangled photons are produced simultaneously in the source, they must be jointly detected by the detectors at $\tau_1 - \tau_2 = 0$. Following this line of thought, can we assume that the entangled signal-idler laser beams are generated simultaneously and periodically by the CW OPO? Not likely! CW OPOs do not work that way.

The contrast of the QGFC is 100%, which means that no joint-detection event occurs whenever the relative delay of $\tau_1 - \tau_2$ falls into the region between the periodic sharp correlation peaks. The non-zero coincidences of sharp correlation peaks only occur within each precise and narrow time windows. From the perspective of classical thinking, it is quite possible for zero-coincidence to occur in the joint detection of single photons, but it is absolutely impossible for zero-coincidence to occur in the measurement of CW laser beams. Even if somehow there are chances to observe zero-coincidences, why are they periodic in such a peculiar manner?

The QGFC is the result of a standard calculation based on the quantum theory of nonlocal superposition, similar to what Einstein–Podolsky–Rosen did in 1935. The periodic QGFC looks like a multi-slit diffraction pattern. The comb-like multi-slit diffraction pattern is the result of coherent superposition of a large number of radiation fields coming from the narrow multi-slits. The "zeros" and periodic sharp peaks of the multi-slit diffraction pattern are the results of destructive and constructive interferences. The QGFC is the result of coherent superposition of a large number of quantum amplitudes, corresponding to the nonlocal joint-measurement of a large number of entangled signal-idler pairs. The "zeros" and periodic peaks of the QGFC are also the results of destructive and constructive interferences, except that this interference occurs at distant space-time coordinates (z_1, t_1) and (z_2, t_2), and the two photodetection events can be easily managed to achieve space-like separation.

Can we trust zero-coincidence observations made by directly measuring CW laser beams? From the perspective of classical theory, the answer is no. From the perspective of quantum mechanics, the answer is yes. Apparently, the observation of QGFC is strong evidence of quantum nonlocal superposition.

It is worth noting that the above discussions are based on idealized conditions. The experimentally observed QGFC contrast may drop below 100% due to imperfect phase matching and partially coherent states caused by photon losses.

6.8   Bell State and Bell's Inequality

In the early 1950s, Bohm simplified the 1935 EPR state from continuous coordinate-momentum space to discrete spin space by introducing a singlet state of two spin $1/2$ particles:

$$|\Psi\rangle = \frac{1}{\sqrt{2}} \left[|\uparrow\rangle_1 |\downarrow\rangle_2 - |\downarrow\rangle_1 |\uparrow\rangle_2 \right], \qquad (6.8.1)$$

where the kets $|\uparrow\rangle$ and $|\downarrow\rangle$ represent states of spin "up" and spin "down", respectively, along an *arbitrary* direction. For the EPR–Bohm state, the spin of neither particle is determined; however, if one particle is measured to be spin up along a certain direction, the other one must be spin down along that direction, despite the distance between the two spin $1/2$ particles. We have shown in Chapter 4, Eq. (6.8.1) is independent of the choice of the spin directions. The introduction of the EPR–Bohm state simplified the physical picture and the discussion about quantum entanglement dramatically.[4]

A more practical example of the EPR–Bohm state concerns the polarization states of photon pairs, such as the spin-zero state of a high energy photon pair disintegrated from the annihilation of positronium. Suppose initially we have a positron and an electron in the spin-zero state with antiparallel spins. The positronium cannot exist very long: it disintegrates into two γ-ray photons within $\sim 10^{-10}$ second of its lifetime. The spin zero state is symmetric under all rotations. Therefore, the photon pair may be disintegrated into any direction in space with equal probability.

---

[4]But, one should keep in mind that it is not a necessary to have "two terms" in an entangled state. In fact, the first entangled state suggested by EPR in 1935 has infinite number of "terms".

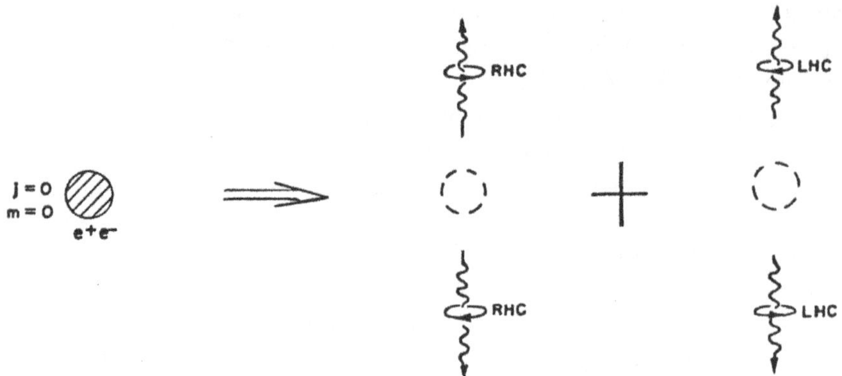

Fig. 6.8.1   Annihilation of Positronium. Due to the conservation of angular momentum, if photon 1 is right-hand circular (RHC) polarized, photon 2 must be right-hand circular polarized. If photon 1 is left-hand circular (LHC) polarized, then photon 2 has to be left-hand circular polarized.

The conservation of linear momentum, however, guarantees that if one of the photon is observed in a certain direction, its twin must be found in the opposite direction (with finite uncertainty $\Delta(\mathbf{p}_1 + \mathbf{p}_1) \neq 0$). The conservation of angular momentum will decide the polarization state of the photon pair. As shown in Fig. 6.8.1, in order to keep spin-zero, if photon 1 is right-hand circular polarized (RHC), photon 2 must be also right-hand circular polarized. The same argument shows that if photon 1 is left-hand circular (LHC) polarized, then photon 2 has to be left-hand circular polarized too. Therefore, the positronium may decay into two RHC photons or two LHC photons with equal probability.

What is the relationship between these two alternatives? If it is just two different ways of disintegration with equal probability, the two photon system may be described by a density matrix,

$$\hat{\rho} = \frac{1}{2} \left[\, | \, R_1 R_2 \, \rangle \langle \, R_1 R_2 \, | + | \, L_1 L_2 \, \rangle \langle \, L_1 L_2 \, | \, \right], \qquad (6.8.2)$$

where $| \, R_j \, \rangle$ and $| \, L_j \, \rangle$ indicate the RHC and LHC states for photon j, respectively. The density matrix specify only the statistics: within N γ-ray photon pairs, 50% of them have RHC–RHC polarization and another 50% have LHC–LHC polarization. The physical process of positronium annihilation, however, is not that simple. The law of parity conservation must be satisfied in the disintegration: the spin-zero ground state of positronium holds an odd parity. Thus, the state of the photon pair must keep its parity odd:

$$|\Psi\rangle = \frac{1}{\sqrt{2}}\big[\,|\,R_1\rangle|\,R_2\rangle - |\,L_1\rangle|\,L_2\rangle\,\big]. \tag{6.8.3}$$

It is the conservation of parity made the two-photon state very special. The two-photon state is a nonfactorizeable pure state of a special superposition between the RHC and LHC states specified with a relative phase of π. Mathematically, "nonfactorizeable" means that the state cannot be written as a product state of photon 1 and photon 2. Physically, it means that photon 1 and photon 2 are not independent anymore. The two γ-ray photons are in an entangled polarization state, or spin state.

The polarization correlation measurement on two γ-ray photons is not an easy job. We do not have suitable polarizers for γ-ray yet. Is it possible to produce a two-photon EPR–Bohm state of Eq. (6.8.3) in lower energy? In 1986, Alley and Shih published an experiment of "New Type of Einstein–Podolsky–Rosen-Bohm Experiment Using Pairs of Light Quanta Produced by Optical Parametric Down Conversion". The entangled signal–idler pairs of light quanta were in visible spectrum. A schematic setup of the Alley-Shih experiment is shown in Fig. 6.8.2. In this experiment, a pair of linearly polarized signal–idler photons at wavelength $\lambda = 532$ nm was created from a type-I SPDC. The linearly polarized signal and idler photons are rotated by two quarter wave-plates to right-hand circular polarization, respectively, and injected onto a 50%-50% beamsplitter from opposite sides. The signal–idler pair has two indistinguishable alternatives to trigger a joint photode-tection event of D_1 and D_2: (1) both are transmitted at the beamsplitter, keeping their right-hand circular polarization; (2) both are reflected at the beamsplitter, changing to left-hand circular polarization. A π phase shift is introduced when the reflection is taken from low index (air) to high index (glass) at the beamsplitter. The coherent superposition of the above two-photon amplitudes produces the EPR–Bohm–Bell state of Eq. (6.8.3).[5]

---

[5](1) The reason to inject the signal–idler pair at the beamsplitter near normally was for achieving 50%–50% transmission and reflection for both S and P polarization components. A near-normal incident beamsplitter was the only choice in the 1980s. In fact, it is unnecessary to adapt the near-normal design for other types of correlation measurement such as demonstrating anti-correlation "dip". (2) Achieving coherent superposition of the two-photon amplitudes, the superposed biphoton wavepackets must be completely overlapped in space–time, which means a balanced interferometer with equal optical path. During the alignment of the beamsplitter, a correlation "peak" and an anti-correlation "dip" were observed. It was interesting to find that the "correlation" can be easily turned into "anti-correlation", and wise versa, by simply rotate one of the polarization analyzers, A_1 or A_2, from θ_j to $\theta_j + 90°$, $j = 1, 2$.

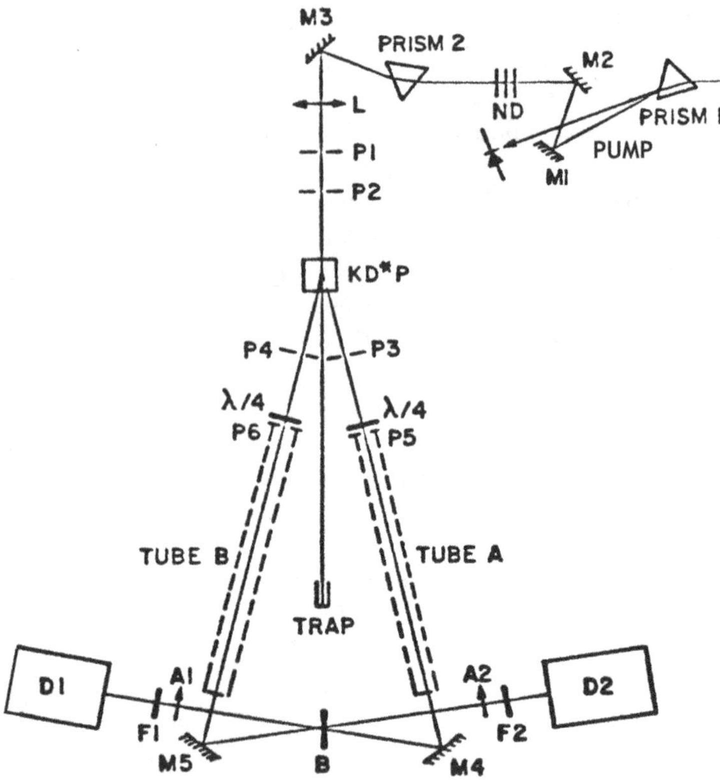

Fig. 6.8.2 Alley–Shih interferometer: successfully produced the EPR–Bohm–Bell state of Eq. (6.8.3) in visible spectrum. This experiment introduced the entangled signal–idler biphoton of SPDC into the experimental study of Bell state and Bell's inequality. In this experiment, Alley-Shih observed a typical Bell correlation of polarization and a violation of a Bell's inequality. In addition, a two-photon "correlation", and an two-photon "anti-correlation" were observed when align the beamsplitter. The state was prepared as that of Eq. (6.8.3), a superposition of two-photon state in circular polarization; however, the correlation measurement was for linear polarization.

The physics behind Eq. (6.8.3) is very interesting. However, exploring the peculiarities of the entangled two-photon state of Eq. (6.8.3) is not so "straightforward" and can even be misleading. One may easily find the following popular statement about this state: the polarization state of photon 1 and photon 2 are both undefined: each has 50%-50% chance to be RHC or LHC; however, the polarization state of photon 1 can be predicted with certainty through the measurement of photon 2: whichever circular polarization state photon 2 is observed, photon 1 must be in the same state, despite the distance between them. In other words, one would always observe two RHC or two LHC polarized photons in a joint detection event:

$$\begin{aligned}
|\langle R_1\,R_2\,|\,\Psi\,\rangle|^2 &= |\langle L_1\,L_2\,|\,\Psi\,\rangle|^2 = 50\% \\
|\langle R_1\,L_2\,|\,\Psi\,\rangle|^2 &= |\langle L_1\,R_2\,|\,\Psi\,\rangle|^2 = 0.
\end{aligned} \tag{6.8.4}$$

The result of Eq. (6.8.4) is interesting, but not surprising. In fact, a statistical ensemble of photons characterized by the density matrix of Eq. (6.8.2) would give the same result. The above statement, which is based on the correlation of circular polarization, has not explored the real interesting physics behind the *pure state* of Eq. (6.8.3).

Can we distinguish the pure state of Eq. (6.8.3) from the statistical mixture of Eq. (6.8.2) through a simple polarization correlation measurement? The answer is positive. In the early 1980s, Alley and Shih measured the linear polarization correlation, instead of measuring circular polarization. In the Alley-Shih experiment, see Fig. 6.8.2, the two-photon state was prepared as that in Eq. (6.8.3) in circular polarization; however, the correlation measurement was for linear polarization. Each linear polarization analyzer has two orthogonal output channels, followed by two detector assemblies, marked as D_1 and D_2, in the figure. Four photodetectors, D_1-D_1' and D_2-D_2', are placed at the four output ports of the two linear polarization analyzers for two-fold joint detections. There are four possible coincidence combinations: D_1 with D_2, D_1 with D_2', D_1' with D_2, and D_1' with D_2'. Suppose we choose the polarization analyzers along $|X\rangle$ and $|Y\rangle$, the measurement on the two-photon pure state of Eq. (6.8.3) yields:

$$\begin{aligned}
|\langle X_1\,Y_2\,|\,\Psi\,\rangle|^2 &= |\langle Y_1\,X_2\,|\,\Psi\,\rangle|^2 = 50\% \\
|\langle X_1\,X_2\,|\,\Psi\,\rangle|^2 &= |\langle Y_1\,Y_2\,|\,\Psi\,\rangle|^2 = 0,
\end{aligned} \tag{6.8.5}$$

i.e., one would always observe two orthogonal polarized photons in a joint detection event. The measurement on the statistical mixture of Eq. (6.8.2), however, gives a completely random result:

$$\begin{aligned}
tr\,\hat{\rho}\,|\,Y_2\,X_1\rangle\langle X_1\,Y_2\,| &= tr\,\hat{\rho}\,|\,X_2\,Y_1\rangle\langle Y_1\,X_2\,| = 25\% \\
tr\,\hat{\rho}\,|\,X_2\,X_1\rangle\langle X_1\,X_2\,| &= tr\,\hat{\rho}\,|\,Y_2\,Y_1\rangle\langle Y_1\,Y_2\,| = 25\%.
\end{aligned} \tag{6.8.6}$$

Furthermore, we may rotate the linear polarization analyzers to any direction in the XY plane, such as θ_1 ($\bar{\theta}_1 = \pi/2 - \theta$ in its orthogonal channel) and θ_2 ($\bar{\theta}_2 = \pi/2 - \theta$ in its orthogonal channel) for the measurement of polarization correlation, where θ_j is the angle between the polarization analyzer and the X axis, and is defined in the right-hand coordinate system following the right-hand convention. The measurement on the two-photon pure state of Eq. (6.8.3) yields

$$| \langle \theta_1\, \theta_2 \,|\, \Psi \rangle |^2 = \frac{1}{2}\, \sin^2(\theta_1 + \theta_2) = \frac{1}{2}\, \sin^2 \varphi, \qquad (6.8.7)$$

where $\varphi = \theta_1 + \theta_2$. Eq. (6.8.7) indicates that the pair must be always orthogonally polarized, independent of the choice of θ_j for each individual polarization analyzer. Considering the other three possible joint detections, we have the following polarization correlation functions in terms of $\varphi = \theta_1 + \theta_2$:

$$| \langle \theta_1\, \theta_2 \,|\, \Psi \rangle |^2 = | \langle \bar{\theta}_1\, \bar{\theta}_2 \,|\, \Psi \rangle |^2 = \frac{1}{2}\, \sin^2 \varphi$$

$$| \langle \theta_1\, \bar{\theta}_2 \,|\, \Psi \rangle |^2 = | \langle \bar{\theta}_1\, \theta_2 \,|\, \Psi \rangle |^2 = \frac{1}{2}\, \cos^2 \varphi. \qquad (6.8.8)$$

On the other hand, the measurement on the statistical mixture of Eq. (6.8.2) gives, again, a completely random result:

$$tr\,\hat{\rho}\,|\,\theta_2\,\theta_1\rangle\langle\theta_1\,\theta_2\,| = tr\,\hat{\rho}\,|\,\bar{\theta}_2\,\bar{\theta}_1\rangle\langle\bar{\theta}_1\,\bar{\theta}_2\,| = 25\%$$

$$tr\,\hat{\rho}\,|\,\bar{\theta}_2\,\theta_1\rangle\langle\theta_1\,\bar{\theta}_2\,| = tr\,\hat{\rho}\,|\,\theta_2\,\bar{\theta}_1\rangle\langle\bar{\theta}_1\,\theta_2\,| = 25\%. \qquad (6.8.9)$$

Now, we are ready to describe the interesting physics behind Eq. (6.8.3) with a better statement: the entangled two-photon polarization state of Eq. (6.8.3) has specified a peculiar two-photon system, in this system, the polarization of neither photon is defined during the course of its propagation; however, if one of the photon is measured to be in a defined polarization state, the polarization state of the other photon is determined with certainty, despite the distance between the pair and *despite the choice of the polarization vector base.*

This statement is similar to the statement we have used for the original EPR state: in an EPR two-particle system, neither *position* nor *momentum* for neither particle is defined during the course of its propagation; however, if one of the particle is measured to be in a defined *position* or *momentum*, the *position* and *momentum* of the other particle is determined with certainty, despite the distance between the two particles.

The independence of the choice for polarization vector base in Eq. (6.8.3) is equivalent to the independence of the choice of position and momentum coordinate for the original EPR state.

Regardless of whether we measure it or not, does a free propagating particle have a defined spin in the Bohm state of Eq. (6.8.1)? Does a free propagating photon have a defined polarization, either circular or linear, in the state of Eq. (6.8.3)? On one hand, the spin of neither independent particle is specified in the Bohm state of Eq. (6.8.1), and the polarization of neither independent photon is specified in the state of Eq. (6.8.3), we may have to believe that the particles do not have any defined spin, and the photons do not have any defined polarization, during the course of their propagation. On the other hand, if the spin of one particle, or the polarization of one photon, uniquely determines the spin of the other distant particle, or the polarization of the other photon, it would be hard for anyone who believes no action-at-a-distance to imagine that the spin of the two particles, or the polarization of the two photons, are not predetermined with defined values before the measurement. The Bohm state thus put us into a paradoxical situation. It seems reasonable for us to ask the same question that EPR had asked in 1935: "Can quantum-mechanical description of physical reality be considered complete?"

Is it possible to have a realistic theory which provides correct predictions of the behavior of a particle similar to quantum theory and, at the same time, respects the description of physical reality by EPR as "complete"? Bohm and his followers have attempted a "hidden variable theory" to formulate the physical reality into the wavefunction of a particle or a pair of particles. The theory seemed to be consistent with quantum mechanics and satisfied the requirements of EPR. The hidden variable theory was successfully applied to many different quantum phenomena until 1964, when Bell proved a theorem to show that an inequality, which is violated by certain quantum mechanical statistical predictions, can be used to distinguish local hidden variable statistics from quantum superpositions. Since then, the testing of Bell's inequalities became a standard instrument for the study of fundamental problems of quantum theory. The experimental testing of Bell's inequality started from the early 1970s. Most of the historical experiments concluded the violation of the Bell's inequalities and thus disproved the local hidden variable theory.[6] The topics of hidden

---

[6]Although Bell's theorem and Bells' inequality involve deep philosophical concerns in general, we should keep in mind that the testing of Bell's inequality in terms of the physical measurement on a physical system is essentially a physical problem. Any proposed statistical model of classical correlation, either local or nonlocal, must obey the basic physical laws of classical theory. One cannot assume a classical correlation that is never allowed in classical theory.

variable theory and Bell's inequality are not the topics of this book. In the end of this chapter, we introduced a simple hidden variable theory for the spin 1/2 Bohm state, and derive the Bell's inequalities of 1964 and 1972 as two appendixes:

Appendix (I) Hidden variable theory for the spin 1/2 Bohm state.

Appendix (II) Bell's theorem and Bell's inequality.

In the rest of this chapter, we discuss Bell states and their preparation.

(I) **Bell States**:

The following set of four two-photon spin states can be generated by manipulating the polarization state of the signal–idler pair of SPDC with beamsplitters, half and/or quarter wave-plates:

$$| \Psi_{RL}^{\pm} \rangle = \frac{1}{\sqrt{2}} \big[| R_1 \rangle | R_2 \rangle \pm | L_1 \rangle | L_2 \rangle \big]$$

$$| \Phi_{RL}^{\pm} \rangle = \frac{1}{\sqrt{2}} \big[| R_1 \rangle | L_2 \rangle \pm | L_1 \rangle | R_2 \rangle \big].$$

$$(6.8.10)$$

The four states in Eq. (6.8.10), named as EPR–Bohm–Bell states, or simply "Bell states", form a complete orthonormal basis in two-particle vector space. In principle, any arbitrary two-photon polarization state can be expressed as an appropriate superposition of Bell states. Bell states have been extensively involved in studying fundamental issues of quantum theory as well as in practical applications of quantum entanglement. In the field of quantum information, Bell states are usually represented in the following form of "qubit":

$$| \Psi_{01}^{(\pm)} \rangle = \frac{1}{\sqrt{2}} \big[| 0_1 1_2 \rangle \pm | 1_1 0_2 \rangle \big]$$

$$| \Phi_{01}^{(\pm)} \rangle = \frac{1}{\sqrt{2}} \big[| 0_1 0_2 \rangle \pm | 1_1 1_2 \rangle \big],$$

$$(6.8.11)$$

where $|0\rangle$ and $|1\rangle$ represent two arbitrary orthogonal polarization bases. The most popular Bell states are prepared from SPDC in linear polarization representation:

$$| \Psi_{XY}^{\pm} \rangle = \frac{1}{\sqrt{2}} \big[| X_1 \rangle | Y_2 \rangle \pm | Y_1 \rangle | X_2 \rangle \big]$$

$$| \Phi_{XY}^{\pm} \rangle = \frac{1}{\sqrt{2}} \big[| X_1 \rangle | X_2 \rangle \pm | Y_1 \rangle | Y_2 \rangle \big],$$

$$(6.8.12)$$

where $|X\rangle$ and $|Y\rangle$, respectively, are defined by the polarization of the o-ray and the e-ray of the nonlinear crystal of SPDC. Treating the circular polarization state as the superposition of the linear polarization states,

$$|R\rangle = |X\rangle + i|Y\rangle$$
$$|L\rangle = -|X\rangle + i|Y\rangle,$$
(6.8.13)

we may find the following states exchangeable between the linear polarization representation and the circular polarization representation:

$$|\Psi_{XY}^{+}\rangle = |\Psi_{RL}^{-}\rangle, \quad |\Psi_{XY}^{-}\rangle = |\Phi_{RL}^{-}\rangle;$$
$$|\Phi_{XY}^{+}\rangle = |\Phi_{RL}^{+}\rangle, \quad |\Phi_{XY}^{-}\rangle = |\Psi_{RL}^{+}\rangle.$$
(6.8.14)

To simplify the notation, we will ignore the subscription for the linear polarization representation to replace $|\Psi_{XY}^{\pm}\rangle$ and $|\Phi_{XY}^{\pm}\rangle$ with $|\Psi^{\pm}\rangle$ and $|\Phi^{\pm}\rangle$, respectively. The polarization correlation functions, in terms of chosen orientations of the polarization analyzers, θ_1 and θ_2, for the Bell states of Eq. (6.8.12) can be easily calculated:

$$|\langle\theta_1\,\theta_2\,|\,\Psi^{\pm}\rangle|^2 = \frac{1}{2}\,\sin^2(\theta_1 \pm \theta_2)$$
$$|\langle\theta_1\,\theta_2\,|\,\Phi^{\pm}\rangle|^2 = \frac{1}{2}\,\cos^2(\theta_1 \mp \theta_2).$$
(6.8.15)

The other correlation functions in terms of $\bar{\theta}$s or in terms of θ and $\bar{\theta}$ can be derived accordingly based on Eq. (6.8.15).

It is noticed that the condition for achieving maximum correlation, in terms of θ_1 and θ_2, in $|\Psi^{+}\rangle$ ($|\Psi^{-}\rangle$) and in $|\Phi^{+}\rangle$ ($|\Phi^{-}\rangle$) are different: $\theta_1 + \theta_2 = \pi/2$ against $\theta_1 - \theta_2 = \pi/2$ ($\theta_1 - \theta_2 = 0$ against $\theta_1 + \theta_2 = 0$). Thus, if one of them implies orthogonal (parallel) polarization of the photon pair, the other one must imply a different geometrical relationship between the two polarization.

Now, we are ready to discuss the rotation symmetry of Bell states. We have mentioned earlier that while preparing Bell states in SPDC, the polarization vector base of $|X\rangle$ and $|Y\rangle$ are usually defined by the chosen orientation of the nonlinear crystal and the associated wave-plates. What happen if we rotate $|X\rangle$-$|Y\rangle$ vector base to $|X'\rangle$-$|Y'\rangle$ with angle θ? It is easy to find that in each set of Eq. (6.8.10) and Eq. (6.8.12), only two of the states reserve the symmetry of rotation. In a right-hand coordinate system with anti-parallel propagation, i.e., the pair propagate to opposite directions, which is illustrated in the upper part of Fig. 6.8.3, the two Bell

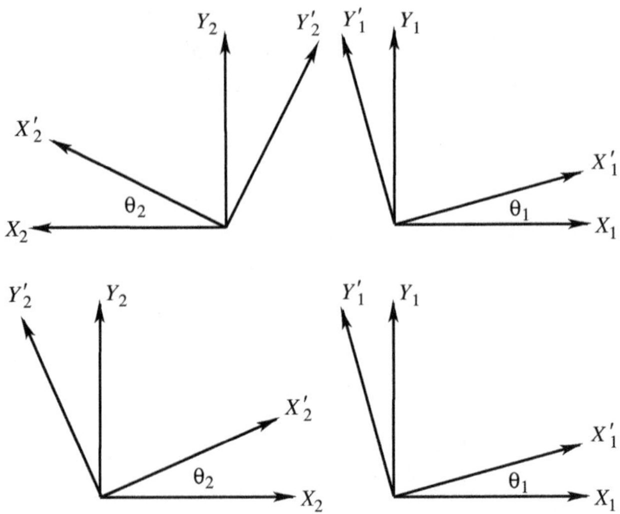

Fig. 6.8.3 Upper: Right hand coordinate system with anti-parallel propagation. The Z_1 (Z_2) axis is point out from (into) the paper. The relative angle between X_1-Y_1 and X_2-Y_2 are $\theta_2 + \theta_1$ after the rotation. Lower: Right hand coordinate system with collinear or near collinear propagation. Both Z_1 and Z_2 axes are point out the paper. The relative angle between X_1-Y_1 and X_2-Y_2 are $\theta_2 - \theta_1$ after the rotation.

states which reserve rotation symmetry are:

$$|\Psi^+\rangle = \frac{1}{\sqrt{2}}\left[\,|X_1'\rangle|Y_2'\rangle + |Y_1'\rangle|X_2'\rangle\,\right]$$
$$|\Phi^-\rangle = \frac{1}{\sqrt{2}}\left[\,|X_1'\rangle|X_2'\rangle - |Y_1'\rangle|Y_2'\rangle\,\right].$$

$$(6.8.16)$$

In a right-hand coordinate system with collinear or near collinear propagation, i.e., the pair propagate to the same direction, which is illustrated in the lower part of Fig. 6.8.3, the two Bell states which reserve rotation symmetry are:

$$|\Psi^-\rangle = \frac{1}{\sqrt{2}}\left[\,|X_1'\rangle|Y_2'\rangle - |Y_1'\rangle|X_2'\rangle\,\right]$$
$$|\Phi^+\rangle = \frac{1}{\sqrt{2}}\left[\,|X_1'\rangle|X_2'\rangle + |Y_1'\rangle|Y_2'\rangle\,\right].$$

$$(6.8.17)$$

It might be easier to figure out the results in Eqs. (6.8.16) and (6.8.17) from the circular polarization based states of Eq. (6.8.10). If the rotation from X-Y to X'-Y' means a rotation of $\theta_1 = \theta$ in arm-1 and $\theta_2 = -\theta$

in arm-2, which is illustrated in the upper part of Fig. 6.8.3, $|\Psi^{\pm}_{RL}\rangle$ will reserve rotation symmetry:

$$|\Psi^{\pm}_{RL}\rangle = \frac{1}{\sqrt{2}}\left[|R_1\rangle|R_2\rangle \pm |L_1\rangle|L_2\rangle\right]$$

$$= \frac{1}{\sqrt{2}}\left[e^{i(\theta_1+\theta_2)}|R'_1\rangle|R'_2\rangle \pm e^{-i(\theta_1+\theta_2)}|L'_1\rangle|L'_2\rangle\right]$$

$$= \frac{1}{\sqrt{2}}\left[|R'_1\rangle|R'_2\rangle \pm |L'_1\rangle|L'_2\rangle\right], \tag{6.8.18}$$

corresponding to Eq. (6.8.16). However, if the rotation from X-Y to X'-Y' means a rotation of $\theta_1 = \theta_2 = \theta$ in both arm-1 and arm-2, which is illustrated in the lower part of Fig. 6.8.3, $|\Phi^{\pm}_{RL}\rangle$ will reserve rotation symmetry:

$$|\Phi^{\pm}_{RL}\rangle = \frac{1}{\sqrt{2}}\left[|R_1\rangle|L_2\rangle \pm |L_1\rangle|R_2\rangle\right]$$

$$= \frac{1}{\sqrt{2}}\left[e^{i(\theta_1-\theta_2)}|R'_1\rangle|L'_2\rangle \pm e^{-i(\theta_1-\theta_2)}|L'_1\rangle|R'_2\rangle\right]$$

$$= \frac{1}{\sqrt{2}}\left[|R'_1\rangle|L'_2\rangle \pm |L'_1\rangle|R'_2\rangle\right], \tag{6.8.19}$$

corresponding to Eq. (6.8.17).

(II) **Preparation of Bell states from biphoton state of SPDC:**

SPDC has been one of the most convenient two-photon sources for the preparation of Bell state since the early 1980s. Although Bell state is for polarization (or spin), the space–time part of the state cannot be ignored. One important "preparation" is to make the two biphoton wavepackes, corresponding to the first and the second terms in the Bell state, completely "overlap" in space–time, or indistinguishable for the joint detection event. This is especially important for type-II SPDC. Type-II SPDC seems easier to use for the preparation of Bell state due to the orthogonal polarization of the signal and idler.

A very interesting situation for type-II SPDC is that of "noncollinear phase matching". The signal–idler pair are emitted from a SPDC crystal, such as BBO, cut in type-II phase matching, into two cones, one ordinarily polarized, the other extraordinarily polarized, see Fig. 6.8.4. Along the intersection, where the cones overlap, two pinholes numbered 1 and 2 are used for defining the direction of the **k** vectors of the signal–idler pair. It is

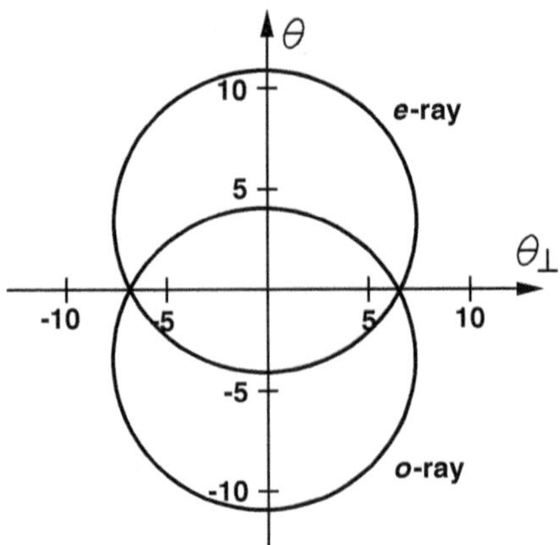

Fig. 6.8.4 Type-II noncolinear phase matching: A cross-section view of the degenerate 702.2 nm cones. The 351.1 nm pump beam is in the center. The numbers along the axes are in degrees.

very reasonable to consider the polarization state of the signal–idler pair as

$$|\Psi\rangle = \frac{1}{\sqrt{2}}(|o_1 e_2\rangle + |e_1 o_2\rangle) = |\Psi_{XY}^+\rangle, \qquad (6.8.20)$$

where o_j and e_j, $j = 1, 2$, are ordinarily and extraordinarily polarization, respectively. It seems straightforward to realize an EPR–Bohm–Bell measurement by simply setting up a polarization analyzer in series with a photon counting detector behind pinholes 1 and 2, respectively, and to expect observe the polarization correlation of Eq. (6.8.15). This is, however, *incorrect!* One can never observe the EPR–Bohm–Bell polarization correlation unless a "compensator" is applied. The "compensator" is a piece of birefringent material. For example, one may place another piece of nonlinear crystal behind the SPDC. It could be the same type of crystal as that of the SPDC, with the same cutting angle, except having half the length and a 90° rotation in respect to that of the SPDC crystal.

What is the role of the "compensator"? There have been naive explanations about the compensator. One suggestion was that the problem comes from the longitudinal "walk-off" of the type-II SPDC. For example, if one uses a type II BBO, which is a negative uni-axis crystal, the extraordinary ray propagates faster than the ordinary-ray inside the BBO. *Suppose* the $o - e \leftrightarrow e - o$ pair is generated in the *middle* of the crystal, the e-polarization will trigger the detector earlier than the o-polarization by a time $\Delta t = (n_o - n_e)L/2c$. This implies that D_2 would be fired first in $|o_1 e_2\rangle$ term; but D_1 would be fired first in $|e_1 o_2\rangle$ term. If Δt is greater than the coherence length of the signal–idler field, one would be able to distinguish which amplitude gave rise to the "click-click" coincidence event. One may compensate the "walk-off" by introducing an additional piece of birefringent material, like the compensator we have suggested above, to delay the e-ray relative to the o-ray by the same amount of time, Δt. If, however, the signal–idler pair is generated in the *front face* or the *back face* of the SPDC, the delay time would be very different: $\Delta t = (n_o - n_e)L/c$ for the *front face* and $\Delta t = 0$ for the *back face*. One can never satisfy all the pairs which are generated at different places along the SPDC crystal. Nevertheless, since SPDC is a *coherent* process, the signal–idler pair is generated in such a way that it is impossible to know the birth place of the pair. So, how is the delay time Δt determined?

Quantum mechanics has provided the correct answer. It is the overlap or indistinguishability of the biphoton wavepackets in space–time.

The Bell state of Eq. (6.8.20) as well as Eqs. (6.8.10) and (6.8.12) are written in the simplified Schrödinger representation. It is sufficient for certain types of discussions of physics. The simplified form, however, has ignored the space–time aspect of the state, which is important here.

Adopting our earlier result, we may rewrite the state of the signal–idler pair of Fig. 6.8.4 in the following form:

$$|\Psi\rangle = \sum_{\mathbf{k}_o, \mathbf{k}_e} \delta(\omega_o + \omega_e - \omega_p)\Phi(\Delta_k L)\,\hat{\mathbf{o}}\,\hat{a}_o^\dagger(\omega(\mathbf{k}_o))\,\hat{\mathbf{e}}\,\hat{a}_e^\dagger(\omega(\mathbf{k}_e))\,|0\rangle, \quad (6.8.21)$$

where $\hat{\mathbf{o}}$ and $\hat{\mathbf{e}}$ are unit vectors along the o-ray and the e-ray polarization direction of the SPDC crystal, and $\Delta_k = k_o + k_e - k_p$. In $\Phi(\Delta_k L)$, the finite length of the nonlinear crystal has been taken into account. Suppose the polarizers of the detectors D_1 and D_2 are set at angles θ_1 and θ_2, relative to the polarization direction of the o-ray of the SPDC crystal, respectively, the field operators can be written as

$$E_j^{(+)}(t_j, r_j) = \int d\omega\,\hat{\theta}_j\,\hat{a}(\omega)\,e^{-i[\omega t_j - k(\omega) r_j]},$$

where $j = 1, 2$, $\hat{\theta}_j$ is the unit vector along the *ith* analyzer direction. Following the standard Glauber formula, the joint detection counting rate of D_1 and D_2 is thus:

$$R_c \propto \int_T dt_1 dt_2\,|(\hat{\theta}_1 \cdot \hat{\mathbf{o}})(\hat{\theta}_2 \cdot \hat{\mathbf{e}})\,\Psi(\tau_1^o, \tau_2^e) + (\hat{\theta}_1 \cdot \hat{\mathbf{e}})(\hat{\theta}_2 \cdot \hat{\mathbf{o}})\,\Psi(\tau_1^e, \tau_2^o)|^2$$

$$= cos^2\theta_1 \sin^2\theta_2 + sin^2\theta_1 \cos^2\theta_2$$

$$+ \cos\theta_1 \sin\theta_2 \sin\theta_1 \cos\theta_2 \int_T dt_1 dt_2\,\Psi^*(\tau_1^o, \tau_2^e)\,\Psi(\tau_1^e, \tau_2^o), \quad (6.8.22)$$

where $\Psi(\tau_1^o, \tau_2^e)$ and $\Psi(\tau_1^e, \tau_2^o)$ are the effective two-photon wavefunctions with the following normalization:

$$\int_T dt_1 dt_2 \, |\, \Psi(\tau_1, \tau_2)\,|^2 = 1.$$

The third term of Eq. (6.8.22) determines the degree of two-photon coherence. Considering degenerate CW laser pumped SPDC, the biphoton wavepacket of Eq. (7.1.9) can be simplified as:

$$\Psi(\tau_1, \tau_2) = \Psi_0 \, e^{-i\omega_p(\tau_1+\tau_2)/2} \, \mathcal{F}_{\tau_-}\{f(\Omega)\}.$$

The coefficient of $\cos\theta_1 \sin\theta_2 \sin\theta_1 \cos\theta_2$ in the third term of Eq. (6.8.22) is thus

$$e^{-i\omega_p(\Delta\tau_1-\Delta\tau_2)/2} \, \mathcal{F}_{\tau_1^o-\tau_2^e}\{f(\Omega)\} \otimes \mathcal{F}_{\tau_1^e-\tau_2^o}\{f(\Omega)\}.$$

Therefore, two important factors will determine the result of the polarization correlation measurement: (1) the phase of $e^{-i\omega_p(\Delta\tau_1-\Delta\tau_2)/2}$; and (2) the value of the convolution between biphoton wavepackets $\Psi^*(\tau_1^o, \tau_2^e)$ and $\Psi(\tau_1^e, \tau_2^o)$. Examining the two wavepackets associated with the $o_1 - e_2$ and $e_1 - o_2$ terms, we found that the convolution of the two dimensional biphoton wavepackets of type II SPDC gives a null result, due to the *asymmetrical* rectangular function of $\Pi(\tau_1 - \tau_2)$ as indicated in Fig. 6.8.5. In order to make the two wavepackets overlap, we may either (1) move both wavepackets a distance of $DL/2$ (case I), or (2) move one of the wavepackets a distance of DL (case II). In both case I and case II, the convolution obtains its maximum value of one. The use of "compensator" is for this purpose. After compensating the two asymmetrical function of $\Pi(\tau_1 - \tau_2)$, we need to further manipulate the phase of $e^{-i\omega_p(\Delta\tau_1-\Delta\tau_2)/2}$ to finalize the desired Bell states. This can be done by means of a retardation plate to introduce

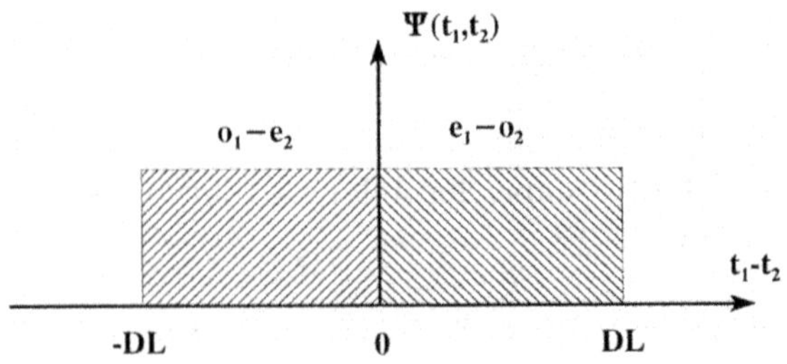

Fig. 6.8.5   Without "compensator", the two dimensional wavepackets of $\Psi(\tau_1^o, \tau_2^e)$ and $\Psi(\tau_1^e, \tau_2^o)$ do not overlap along $\tau_1 - \tau_2$ axis.

phase delay of 2π (+1) or π (-1) between the o-ray and the e-ray in either arm 1 or arm 2. The EPR–Bohm–Bell polarization correlation

$$R_c \propto \sin^2(\theta_1 \pm \theta_2)$$

is expected only when the above two conditions are satisfied. We can simplify the polarization state of the signal–idler photon pair in the form of Bell states $|\Psi_{XY}^{(\pm)}\rangle$ in this situation only.

The biphoton wavepacket looks very different in the case of femtosecond laser pumped SPDC. As an example, a calculation of type-II biphoton wavepacket is given in the following.

We start with the quantum state of type-II SPDC in the following format without losing generality:

$$|\Psi\rangle = \int dk_o \int dk_e \int d\omega_p \int_0^L dz e^{-4\ln 2\left[\frac{\omega_p-\Omega_p}{\sigma_p}\right]^2} e^{i\Delta z}\delta(\omega_o + \omega_e - \omega_p)a_o^\dagger a_e^\dagger|0\rangle,$$

$$(6.8.23)$$

where $\Delta = k_p - k_o - k_e$, L is the length of the nonlinear crystal. We have considered a Gaussian like pump pulse with bandwidth σ_p.

To generalize the result to a realistic experimental setup, we consider placing a Gaussian like spectral filter in front of the detector $D_j, j = 1, 2$. The field operator is written as

$$E_j^{(+)}(t_j, r_j) = \int d\omega e^{-4\ln 2\left[\frac{\omega-\Omega_j}{\sigma_j}\right]^2} a_o(\omega)e^{-i[\omega t_j - k(\omega)r_j]}. \qquad (6.8.24)$$

Define

$$\omega_o = \Omega_o + \nu_o,$$

$$\omega_e = \Omega_e + \nu_e,$$

$$\omega_p = \Omega_p + \nu_p,$$

where $\Omega_o \equiv \Omega_1$ and $\Omega_e \equiv \Omega_2$. ν_j is the detuning from the central frequency Ω_j. Note that $\Omega_o + \Omega_e = \Omega_p$ and $\nu_o + \nu_e = \nu_p$. The type-II phase mismatch Δ is now expanded to the first leading order. Assume $k(\nu) = [\Omega + \nu]n(\Omega + \nu)/c$, where $n(\Omega + \nu)$ is the index of refraction of the crystal at frequency $\Omega + \nu$. $k(\nu)$ can then be expanded to

$$k_j = K_j + \frac{\nu_j}{u_j(\Omega_j)}, \qquad (6.8.25)$$

where the subscripts $j = o, e, p$ and $K_j = \frac{n_j\Omega_j}{c}$, and where n_j is the index of refraction inside the nonlinear crystal. Therefore, Δ can now be written as

$$\Delta = k_p - k_o - k_e$$

$$= \frac{\nu_p}{u_p(\Omega_p)} - \frac{\nu_o}{u_o(\Omega_o)} - \frac{\nu_e}{u_e(\Omega_e)}$$

$$= -\nu_p D_+ - \frac{1}{2}\nu_- D, \tag{6.8.26}$$

where $\nu_- = \nu_o - \nu_e$, $D_+ \equiv \frac{1}{2}\left[\frac{1}{u_o(\Omega_o)} + \frac{1}{u_e(\Omega_e)}\right] - \frac{1}{u_p(\Omega_p)}$, and $D \equiv \frac{1}{u_o(\Omega_o)} - \frac{1}{u_e(\Omega_e)}$. To make the calculation easier, we assume $\sigma_1 = \sigma_2 = \sigma$; i.e., the same bandwidths for both filters. This assumption makes the ν_- and ν_+ integrals uncoupled. We also define $\Omega_o = \Omega_p/2 + \Omega_d$ and $\Omega_e = \Omega_p/2 - \Omega_d$. Note that in the case of degenerate type-II, $\Omega_d = 0$.

By combining Eqs. (6.8.24), (6.8.23), and (6.8.26), the biphoton wavefunction $\Psi(\tau_+, \tau_-)$ is given by

$$\Psi(\tau_+, \tau_-) = e^{-i\Omega_p \tau_+} e^{-i\Omega_d \tau_-} \Pi(\tau_+, \tau_-), \tag{6.8.27}$$

where

$$\Pi(\tau_+, \tau_-) = \int d\nu_p \int d\nu_- \int_0^L dz\, e^{-[\nu_p/\delta]^2} e^{-2\ln 2[\nu_-/\sigma]^2}$$

$$\times e^{-i\nu_p \tau_+} e^{-\frac{i\nu_- \tau_-}{2}} e^{-i[\nu_p D_+ + \frac{\nu_-}{2}D]z} \tag{6.8.28}$$

with $\frac{1}{\delta^2} = \frac{2\ln 2}{\sigma^2} + \frac{4\ln 2}{\sigma_p^2}$. If the bandwidths of the filters are taken to be infinite, Eq. (6.8.28) can be simplified as

$$\Pi(\tau_+, \tau_-) = \begin{cases} e^{-\sigma_p^2[\tau_+ - [D_+/D]\tau_-]^2/16\ln 2} & 0 < \tau_- < DL \\ 0 & \text{otherwise.} \end{cases}$$

The Π function for pulsed type-II SPDC looks quite different from a CW-pumped type-II SPDC. In a CW-pumped type-II SPDC, the Π function is approximately independent of τ_+. However, in short pulse pumped type-II SPDC, it must be treated as a function of both τ_- and τ_+.

As an example, we consider a type-II BBO crystal with 2mm thickness pumped by a laser pulse which has a 400 nm central wavelength, and for the degenerate case, $\lambda_o = \lambda_e = 800$ nm. Figure 6.8.6(a) shows the biphoton wavepacket generated by a 80 fs pump pulse with no spectral filtering. Strong dependence on τ_+ is observed. The Π function is "tilted" on the τ_- - τ_+ plane. In Fig. 6.8.6(b), the biphoton wavepacket is generated by $3ps$ pump pulse with no spectral filtering. Note that the biphoton is now

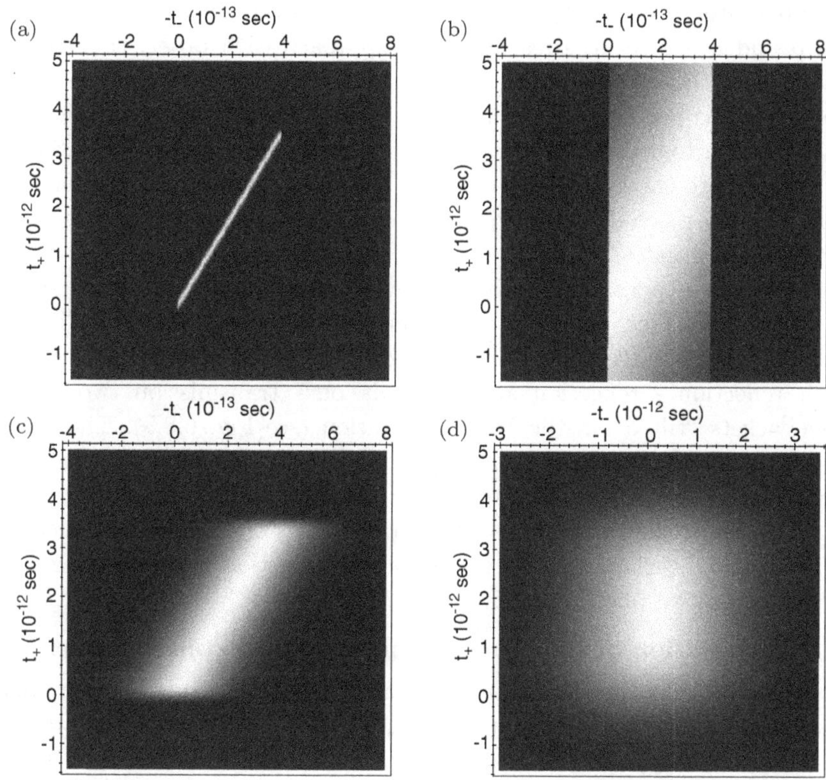

Fig. 6.8.6 Type-II biphoton as a function of the pump bandwidth and the filter bandwidth. $\Pi(\tau_+, \tau_-)$ is shown in a density plot. (a) With no spectral filters, pump pulse duration $\delta t_p = 80$ fs. (b) With no spectral filters, $\delta t_p = 3$ ps. (c) With 10 nm FWHM spectral filters, $\delta t_p = 80$ fs. (d) With 1 nm FWHM spectral filters, $\delta t_p = 80$ fs.

stretched in τ_+ direction and starts to resemble the CW-pumped type-II biphoton. In Figs. 6.8.6(c) and 6.8.6(d), we show the effects of spectral filtering. The pump pulse duration is fixed at 80 fs, but the bandwidth of the spectral filters are varied. In Fig. 6.8.6(c), we use 10 nm FWHM spectral filters and in Fig. 6.8.6(d), a 1 nm FWHM spectral filter is applied. As expected, spectral filtering broadens the biphoton in both τ_+ and τ_- directions. The effects in τ_- is greater than that in τ_+.

Femtosecond pulse pumped type-II SPDC has a very peculiar space–time structure. The "tilted" biphoton wavepacket requires special efforts for preparing Bell states. Differing from CW pumped SPDC, applying a "compensator" in this situation is not enough. Although we could shift the $e - o$ and $o - e$ terms to make them approach each other, the "tilted" wavepackets cannot be overlapped due to the different orientation. To overcome this problem, basically, we need to either (1) use a narrow band spectral filter to broaden the wavepacket as we have learned in Fig. 6.8.6(d); or (2) prepare the two wavepackets in the same "tilted" orientation, i.e., making e-ray reach detector D_1 and o-ray reach detector D_2, or vise versa, for both amplitudes.

A clever experimental scheme developed by Kim *et al.* is shown in Fig. 6.8.7. This design uses a noncollinear type-II SPDC. The use of half-waveplate (HWP) and polarization beamsplitter (PBS) makes both reflection - reflection and transmission - transmission two-photon wavepackets "tilted" in the same orientation (see Fig. 6.8.8). The e-rays and o-rays, respectively, are always go to detectors D_1 and D_2. The Bell state measurements have been tested for both CW and Femtosecond pumps. In both cases, high visibilities of interferences were achieved, indicating high degree overlapping of the reflection–reflection and transmission–transmission two-photon amplitudes.

(III) Preparation of Bell states from entangled coherent state:

Consider an experimental setup similar to the Bell state measurement discussed earlier. In this setup, entangled laser beams are generated by an OPO with type-II phase matching. The orthogonal polarized signal

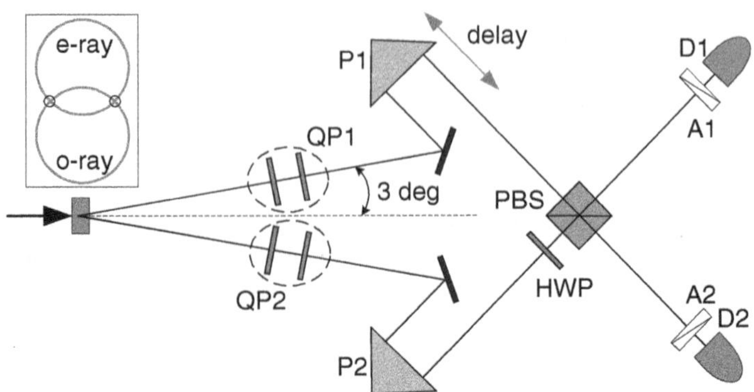

Fig. 6.8.7 Bell states generation by ultrashort laser pulse. The nonoverlapping $o_1 - e_2$ and $e_1 - o_2$ amplitudes are generated via noncollinear type-II SPDC. The use of half-waveplate (HWP) and polarization beamsplitter (PBS) make the reflection — reflection and transmission — transmission two-photon wavepackets "tilted" in the same orientation. QP1 and QP2 are quartz plates for fine tuning of the relative phase between the biphoton amplitudes.

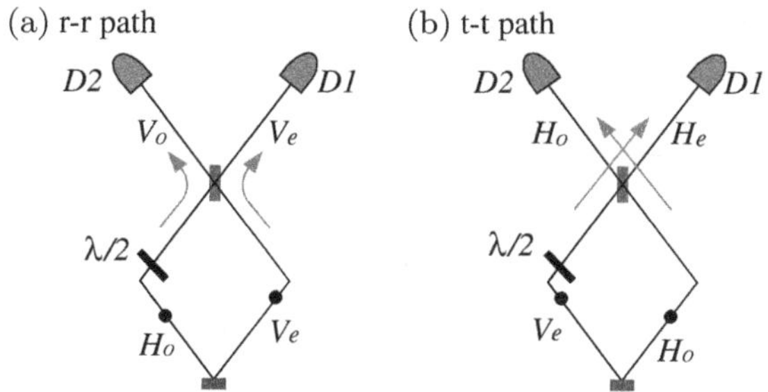

Fig. 6.8.8   Two alternative ways for a photon pair trigging a joint detection event. This figure further clarifies the role of the half-waveplate (HWP) and the polarization beamsplitter (PBS): How to make the reflection — reflection and transmission — transmission two-photon wavepackets "tilted" in the same orientation. The e-rays and o-rays are always go to detectors D_1 and D_2, respectively. Bell state measurements were tested for CW and femtosecond pump. In both cases, high interference visibilities were observed.

beam and idler beam both have 50%-50% chance to trigger two point-like photodetectors D_1 and D_2 for joint-photodetection. Two polarization analyzers oriented at $\hat{\theta}_1$ and $\hat{\theta}_2$, are coupled with D_1 and D_2, respectively. The field operators at D_j, $j = 1, 2$, can be formally written

$$\hat{E}^{(+)}(\mathbf{r}_j, t_j) = \hat{\theta}_j\, \hat{E}_s^{(+)}(\mathbf{r}_j, t_j) + \hat{\theta}_j\, \hat{E}_i^{(+)}(\mathbf{r}_j, t_j), \tag{6.8.29}$$

where $\hat{\theta}_j$ specify the direction of the polarizer coupled with D_j. The second-order coherence function, or correlation function, is thus

$$\begin{aligned}
G^{(2)}(\tau_1, \tau_2) &= \langle \Psi | \hat{E}^{(-)}(\tau_1) \hat{E}^{(-)}(\tau_2) \hat{E}^{(+)}(\tau_2) \hat{E}^{(+)}(\tau_1) | \Psi \rangle \\
&= \langle \Psi | [\hat{\theta}_1 \hat{E}_s^{(-)}(\tau_1) + \hat{\theta}_1 \hat{E}_i^{(-)}(\tau_1)][\hat{\theta}_2 \hat{E}_s^{(-)}(\tau_2) + \hat{\theta}_2 \hat{E}_i^{(-)}(\tau_2)] \\
&\quad \times [\hat{\theta}_2 \hat{E}_s^{(+)}(\tau_2) + \hat{\theta}_2 \hat{E}_i^{(+)}(\tau_2)][\hat{\theta}_1 \hat{E}_s^{(+)}(\tau_1) + \hat{\theta}_1 \hat{E}_i^{(+)}(\tau_1)] | \Psi \rangle.
\end{aligned} \tag{6.8.30}$$

To simplify a lengthy but straightforward calculation, we find the only surviving terms among the above sixteen expectations are the following three groups of second-order coherence or correlation functions:

$$\begin{aligned}
G_{s,i}^{(2)}(\tau_1, \tau_2) &= \langle \Psi | [\hat{\theta}_1 \hat{E}_s^{(-)}(\tau_1)\, \hat{\theta}_2 \hat{E}_i^{(-)}(\tau_2) \hat{\theta}_2 \hat{E}_i^{(+)}(\tau_2) \hat{\theta}_1 \hat{E}_s^{(+)}(\tau_1) | \Psi \rangle \\
&\quad + \langle \Psi | [\hat{\theta}_1 \hat{E}_i^{(-)}(\tau_1) \hat{\theta}_2 \hat{E}_s^{(-)}(\tau_2) \hat{\theta}_2 \hat{E}_s^{(+)}(\tau_2) \hat{\theta}_1 \hat{E}_i^{(+)}(\tau_1) | \Psi \rangle \\
&\quad + \langle \Psi | [\hat{\theta}_1 \hat{E}_s^{(-)}(\tau_1) \hat{\theta}_2 \hat{E}_i^{(-)}(\tau_2) \hat{\theta}_2 \hat{E}_s^{(+)}(\tau_2) \hat{\theta}_1 \hat{E}_i^{(+)}(\tau_1) | \Psi \rangle \\
&\quad + \langle \Psi | [\hat{\theta}_1 \hat{E}_i^{(-)}(\tau_1) \hat{\theta}_2 \hat{E}_s^{(-)}(\tau_2) \hat{\theta}_2 \hat{E}_i^{(+)}(\tau_2)\, \hat{\theta}_1 \hat{E}_s^{(+)}(\tau_1) | \Psi \rangle
\end{aligned} \tag{6.8.31}$$

in which the signal and idler beams are measured by D_1 and D_2, respectively and jointly;

$$G_{s,s}^{(2)}(\tau_1,\tau_2) = \langle\Psi|\,[\hat{\theta}_1\hat{E}_s^{(-)}(\tau_1)\,\hat{\theta}_2\hat{E}_s^{(-)}(\tau_2)\,\hat{\theta}_2\hat{E}_s^{(+)}(\tau_2)\,\hat{\theta}_1\hat{E}_s^{(+)}(\tau_1)|\Psi\rangle$$

$$(6.8.32)$$

in which the signal beam is measured by D_1 and D_2 respectively and jointly; and

$$G_{i,i}^{(2)}(\tau_1,\tau_2) = \langle\Psi|\,[\hat{\theta}_1\hat{E}_i^{(-)}(\tau_1)\,\hat{\theta}_2\hat{E}_i^{(-)}(\tau_2)\,\hat{\theta}_2\hat{E}_i^{(+)}(\tau_2)\hat{\theta}_1\hat{E}_i^{(+)}(\tau_1)|\Psi\rangle$$

$$(6.8.33)$$

in which the idler beam is measured by D_1 and D_2 respectively and jointly.

First, we consider the second-order coherence function of the first group $G_{s,i}^{(2)}(\tau_1,\tau_2)$. Applying the concept of effective wavefunction of the entangled signal–idler beams

$$\Psi_{s,i}(\tau_2;\tau_1) = \langle\Psi|\hat{\theta}_2\hat{E}_i^{(+)}(\tau_2)\,\hat{\theta}_1\hat{E}_s^{(+)}(\tau_1) + \hat{\theta}_2\hat{E}_s^{(+)}(\tau_2)\,\hat{\theta}_1\hat{E}_i^{(+)}(\tau_1)|\Psi\rangle$$

$$= \cos\theta_1\sin\theta_2\Psi(\tau_{s1},\tau_{i2}) + \sin\theta_1\cos\theta_2\Psi(\tau_{i1},\tau_{s2}), \qquad (6.8.34)$$

the second-order coherence function of the first group is

$$G_{s,i}^{(2)}(\tau_1,\tau_2) = \big|\cos\theta_1\sin\theta_2\Psi(\tau_{s1},\tau_{i2}) + \sin\theta_1\cos\theta_2\Psi(\tau_{i1},\tau_{s2})\big|^2,$$

where we have normalized the effective wavefunction. The second-order coherence function $G_{s,i}^{(2)}(\tau_1,\tau_2)$ is thus the result of quantum interference which involves the superposition of two different yet indistinguishable "two-photon" probability amplitudes: (1) one or more $\hat{x}$-polarized identical photons from the signal beam triggers D_1 at (r_1,t_1) while one or more $\hat{y}$-polarized identical photons from the idler beam triggers D_2 at (r_2,t_2); and (2) one or more $\hat{x}$-polarized identical photons from the signal beam triggers D_2 at (r_2,t_2) while one or more $\hat{y}$-polarized identical photons from the idler beam triggers D_1 at (r_1,t_1).

When different experimental conditions are achieved, we will be able to observe the following nontrivial second-order correlations:

(A) EPR–Bohm–Bell polarization correlation when choosing $\tau_1 - \tau_2 = 0$ or any of the nth correlation peaks $\tau_1 - \tau_2 = n/\nu_b$:

$$G^{(2)}(\theta_1,\theta_2) = \big|\cos\theta_1\sin\theta_2 + \sin\theta_1\cos\theta_2\big|^2 = \sin^2(\theta_1 + \theta_2) \qquad (6.8.35)$$

which means: although the polarization of neither beam-1 at D_1 nor beam-2 at D_2 is defined before its measurement, if beam-1 is measured with a polarization along a certain orientation θ_1 the polarization of the beam-2 is uniquely determined to be polarized at $\theta_2 = \pi/2 - \theta_1$, despite the distance between the two detectors.

(B) "Correlation peaks" (QGFC) when choosing $\theta_1 = \theta_2 = \pi/4$ and $t_1 = t_2$, but scanning $r_1 - r_2$:

$$G^{(2)}(\delta) = \left| \Psi_{s,i}(\tau_1; \tau_2) + \Psi_{i,s}(\tau_1; \tau_2) \right|^2 = 1 + \Psi_{s,i}^*(\tau)\Psi_{i,s}(\tau - \delta), \quad (6.8.36)$$

where $\delta = (r_1 - r_2)/c$ is the relative delay between the two wavepackets along $\tau_1 - \tau_2$ axis, which can be easily introduced by inserting birefringence material at the input port of the beamsplitter. The width of each "peak" is approximately $\Delta t \simeq 1/\nu_b N$.

(C) "Anti-correlation dips" (anti-QGFC) when choosing $\theta_1 = \pi/4, \theta_2 = -\pi/4$, and $t_1 = t_2$, but scanning $r_1 - r_2$:

$$G^{(2)}(\delta) = \left| \Psi_{s,i}(\tau_1; \tau_2) + \Psi_{i,s}(\tau_1; \tau_2) \right|^2 = 1 - \Psi_{s,i}^*(\tau)\Psi_{i,s}(\tau - \delta), \quad (6.8.37)$$

where, again, $\delta = (r_1 - r_2)/c$ is the relative delay between the two wavepackets along $\tau_1 - \tau_2$ axis, which can be easily introduced by inserting birefringence material at the input port of the beamsplitter. The width of each "dip" is approximately $\Delta t \simeq 1/\nu_b N$.

The above results are similar to that of the Alley-Shih interferometer in which a pair of entangled signal–idler photons, created from SPDC, was measured jointly at distant space–time coordinates. Superior to entangled photon pair, measurements of entangled laser beams do not rely on photon counting and can be performed over much greater distances and in shorter time.

Next, we discuss the second-order coherence function, or correlation functions $G_{s,s}^{(2)}(\tau_1, \tau_2)$ and $G_{i,i}^{(2)}(\tau_1, \tau_2)$. The $G_{ss}^{(2)}$ and $G_{ii}^{(2)}$ terms simply involve only the signal beam or only the idler beam. In fact, the second-order coherence or correlation function of the signal or idler beam has been calculated in Chapter 5. Although the state of the entangled signal–idler pair is pure, the signal and idler as subsystems are both in mixed coherent states. The normalized second-order coherence function of the signal beam or idler beam, respectively, is

$$g^{(2)}(\tau) = 1 + \operatorname{sinc}^2\left(\frac{\Delta\omega\tau}{2}\right) \frac{\sin^2\left(N\omega_b\tau/2\right)}{N\sin^2\left(\omega_b\tau/2\right)}, \quad (6.8.38)$$

representing a similar GFC, except a reduced contrast of 50%.

The measured second-order coherence function contains three contributions

$$G^{(2)}(\tau_1, \tau_2) = G^{(2)}_{s,i}(\tau_1, \tau_2) + G^{(2)}_{s,s}(\tau_1, \tau_2) + G^{(2)}_{i,i}(\tau_1, \tau_2). \qquad (6.8.39)$$

The experimental observation is different for different experimental setups. If the experimental setup is able to distinguish $G^{(2)}_{s,i}(\tau_1, \tau_2)$ from $G^{(2)}_{s,s}(\tau_1, \tau_2)$ and $G^{(2)}_{i,i}(\tau_1, \tau_2)$ the measurement will give similar result as that of SPDC. In fact, $G^{(2)}_{s,i}(\tau_1, \tau_2)$ can be easily distinguished from $G^{(2)}_{s,s}(\tau_1, \tau_2)$ and $G^{(2)}_{i,i}(\tau_1, \tau_2)$ by manipulating the optical delays for different polarizations.

Appendix (I) Hidden Variable Theory for the Spin 1/2 Bohm State

Consider two spin 1/2 particles, in the Bohm state of Eq. (6.8.1). The particles are well separated at distance. Two groups of observers decide to measure their spins by using Stern-Gerlach analyzers. Three sets of Stern-Gerlach apparatus whose z-directions are defined by unit vectors $\hat{\mathbf{a}}$, $\hat{\mathbf{b}}$ and $\hat{\mathbf{c}}$ as shown in Fig. A1.1. What is the probability to find the spin of particle-1 to be $+1/2$ in the direction $\hat{\mathbf{a}}$ and the spin of particle-2 to be $+1/2$ in the direction of $\hat{\mathbf{b}}$? The following is a simple classical hidden variable model

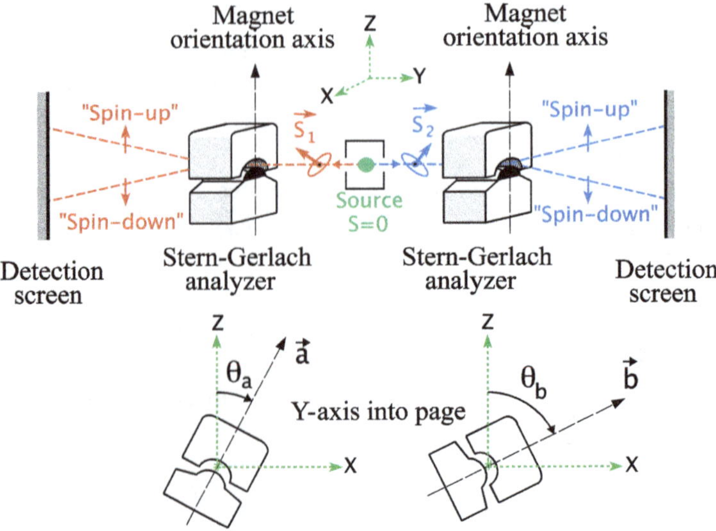

Fig. A1.1   Bohm's version of the EPR *gedankenexperiment*.

addressing this question. Before giving the detail, we first introduce the following notation P_{ab} to specify the classical probability,

$$P_{ab} = P(+_a \ d_b \ d_c \,|\, d_a \ +_b \ d_c). \tag{A1.1}$$

In this notation, we chose six-elements to specify the observation. The left-side of the partition refers particle one (in channel 1) and the right-side to particle two (in channel 2). Each element represents wether a measurement of the component of the spin of the jth particle, $j = 1, 2$, gives $+$ or $-$ in the corresponding direction $\hat{a}$, $\hat{b}$, or $\hat{c}$. d_a, d_b and d_c indicate no determination in the direction $\hat{a}$, $\hat{b}$ and $\hat{c}$, respectively. Note, P_{ab} is defined as to have $+$ in direction $\hat{a}$ which is observed in channel 1 (left-side), and to have $+$ in direction $\hat{b}$ which is observed in channel 2 (right-side).

In a classical hidden variable theory, if the spin component of particle-1 is measured with $+$ in the direction $\hat{a}$, then a measurement of the spin of particle-2 in the direction $\hat{a}$ must be $-$ with certainty. The probability P_{ab} in which particle-1 is measured with $+$ in the direction $\hat{a}$ (in channel 1), and particle-2 is measured with $-$ in the direction $\hat{b}$ (in channel 2) is written as

$$P_{ab} = P(+ \ - \ d_c \,|\, - \ + \ d_c). \tag{A1.2}$$

Similarly, we have

$$\begin{aligned}
P_{bc} &= P(d_a \ + \ - \,|\, d_a \ - \ +) \\
P_{ac} &= P(+ \ d_b \ - \,|\, - \ d_b \ +).
\end{aligned} \tag{A1.3}$$

In a classical probability theory, the undetermined spin components of particle-1 and particle-2, d_a, d_b, or d_c, must have specific values, either $+$ or $-$ in any direction. Consequently, we may write P_{ab} as a sum of probabilities for these two possibilities

$$\begin{aligned}
P_{ab} &= P(+ \ - \ + \,|\, - \ + \ -) + P(+ \ - \ - \,|\, - \ + \ +) \\
&\equiv P(+ \ - \ +) + P(+ \ - \ -),
\end{aligned} \tag{A1.4}$$

where on the second line we have dropped the redundant information involving the second particle by writing $P(+ \ - \ +)$ for $P(+ \ - \ +\,|\,- \ + \ -)$. Thus, we write Eqs. (A1.2)–(A1.3) as

$$\begin{aligned}
P_{ab} &= P(+ \ - \ +) + P(+ \ - \ -), \\
P_{bc} &= P(+ \ + \ -) + P(- \ + \ -), \\
P_{ac} &= P(+ \ + \ -) + P(+ \ - \ -).
\end{aligned} \tag{A1.5}$$

It should be emphasized that the left side quantities in Eq. (A1.5), P_{ab}, *etc.*, are measurable probabilities, but we have no procedure for measuring the separate classical probabilities $P(+ - +)$, $P(+ - -)$, *etc.* on the right side of Eq. (A1.5). Adding P_{ab} and P_{bc} and comparing with P_{ac}, it is easy to see that

$$P_{ab} + P_{bc} = P_{ac} + P(+ - +) + P(- + -). \tag{A1.6}$$

Assuming the classical probabilities $P(+ - +)$ and $P(- + -)$ are positive, which is true in any classical probability theory, we then obtain the Bell-Wigner inequality

$$P_{ab} + P_{bc} \geq P_{ac}. \tag{A1.7}$$

The seemingly obvious inequality of Eq. (A1.7) has been experimentally violated.

We next present a simple quantum mechanical calculation for the same measurement. Let us rewrite the state vector of Eq. (6.8.1) into the following form

$$|\Psi\rangle = \frac{1}{\sqrt{2}}\left[\begin{pmatrix} 1 \\ 0 \end{pmatrix}_1 \begin{pmatrix} 0 \\ 1 \end{pmatrix}_2 - \begin{pmatrix} 0 \\ 1 \end{pmatrix}_1 \begin{pmatrix} 1 \\ 0 \end{pmatrix}_2 \right]. \tag{A1.8}$$

Recalling that for a spin 1/2 particle in a state $|\psi\rangle$ the probability of observing spin up in a direction at angle θ to the z-axis is

$$|\langle\theta|\Psi\rangle\}^2 = \langle\Psi|\theta\rangle\langle\theta|\Psi\rangle = \langle\Psi|\hat{\pi}_\theta|\Psi\rangle, \tag{A1.9}$$

where $\hat{\pi}_\theta = |\theta\rangle\langle\theta|$ is the projection operator, and $|\theta\rangle = e^{-(i/2)\,\sigma_y\theta}|\uparrow\rangle$. The projection operator for spin up in the direction of a unit vector $\hat{\mathbf{r}}$ at angle θ to the z-axis is

$$\hat{\pi}_\theta = |\theta\rangle\langle\theta| = \frac{1}{2}(1 + \sigma_z \cos\theta + \sigma_x \sin\theta) \equiv \frac{1}{2}(\mathbf{I} + \vec{\sigma}\cdot\hat{\mathbf{r}}), \tag{A1.10}$$

where $\vec{\sigma} = (\sigma_x, \sigma_y, \sigma_z)$ are the Pauli matrices and $\mathbf{I}$ is the identity matrix. We can also explicitly rewrite $\hat{\pi}_\theta$ as a matrix,

$$\hat{\pi}_\theta = \frac{1}{2}\begin{pmatrix} 1 + \cos\theta & \sin\theta \\ \sin\theta & 1 - \cos\theta \end{pmatrix}. \tag{A1.11}$$

The physical observable P_{ab} is defined as the joint probability for particle-1 to be spin up along an SGA oriented in direction $\hat{\mathbf{a}}$ and particle-2 to be spin up along another SGA oriented in direction $\hat{\mathbf{b}}$

$$P_{ab} = \langle \Psi | \hat{\pi}_{1\hat{\mathbf{a}}} \, \hat{\pi}_{2\hat{\mathbf{b}}} | \Psi \rangle, \tag{A1.12}$$

where each projector has two subscripts defining the particles (and channel) and the measurement direction in the $x - z$ plane. Similarly,

$$P_{bc} = \langle \Psi | \hat{\pi}_{1\hat{\mathbf{b}}} \, \hat{\pi}_{2\hat{\mathbf{c}}} | \Psi \rangle,$$
$$P_{ac} = \langle \Psi | \hat{\pi}_{1\hat{\mathbf{a}}} \, \hat{\pi}_{2\hat{\mathbf{c}}} | \Psi \rangle. \tag{A1.13}$$

Evaluation of P_{ab}, P_{bc}, and P_{ac} using the operator in Eq. (A1.11) and the state in Eq. (A1.8) gives

$$P_{ab} = \frac{1}{4}\left[1 - \cos\left(\theta_a - \theta_b\right)\right] = \frac{1}{4}\left[1 - \hat{\mathbf{a}} \cdot \hat{\mathbf{b}}\right],$$
$$P_{bc} = \frac{1}{4}\left[1 - \cos\left(\theta_b - \theta_c\right)\right] = \frac{1}{4}\left[1 - \hat{\mathbf{b}} \cdot \hat{\mathbf{c}}\right], \tag{A1.14}$$
$$P_{ac} = \frac{1}{4}\left[1 - \cos\left(\theta_a - \theta_c\right)\right] = \frac{1}{4}\left[1 - \hat{\mathbf{a}} \cdot \hat{\mathbf{c}}\right].$$

It is easy to find that the Bell-Wigner inequality in Eq. (A1.7) achieves its maximum violation when $\theta_a = 0$, $\theta_b = \pi/3$, and $\theta_c = 2\pi/3$. The experimental observations have confirmed the violation.

Appendix (II) Bell's Theorem and Bell's Inequality

Bell first proved his theorem in 1964 under idealized experimental condition of 100% detection efficiency. It was soon realized that the Bell's inequality under idealized condition cannot be tested experimentally. Any realistic particle detector cannot have quantum efficiency of 100%. Bell improved his theorem and inequality of 1964 in 1972 by including realistic nonperfect particle detection.

(A) Bell's theorem of 1964.

In his 1964 paper, based on the EPR–Bohm state of Eq. (6.8.1) and a hidden parameter λ, Bell derived an inequality to distinguish quantum mechanics from local realistic probability theory of hidden variable. In his pioneer work Bell introduced a "more complete specification effected by means of parameter λ" with probability distribution $\rho(\lambda)$ for the classical statistical estimation of the expectation value of the joint measurement $\langle \Psi | (\vec{\sigma}_1 \cdot \hat{\mathbf{a}}) (\vec{\sigma}_2 \cdot \hat{\mathbf{b}}) | \Psi \rangle$ of particle-1 and particle-2 in the directions $\hat{\mathbf{a}}$ and $\hat{\mathbf{b}}$, simultaneously and respectively. The quantum mechanical result of this measurement gives

$$E_{ab} = \langle \Psi | (\vec{\sigma}_1 \cdot \hat{\mathbf{a}})\,(\vec{\sigma}_2 \cdot \hat{\mathbf{b}}) | \Psi \rangle = -\hat{\mathbf{a}} \cdot \hat{\mathbf{b}}. \tag{A2.1}$$

A special case of this result contains the determinism implicit in this idealized system, When the SGA analyzers are parallel, we have

$$E_{ab} = \langle \Psi | (\vec{\sigma}_1 \cdot \hat{\mathbf{a}})\,(\vec{\sigma}_2 \cdot \hat{\mathbf{a}}) | \Psi \rangle = -1 \tag{A2.2}$$

for all λ and all $\hat{\mathbf{a}}$. Thus, we can predict with certainty the result B by obtaining the result of A. Since the quantum mechanical state $|\Psi\rangle$ does not determine the result of an individual measurement, this fact (via EPR's argument) suggests that there exits a more complete specification of the state by a single symbol λ it may have many dimensions, discrete and/or continuous parts, and different parts of it interacting with either apparatus, *etc.* Let Λ be the space of λ for an ensemble composed of a very large number of the particle systems. Bell represented the distribution function for the state λ on the space Λ by the symbol $\rho(\lambda)$ and take $\rho(\lambda)$ to be normalized

$$\int_{\Lambda} \rho(\lambda)\,d\lambda = 1. \tag{A2.3}$$

In a deterministic hidden variable theory the observable $[A(\hat{\mathbf{a}})B(\hat{\mathbf{b}})]$ has a defined value $[A(\hat{\mathbf{a}})B(\hat{\mathbf{b}})](\lambda)$ for the state λ.

The locality is defined as follows: a deterministic hidden variable theory is local if for all $\hat{\mathbf{a}}$ and $\hat{\mathbf{b}}$ and and all $\lambda \in \Lambda$

$$\left[A(\hat{\mathbf{a}})B(\hat{\mathbf{b}})\right](\lambda) = A(\hat{\mathbf{a}}, \lambda)\,B(\hat{\mathbf{b}}, \lambda). \tag{A2.4}$$

This is, once λ is specified and the particle have separated, measurements of A can depend only upon λ and $\hat{\mathbf{a}}$ but not $\hat{\mathbf{b}}$. Likewise measurements of B depend only upon λ and $\hat{\mathbf{b}}$. Any reasonable physical theory that is realistic and deterministic and that denies action-at-a-distance is local in this sense. For such theories the expectation value of $[A(\hat{\mathbf{a}})B(\hat{\mathbf{b}})]$ is given by

$$E(\hat{\mathbf{a}}, \hat{\mathbf{b}}) = \int_{\Lambda} d\lambda\, \rho(\lambda)\,[A(\hat{\mathbf{a}})B(\hat{\mathbf{b}})](\lambda)$$

$$= \int_{\Lambda} d\lambda\, \rho(\lambda)\, A(\hat{\mathbf{a}}, \lambda)\, B(\hat{\mathbf{b}}, \lambda), \tag{A2.5}$$

where $E(\hat{\mathbf{a}}, \hat{\mathbf{b}}) \equiv E_{ab}$, corresponding to our previous notation. It is clear that Eq. (A2.2) can hold if only if

$$A(\hat{\mathbf{a}}, \lambda) = -B(\hat{\mathbf{b}}, \lambda) \tag{A2.6}$$

hold for all $\lambda \in \Lambda$.

Using Eq. (A2.6) we calculate the following expectation values, which involves three different orientations of the SGA analyzers:

$$E(\hat{\mathbf{a}}, \hat{\mathbf{b}}) - E(\hat{\mathbf{a}}, \hat{\mathbf{c}})$$

$$= \int_\Lambda d\lambda\, \rho(\lambda) \left[A(\hat{\mathbf{a}}, \lambda)\, B(\hat{\mathbf{b}}, \lambda) - A(\hat{\mathbf{a}}, \lambda)\, B(\hat{\mathbf{c}}, \lambda) \right]$$

$$= - \int_\Lambda d\lambda\, \rho(\lambda) \left[A(\hat{\mathbf{a}}, \lambda)\, A(\hat{\mathbf{b}}, \lambda) - A(\hat{\mathbf{a}}, \lambda)\, A(\hat{\mathbf{c}}, \lambda) \right]$$

$$= - \int_\Lambda d\lambda\, \rho(\lambda)\, A(\hat{\mathbf{a}}, \lambda)\, A(\hat{\mathbf{b}}, \lambda) \left[1 - A(\hat{\mathbf{b}}, \lambda)\, A(\hat{\mathbf{c}}, \lambda) \right]. \tag{A2.7}$$

Since $A(\hat{\mathbf{a}}, \lambda) = \pm 1$, $A(\hat{\mathbf{b}}, \lambda) = \pm 1$, this expression can be written as

$$\left| E(\hat{\mathbf{a}}, \hat{\mathbf{b}}) - E(\hat{\mathbf{a}}, \hat{\mathbf{c}}) \right| \leq \int_\Lambda d\lambda\, \rho(\lambda) \left[1 - A(\hat{\mathbf{b}}, \lambda)\, A(\hat{\mathbf{c}}, \lambda) \right], \tag{A2.8}$$

and consequently,

$$\left| E(\hat{\mathbf{a}}, \hat{\mathbf{b}}) - E(\hat{\mathbf{a}}, \hat{\mathbf{c}}) \right| \leq 1 + E(\hat{\mathbf{b}}, \hat{\mathbf{c}}). \tag{A2.9}$$

This inequality is the first of a family of inequalities which are collectively called "Bell's inequalities".

It is easy to find a disagreement between the quantum mechanics prediction of Eq. (A2.1) and the inequality of Eq. (A2.9). When we choose $\hat{\mathbf{a}}$, $\hat{\mathbf{b}}$ and $\hat{\mathbf{c}}$ to be coplanar with $\hat{\mathbf{c}}$ making an angle of $2\pi/3$ with $\hat{\mathbf{a}}$, and $\hat{\mathbf{b}}$ making an angle of $\pi/3$ with both $\hat{\mathbf{a}}$ and $\hat{\mathbf{c}}$, the quantum prediction gives

$$\left| \left[E(\hat{\mathbf{a}}, \hat{\mathbf{b}}) - E(\hat{\mathbf{a}}, \hat{\mathbf{c}}) \right]_{QM} \right| = 1, \tag{A2.10}$$

while

$$1 + \left[E(\hat{\mathbf{b}}, \hat{\mathbf{c}}) \right]_{QM} = \frac{1}{2}. \tag{A2.11}$$

It does not satisfy inequality of Eq. (A2.9).

In summary, Bell' theorem has proved that no deterministic hidden variable theory satisfying Eqs. (A2.2) and (A2.4) can agree with all of the predications of quantum mechanics concerning the spin of a pair of spin-1/2 particles in the singlet state of Eq. (6.8.1).

Since the space of Λ in Eq. (A2.5) is spanned into four regions with classical probabilities P_{ab}, P_{-ab}, P_{a-b}, P_{-a-b} in which A and B have values ± 1, the expectation value evaluation of Eq. (A2.5) can be explicitly calculated as

$$E_{ab} = (+1)(+1)P_{ab} + (-1)(+1)P_{-ab} + (+1)(-1)P_{a-b} + (-1)(-1)P_{-a-b}$$
$$= P_{ab} - P_{-ab} - P_{a-b} + P_{-a-b}. \tag{A2.12}$$

Consider three directions, $\hat{a}$, $\hat{b}$, $\hat{c}$, using the results of Eqs. (A1.5) and (A2.12) gives

$$E_{ab} = P(+ \ - \ +) + P(+ \ - \ -) - P(- \ - \ +) - P(- \ - \ -)$$
$$- P(+ \ + \ +) - P(+ \ + \ -) + P(- \ + \ +) + P(- \ + \ -).$$

Together with the normalization condition

$$\sum P(\pm \ \pm \ \pm) = 1,$$

E_{ab} becomes

$$E_{ab} = 2\big[P(+ \ - \ +) + P(+ \ - \ -) + P(- \ + \ +) + P(- \ + \ -)\big] - 1.$$

Similarly, we have

$$E_{bc} = 2\big[P(+ \ + \ -) + P(- \ + \ -) + P(+ \ - \ +) + P(- \ - \ +)\big] - 1.$$

$$E_{ac} = 2\big[P(+ \ + \ -) + P(+ \ - \ -) + P(- \ + \ +) + P(- \ - \ +)\big] - 1.$$

From the above three equations, it is not difficult to find that

$$E_{ab} - E_{ac} = 1 + E_{bc} - 4P(+ \ + \ -) - 4P(- \ - \ +)$$
$$E_{ac} - E_{ab} = 1 + E_{bc} - 4P(- \ + \ -) - 4P(+ \ - \ +). \tag{A2.13}$$

Considering the positive values of the probabilities, Eq. (A2.13) implies

$$E_{ab} - E_{ac} \le 1 + E_{bc}$$
$$E_{ac} - E_{ab} \le 1 + E_{bc}, \tag{A2.14}$$

and consequently we obtain the original Bell's inequality,

$$\big|E_{ab} - E_{ac}\big| \le 1 + E_{bc}. \tag{A2.15}$$

(B) Bell's theorem of 1971.

It was soon realized that the Bell's inequality of Eq. (A2.9) cannot be tested in a real experiment. Because Eq. (A2.2) cannot be hold exactly in an realistic measurement. Any real detector cannot have a perfect quantum efficiency of 100%, and any real analyzer cannot have a perfect distinguish ration between orthogonal channels. In 1971, Bell proved a new inequality which includes these concerns by assuming the outcomes of measurement A or B may take one of the following possible results

$$A(\hat{\mathbf{a}}, \lambda) \quad \text{or} \quad B(\hat{\mathbf{b}}, \lambda) = \begin{cases} +1 & \text{``spin-up''} \\ -1 & \text{``spin-down''} \\ 0 & \text{particle not detected} \end{cases} \tag{A2.16}$$

For a given state λ, we define the measured values for these quantities by the symbols $\bar{A}(\hat{\mathbf{a}}, \lambda)$ and $\bar{B}(\hat{\mathbf{b}}, \lambda)$, which satisfy

$$\left|\bar{A}(\hat{\mathbf{a}}, \lambda)\right| \le 1 \quad \text{and} \quad \left|\bar{B}(\hat{\mathbf{b}}, \lambda)\right| \le 1. \tag{A2.17}$$

Following the same definition of locality, the expectation value of $A(\hat{\mathbf{a}})B(\hat{\mathbf{b}})$ is calculated as

$$E(\hat{\mathbf{a}}, \hat{\mathbf{b}}) = \int_\Lambda d\lambda\, \rho(\lambda)\, \bar{A}(\hat{\mathbf{a}}, \lambda)\, \bar{B}(\hat{\mathbf{b}}, \lambda). \tag{A2.18}$$

Consider a measurement which involves $E(\hat{\mathbf{a}}, \hat{\mathbf{b}})$ and $E(\hat{\mathbf{a}}, \hat{\mathbf{b}}')$

$$\begin{aligned} &E(\hat{\mathbf{a}}, \hat{\mathbf{b}}) - E(\hat{\mathbf{a}}, \hat{\mathbf{b}}') \\ &= \int_\Lambda d\lambda\, \rho(\lambda)\, \left[\bar{A}(\hat{\mathbf{a}}, \lambda)\, \bar{B}(\hat{\mathbf{b}}, \lambda) - \bar{A}(\hat{\mathbf{a}}, \lambda)\, \bar{B}(\hat{\mathbf{b}}', \lambda)\right], \end{aligned} \tag{A2.19}$$

which can be written in the following form:

$$\begin{aligned} &E(\hat{\mathbf{a}}, \hat{\mathbf{b}}) - E(\hat{\mathbf{a}}, \hat{\mathbf{b}}') \\ &= \int_\Lambda d\lambda\, \rho(\lambda)\, \bar{A}(\hat{\mathbf{a}}, \lambda)\, \bar{B}(\hat{\mathbf{b}}, \lambda)\left[1 \pm \bar{A}(\hat{\mathbf{a}}', \lambda)\, \bar{B}(\hat{\mathbf{b}}', \lambda)\right] \\ &\quad - \int_\Lambda d\lambda\, \rho(\lambda)\, \bar{A}(\hat{\mathbf{a}}, \lambda)\, \bar{B}(\hat{\mathbf{b}}', \lambda)\left[1 \pm \bar{A}(\hat{\mathbf{a}}', \lambda)\, \bar{B}(\hat{\mathbf{b}}, \lambda)\right]. \end{aligned} \tag{A2.20}$$

Using inequality in Eq. (A2.17), we then have

$$\begin{aligned} \left|E(\hat{\mathbf{a}}, \hat{\mathbf{b}}) - E(\hat{\mathbf{a}}, \hat{\mathbf{b}}')\right| &\le \int_\Lambda d\lambda\, \rho(\lambda)\, \left[1 \pm \bar{A}(\hat{\mathbf{a}}', \lambda)\, \bar{B}(\hat{\mathbf{b}}', \lambda)\right] \\ &\quad + \int_\Lambda d\lambda\, \rho(\lambda)\, \left[1 \pm \bar{A}(\hat{\mathbf{a}}', \lambda)\, \bar{B}(\hat{\mathbf{b}}, \lambda)\right], \end{aligned} \tag{A2.21}$$

or

$$\left|E(\hat{\mathbf{a}}, \hat{\mathbf{b}}) - E(\hat{\mathbf{a}}, \hat{\mathbf{b}}')\right| \le \pm\left[E(\hat{\mathbf{a}}', \hat{\mathbf{b}}') + E(\hat{\mathbf{a}}', \hat{\mathbf{b}})\right] + 2\int_\Lambda d\lambda\, \rho(\lambda). \tag{A2.22}$$

We thus derive a measurable inequality

$$-2 \leq E(\hat{\mathbf{a}}, \hat{\mathbf{b}}) - E(\hat{\mathbf{a}}, \hat{\mathbf{b}}') + E(\hat{\mathbf{a}}', \hat{\mathbf{b}}) + E(\hat{\mathbf{a}}', \hat{\mathbf{b}}') \leq 2. \tag{A2.23}$$

The quantum mechanical prediction of the Bohm state in a realistic measurement with in-perfect detectors, analyzers *etc.*, can be written as

$$\left[E(\hat{\mathbf{a}}, \hat{\mathbf{b}}) \right]_{QM} = \mathrm{C}\,\hat{\mathbf{a}} \cdot \hat{\mathbf{b}}, \tag{A2.24}$$

where $|\mathrm{C}| \leq 1$. Suppose we take $\hat{\mathbf{a}}$, $\hat{\mathbf{a}}'$, $\hat{\mathbf{b}}$, $\hat{\mathbf{b}}'$ to be coplanar with $\phi = \pi/4$, we can easily find a disagreement between the quantum mechanics prediction and the inequality of Eq. (A2.23):

$$\left[E(\hat{\mathbf{a}}, \hat{\mathbf{b}}) - E(\hat{\mathbf{a}}, \hat{\mathbf{b}}') + E(\hat{\mathbf{a}}', \hat{\mathbf{b}}) + E(\hat{\mathbf{a}}', \hat{\mathbf{b}}') \right]_{QM} = 2\sqrt{2}\,\mathrm{C}. \tag{A2.25}$$

Although Bell derived his inequalities based on the measurement of EPR–Bohm state of Eqs. (6.8.1), (A2.9) and (A2.23) are not restricted to the measurement of spin $1/2$ particle pairs. In fact, most of the historical experimental testing have been the polarization measurements of photon pairs. The photon pairs are prepared in similar states which have been called EPR–Bohm–Bell states, or Bell states in short. Most of the experimental observations violated Bell's inequalities which may have different forms and have their violation occur at different orientations of the polarization analyzers. However, the physics behind the violations are all similar to that of Bell's theorem.

Bibliography

Alley C.O. and Y.H. Shih, *Foundations of Quantum Mechanics in the Light of New Technology*, in M. Namiki (Ed.), Vol. 47, Physical Society of Japan, Tokyo, 1986; Y.H. Shih and C.O. Alley, *Phys. Rev. Lett.* **61**, 2921 (1988).

Aspect A., *et al.*, *Phys. Rev. Lett.* **47**, 460 (1981); *Phys. Rev. Lett.* **49**, 91 (1982); *Phys. Rev. Lett.* **49**, 1804 (1982).

Bell J.S., *Physics* **1**, 195 (1964).

Bell J.S., *Speakable and Unspeakable in Quantum Mechanics*, Cambridge University Press, Cambridge, 1993.

Bohm D., *Quantum Theory*, Prentice Hall Inc., New York, 1951.

Clauser J.F., M.A. Horne, A.Shimony, and R.A. Holt, *Phys. Rev. Lett.* **23**, 880 (1969).

Clauser J.F. and M.A. Horne, *Phys. Rev. D* **10**, 526 (1974).

Clauser J.F. and A. Shimony, *Rep. Prog. Phys.* **41**, 1881 (1978).

Einstein A., B. Podolsky, and N. Rosen, *Phys. Rev.* **47**, 777 (1935).

Franson J.D., *Phys. Rev. A* **45**, 3126 (1992); **80**, 032119 (2009); **81**, 023825 (2010).

Greenberger D.M., M. Horne, and A. Zeilinger, *Bell's Theorem, Quantum Theory, and Conceptions of the Universe*, in M. Kafatos (Ed.), Kluwer Academic, Dordrecht, The Netherlands, 1989.

Greenberger D.M., M. Horne, A. Shimony, and A. Zeilinger, *Am. J. Phys.* **58**, 1131 (1990).

Kim Y.-H., M.V. Chekhova, S.P. Kulik, M.H. Rubin, and Y.H. Shih, *Phys. Rev. A* (2001).

Kim Y.-H. S.P. Kulik, and Y.H. Shih, *Phys. Rev. A* **63**, 060301(R) (2001).

Kim Y.-H., M.V. Chekhova, S.P. Kulik, and Y.H. Shih, *Phys. Rev. A* **67**, 010301(R) (2003).

Kim Y.-H., S.P. Kulik, and Y.H. Shih, *Phys. Rev. Lett.* **86**, 1370 (2001).

Klyshko D.N., *Photon and Nonlinear Optics*, Gordon and Breach Science, New York, 1988; A. Yariv, *Quantum Electronics*, John Wiley and Sons, New York, 1989; R.W. Boyd, *Nonlinear Optics*, Academic Press, San Diego, 1992.

Kwiat P.G. *et al.*, *Phys. Rev. A.* **60**, R773 (1999).

Ou Z.Y. and L. Mandel, *Phys. Rev. Lett.* **62**, 50 (1988).

Pittman T.B., Y.H. Shih, A.V. Sergienko, and M.H. Rubin, *Phys. Rev. A* **51**, 3495 (1995).

Rubin M.H., D.N. Klyshko, Y.H. Shih, and A.V. Sergienko, *Phys. Rev. A* **50**, 5122 (1994).

Schrödinger E., *Naturwissenschaften* **23**, 807, 823, 844 (1935); translations appear in *Quantum Theory and Measurement*, in J.A. Wheeler and W.H. Zurek (Eds.), Princeton University Press, New York, 1983.

Scully M.O., N. Erez, and E.S. Fry, *Phys Lett. A* **347** 56 (2005).

Sergienko A.V., Y.H. Shih, and M.H. Rubin, *JOSAB* **12**, 859 (1995).

Shih Y.H., Two-photon entanglement and quantum reality, in B. Bederson and H. Walther (Eds.), *Advances in Atomic, Molecular, and Optical Physics*, Academic Press, Cambridge, 1997; Entangled photons, *IEEE J. Sel. Top. Quantum Electron.* **9**(6) (2003).

Strekalov D.V., T.B. Pittman, and Y.H. Shih, *Phys. Rev. A* **57**, 567 (1998).

Strekalov D.V., Y.H. Kim, and Y.H. Shih, *Phys. Rev. A* 60, 2685 (1999).

Strekalov D.V., T.B. Pittman, A.V. Sergienko, Y.H. Shih, and P.G. Kwiat, *Phys. Rev. A* **54**, R1 (1996).

Valencia A., M.V. Chekhova, A. Trifonov, and Y.H. Shih, *Phys. Rev. Lett.* **88**, 183601 (2002).

Chapter 7

Two-Photon Interferometry I:
Biphoton Interference

Two-photon interferometry started to play an important role in quantum optics since 1980's. In that time, Dirac was criticized to be mistaken because he stated in his book, *The Principles of Quantum Mechanics*, that "... photon ... only interferes with itself. Interference between two different photons never occurs". Two-photon interference was considered the interference between two photons.

Is two-photon interference the interference of two photons? In this chapter, we provide a negative answer to this question: No! Two-photon interference is not the interference between two photons. Two-photon interference is the result of the superposition of two-photon amplitudes, a nonclassical entity corresponding to different yet indistinguishable alternatives which lead to a joint-photodetection event. In Dirac's language, "... a pair of photon only interferes with the pair itself ...".

In fact, neither classical theory nor quantum theory suggested the interference between different photons. Classical theory views optical interference the result of coherent superposition between electromagnetic waves. In quantum theory, the superposition occurs between quantum amplitudes, which corresponding to alternative ways of annihilating a photon in a photoelectron event, or annihilating a pair of photons in a joint-detection event between two individual photodetectors, or annihilating a group of N-photons in a N-fold joint-detection event between N individual photodetectors. Perhaps, the idea of interference between different photons came from the successful experimental observation of the first-order interference between radiations of independent light sources. The concept of interference between different photons was introduced in the middle of 1960's when Mandel demonstrated the interferences between two individual CW laser beams and between two laser pulses produced from two synchronized lasers.[1] Since then, interference between two photons became a hot and controversial topic in the study of optical interferometry. Before 1980's most experimental and theoretical studies focused on the first-order phenomena between coherent laser sources, which involve the measurement of a large number of indistinguishable photons. It is hard to distinguish the behavior of one photon, two photons, or a few photons from a bright laser beam or coherent electromagnet laser field, unless reducing the intensity to single-photon's level. However, at single-photon's level, on one hand, one "exposure" of two individual photons cannot provide us an observable interference pattern either spatially or temporally. On the other hand, a statistical measurement based on a large number of photons averages out any possible observable interference pattern due to the random phases between "patterns" from one "exposure" to another "exposure". We may never be able to draw a definite conclusion to support or to reject the idea of interference of two photons.

The situation changed in the middle of 1980's since the observation of two-photon interference of an entangled photon pair of SPDC. In a two-photon interferometer, Alley and Shih successfully demonstrated the interference of an orthogonally polarized signal–idler photon pair and reported in 1986. In that experiment, the signal photon and the idler photon were prepared with well-defined orthogonal polarization, either in the linear $|X\rangle|Y\rangle$ base or in the circular $|R\rangle|L\rangle$ base. With the help of a beamsplitter the photon pair seemed lost their original polarization and exhibited a typical EPR–Bohm–Bell correlation with the violation of a Bell's inequality. How cold a pair of well polarized photon turn into an entangled Bell state in which no polarization for either photon is specified? Alley and Shih provided an interpretation of the phenomenon as the result of a coherent superposition between two different yet indistinguishable biphoton amplitudes:[2] In addition to the Bell-type polarization correlation,

---

[1](1) In the CW case, in order to obtain an observable interference pattern, the exposure time of the graphic film has to be less than the coherent time of the laser radiation. Although the observed interference patterns may differ from one exposure to another significantly, there is no doubt about interference in each individual exposure. (2) In the case of pulsed lasers, the two pulses have to be synchronized to overlap in time. The interference pattern may differ from pulse to pulse with random phases, the experiment of Mandel *et al.* observed clear interference pattern in each exposure of a pair of overlapped pulses.

[2]The terminology *biphoton* was introduced later by Klyshko.

a two-photon correlation ($g^{(2)}(0) = 2$) and anti-correlation ($g^{(2)}(0) = 0$) were also observed with different chosen set of orientations of the polarization analyzers. The two-photon interference picture provided reasonable interpretation to that observation too: under the condition of complete overlapping, when the two biphoton amplitudes superpose in phase (with a "+" sign in between) the joint-detection of the two distant photodetectors reaches its maximum counting rate as the sign of "correlation", i.e., $g^{(2)}(0) = 2$; when the two biphoton amplitudes superpose out of phase (with a "−" sign in between) the joint-detection counting rate achieves its minimum value as the signature of "anti-correlation", i.e., $g^{(2)}(0) = 0$; when the two biphoton amplitudes superpose with other relative phases, a sinusoidal joint-detection counting rate comes out as the sign of two-photon interference. In 1987, Hong, Ou, Mandel, using a similar two-photon interferometer, demonstrated an anti-correlation "dip" by selecting one polarization, i.e., HOM dip. Hong, Ou, and Mandel provided a different interpretation of the phenomenon as the interference between two different photons: the observation of anti-correlation requires the signal photon and the idler photon to "meet" at the beamsplitter, indicating the interference between the signal photon and the idler photon. The debate took a decay to draw its conclusion. We address this fundamentally important debate in the following section. The physics of two-photon interferometry will be discussed in the rest of this chapter in the process of analyzing a few historical two-photon interference experiments.

7.1 Is Two-Photon Interference the Interference of Two Photons?

Figure 7.1.1 illustrates a typical historical two-photon interference experiment for the observation of biphoton anti-correlation.[3] The entangled signal–idler photon pair generated in SPDC is mixed by a 50%–50% near-normal incident beamsplitter,[4] BS, and detected by two detectors, D_1 and D_2, for coincidences. Balancing the signal and idler optical paths

---

[3]In fact, both biphoton anti-correlation and correlation are observable from the Alley–Shih interferometer.

[4]The first a few historical "dip" experiments adopted this near-normal incidence configuration from Alley and Shih's two-photon interferometer which used orthogonal polarization for observing both anti-correlation and correlation. Near-normal incidence is the only configuration to achieve 50%–50% transmission-reflection for both S and P polarization. In fact, it is unnecessary to choose near-normal incident beamsplitter

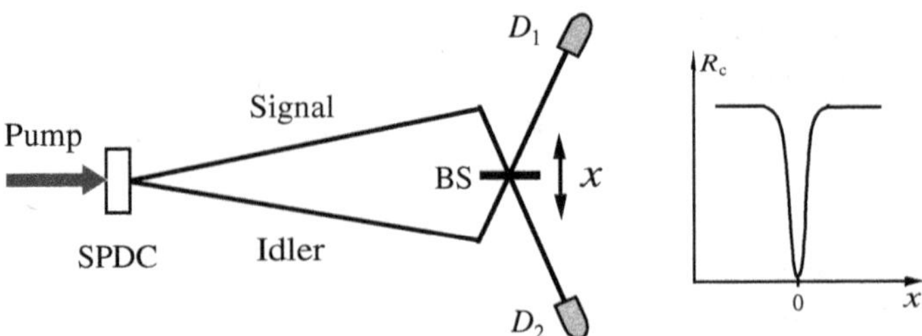

Fig. 7.1.1 Schematic of a typical two-photon interferometer. A anti-correlation is observable in coincidences of D_1 and D_2 when scanning the beamsplitter around its balanced position $x = 0$. It is quite tempting to rely on a incorrect picture which somehow envisions the anti-correlation as arising from two individual photons of a given signal–idler pair meeting at the beamsplitter: interference between the signal photon and the idler photon.

by positioning the beamsplitter, one can observe a "null" in coincidences which indicates a complete destructive interference. When the optical path difference is increased from zero to unbalanced values, a coincidence curve of "dip" is observed. The width of the "dip" equals the coherence length of the signal and idler, which is mainly determined by the spectral bandwidth of the filters placed in front of D_1 and D_2.

It is quite tempting to rely on a picture which somehow envisions the interference as arising from two individual photons of a given signal–idler pair "meeting" at the beamsplitter. Loosely speaking, indistinguishability leads to interference, for one sees that when the condition for total destructive interference is held, the two optical paths of the interferometer are of exactly the same length. It appears the signal and idler photons "meet" at the beamsplitter, and then it becomes impossible to distinguish which photon caused which single detector detection event. Destructive interference between the signal and idler photons occurs.

The picture of "interference between two photons" is further reinforced by the fact that changing the position of the beamsplitter from its balanced position, which begins to make these paths distinguishable, will bring about a degradation of interference. The coincidence counting rate seems to depend on the amount of overlap of the signal "wavepacket" and the idler "wavepacket" that is achieved. The shape of the "dip" is determined by the temporal convolution of the signal "wavepacket" and the idler

---

for an interferometer which uses one polarization for "dip" measurement only. For one polarization, 50%–50% can be achieved at any incident angle.

"wavepacket", and therefore provides information about them. The "dip" was interpreted as the result of "photon bunching" or "anti-bunching" effects as well.

Is the above explanation correct? Does the observation of the anti-correlation "dip" mean the interference of the signal photon and the idler photon at the beamsplitter? Is the statement of Dirac failed in this experiment? We provide an answer to these questions in the following way.

Let us examine a slightly modified experiment that is illustrated in Fig. 7.1.2. The experimental set up is similar to Fig. 7.1.1, except that the signal has two paths: one path length is L_l (longer path), the other is L_s (short path) with $L_l - L_s \equiv 2\Delta L$, such that $l_c^{s,i} \ll \Delta L \ll l_c^p$, where $l_c^{s,i}$ and l_c^p are the coherence length of the signal–idler field and the pump field, respectively. Experimentally, $l_c^{s,i}$ is usually determined by the chosen spectral filters for the photodetectors, l_c^p is determined by the spectral bandwidth of the pump laser. Due to the condition of $\Delta L > l_c^{s,i}$, although the setup provides two paths for the signal, there is no observable first-order interference of the signal photon itself. The counting rates of the single detectors, D_1 and D_2, respectively, remain constant. When the position of the beamsplitter BS is chosen to be $x = 0$, the path length of the idler arm takes its value of L_0 such that $L_0 = (L_l + L_s)/2$, i.e., L_0 takes the middle value of L_l and L_s. In this case, $L_l - L_0 = L_0 - L_s = \Delta L$, so that the signal photon and the idler photon are unable to "meet" at the beamsplitter when $x = 0$ due to the experimental condition of $\Delta L > l_c^{s,i}$.

Based on the idea of "distinguishability of two photons", the interference arising from the overlap of the signal and idler "wavepackets", "dips"

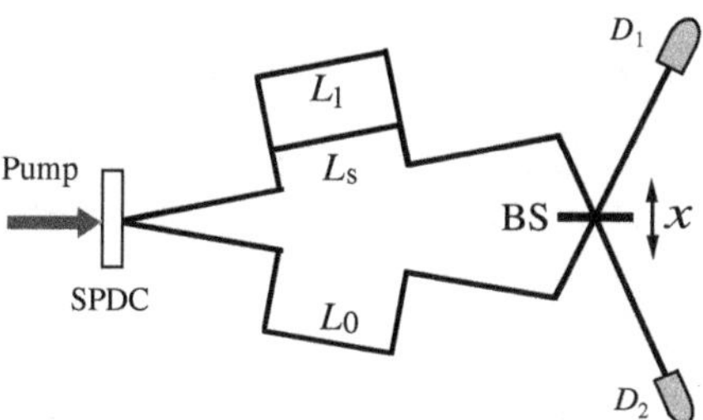

Fig. 7.1.2 Schematic of a modified two-photon interferometer. In contrast with Fig. 7.1.1, there exists two optical paths, L_l (long) and L_s (short), for the signal while the idler takes optical paths $L_0 = (L_l + L_s)/2$ when BS is placed at its balanced position $x = 0$. The optical paths are chosen to satisfy $l_c^{s,i} \ll \Delta L \ll l_c^p$, where $\Delta L \equiv (L_l - L_s)/2$. In this setup, the signal photon and the idler photon never "meet" at the beamsplitter when taking $x = 0$. Do we expect an anti-correlation "dip" at $x = 0$?

are expected to appear at two positions of the beamsplitter only, i.e., $x = \pm\Delta L/2$. In these two cases, the idler photon has a 50% chance of overlapping with the signal photon. This partial distinguishability results in that the contrast of these two dips should be at most 50%.

When $x = 0$, however, the signal photon and the idler photon do not "meet": there is no overlap of the signal and idler photon "wavepackets" because of $\Delta L > l_c^{s,i}$. Moreover, the detectors fire at random: in 50% of the joint detections D_1 fires ahead of D_2 by $\tau = \Delta L/c$; in the other 50% the opposite happens. So, no interference is expected.

Figure 7.1.3 shows the experimental result of Strekalov *et al* which tells quite a different story. We observe a high contrast interference "dip" in the middle ($x = 0$). In addition, the "dip" can turn into a "peak", or any Gaussian-like function between the "peak" and "dip", if the experimental conditions are slightly changed. The transition from "dip" to "peak" depends on $\phi = \omega_p\tau$, where $\tau = \Delta L/c$ is the time delay between the long path and the short path. Fixing $x = 0$ and varying ϕ, by slightly increasing or decreasing the value of ΔL, we observe a sinusoidal fringe in the joint detection counting rate, which is shown in Fig. 7.1.4, corresponding to a transition from "dip" to "peak" shown in Fig. 7.1.3. The experimental data indicates that it is not a necessary condition to have the signal photon and the idler photon "meeting" each other at the beamsplitter for observing

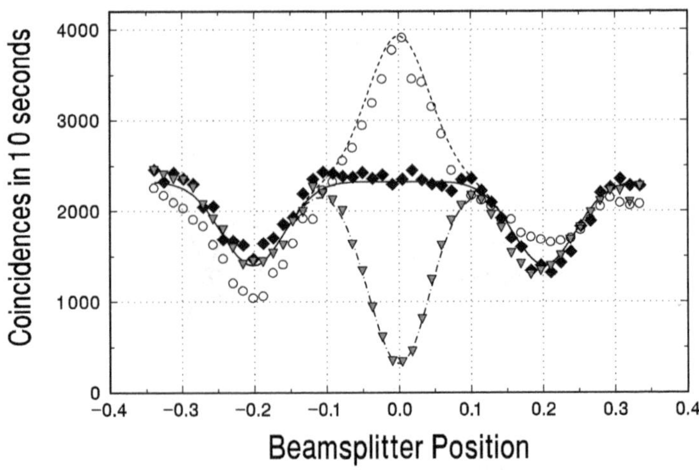

Fig. 7.1.3   A high contrast anti-correlation "dip" is observed at $x = 0$, where the signal photon is unable to "meet" with the idler photon. In addition, the destructive "dip" can turn to a constructive "peak" when L_l-L_s is slightly changed.

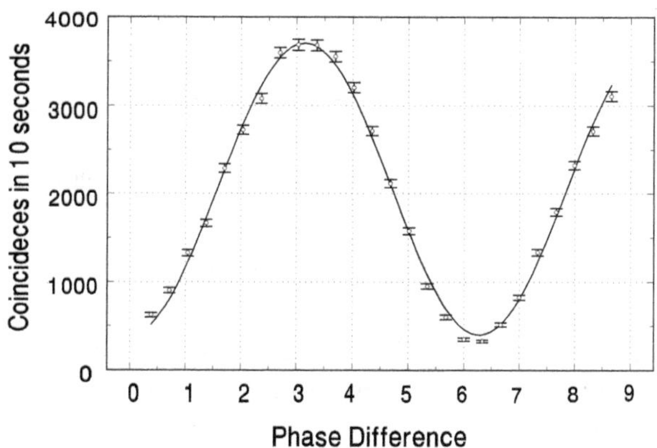

Fig. 7.1.4   The dip-peak transition is measured as a sinusoidal function of ϕ, where $\phi = \omega_p \tau$ and $\tau \equiv \Delta L/c$.

two-photon correlation, anti-correlation, or two-photon interference. The idea of "destructive interference between signal and idler photons" has failed to give a correct prediction. Thus, the "dip" or "peak" may not be considered as the interference between the signal and idler photons. Indeed, two-photon interference is not the interference between two individual photons. Two-photon interference arises from the superposition of two-photon amplitudes, different yet indistinguishable alternatives that result in a click-click joint-detection event between two photodetectors. In this regard, Dirac is correct. His statement is still valid if we modify it slightly: "...*biphoton*... only interferes with itself. Interference between two different *biphotons* never occurs". Probably, Dirac's statement "...photon... only interferes with itself" is confusing from the beginning, we may modify his statement as follows:

> "Interference is the result of the superposition of quantum amplitudes, a nonclassical entity corresponding to different yet indistinguishable alternatives which lead to a photodetection event or a joint-photodetection event. Interference between different photons or photon pairs never occurs."

(I) Analysis of the historical anti-correlation experiment of Fig. 7.1.1:

Let us first analyze the historical experiments of Fig. 7.1.1. As we have discussed earlier, the joint detection counting rate, R_c, of detectors D_1 and D_2 on the time interval T is given by the Glauber theory in a general form:

$$R_c \propto \int_T dt_1 dt_2 \, G^{(2)}(\mathbf{r}_1, t_1; \mathbf{r}_2, t_2)$$

$$= \int_T dt_1 dt_2 \, \langle \Psi | \hat{E}_1^{(-)} \hat{E}_2^{(-)} \hat{E}_2^{(+)} \hat{E}_1^{(+)} | \Psi \rangle$$

$$= \int_T dt_1 dt_2 \, \left| \langle 0 | \hat{E}_2^{(+)} \hat{E}_1^{(+)} | \Psi \rangle \right|^2, \tag{7.1.1}$$

where $\hat{E}_j^{(\pm)}$, $j = 1, 2$, are positive and negative-frequency components of the field at detectors D_1 and D_2, respectively, and $|\Psi\rangle$ is the state of the signal–idler photon pair:

$$|\Psi\rangle \simeq \int d\omega_p \, g(\omega_p) \int d\omega_s \, f(\omega_p, \omega_s) a_s^\dagger(\omega_s) \, a_i^\dagger(\omega_p - \omega_s) \, | 0 \rangle, \tag{7.1.2}$$

where, again, we concentrate to the temporal part of the state by selecting a pair of conjugate mode $\mathbf{k}_s$ and $\mathbf{k}_i$ for the measurement. Different from earlier discussions, here, $\omega_p = $ constant has been released. We assume a well-collimated pump beam ($\Delta \vec{\kappa} \sim 0$) with longitudinal mode distribution function $g(\omega_p)$. In the study of two-photon interference, we need to deal with finite bandwidth of pump, especially in the case of pulse pumped SPDC.

The fields $\hat{E}_1^{(+)}$ and $\hat{E}_2^{(+)}$ both have two contributions. Propagating the field operators from the source to the photodetector, and ignoring the transverse part of Green's function:

$$\hat{E}_1^{(+)} = \frac{1}{\sqrt{2}} \left[i \int d\omega \, E_0(\omega) \, e^{-i\omega\tau_1^R} a_s(\omega) + \int d\omega \, E_0(\omega) \, e^{-i\omega\tau_1^T} a_i(\omega) \right],$$

$$\hat{E}_2^{(+)} = \frac{1}{\sqrt{2}} \left[\int d\omega \, E_0(\omega) \, e^{-i\omega\tau_2^T} a_s(\omega) + i \int d\omega \, E_0(\omega) \, e^{-i\omega\tau_2^R} a_i(\omega) \right],$$

$$\tag{7.1.3}$$

where the superscript R and T stand for reflection and transmission, again, only one polarization is considered. $E_0(\omega) = \sqrt{\hbar\omega / 2\epsilon_0 V}$, V is the quantization volume, $\tau \equiv t - z/c$, again, z is the longitudinal coordinate along the optical path. Similar to earlier calculations, we treat $E_0(\omega)$ as a constant.

Applying the biphoton state of the signal–idler pair to Eq. (7.1.1), it is easy to find that the effective two-photon wavefunction has two amplitudes:

$$\Psi_{21} = \langle 0 | E_2^{(+)} E_1^{(+)} | \Psi \rangle = \Psi(\tau_2^T, \tau_1^T) - \Psi(\tau_2^R, \tau_1^R), \tag{7.1.4}$$

where $\Psi(\tau_2^T, \tau_1^T) = \langle 0|E(\tau_2^T)E(\tau_1^T)|\Psi\rangle$ corresponds to the case when the signal and idler are both transmitted at the beamsplitter, while $\Psi(\tau_2^R, \tau_1^R) = \langle 0|E(\tau_2^R)E(\tau_1^R)|\Psi\rangle$ corresponds to their reflection. The normalization constant has been absorbed into each of the amplitudes. The superposition of these two different yet indistinguishable two-photon amplitudes, or biphoton wavepackets, which contribute to a click-click joint-detection event between photodetectors D_1 and D_2, determine the probability of having a joint-detection at space–time $(\mathbf{r}_1, t_1; \mathbf{r}_2, t_2)$:

$$G^{(2)}(\mathbf{r}_1, t_1; \mathbf{r}_2, t_2)$$

$$= \left|\Psi(\tau_2^T, \tau_1^T) - \Psi(\tau_2^R, \tau_1^R)\right|^2$$

$$= \left|\Psi(\tau_2^T, \tau_1^T)\right|^2 + \left|\Psi(\tau_2^R, \tau_1^R)\right|^2$$

$$- \Psi^*(\tau_2^T, \tau_1^T)\Psi(\tau_2^R, \tau_1^R) - \Psi(\tau_2^T, \tau_1^T)\Psi^*(\tau_2^R, \tau_1^R). \tag{7.1.5}$$

The biphoton interference is thus observable in the coincidence counting rate

$$R_c \propto \int_T dt_1\, dt_2 \left| \Psi(\tau_2^T, \tau_1^T) - \Psi(\tau_2^R, \tau_1^R) \right|^2. \tag{7.1.6}$$

Examining Eqs. (7.1.5) and (7.1.6), when $\Psi(\tau_2^T, \tau_1^T)$ and $\Psi(\tau_2^R, \tau_1^R)$ are completely "overlap" in space–time, or are completely indistinguishable in the joint detection events, the coincidence counting rate is expected to be "null". Shifting the position of the beamsplitter from its balanced position, which begins to make these wavepackets nonoverlapping, or distinguishable, will bring about a degradation of interference, i.e., observing a "dip" in coincidences. This is the result of the convolution of the biphoton wavepackets. In addition, if one could change the "$-$" to "$+$", for example, by playing with the polarization of the photon pair, one can make a "peak" instead of a "dip". Both "dip" and "peak" can be easily observed in a polarization two-photon interferometer, which will be discussed latter.

Further computing the interference as a function of the optical path difference of the two-photon interferometer, we need to calculate the biphoton wavepackets and their convolution. The biphoton wavepacket of SPDC has been calculated earlier in the case of monochromatic plane wave pump. Similar to the earlier calculation, except taking into account of the pump distribution function $g(\omega_p)$, the biphoton wavepacket is thus

$$\Psi(\tau_2, \tau_1) = \Psi_0 \int d\omega_p \, g(\omega_p) \, e^{-i\frac{1}{2}\omega_p(\tau_2+\tau_1)}$$

$$\times \int d\omega_s \, f(\omega_p, \omega_s) \, e^{-i\frac{1}{2}(\omega_s-\omega_i)(\tau_2-\tau_1)}, \qquad (7.1.7)$$

where Ψ_0 absorbs all constants from the field and the state. To simplify the mathematics, we start with a factorize-able integral by imposing the following approximations:

$$\omega_s = \omega_s^0 + \nu, \quad \omega_i = \omega_i^0 - \nu,$$

$$\omega_s^0 + \omega_i^0 \simeq \omega_p^0, \quad \omega_p = \omega_p^0 + \nu_p, \qquad (7.1.8)$$

where ω_s^0, ω_i^0, and ω_p^0 are the center frequencies for the signal, idler, and pump, respectively. ω_s^0, ω_i^0, and ω_p^0 are considered as constants. Equation (7.1.7) is then simplified to a product of two functions:

$$\Psi(\tau_2, \tau_1) = \Psi_0 \, v(\tau_2 + \tau_1) \, u(\tau_2 - \tau_1), \qquad (7.1.9)$$

with

$$v(\tau_2 + \tau_1) = e^{-i\omega_p^0(\tau_2+\tau_1)/2} \int_{-\infty}^{\infty} d\nu_p \, g(\nu_p) \, e^{-i\nu_p(\tau_2+\tau_1)/2} \qquad (7.1.10)$$

and

$$u(\tau_2 - \tau_1) = e^{-i\frac{1}{2}(\omega_s^0-\omega_i^0)(\tau_2-\tau_1)} \int_{-\infty}^{\infty} d\nu \, f(\nu) \, e^{-i\nu(\tau_2-\tau_1)}. \qquad (7.1.11)$$

Basically, we have assumed the integral of $u(\tau_2-\tau_1)$ independent of ω_p. This approximation is valid only for narrow bandwidth of $g(\nu_p)$ such as that of a CW laser pump. This approximation cannot be used for short pulse pump, especially in the case of femtosecond laser pumped SPDC. The discussion for ultra-short pulse pumped SPDC is given later.

The functions $v(\tau_2 + \tau_1)$ and $u(\tau_2 - \tau_1)$ can be written in terms of the Fourier transforms of $g(\nu_p) \to \mathcal{F}_{\tau_+}\{g(\nu_p)\}$ and $f(\nu) \to \mathcal{F}_{\tau_-}\{f(\nu)\}$, where $\tau_+ \equiv (\tau_2 + \tau_1)/2$ and $\tau_- \equiv \tau_2 - \tau_1$. The effective two-photon wavefunction, or biphoton wavepacket, is given by

$$\Psi(\tau_2, \tau_1) = \Psi_0 \, \mathcal{F}_{\tau_+}\{g(\nu_p)\} \, e^{-i\omega_p^0\tau_+} \, \mathcal{F}_{\tau_-}\{f(\nu)\} \, e^{-i\omega_d^0\tau_-}$$

$$= \Psi_0 \, \mathcal{F}_{\tau_+}\{g(\nu_p)\} \, \mathcal{F}_{\tau_-}\{f(\nu)\} \, e^{-i\omega_s^0\tau_2} \, e^{-i\omega_i^0\tau_1}, \qquad (7.1.12)$$

where $\omega_d^0 \equiv \frac{1}{2}(\omega_s^0 - \omega_i^0)$.

The cross-interference term in Eq. (7.1.6) is calculated as follows:

$$\int dt_1 dt_2 \, \Psi^*(\tau_2^T, \tau_1^T) \, \Psi(\tau_2^R, \tau_1^R)$$

$$= |\Psi_0|^2 \int dt_+ \, \mathcal{F}_{\tau_+^T}^* \{g(\nu_p)\} \, \mathcal{F}_{\tau_+^R} \{g(\nu_p)\} \, e^{i\omega_p^0(\tau_+^T - \tau_+^R)}$$

$$\times \int dt_- \, \mathcal{F}_{\tau_-^T}^* \{f(\nu)\} \, \mathcal{F}_{\tau_-^R} \{f(\nu)\} \, e^{i\omega_d^0(\tau_-^T - \tau_-^R)}$$

$$\simeq |\Psi_0|^2 \int dt_- \, \mathcal{F}_{t_-}^* \{f(\nu)\} \, \mathcal{F}_{t_- - \delta} \{f(\nu)\}$$

$$= |\Psi_0|^2 \, \mathcal{F}_{t_-}^* \{f(\nu)\} \otimes \mathcal{F}_{t_- - \delta} \{f(\nu)\}, \tag{7.1.13}$$

where $t_+ \equiv t_2 + t_1$, $t_- \equiv t_2 - t_1$, and $\delta = [(z_2^R - z_1^R) - (z_2^T - z_1^T)]/c$ is the optical path difference introduced by moving the beamsplitter upward from its balanced position in the two-photon interferometer of Fig. 7.1.1. We have assumed degenerate ($\omega_d^0 = 0$) type-I SPDC in the above calculation. The coincidence counting rate R_c is therefore

$$R_c(\delta) = R_0 \left[1 - \mathcal{F}_{t_-}^* \{f(\nu)\} \otimes \mathcal{F}_{t_- - \delta} \{f(\nu)\} \right], \tag{7.1.14}$$

where R_0 is a constant. Assuming Gaussian wavepackets, the convolution yields a Gaussian function $|\Psi_0|^2 e^{-\delta^2/\tau_c^2}$. The coincidence counting rate R_c is approximately

$$R_c(\delta) = R_0 \left[1 - e^{-\delta^2/\tau_c^2} \right].$$

It is now clear that the observed "dip" is a biphoton "destructive interference" phenomenon. Mathematically, the "dip" is the result of a convolution, or cross-correlation, of the biphoton wavepackets, $\Psi(\tau_2^T, \tau_1^T)$ and $\Psi(\tau_2^R, \tau_1^R)$, along τ_- axis. Figure 7.1.5 is a conceptual Feynman diagram for the two-photon interference experiment of Fig. 7.1.1. While the beamsplitter is in its balanced position, the two Feynman alternatives (reflect-reflect vs transmit-transmit) are indistinguishable. Moving the beamsplitter away from its balanced position, the optical path difference of the two Feynman paths are no longer the same, corresponds to the moving away of the 2-D wavepacket along the τ_- axis. The superposition takes place between the transmit-transmit and the reflect-reflect biphoton amplitudes, instead of the signal photon and the idler photon. In Dirac's language: it is the interference of biphoton itself, but not the interference between the signal and the idler photons.

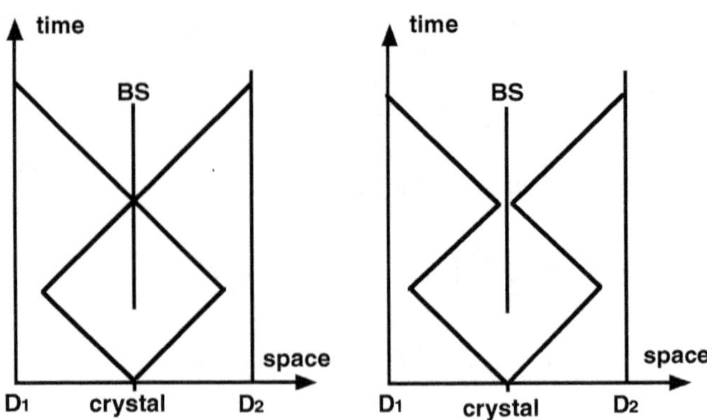

Fig. 7.1.5   Conceptual Feynman diagrams. The beamsplitter is represented by the thin vertical lines. The biphoton amplitudes, or biphoton wavepackets, are represented by "straight lines". Left: $\Psi(z_2^R, t_2; z_1^R, t_1)$ (reflect-reflect); Right: $\Psi(z_2^T, t_2; z_1^T, z_1)$ (transmit-transmit). The two Feynman alternatives both contribute to a "click-click" joint-detection event of D_1 and D_2.

In general, the biphoton interference can occur in two different ways: (1) The convolution takes place along τ_- direction; (2) the convolution takes place along τ_+ direction. In the experiment shown in Fig. 7.1.1, the biphoton wavepackets, $\Psi(\tau_2^T, \tau_1^T)$ and $\Psi(\tau_2^R, \tau_1^R)$ completely "overlap" along τ_+, since $\tau_+^T = \tau_+^R$ in any position of the beamsplitter. $\Psi(\tau_2^R, \tau_1^R)$, however, moves away from $\Psi(\tau_2^T, \tau_1^T)$ along τ_-, when scanning the beamsplitter from its balanced position.

(II) Analysis of the modified "dip" experiment in Fig. 7.1.2:

In the view of biphoton interference, we now present an interpretation for the experiment of Fig. 7.1.2. Differing from that of the experiment of Fig. 7.1.1, the special experimental setup in Fig. 7.1.2 achieves four alternatives of producing a joint-detection event between D_1 and D_2. The effective biphoton wavefunction thus consists of four amplitudes:

$$\Psi_{21} = \Psi(\tau_2^{LT}, \tau_1^{0T}) - \Psi(\tau_2^{0R}, \tau_1^{LR}) + \Psi(\tau_2^{ST}, \tau_1^{0T}) - \Psi(\tau_2^{0R}, \tau_1^{SR}),$$

where the superscript L, S, and 0 represent the long path, the short path and the middle path of Fig. 7.1.2, respectively. Consequently, $G^{(2)}$ has sixteen terms contributing to the coincidence photon-counting:

$$G^{(2)} = \left| \Psi(\tau_2^{LT}, \tau_1^{0T}) - \Psi(\tau_2^{0R}, \tau_1^{LR}) + \Psi(\tau_2^{ST}, \tau_1^{0T}) - \Psi(\tau_2^{0R}, \tau_1^{SR}) \right|^2.$$

However, due to the experimental condition that we have chosen, $L_l - L_0 = L_0 - L_s \equiv \Delta L \gg l_c^{s,i}$, where, again, $l_c^{s,i}$ is the coherence length of the signal and idler, only four cross terms are nonzero. We have the following eight nonzero contributions to the coincidence counting rate of D_1 and D_2:

$$R_c \propto \int_T dt_1 dt_2 \left\{ |\Psi(\tau_2^{LT}, \tau_1^{OT})|^2 + |\Psi(\tau_2^{OR}, \tau_1^{LR})|^2 \right.$$

$$+ |\Psi(\tau_2^{ST}, \tau_1^{OT})|^2 + |\Psi(\tau_2^{OR}, \tau_1^{SR})|^2$$

$$- \Psi^*(\tau_2^{LT}, \tau_1^{OT})\Psi(\tau_2^{OR}, \tau_1^{SR}) - \Psi(\tau_2^{LT}, \tau_1^{OT})\Psi^*(\tau_2^{OR}, \tau_1^{SR})$$

$$\left. - \Psi^*(\tau_2^{ST}, \tau_1^{OT})\Psi(\tau_2^{OR}, \tau_1^{LR}) - \Psi(\tau_2^{ST}, \tau_1^{OT})\Psi^*(\tau_2^{OR}, \tau_1^{LR}) \right\}. \quad (7.1.15)$$

The interference cross terms are calculated as

$$\Psi^*(\tau_2^{LT}, \tau_1^{OT})\Psi(\tau_2^{OR}, \tau_1^{SR})$$

$$= |\Psi_0|^2\, e^{-i\omega_p^0 \frac{\Delta L}{c}} \mathcal{F}_{t_+}^* \{g(\nu_p)\}\, \mathcal{F}_{t_+ + \frac{\Delta L}{c}}\{g(\nu_p)\}\, \mathcal{F}_{t_-}^*\{f(\nu)\}\, \mathcal{F}_{t_- - \delta}\{f(\nu)\}$$

$$\Psi^*(\tau_2^{ST}, \tau_1^{OT})\,\Psi(\tau_2^{OR}, \tau_1^{LR})$$

$$= |\Psi_0|^2\, e^{i\omega_p^0 \frac{\Delta L}{c}} \mathcal{F}_{t_+}^* \{g(\nu_p)\}\, \mathcal{F}_{t_+ - \frac{\Delta L}{c}}\{g(\nu_p)\}\, \mathcal{F}_{t_-}^*\{f(\nu)\}\, \mathcal{F}_{t_- - \delta}\{f(\nu)\},$$

where $\delta = [(r_2^{OR} - r_1^{SR}) - (r_2^{LT} - r_1^{OT})]/c$ is the additional optical path difference introduced by moving the beamsplitter upward from its "balanced" position $x = 0$.

The Feynman paths for this experiment are illustrated in Fig. 7.1.6. The upper two correspond to $\Psi(\tau_2^{LT}, \tau_1^{OT})$ and $\Psi(\tau_2^{OR}, \tau_1^{SR})$; the lower two correspond to $\Psi(\tau_2^{ST}, \tau_1^{OT})$ and $\Psi(\tau_2^{OR}, \tau_1^{LR})$. Note that if $\Delta L \ll l_c^p$, the upper two and the lower two Feynman paths, respectively, are indistinguishable by means of the click-click joint-photodetection of D_1 and D_2.

By increasing or decreasing ΔL or δ, we have two freedom to "shift" the 2-D biphoton wavepackets, independently, along τ_+ and τ_- axes:

$$\left[\mathcal{F}_{t_+}^* \{g(\nu_p)\} \otimes \mathcal{F}_{t_+ \pm \frac{\Delta L}{c}}\{g(\nu_p)\}\right]\left[\mathcal{F}_{t_-}^*\{f(\nu)\} \otimes \mathcal{F}_{t_- - \delta}\{f(\nu)\}\right].$$

For a chosen value of $\delta \ll l_c^p$, the upper pair and the lower pair of 2-D wavepackets illustrated in Fig. 7.1.6 are almost 100% overlapped along the τ_+ axis, respectively, yields $\mathcal{F}_{t_+}^* \{g(\nu_p)\} \otimes \mathcal{F}_{t_+ \pm \Delta L/c}\{g(\nu_p)\} \sim 1$. The joint

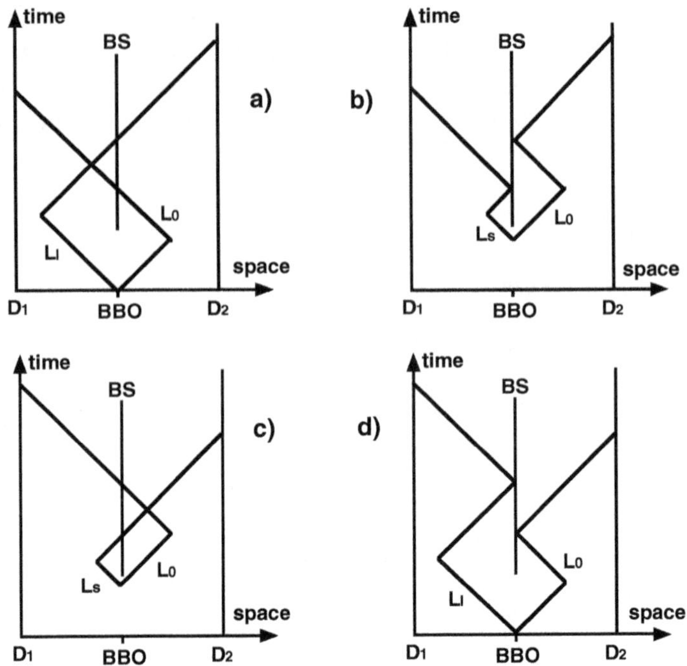

Fig. 7.1.6   Conceptual Feynman diagrams. (a) and (b) are two amplitudes for a joint detection such that D_1 fires ahead of D_2; (c) and (d) are two amplitudes in the reversed order. If $\Delta L \ll l_p^{\mathrm{coh}}$, the upper two and the lower two, respectively, are indistinguishable.

detection counting rate in the neighborhood of $x = 0$ is thus

$$R_c(\delta) = R_0 \left[1 - \cos\phi \, \mathcal{F}_{t_-}^* \{ f(\nu) \} \otimes \mathcal{F}_{t_- - \delta} \{ f(\nu) \} \right]. \qquad (7.1.16)$$

For Gaussian wavepacket along the t_- axis, R_c is approximately

$$R_c(\delta) = R_0 \left[1 - \cos\phi \, e^{-\delta^2/\tau_c^2} \right]. \qquad (7.1.17)$$

Equation (7.1.16) indicates a $\sim$100% interference modulation while scanning the beamsplitter in the neighborhood of $x = 0$. It is noted that the phase factor $\phi = \omega_p^0(\Delta L/c)$ plays an important role in this measurement. Subsequently setting the phase ϕ to be equal to π, 0, or $\pi/2$ we observe respectively a peak, dip, or flat coincidence rate R_c distribution centered at the "balanced" position of the beamsplitter, agreeing with the experimental results shown in Figs. 7.1.3 and 7.1.4. The mechanism of manipulating phase ϕ along τ_+ axis is very useful for the preparation of Bell states. We learn more about it in the following section.

7.2 Two-Photon Interference with Orthogonal Polarization

In the history of two-photon interferometry, a great driving force was the experimental testing of Bell's inequality. In fact, the first historical two-photon interferometer was designed for that purpose. Before the discussions of Bell's states and Bell's inequality, we analyze a simple two-photon interferometer with a pair of orthogonal polarized photons and a pair of independent polarization analyzers that is schematically illustrated in Fig. 7.2.1.

Assuming an idealized biphoton source generates an orthogonal polarized signal–idler pair in the following state:

$$|\Psi\rangle \simeq \int d\nu \, f(\nu) \, \hat{\mathbf{o}} \, a_s^\dagger(\omega_s^0 + \nu) \, \hat{\mathbf{e}} \, a_i^\dagger(\omega_i^0 - \nu) \, |\, 0\rangle, \qquad (7.2.1)$$

where $\hat{\mathbf{o}}$ and $\hat{\mathbf{e}}$ are unit vectors along the o-ray and the e-ray polarization direction of the SPDC crystal. In Eq. (7.2.1), we have assumed perfect phase matching $\omega_s + \omega_i - \omega_p = 0$ and $k_{s,o} + k_{i,e} - k_p = 0$. Suppose the polarizers of the detectors D_1 and D_2 are set at angles θ_1 and θ_2, relative to the polarization direction of the o-ray of the SPDC crystal, respectively, the field operators can be written as

$$E_1^{(+)} = \frac{1}{\sqrt{2}} \left[i \int d\omega \, E_0(\omega) \, e^{-i\omega\tau_1^R} \, \hat{\theta}_1 a_s(\omega) + \int d\omega \, E_0(\omega) \, e^{-i\omega\tau_1^T} \, \hat{\theta}_1 a_i(\omega) \right],$$

$$E_2^{(+)} = \frac{1}{\sqrt{2}} \left[\int d\omega \, E_0(\omega) \, e^{-i\omega\tau_2^T} \, \hat{\theta}_2 a_s(\omega) + i \int d\omega \, E_0(\omega) \, e^{-i\omega\tau_2^R} \, \hat{\theta}_2 a_i(\omega) \right],$$

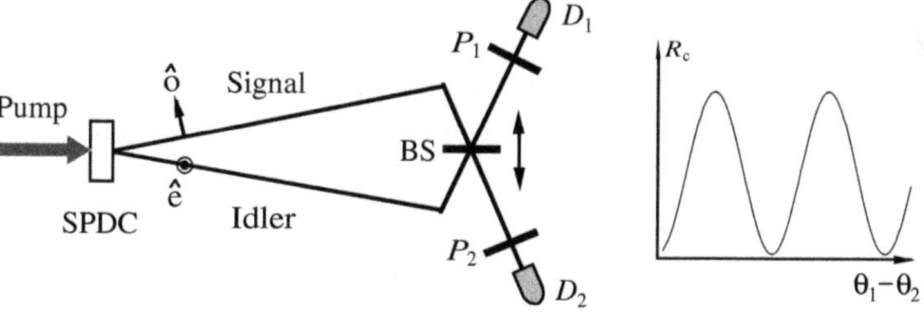

Fig. 7.2.1 Schematic of an Alley–Shih two-photon interferometer. The type-II SPDC produces an orthogonally polarized signal–idler pair. BS is a 50%–50% beamsplitter for both $\hat{o}$ and $\hat{e}$ polarized signal and idler photons. The polarization analyzers P_1 and P_2 are oriented at any chosen angles θ_1 and θ_2 for polarization correlation measurement. Fixing BS at $x = 0$ by examining the correlation "peak" and the anti-correlation "dip", a sinusoidal polarization correlation of $\sin^2(\theta_1 - \theta_2)$ is observed in the coincidences of D_1 and D_2.

where $\hat{\theta}_j$, $j = 1, 2$, is the unit vector along the jth analyzer direction. The effective wavefunction that contribute to the joint-detection events of D_1 and D_2 is calculated to be

$$\Psi_{21} = (\hat{\theta}_1 \cdot \hat{\mathbf{e}})(\hat{\theta}_2 \cdot \hat{\mathbf{o}}) \, \Psi(\tau_2^T, \tau_1^T) - (\hat{\theta}_1 \cdot \hat{\mathbf{o}})(\hat{\theta}_2 \cdot \hat{\mathbf{e}}) \, \Psi(\tau_2^R, \tau_1^R). \qquad (7.2.2)$$

The joint detection counting rate of D_1 and D_2 is thus

$$R_c \propto \int_T dt_1 dt_2 \left| (\hat{\theta}_1 \cdot \hat{\mathbf{e}})(\hat{\theta}_2 \cdot \hat{\mathbf{o}}) \, \Psi(\tau_2^T, \tau_1^T) - (\hat{\theta}_1 \cdot \hat{\mathbf{o}})(\hat{\theta}_2 \cdot \hat{\mathbf{e}}) \, \Psi(\tau_1^R, \tau_2^R) \right|^2$$

$$= R_0 \left\{ \sin^2\theta_1 \cos^2\theta_2 + \cos^2\theta_1 \sin^2\theta_2 - \sin\theta_1 \cos\theta_2 \cos\theta_1 \sin\theta_2 \right.$$

$$\left. \times \frac{1}{R_0} \int_T dt_1 dt_2 \left[\Psi^*(\tau_2^T, \tau_1^T) \, \Psi(\tau_2^R, \tau_2^R) + \Psi(\tau_2^T, \tau_1^T) \, \Psi^*(\tau_2^R, \tau_2^R) \right] \right\},$$

$$(7.2.3)$$

where $\Psi(\tau_2^T, \tau_1^T)$ and $\Psi(\tau_2^R, \tau_1^R)$ are the transmitted-transmitted and reflected-reflected biphoton wavepackets with the following time averaging:

$$\int_T dt_1 dt_2 \, | \, \Psi(\tau_1, \tau_2) \, |^2 = R_0.$$

The third term of Eq. (7.2.3) determines the degree of two-photon coherence. Considering degenerate CW laser pumped SPDC, the biphoton wavepacket of Eq. (7.1.9) can be simplified as

$$\Psi(\tau_2, \tau_1) = \Psi_0 \, \mathcal{F}_{\tau_-} \{ f(\nu) \}.$$

Where we have absorbed the phase factor $e^{-i\omega_p(\tau_1 + \tau_2)/2}$ into Ψ_0. The coefficient of $(\sin\theta_1 \cos\theta_2 \cos\theta_1 \sin\theta_2)$ in the third term of Eq. (7.2.3) is thus

$$\int_T dt_1 dt_2 \, \mathcal{F}_{\tau_2^T - \tau_1^T} \{ f(\nu) \} \mathcal{F}_{\tau_2^R - \tau_1^R} \{ f(\nu) \}$$

$$= \mathcal{F}_{t_-} \{ f(\nu) \} \otimes \mathcal{F}_{t_- - \delta} \{ f(\nu) \},$$

where, again, δ is the optical path delay introduced by moving the beamsplitter from its balanced position of $x = 0$.

In a polarization two-photon interferometer, we are able to observe two types of biphoton interference effects:

(1) **Anti-correlation "dip" and correlation "peak"**

In this measurement, we fix θ_1 and θ_2, such that $\theta_1 = 45°$ with $\theta_2 = 45°$ or $\theta_1 = 45°$ with $\theta_2 = -45°$, an anti-correlation-"dip" or a correlation-"peak" as function of δ will be observed in the coincidence counting rate R_c when scanning δ in the neighborhood of $x = 0$,

$$R_c(\delta) = R_0 \left[1 \mp \mathcal{F}_{t_-}^* \{ f(\nu) \} \otimes \mathcal{F}_{t_- - \delta} \{ f(\nu) \} \right]. \tag{7.2.4}$$

(2) **Polarization correlation**

In this measurement, we make $\delta = 0$ to achieve complete overlapping between biphoton wavepackets $\Psi(\tau_2^T, \tau_1^T)$ and $\Psi(\tau_2^R, \tau_1^R)$. The coefficient of $(\sin\theta_1 \cos\theta_2 \cos\theta_1 \sin\theta_2)$ in the third term of Eq. (7.2.3) achieves its maximum value of 2. The coincidence counting rate will be a function of $\theta_1 - \theta_2$ when manipulating the relative angle of the two polarization analyzers:

$$R_c(\theta_1, \theta_2) = R_0 \sin^2(\theta_1 - \theta_2). \tag{7.2.5}$$

This result is equivalent to the polarization correlation measurement for Bell's state

$$|\Psi\rangle = \frac{1}{\sqrt{2}} \left[|X_1\rangle |Y_2\rangle - |Y_1\rangle |X_2\rangle \right],$$

where $|X_1\rangle$ and $|X_1\rangle$ are defined as the polarization state that are respectively coincide with the o-ray and e-ray polarization direction of SPDC. Bell's states and polarization correlation will be discussed in detail in Chapter 14.

Since Einstein–Podolsky–Rosen published their 1935 paper, the concept of "physical reality" became an important subject for physicists and philosophers to study. In the early 1950's, Bohm simplified the Einstein–Podolsky–Rosen state of 1935 to discrete spin variables by introducing the singlet state of two spin $1/2$ particles:

$$|\Psi\rangle = \frac{1}{\sqrt{2}} \left[|\uparrow\rangle_1 |\downarrow\rangle_2 - |\downarrow\rangle_1 |\uparrow\rangle_2 \right], \tag{7.2.6}$$

where the kets $|\uparrow\rangle$ and $|\downarrow\rangle$ represent states of spin "up" and spin "down", respectively, along an *arbitrary* direction. For the EPR–Bohm state, the spin of neither particle is determined; however, if one particle is measured to be spin up along a certain direction, the other one must be spin down along that direction, despite the distance between the two spin $1/2$ particles and the orientation of the Stern–Gerlach analyzers (SGA). The nonlocal behavior of this two-particle system leads to the questions of EPR–Bohm: Are the two spin 1/2 particles prepared with defined spins at the source and in the course of their propagation? Is spin a physical reality of a particle independent of the observation?

In the Alley–Shih experiment, the same question was asked in a slightly different way: If two particles each is prepared with well-defined spin, can we expect similar nonlocal behavior? This question lead to their 1986 experiment. With the help of a two-photon interferometer, Alley and Shih discovered that a pair of photons with well-defined polarization can give similar EPR–Bohm-type correlation. Since than, the complete set of Bell states have been experimentally observed:

$$|\Psi^{(\pm)}\rangle = \frac{1}{\sqrt{2}}\big[|H_1\rangle|V_2\rangle \pm |V_1\rangle|H_2\rangle\big],$$

$$|\Phi^{(\pm)}\rangle = \frac{1}{\sqrt{2}}\big[|H_1\rangle|H_2\rangle \pm |V_1\rangle|V_2\rangle\big], \tag{7.2.7}$$

where $|H\rangle$ and $|V\rangle$, respectively, indicate well-defined horizontal and vertical polarization. In fact, any set of orthogonal polarization can be used to construct Bell states. In general, we use polarization state vector $|X\rangle$ and $|Y\rangle$, which can be defined in any orthogonal orientation, to replace $|H\rangle$ and $|V\rangle$.

This observation has been puzzling us for two decays: (1) There seems nothing "hidden" in this experiment. The signal photon and the idler photon both have well-defined polarization before entering into the interferometer. (2) The signal field and the idler field are first-order incoherent, the incoherent superposition of the signal–idler fields cannot change the polarization of the signal and idler, either during the course of their propagation or in the process of their annihilation. (3) It seems neither "correlation" nor "anti-correlation" is the intrinsic property of the photon source, or the state of the signal photon and idler photon, since Alley–Shih observed both "correlation" and "anti-correlation" from one experiment utilizing the same signal–idler photon source.

What is the cause of the nonlocal EPR–Bohm–Bell correlation for a pair of photons with well-defined polarization? We have attempted to introduce the concept of two-photon (two-particle) interference since 1986. In fact, this concept has been applied in the above analysis of the Alley–Shih experiment. In this regard, the nonlocal behavior of the EPR–Bohm spin 1/2 particles is a two-particle interference phenomenon. The EPR–Bohm state specifies a coherent superposition of two-particle amplitudes, corresponding to two different yet indistinguishable alternative ways for the two spin $1/2$ particles to trigger a joint-detection event through the two distant SGAs. We continue our discussion on the concept of physical reality and the physics behind this interesting observation.

7.3 Two-photon Double-slit Interferometer

A two-photon double-slit, or double-pinhole, interferometer is schematically shown in Fig. 7.3.1. The input light source is a Type-II degenerate noncollinear SPDC, which has been introduced in section 6.3. The orthogonally polarized signal-idler pair has two different yet indistinguishable alternatives to pass the double-slit or double-pinhole: (1) the signal passes the upper slit-A while the idler passes the lower slit-B; (2) the idler passes the upper slit-A while the signal passes the lower slit-B. A $45°$ oriented polarizers is placed in-front of the double-slit to select the $45°$ polarization from the orthogonally polarized signal and idler. The separation between the upper slit-A and the lower slit-B is much greater than the spatial coherence of the signal field and the idler field, $d \gg l_c$. No first-order

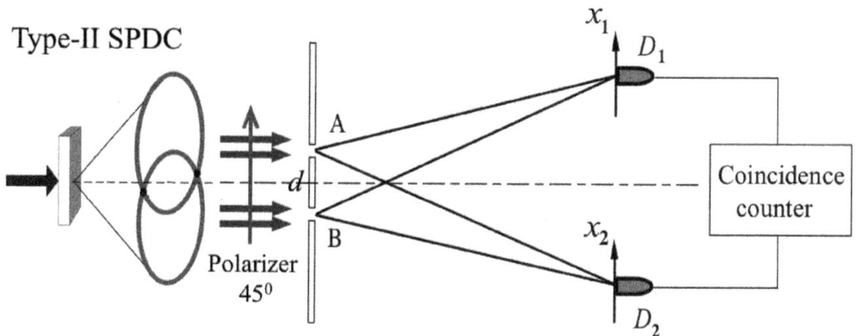

Fig. 7.3.1 Schematic of a two-photon Young's double-slit, or double-pinhole, interferometer. The input source is a Type-II degenerate noncollinear SPDC, $\omega_{s0} = \omega_{i0}$. The orthogonally polarized signal-idler pair has two alternatives to pass the double-slit or double-pinhole: (1) the signal passes the upper slit-A while the idler passes the lower slit-B; (2) the idler passes the upper slit-A while the signal passes the lower slit-B. A $45°$ oriented polarizers is placed in-front of the double-slit. The separation between the upper slit-A and the lower slit-B is much greater than the spatial coherence of the signal field and the idler field, $d \gg l_c$. Two point-like photodetectors, D_1 and D_2, are scannable along their x-axises in the far-field, $z = z_0$, of the interferometer. The single-detector counting rates of D_1 and D_2 are monitored respectively; and the coincidence counting rate of D_1 and D_2 is monitored jointly, during the scanning of D_1 and D_2. To simplify the mathematics, we assume line-like slits, or pint-like pinholes.

interferences are observable. Two point-like photodetectors, D_1 and D_2, are scannable along their x-axises in the far-field, $z = z_0$, of the interferometer. The single-detector counting rates of D_1 and D_2 are monitored respectively; and the coincidence counting rate of D_1 and D_2 is monitored jointly, during the scanning of D_1 and D_2. There are no observable interferences from the single-detector counting rate of D_1 and D_2. Do we expect observing interference from the joint-photodetection of D_1 and D_2?

Taking previous result, we find the effective wavefunction measured by D_1 and D_2 is:

$$\Psi(\tau_1, \tau_2) = \Psi_{si}^{(1)}(\tau_{A1}, \tau_{B2}) + \Psi_{is}^{(1)}(\tau_{B1}, \tau_{A2}) + \Psi_{is}^{(2)}(\tau_{A1}, \tau_{B2})$$

$$+ \Psi_{si}^{(2)}(\tau_{B1}, \tau_{A2})$$

$$= \Psi_{si}(\tau_{A1}, \tau_{B2}) + \Psi_{is}(\tau_{B1}, \tau_{A2}) \tag{7.3.1}$$

corresponding to four different yet indistinguishable alternatives for the signal-idler pair to produce a joint-photodetection event; where the superscripts (1) and (2) label the alternatives (1) and (2), respectively. Obviously, we have $\Psi_{si}^{(1)}(\tau_{A1}, \tau_{B2}) = \Psi_{si}^{(2)}(\tau_{A1}, \tau_{B2})$ and $\Psi_{is}^{(1)}(\tau_{B1}, \tau_{A2}) = \Psi_{is}^{(2)}(\tau_{B1}, \tau_{A2})$. The second-order coherence function is thus

$$G^{(2)}(\tau_1, \tau_2)$$

$$= \left| \Psi_{si}(\tau_{A1}, \tau_{B2}) + \Psi_{is}(\tau_{B1}, \tau_{A2}) \right|^2$$

$$= \left| e^{-i\omega_p(\tau_{A1}+\tau_{B2})/2} \mathcal{F}_{\tau_{A1}-\tau_{B2}}\{f(\nu)\} e^{-i(\omega_{s0}-\omega_{i0})(\tau_{A1}-\tau_{B2})/2} \right.$$

$$\left. + e^{-i\omega_p(\tau_{B1}+\tau_{A2})/2} \mathcal{F}_{\tau_{B1}-\tau_{A2}}\{f(\nu)\} e^{-i(\omega_{s0}-\omega_{i0})(\tau_{B1}-\tau_{A2})/2} \right|^2 \tag{7.3.2}$$

Examining the cross interference term

$$e^{i\omega_p(\tau_{A1}+\tau_{B2})/2} \mathcal{F}_{\tau_{A1}-\tau_{B2}}^*\{f(\nu)\} e^{i(\omega_{s0}-\omega_{i0})(\tau_{A1}-\tau_{B2})/2}$$

$$\times\, e^{-i\omega_p(\tau_{B1}+\tau_{A2})/2} \mathcal{F}_{\tau_{B1}-\tau_{A2}}\{f(\nu)\} e^{-i(\omega_{s0}-\omega_{i0})(\tau_{B1}-\tau_{A2})/2}$$

$$= e^{-i\omega_{s0}[(\tau_{B1}-\tau_{A1})+(\tau_{A2}-\tau_{B2})]} \mathcal{F}_{\tau_{A1}-\tau_{B2}}^*\{f(\nu)\} \mathcal{F}_{\tau_{B1}-\tau_{A2}}\{f(\nu)\}, \tag{7.3.3}$$

we found the cross interference term will contribute a sinusoidal modulation to the measurement of $G^{(2)}(\tau_1, \tau_2)$, if the two Fourier transforms overlap in space-time during the scanning of D_1 and D_2.

Obviously, the optimal experimental practice is to move D_1 and D_2 symmetrically about $x = 0$ with $x_1 = -x_2$. In this case, we have $\tau_{A1} - \tau_{B2} = \tau_{B1} - \tau_{A2} = 0$, the second-order coherence function

$G^{(2)}(\tau_1, \tau_2)$ becomes:

$$G^{(2)}(\tau_1, \tau_2) = \left| e^{-i\omega_p(\tau_{A1}+\tau_{B2})/2} + e^{-i\omega_p(\tau_{B1}+\tau_{A2})/2} \right|^2$$

$$\propto 1 + Re \, e^{-i\omega_{s0}[(\tau_{B1}-\tau_{A1})+(\tau_{A2}-\tau_{B2})]}$$

$$\simeq 1 + \cos\frac{2\pi d}{\lambda_{s0} z_0}(x_1 - x_2). \tag{7.3.4}$$

The coincidence counting rate is thus:

$$R_c(x_1, x_2) \propto \int_T dt_1 dt_2 \left| e^{-i\omega_p(\tau_{A1}+\tau_{B2})/2} + e^{-i\omega_p(\tau_{B1}+\tau_{A2})/2} \right|^2$$

$$\simeq 1 + \cos\frac{2\pi d}{\lambda_{s0} z_0}(x_1 - x_2). \tag{7.3.5}$$

A two-photon interference pattern of 100% visibility is than observable when scanning D_1 and D_2, even when the separation between the double-slit, or double-pinhole, is much greater than the coherence length of the signal field and the idler field.

It is not difficult to find that the scanning of D_1 and D_2 realized a relative shift between two biphoton amplitudes, $\Psi_{si}(\tau_{A1}, \tau_{B2})$ and $\Psi_{is}(\tau_{B1}, \tau_{A2})$, along their $\tau_1 + \tau_2$ axis. This unusual phenomenon has been observed experimentally in quite a number different experimental configurations.

7.4 Quantum Beats

In the two-photon double-slit, or double-pinhole, interferometer, we have realized a relative shift between two biphoton amplitudes, i.e., the nonfacterizeable 2D wavepackets, along their $\tau_1 + \tau_2$ axis by scanning D_1 and D_2. Is it possible to manipulate a relative shift along the $\tau_1 - \tau_2$ axis of the wavepackets? The answer is positive. Shih and Sergienko observed a quantum beats experimentally in 1994. The quantum beats is the result of the relative shit between two biphoton wavepackets along their $\tau_1 - \tau_2$ axis.

The 1994 experiment of quantum beats is schematically illustrated in Fig. 7.4.1. An entangled pair of signal-idler photon is generated from a nondegenerate but collinear Type-II SPDC. The signal photon, $\lambda_0 = 700.7\,\text{nm}$, and the idler photon, $\lambda_0 = 703.7\,\text{nm}$, are orthogonal polarized, horizontally and vertically, respectively. A set of 15 crystal quartz plates is insered in the incident biphoton beam one-by-one or changing the optical

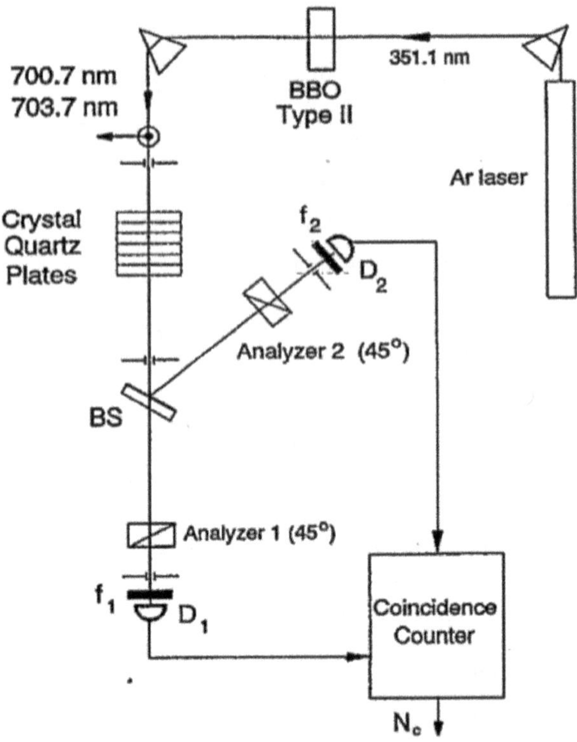

Fig. 7.4.1   Schematic experimental setup to demonstrate Quantum Beats. An entangled pair of signal-idler photon is generated from a nondegenerate but collinear Type-II SPDC. The signal photon, $\lambda_0 = 700.7\,$nm, and the idler photon, $\lambda_0 = 703.7\,$nm, are orthogonal polarized, horizontally and vertically, respectively. A set of crystal quartz plates is inserted in the incident biphoton beam one-by-one for changing the optical delay, δ, between the orthogonally polarized signal and idler photons. A 50%–50% beamsplitter at near-normal incident angle is used to split the signal and idler beam into two for the joint-photodetector of D_1 and D_2.

delay, δ, between the orthogonally polarized signal and idler photons. The fast axes of the quartz plates were carefully aligned to match the o-ray or e-ray polarization planes of the SPDC nonlinear crystal. A 50%–50% beamsplitter at near-normal incident angle is followed to split the signal and idler beam into two arms, followed by two polarization analyzers, both oriented at 45°, and two photon counting detectors D_1 and D_2, respectively. The experiment was designed in such a way that the signal photon and the idler photon have equal chance to trigger D_1 and D_2.

In this experiment, the signal-idler photon pair has two different yet indistinguishable alternatives to produce a joint-photdetection event of D_1 and D_2: (1) the signal photon, $\lambda_0 = 700.7\,\text{nm}$, is transmitted at BS and triggers D_1, while the idler, $\lambda_0 = 703.7\,\text{nm}$, is reflected at BS and triggers D_2; (2) the signal photon, $\lambda_0 = 700.7\,\text{nm}$, is reflected at BS and triggers D_2, while the idler, $\lambda_0 = 703.7\,\text{nm}$, is transmitted at BS and triggers D_1. Following our discussions in previous section, we find the effective wavefunction measured by D_1 and D_2 is the coherent superposition of the above two indistinguishable two-photon amplitudes:

$$\Psi(\tau_1, \tau_2) = (\hat{e}_1 \cdot \hat{e}_o)(\hat{e}_2 \cdot \hat{e}_e)\,\Psi(\tau_{s1}, \tau_{i2}) + (\hat{e}_1 \cdot \hat{e}_e)(\hat{e}_2 \cdot \hat{e}_0)\,\Psi(\tau_{i1}, \tau_{s2}).$$

$$(7.4.1)$$

The second-order coherence function is thus

$$G^{(2)}(\tau_1, \tau_2)$$

$$= \left|\, \Psi(\tau_{s1}, \tau_{i2}) - \Psi(\tau_{i1}, \tau_{s2}) \,\right|^2$$

$$= \left|\, e^{-i\omega_p(\tau_{s1}+\tau_{i2})/2}\, \mathcal{F}_{\tau_{s1}-\tau_{i2}}\{f(\nu)\}e^{-i(\omega_{s0}-\omega_{i0})(\tau_{s1}-\tau_{i2})/2} \right.$$

$$\left. - e^{-i\omega_p(\tau_{i1}+\tau_{s2})/2}\, \mathcal{F}_{\tau_{i1}-\tau_{s2}}\{f(\nu)\}e^{-i(\omega_{s0}-\omega_{i0})(\tau_{i1}-\tau_{s2})/2} \right|^2$$

$$= \left|\, \mathcal{F}_{\tau_{s1}-\tau_{i2}}\{f(\nu)\}e^{-i\omega_b(\tau_{s1}-\tau_{i2})/2} - \mathcal{F}_{\tau_{i1}-\tau_{s2}}\{f(\nu)\}e^{-i\omega_b(\tau_{i1}-\tau_{s2})/2} \right|^2$$

$$(7.4.2)$$

where $\omega_b = \omega_{s0} - \omega_{i0}$ is the beating frequency. It is easy to find that $\tau_{s1} + \tau_{i2} = \tau_{i1} + \tau_{s2}$, therefore, we can move $e^{-i\omega_p(\tau_{s1}+\tau_{i2})/2}$ term outside $|...|^2$ and normalized it to 1. Let us examine the cross interference term

$$\mathcal{F}^*_{\tau_{s1}-\tau_{i2}}\{f(\nu)\}\,\mathcal{F}_{\tau_{i1}-\tau_{s2}}\{f(\nu)\}e^{-i\omega_b[(\tau_{i1}-\tau_{s2})-(\tau_{s1}-\tau_{i2})]/2}, \qquad (7.4.3)$$

we found the cross interference term will contribute a sinusoidal modulation of beating frequency ω_b to the measurement of $G^{(2)}(\tau_1, \tau_2)$, if we keep the two Fourier transforms overlap in space-time while introducing the time delay, δ, of the quartz plates.

The coincidence counting rate becomes:

$$R_c(\delta) = \int_T dt_1 dt_2 \left|\, \Psi(\tau_{s1}, \tau_{i2}) - \Psi(\tau_{i1}, \tau_{s2}) \,\right|^2$$

$$= R_{c0}\left[1 - \cos\omega_B\delta\; \mathcal{F}^*_{\tau_{s1}-\tau_{i2}}\{f(\nu)\} \otimes \mathcal{F}_{\tau_{i1}-\tau_{s2}}\{f(\nu)\} \right]$$

$$\simeq R_{c0}\left[1 - \cos\omega_b\delta \right] \qquad (7.4.4)$$

when the convolution yields a value close to one.

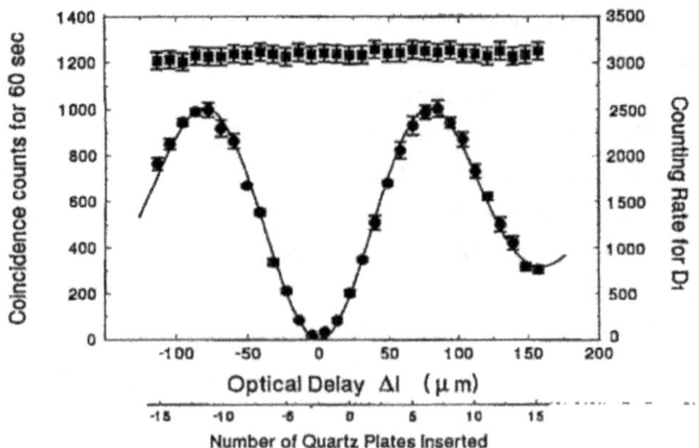

Fig. 7.4.2   Lower curve: Coincidence counts in 60 second as a function of δ, which corresponds to a certain number of quartz plates. The solid curve is a fitting of sinusoidal modulation of frequency $\omega_b = \omega_{s0} - \omega_{i0}$. The observed beating frequency is 1.83×10^{10} Hz, agreeing well with the expected beating frequency. The modulation visibility is $(97 \pm 2)\%$. Upper curve: Single detector counting rate of D_1 as a function of δ. The counting rates of both D_1 and D_2 are constants during the insertion of the crystal quartz plates.

Two-photon beats is therefore observable when the number of crystal quartz plates, i.e., the optical delay δ, is manipulated. The visibility of the sinusoidal modulation is mainly determined by the degree of overlapping between the two Fourier transforms. Figure 7.4.2 reports a measurement of the 1994 experiment of Shih and Sergienko. This experiment observed a sinusoidal modulation of 1.83×10^{10} Hz, agreeing well with the expected beating frequency. The modulation visibility was $(97 \pm 2)\%$ while the single detector counting rates of D_1 and D_2 both kept flat during the insertion of the crystal quartz plates.

7.5   Franson Interferometer

In 1989, Franson proposed an interferometer to explore the surprising behavior of entangled photon pairs. Figure 7.5.1 is a schematic setup of a Franson interferometer, which consists of an entangled biphoton source, and a pair of classic unbalanced interferometers with photon counting detectors coupled at their output ports. A pair of entangled photons, such as the signal photon and the idler photon of SPDC are sent into the unbalanced interferometers 1 and 2, respectively. The photon counting detectors are

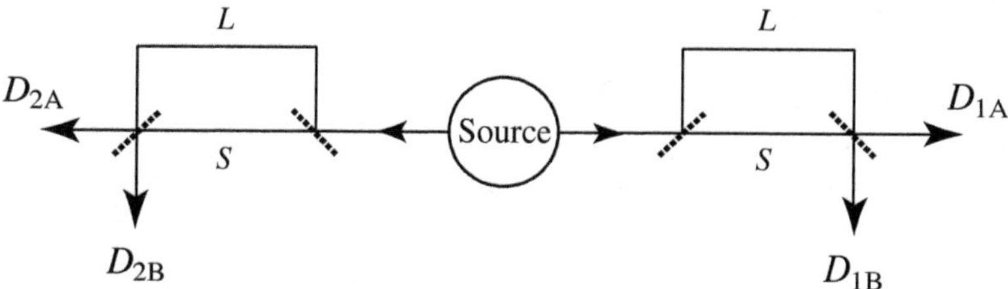

Fig. 7.5.1   Schematic setup of a Franson interferometer.

used for monitoring the single-detector counting rates, independently, and for observing the joint-detection counting rate, coincidentally. The optical path differences of the two interferometers, ΔL_1 and ΔL_1 are both chosen to be much greater than the coherence length, $l_c^{s,i}$, of the signal–idler field, thus, there is no observable first-order interference in the single-detector counting rates of D_1 and D_2 when increasing or decreasing the vales of ΔL_1 and ΔL_2 either individually or simultaneously. The joint-detection counting rate of D_1 and D_2, however, shows $\sim 100\%$ interference if the operation of the two interferometers satisfy the following conditions: (1) the photon pair only passes through the long-long and the short-short paths of the interferometers; (2) $|\Delta L_1 - \Delta L_2| \ll l_c^{s,i}$; (3) $\Delta L_1 + \Delta L_2 \ll l_c^p$ where l_c^p is the coherence length of the pump for SPDC. The surprising observation is the result of a biphoton interference phenomenon.

In the following calculation, we assume condition (1) is satisfied, i.e., there are only two alternatives, $\Psi(\tau_1^L, \tau_2^L)$ and $\Psi(\tau_1^S, \tau_2^S)$, contribute to a joint-photodetection event of D_1 and D_2. The coincidence counting rate of D_1 and D_2 is thus

$$
R_c \propto \int_T dt_1\, dt_2 \left| \Psi(\tau_1^L, \tau_2^L) + \Psi(\tau_1^S, \tau_2^S) \right|^2
$$

$$
= \int_T dt_1\, dt_2 \left[\left| \Psi(\tau_1^L, \tau_2^L) \right|^2 + \left| \Psi(\tau_1^S, \tau_2^S) \right|^2 \right.
$$

$$
\left. + \Psi^*(\tau_1^L, \tau_2^L)\, \Psi(\tau_1^S, \tau_2^S) + \Psi(\tau_1^L, \tau_2^L)\, \Psi^*(\tau_1^S, \tau_2^S) \right], \tag{7.5.1}
$$

where the subscripts L and S of τ label the long path and the short path of the jth classic interferometer, $j = 1, 2$. The cross term is the nontrivial term that determines the interference. Now, we further assume a biphoton wavepacket of SPDC, $\Psi(\tau_1, \tau_2) \sim \Psi_0 v(\tau_1 + \tau_2) u(\tau_1 - \tau_2)$, as shown in

Eq. (7.1.9). The interference term can be written as

$$\int_T dt_1\, dt_2\, \Psi^*(\tau_1^L, \tau_2^L)\, \Psi(\tau_1^S, \tau_2^S)$$

$$= e^{i\,\omega_p^0(\Delta L_1 + \Delta L_2)/2c}\left[\mathcal{F}^*_{\tau_1^L + \tau_2^L}\{g(\nu_p)\} \otimes \mathcal{F}_{\tau_1^S + \tau_2^S}\{g(\nu_p)\}\right]$$

$$\times\, e^{i(\omega_s^0 - \omega_i^0)(\Delta L_1 - \Delta L_2)/2c}\left[\mathcal{F}^*_{\tau_1^L - \tau_2^L}\{f(\nu)\} \otimes \mathcal{F}_{\tau_1^S - \tau_2^S}\{f(\nu)\}\right].$$

$$(7.5.2)$$

It is easy to see that the two convolutions in the brackets require the satisfaction of conditions (2) and (3) for observing interference from a Franson interferometer. The interference pattern has two parts of sinusoidal modulation: the sum frequency $\omega_s^0 + \omega_i^0 = \omega_p^0$ and the beating frequency $\omega_s^0 - \omega_i^0$. If degenerate SPDC is applied, and if one manipulates the the optical path difference of the interferometers simultaneously with $\Delta L_1 = \Delta L_2 = \Delta L$, the interference pattern keeps the sum frequency only as predicated by Franson in 1989:

$$R_c \propto 1 + V\cos(\omega_p\tau), \qquad (7.5.3)$$

where $\tau \equiv \Delta L/c$ is the time delay between the long and short paths of the interferometer and V is the interference visibility, which is evaluated from the two convolutions in Eq. (7.5.2).

Franson interferometer has been studied intensively in the 1990s with the use of entangled two-photon source of SPDC. Most of the interesting physics associated with Franson interferometer have been experimentally observed.

The implementation of condition (1) is not that straightforward even if we are given an entangled biphoton source of SPDC, two standard Mach–Zehnder interferometers, and proper joint-photodetection electronics. Naturally, there are four alternative ways in which the signal–idler photon pair may contribute to a joint-photodetection event. Besides $\Psi(\tau_1^L, \tau_2^L)$ and $\Psi(\tau_1^S, \tau_2^S)$, the other two alternatives, or biphoton amplitudes, $\Psi(\tau_1^L, \tau_2^S)$ and $\Psi(\tau_1^S, \tau_2^L)$ do not despair automatically in a realistic "coincidence" measurement with finite coincidence time window. The joint-photodetection counting rate of D_1 and D_2 is the result of a superposition that contains four alternatives:

$$R_c \propto \int_T dt_1\, dt_2\, \left|\, \Psi(\tau_1^L, \tau_2^L) + \Psi(\tau_1^S, \tau_2^S) + \Psi(\tau_1^L, \tau_2^S) + \Psi(\tau_1^S, \tau_2^L)\,\right|^2.$$

$$(7.5.4)$$

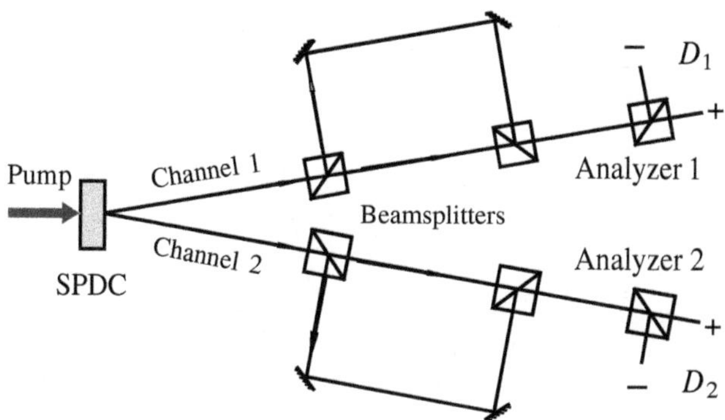

Fig. 7.5.2 Scheme setup of a Franson interferometer. The entangled biphoton source is a type II noncollinear SPDC. The clever use of polarization guarantees the implementation of condition (1): only $\Psi(\tau_1^L, \tau_2^L)$ and $\Psi(\tau_1^S, \tau_2^S)$ contribute to a joint-photodetection event.

Due to the operation condition of the interferometer, $\Delta L_{1,2} > l_c^{s,i}$, however, only one cross term has nonzero contribution to the interference, which is the same as shown in Eq. (7.5.2). In this case, one would observe the same interference pattern as that of Eq. (7.5.1), except the maximum interference visibility is reduced from 100% to 50%:

$$R_c \propto 1 + \frac{1}{2} V \cos\left(\omega_p \tau\right). \tag{7.5.5}$$

Figure 7.5.2 schematically illustrates a clever realization of Franson interference by Strekalov *et al.* from which $\sim 100\%$ interference visibility was observed. The entangled two-photon source is a noncollinear type II SPDC similar to that shown in Fig. 6.8.4. The unbalanced Mach–Zehnder interferometer in channel 1 and 2 is implemented by a long quartz rod followed with a Pockels cell. The quartz rods delay the slow polarization component relative to the fast one due to their birefringence. The Pockels cell, by applying an adjustable DC voltage, is for "fine-tuning" of the optical path difference, ΔL, of the interferometer. The birefringent delay of the interferometer is carefully chosen to be greater than the coherence time of the measured signal–idler field, which is mainly determined by the bandwidth of the spectral filters placed in front of D_1 and D_2. The fast-slow axes of the quartz rods as well as that of the Pockels cell are both oriented carefully to provide the $o_1 - e_2$ and $e_1 - o_2$ amplitudes with long-long and short-short optical paths, thus satisfy condition (1) of the Franson interferometer. Following the quartz rods and Pockels cells, in channel 1 and 2, are two polarization analyzers, A_1 and A_2. The axes of the analyzers

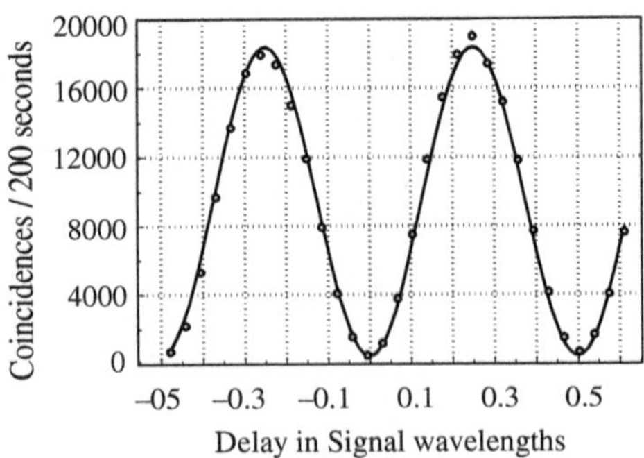

Fig. 7.5.3 An earlier experimental data of Strekalov *et al.* reported $(95.0 \pm 1.4)\%$ interference visibility in joint detection counting rate. The single detector counting rates of D_1 and D_2, however, both kept constants while tuning the optical path differences of ΔL.

are oriented at $45°$ relative to that of the quartz rod and the Pockels cell. The joint photodetection events are recorded as a function of the optical path difference ΔL with the help of a coincidence circuits in nanosecond time window. A $(95.0 \pm 1.4)\%$ visibility of interference pattern specified by Eq. (7.5.3) was reported in an earlier publication of Strekalov *et al.*, see Fig. 7.5.3. Recent measurements of Franson interferometer have observed $\sim 100\%$ interference visibility with statistical errors a few orders smaller.

The high degree two-photon coherence observed in Franson interferometer is considered as a demonstration of the nonlocal EPR inequality in energy. As we know, the loss of first-order interference in each of the Mach–Zehnder interferometer indicates a considerable large uncertainty in $\Delta \hbar \omega_{s,i}$, at least $\Delta \hbar \omega_{s,i} > 2\pi \hbar c / \Delta L$. In contrast, the two-photon interference pattern has shown quiet a high degree of visibility, which indicates

$$\Delta(\hbar \omega_s + \hbar \omega_i) \ll \min(\Delta \hbar \omega_s, \Delta \hbar \omega_i).$$

In EPR's language, the energy of neither signal photon nor idler photon is defined in the course of their preparation and propagation; however, if one is measured with a certain value the other one must be measured with a unique value.

7.6 Two-Photon Ghost Interference

A two-photon interference experiment reported by Strekalov *et al.* in 1995 surprised the physics community. The experiment was named as "ghost" interference soon after the publication. The experiment itself is quite simple. The signal photon and the idler photon of SPDC are propagated to different directions to trigger two distant point-like photon counting detectors D_1 and D_2, respectively. On the way of its propagation, the signal passed a standard Young's double-slit, while the idler propagated freely to reach D_2. Due to the poor spatial coherence of the signal field, there is no observable standard first-order Young's interference when scanning D_1 transversely behind the double-slit. A high visibility second-order double-slit interference–diffraction pattern, however, was observed in the joint photodetection counting rate of D_1 and D_2 when D_2 was scanned across the "empty" idler beam while D_1 was placed in a fixed position behind the double-slit. The name of "ghost" was given because of the surprising nonlocal feature of the phenomenon.

We now understood the observation is a two-photon interference phenomenon. The very special physics explored in the ghost interference experiment might have been its apparent nonlocal behavior: By scanning D_2 across the idler beam, how could one observe the interference pattern produced by the signal beam in distance?

The schematic experimental setup of the historical ghost interference experiment is illustrated in Fig. 7.6.1. A pair of orthogonal polarized signal–idler photon is prepared by a near-collinear degenerate type-II SPDC. The signal and the idler are separated by a polarization beamsplitter. The signal passes through a Young's double-silt (or single-slit) aperture and then travels about $1m$ to meet a point-like photon counting detector D_1. The idler travels to the far-field zone to feed into an optical fiber which

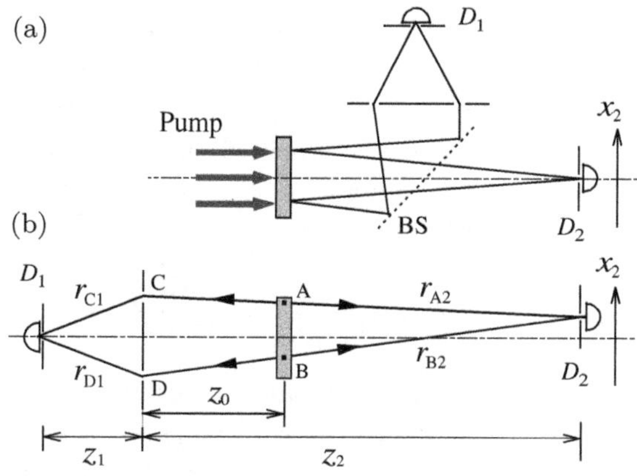

Fig. 7.6.1 Simplified experimental scheme (a) and the "unfolded" version (b).

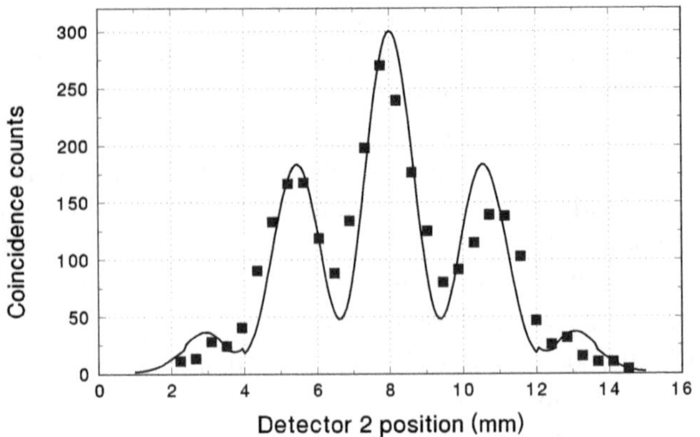

Fig. 7.6.2 Typical observed interference–diffraction pattern. The solid curve is a theoretical fitting. The calculation has taken into account the finite size of D_1 and D_2, resulting in less than 100% interference visibility. In this measurement, D_1 was fixed in a symmetrical position behind the double-slit in the signal beam, while D_2 was scanned across the "empty" idler beam. If D_1 is moved to an asymmetrical point, which results in unequal distance to the two slits, the interference–diffraction pattern is observed to be simply shifted to one side.

is mated with a photon counting detector D_2. During the joint-detection measurement, D_1 is fixed at a point behind the double-slit while the horizontal transverse coordinate, x_2, of the fiber input tip, which is equivalent to that of D_2, is scanned by a step motor.

Figure 7.6.2 is the ghost interference–diffraction pattern published by Strekalov *et al.* The coincidence counting rate is reported as a function of x_2, which is obtained by scanning D_2 (the fiber tip) across the idler beam, whereas the double-slit is in the signal beam. Young's double-silt has a slit-width of $a = 0.15$ mm and slit-distance of $d = 0.47$ mm. The interference period is measured to be 2.7 ± 0.2 mm and the half-width of the envelope is estimated to be about 8 mm. By curve fittings, it is easy to find that the observation is a standard Young's interference pattern, i.e., a sinusoidal oscillation with a sinc-function envelope:

$$R_c \propto \mathrm{sinc}^2 \left(\frac{\pi a x_2}{\lambda z_2} \right) \cos^2 \left(\frac{\pi d x_2}{\lambda z_2} \right). \tag{7.6.1}$$

The interference pattern in Fig. 7.6.2 which is described by Eq. (7.6.1) was taken when D_1 was placed in a symmetrical point between the double-slit. If D_1 is moved to an asymmetrical point, which results in unequal distances to the two slits, the interference–diffraction pattern is shifted from the current symmetrical position to one side of x_2. Similar to the ghost imaging experiment, the remarkable feature here is that z_2 is the distance from the slits' plane, which is in the signal beam, back through BS to the SPDC crystal and then along the idler beam to the scanning fiber tip of detector D_2 (see Fig. 7.6.1). The calculated interference period and half-width of the sinc-function from Eq. (7.6.1) are 2.67 mm and 8.4 mm, respectively.

Although the interference–diffraction pattern is observed in coincidences, *the single detector counting rates are both observed to be constant* when scanning detector D_1 and D_2. It seems reasonable not to have any interference modulation in the single counting rate of D_2, which is located in the "empty" idler beam. Of interest, however, is that the absence of the interference–diffraction structure in the single counting rate of D_1, which is behind the double-slit, is mainly due to the poor spatial coherence or the considerable large divergence of the signal beam, $\Delta\theta \gg \lambda/d$.

To explain the ghost interference, a simple model is presented in the following. The basic concept of the model is, again, two-photon interference of entangled state. In Chapter 9, we have introduced two EPR δ-functions $\delta(\vec{\rho}_s - \vec{\rho}_i)$ and $\delta(\vec{k}_s + \vec{k}_i)$ for near-collinear degenerate SPDC, which means (1) the signal–idler pair may come out from any point on the output plane of the SPDC, however, if the signal is measured at a certain position the idler must be emitted from the same position and (2) the signal and idler may propagate to any directions around the pump beam, however, if the signal is observed in a certain direction the idler must be emitted to the opposite direction with equal angle relative to the pump. This peculiar entanglement nature of the signal–idler two-photon system determines the only two possible two-photon amplitudes in Fig. 7.6.1, when signal passes through the double-slit aperture while the idler triggers D_2. The coherent superposition is taken between these two-photon amplitudes.

Encouraged by the EPR δ-functions, Klyshko suggested an "Klyshko picture" to treat the SPDC crystal as a mirror in terms of "usual" geometrical optics in the following manner: we envision the output plane as a "hinge point" and "unfold" the schematic of Fig. 7.6.1(a) into that shown in Fig. 7.6.1(b). Based on the unfolded Klyshko picture of Fig. 7.6.1(b), we now give an quantitative calculation of the experiment.

The joint detection counting rate R_c is proportional to the probability of jointly detecting the signal–idler pair by detectors D_1 and D_2:

$$R_c \propto G^{(2)} = \langle\Psi|\,\hat{E}_1^{(-)}\hat{E}_2^{(-)}\hat{E}_2^{(+)}\hat{E}_1^{(+)}\,|\Psi\rangle = \left|\langle 0|\hat{E}_2^{(+)}\hat{E}_1^{(+)}|\Psi\rangle\right|^2, \qquad (7.6.2)$$

where $|\Psi\rangle$ is the two-photon state of SPDC. Let us simplify the mathematics by using the following "two-mode" expression for the state, bearing in mind that the EPR δ-functions have been taken into account based on the "straight line" picture of Fig. 7.6.1:

$$|\Psi\rangle = \epsilon\,[a_s^\dagger a_i^\dagger e^{i\varphi_A} + b_s^\dagger b_i^\dagger\, e^{i\varphi_B}]\,|0\rangle\,, \tag{7.6.3}$$

where ϵ is a normalization constant that is proportional to the pump field (classical) and the nonlinearity of the crystal, φ_A and φ_B are the phases of the pump field at A and B, and $a_j^\dagger$ $(b_j^\dagger)$ are the photon creation operators for the upper (lower) mode in Fig. 7.6.1 $(j = s, i)$. In terms of the Copenhagen interpretation, one may say that the interference is due to the uncertainty in the birth-place (A or B in Fig. 7.6.1) of a signal–idler pair.

In Eq. (7.6.2), the fields at the detectors are given by

$$\begin{aligned}
\hat{E}_1^{(+)} &= \hat{a}_s\,\exp(ik\,r_{A1}) + \hat{b}_s\exp(ik\,r_{B1}), \\
\hat{E}_2^{(+)} &= \hat{a}_i\,\exp(ik\,r_{A2}) + \hat{b}_i\,\exp(ik\,r_{B2}),
\end{aligned} \tag{7.6.4}$$

where r_{Ai} (r_{Bi}) are the optical path lengths from region A (B) along the upper (lower) path to the ith detector. Substituting Eqs. (7.6.3) and (7.6.4) into Eq. (7.6.2),

$$G^{(2)} \propto \left| e^{i(kr_A+\varphi_A)} + e^{i(kr_B+\varphi_B)} \right|^2 \propto 1 + \cos[k\,(r_A - r_B)], \tag{7.6.5}$$

where we have assumed $\varphi_A = \varphi_B$ in the second line of Eq. (7.6.5). We have also defined the overall optical path lengths between the detectors D_1 and D_2 along the upper and lower paths (see Fig. 7.6.1): $r_A \equiv r_{A1} + r_{A2} = r_{C1} + r_{C2}$, $r_B \equiv r_{B1} + r_{B2} = r_{D1} + r_{D2}$, where r_{Ci} and r_{Di} are the respective path lengths from the slits C and D to the ith detector.

If the optical paths from the fixed detector D_1 to the two slits are equal, i.e., $r_{C1} = r_{D1}$, and if $z_2 \gg d^2/\lambda$ (far field), then $r_A - r_B = r_{C2} - r_{D2} \cong x_2 d/z_2$, and Eq. (7.6.5) can be written as

$$R_c \propto \cos^2\left(\frac{\pi d x_2}{\lambda z_2}\right). \tag{7.6.6}$$

Equation (7.6.6) has the form of standard Young's double-slit interference pattern. Here, again, z_2 is the unusual distance from the slits plane, which is in the signal beam, back through BS to the crystal and then along the idler beam to the scanning fiber tip of detector D_2.

If the optical paths from the fixed detector D_1 to the two slits are unequal, i.e., $r_{C1} \neq r_{D1}$, the interference pattern will be shifted from the symmetrical position of Eq. (7.6.6) to an asymmetrical position of Eq. (7.6.5). This interesting phenomenon has been observed and discussed following the discussion of Fig. 7.6.2.

To calculate the "ghost" diffraction effect of a single-slit, we need an integral of the effective two-photon wavefunction over the slit width (the superposition of infinite number of probability amplitudes results in a click-click coincidence detection event):

$$R_c \propto \left| \int_{-a/2}^{a/2} dx_0 \exp[-ik\, r(x_0, x_2)] \right|^2 \cong \mathrm{sinc}^2 \left(\frac{\pi a x_2}{\lambda z_2} \right), \qquad (7.6.7)$$

where $r(x_0, x_2)$ is the distance between points x_0 and x_2, x_0 belongs to the slit's plane, and the inequality $z_2 \gg a^2/\lambda$ is applied (far field approximation).

Repeating the above calculations, the combined interference–diffraction joint detection counting rate for the double-slit case is given by

$$R_c \propto \mathrm{sinc}^2 \left(\frac{\pi a x_2}{\lambda z_2} \right) \cos^2 \left(\frac{\pi d x_2}{\lambda z_2} \right), \qquad (7.6.8)$$

which is the same function as that of Eq. (7.6.1) obtained from experimental data fittings. If the finite size of the detectors and the divergence of the pump are taken into account by a convolution, the interference visibility will be reduced. These factors have been considered in the theoretical plots of Fig. 7.6.2.

Similar to Franson interferometer which demonstrated the nonlocal EPR inequality in energy, the ghost interference experiment has explored another nonlocal EPR inequality in momentum. As we know, the loss of first-order spatial coherence of the signal (idler) indicates a considerable large uncertainty in the transverse component of its momentum, $\Delta \vec{\kappa}_{s,i}$. In contrast, the two-photon interference pattern has shown quiet a high degree of two-photon spatial coherence, which indicates

$$\Delta(\vec{\kappa}_s + \vec{k}_i) \ll \min(\Delta \vec{k}_s, \Delta \vec{\kappa}_i).$$

In EPR's language, the transverse momentum of neither signal photon nor idler photon is defined in the course of their preparation and propagation; however, if one is measured with a certain value the other one must be measured with a unique value despite the distance between the two measurements.

7.7 Biphoton Ghost Imaging

The *nonlocal* position–position and momentum–momentum correlation of the entangled biphoton system of SPDC was successfully demonstrated in 1995. The experiment received an interesting name "ghost" imaging immediately in the physics community. The important physics demonstrated in that experiment, however, may not be the so called "ghost". Indeed, the original purpose of the experiment was to study the EPR correlation in position and in momentum for an entangled biphoton system.

The schematic setup of the ghost imaging experimental is shown in Fig. 7.7.1. A CW laser is used to pump a nonlinear crystal, which is cut for degenerate type-II phase matching to produce a pair of orthogonally polarized signal (e-ray of the crystal) and idler (o-ray of the crystal) photon. The pair emerges from the crystal as collinear, with $\omega_s \cong \omega_i \cong \omega_p/2$. The pump is then separated from the signal–idler pair by a dispersion prism, and the remaining signal and idler beams are sent in different directions by a polarization beam splitting (Thompson prism). The signal beam passes through a convex lens with a 400 mm focal length and illuminates a chosen aperture (mask). As an example, one of the demonstrations used letters "UMBC" for the object mask. Behind the aperture is the "bucket" detector package D_1, which consists of a short focal length collection lens in whose

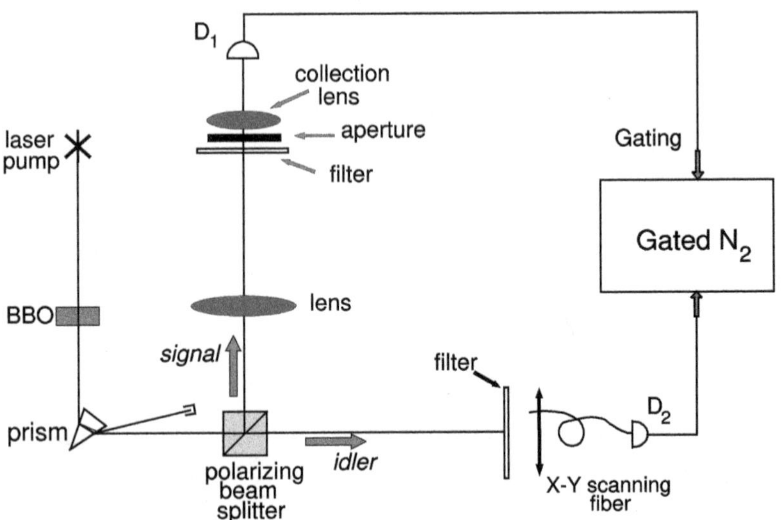

Fig. 7.7.1 Schematic set-up of the "ghost" image experiment.

focal spot is an avalanche photodiode. D_1 is mounted in a fixed position during the experiment. The idler beam is met by detector package D_2, which consists of an optical fiber whose output is mated with another avalanche photodiode. The input tip of the fiber is scannable in the transverse plane by two step motors. The output pulses of each detector, which are operated in photon counting mode, are sent to a coincidence circuit for counting the joint-detection event of the signal–idler pair.

By recording the coincidence counts as a function of the fiber tip's transverse plane coordinates, the image of the chosen aperture (for example, "UMBC") is observed, as reported in Fig. 7.7.2, if the following experimental condition is satisfied: the focal length of the imaging lens f, the aperture's optical distance from the lens s_o, and the image's optical distance from the lens s_i (which is from the imaging lens going backward along the signal photon path to the biphoton source of SPDC crystal then going forward along the path of idler photon to the image), satisfy the Gaussian thin lens equation. When choosing $s_i/s_o = 2$, it is interesting to

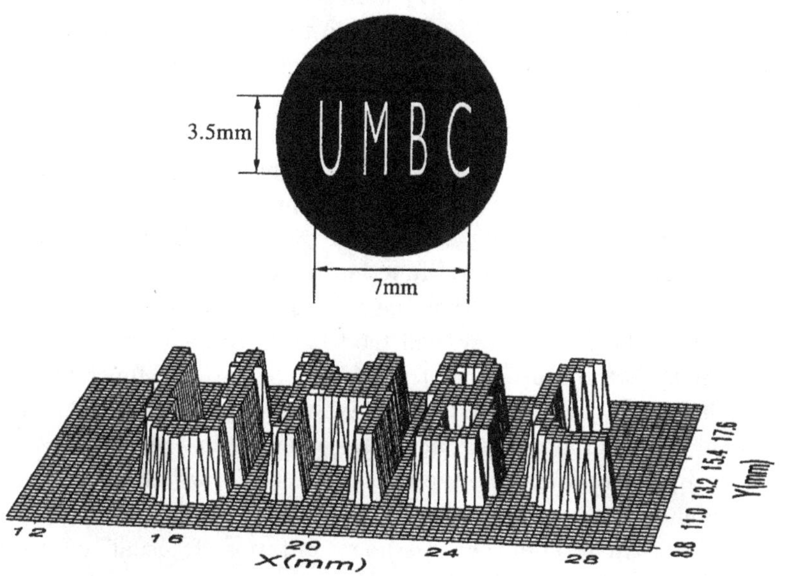

Fig. 7.7.2 (a) A reproduction of the actual aperture "UMBC" placed in the signal beam. (b) The image of "UMBC": Coincidence counts as a function of the fiber tip's transverse plane coordinates. The step size is 0.25 mm. The image shown is a "slice" at the half-maximum value.

find that the observed image measures 7 mm $\times$ 14 mm while the size of the "UMBC" aperture inserted in the signal beam is only about 3.5 mm $\times$ 7 mm. The observed image in the joint detection of D_1 and D_2 is magnified exactly a factor of $m = s_i/s_o = 2$. In the measurement of Fig. 7.7.2, s_o was chosen to be $s_o = 600$ mm, and the twice magnified clear image was found when the fiber tip was on the plane of $s_i = 1200$ mm. While D_2 was scanned on other transverse planes other than that of $s_i = 1200$ mm, the images blurred out.

The first reaction of many people may be negative: "absolutely impossible!" Examine the two observers D_1 and D_2: D_1 is in the signal beam behind the object, however, it is a "bucket" detector, it does not have any ability to image the object. What D_1 can do is simply counting the signal photons; whenever a photoelectron is created by a signal photon, D_1 sends an electronic pulse to open the "gate" of the coincidence circuit for D_2. D_2, however, is placed in the idler beam, how could D_2 "see" the object that is in the signal beam? The measurement of the signal and the idler subsystem themselves may further strengthen the negative reaction. In fact, the single photon counting rate of D_1 and D_2 were recorded during the scanning of the image and was found fairly constant in the entire region of the image. The counting rate of neither D_1 nor D_2 is a function of the fiber tips transverse coordinates.

The EPR-correlation δ-functions, $\delta(\vec{\rho}_s - \vec{\rho}_i)$ and $\delta(\vec{\kappa}_s + \vec{\kappa}_i)$ in transverse dimension, are the key to understand this interesting phenomenon. In degenerate SPDC, although the signal–idler photon pair has equal probability to be emitted from any points on the output surface of the nonlinear crystal, the transverse position δ-function indicates that if one of them is observed at one position, the other one must be found at the same position. In other words, the pair is always emitted from the same point on the output plane of the biphoton source. The transverse momentum δ-function, defines the angular correlation of the signal–idler pair: the transverse momenta of a signal–idler amplitude are equal but pointed in opposite directions: $\vec{\kappa}_s = -\vec{\kappa}_i$. In other words, the biphoton amplitudes are always existing at roughly equal yet opposite angles relative to the pump. This then allows for a simple explanation of the experiment in terms of "usual" geometrical optics in the following manner: We envision the nonlinear crystal as a "hinge point" and "unfold" the schematic of Fig. 7.7.1 into that shown in Fig. 7.7.3. The signal–idler biphoton amplitudes can then be represented by straight lines (but keep in mind the different propagation directions) and therefore, the image is well produced in coincidences when the aperture, lens, and fiber

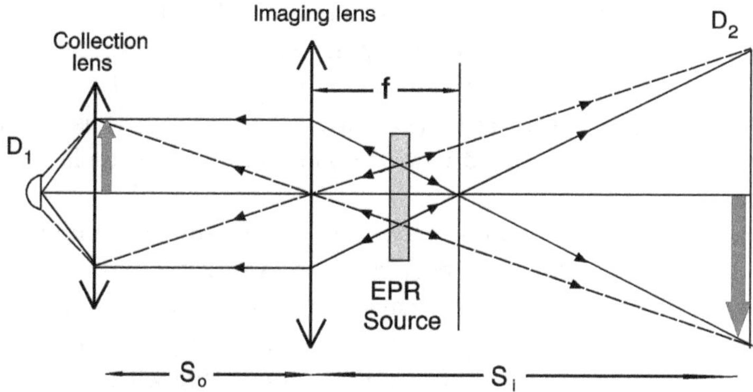

Fig. 7.7.3   An unfolded setup of the "ghost" imaging experiment, which is helpful for understanding the physics. Since the biphoton "light" propagates along "straight-lines", it is not difficult to find that any geometrical light point on the subject plane corresponds to an unique geometrical light point on the image plane. Thus, a "ghost" image of the subject is made nonlocally in the image plane. Although the placement of the lens, the object, and detector D_2 obeys the Gaussian thin lens equation, it is important to remember that the geometric rays in the figure actually represent the biphoton amplitudes of an entangled signal–idler pair. The point to point correspondence is the result of the superposition of these biphoton amplitudes.

tip are located according to the Gaussian thin lens equation. The image is exactly the same as one would observe on a screen placed at the fiber tip if detector D_1 were replaced by a point-like light source and the nonlinear crystal by a reflecting mirror.

Following a similar analysis in geometric optics, it is not difficult to find that any geometrical "light spot" on the subject plane, which is the intersection point of all possible signal–idler amplitudes coming from the entangled biphoton source, corresponds to an unique geometrical "light spot" on the image plane, which is another intersection point of all the possible signal–idler amplitudes. This point to point correspondence made the "ghost" image of the subject-aperture possible. Despite the completely different physics from classical geometrical optics, the remarkable feature is that the relationship between the focal length of the lens f, the aperture's optical distance from the lens S_o, and the image's optical distance from the lens S_i, satisfy the Gaussian thin lens equation:

$$\frac{1}{s_o} + \frac{1}{s_i} = \frac{1}{f}.$$

Although the placement of the lens, the object, and the detector D_2 obeys the Gaussian thin lens equation, it is important to remember that the geometric rays in the figure actually represent the biphoton amplitudes of a signal–idler photon pair and the point to point correspondence is the result of the superposition of these biphoton amplitudes. It is the imaging lens made these biphoton paths (amplitudes) equal and thus superposed constructively at these unique pair of points on the object plane and the ghost image plane.

The signal and idler photons may propagate to any points on the object and the ghost image planes, respectively, however, if the signal is passing a point on the object plane, the idler must stopped at an unique point on the ghost image plane. In other words, the transverse coordinate uncertainty of either signal or idler is considerably greater comparing with the correlation of the transverse coordinates of the entangled signal–idler photon pair: Δx_1 (Δy_1) and Δx_2 (Δy_2) are much greater than $\Delta(x_1 - x_2)$ $(\Delta(y_1 - y_2))$. The "ghost" image is a realization of the 1935 EPR *gedankenexperiment*.

Now we calculate $G^{(2)}(\vec{\rho}_o, \vec{\rho}_i)$ for the "ghost" imaging experiment, where $\vec{\rho}_o$ and $\vec{\rho}_i$ are the transverse coordinates on the object plane and the image plane. We show that there exists a δ-function like point-to-point relationship between the object plane and the image plane, i.e., if one observes the signal photon at a position of $\vec{\rho}_o$ on the object plane the idler photon can be observed only at a certain unique position of $\vec{\rho}_I$ on the image plane satisfying $\delta(m\vec{\rho}_o - \vec{\rho}_I)$, where $m = -(s_i/s_o)$ is the image-object magnification factor. After demonstrating the δ-function, we show how the object function of $A(\vec{\rho}_o)$ is transferred to the image plane as a magnified image $A(\vec{\rho}_i/m)$. Before showing the calculation, it is worth to emphasize again that the "straight lines" in Fig. 7.7.3 schematically represent the biphoton amplitudes all belong to a pair of signal–idler photon. A "click-click" joint measurement at $(\mathbf{r}_1, t_1)$, which is on the object plane, and $(\mathbf{r}_2, t_2)$, which is on the image plane, in the form of EPR δ-function, is the result of the coherent superposition of all these biphoton amplitudes.

We follow the unfolded experimental setup shown in Fig. 7.7.4 to establish Green's functions $g(\vec{\kappa}_s, \omega_s, \vec{\rho}_o, z_o)$ and $g(\vec{\kappa}_i, \omega_i, \vec{\rho}_2, z_2)$. In arm-1, the signal propagates freely over a distance d_1 from the output plane of the source to the imaging lens, then passes an object aperture at distance s_o, and then is focused onto photon counting detector D_1 by a collection lens. We evaluate $g(\vec{\kappa}_s, \omega_s, \vec{\rho}_o, z_o)$ by propagating the field from the output plane of the biphoton source to the object plane. In arm-2, the idler propagates freely over a distance d_2 from the output plane of the biphoton source to a point-like detector D_2. $g(\vec{\kappa}_i, \omega_i, \vec{\rho}_2, z_2)$ is thus a free propagator.

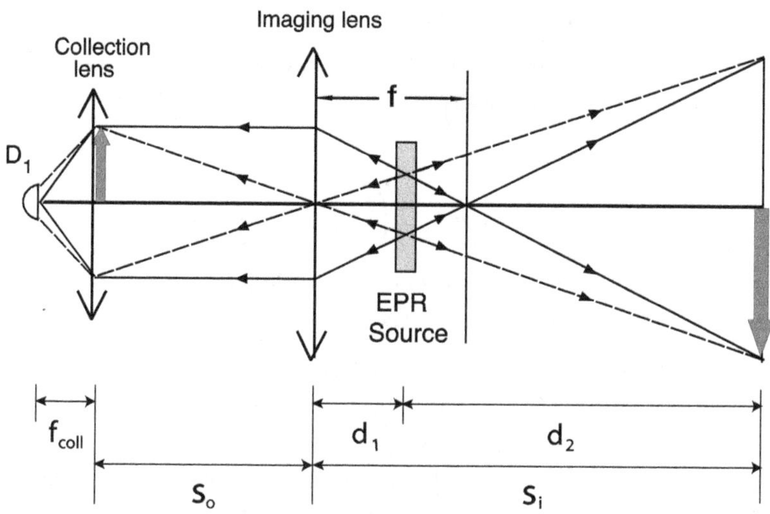

Fig. 7.7.4 In arm-1, the signal propagates freely over a distance d_1 from the output plane of the source to the imaging lens, then passes an object aperture at distance s_o, and then is focused onto photon counting detector D_1 by a collection lens. In arm-2, the idler propagates freely over a distance d_2 from the output plane of the source to a point-like photon counting detector D_2.

(I) **Arm-1 (from source to object):**

The optical transfer function or Green's function in arm-1, which propagates the field from the source plane to the object plane, is given by

$$g(\vec{\kappa}_s, \omega_s; \vec{\rho}_o, z_o = d_1 + s_o)$$

$$= e^{i\frac{\omega_s}{c}z_o} \int_{\text{lens}} d\vec{\rho}_l \int_{\text{source}} d\vec{\rho}_S \left\{ \frac{-i\omega_s}{2\pi c d_1} e^{i\vec{\kappa}_s \cdot \vec{\rho}_S} G\left(|\vec{\rho}_S - \vec{\rho}_l|, \frac{\omega_s}{cd_1} \right) \right\}$$

$$\times \left\{ G\left(|\vec{\rho}_l|, \frac{\omega_s}{cf} \right) \right\} \left\{ \frac{-i\omega_s}{2\pi c s_o} G\left(|\vec{\rho}_l - \vec{\rho}_o|, \frac{\omega_s}{cs_o} \right) \right\}, \tag{7.7.1}$$

where $\vec{\rho}_S$ and $\vec{\rho}_l$ are the transverse vectors defined, respectively, on the output plane of the source and on the plane of the imaging lens. The terms in the first and third curly brackets in Eq. (7.7.1) describe free space propagation from the output plane of the source to the imaging lens and from the imaging lens to the object plane, respectively. The function $G(|\vec{\rho}_l|, \frac{\omega}{cf})$ in the second curly brackets is the transformation function of the imaging lens. Here, we treat it as a thin lens: $G(|\vec{\rho}_l|, \frac{\omega}{cf}) \cong e^{-i\frac{\omega}{2cf}|\vec{\rho}_l|^2}$.

(II) **Arm-2 (from source to ghost image):**

In arm-2, the idler propagates freely from the source to the plane of D_2, which is also the plane of the image. Green's function is thus

$$g(\vec{\kappa}_i, \omega_i; \vec{\rho}_2, z_2 = d_2)$$

$$= \frac{-i\omega_i}{2\pi c d_2} \, e^{i\frac{\omega_i}{c} d_2} \int_{\text{source}} d\vec{\rho}_S \, G\left(|\vec{\rho}_S - \vec{\rho}_2|, \frac{\omega_i}{c d_2}\right) e^{i\vec{\kappa}_i \cdot \vec{\rho}_S}, \qquad (7.7.2)$$

where $\vec{\rho}_S$ and $\vec{\rho}_2$ are the transverse vectors defined, respectively, on the output plane of the source, and on the plane of the photodetector D_2.

(III) $\Psi(\tilde{\rho}_o, \tilde{\rho}_i)$ (object plane — ghost image plane):

To simplify the calculation and to focus on the transverse correlation, in the following calculation, we assume degenerate ($\omega_s = \omega_i = \omega$) and collinear SPDC. The transverse biphoton effective wavefunction $\Psi(\vec{\rho}_o, \vec{\rho}_2)$ is then evaluated by substituting Green's functions $g(\vec{\kappa}_s, \omega; \vec{\rho}_o, z_o)$ and $g(\vec{\kappa}_i, \omega; \vec{\rho}_2, z_2)$ into the expression given in Eq. (6.4.7):

$$\Psi(\vec{\rho}_o, \vec{\rho}_2) \propto \int d\vec{\kappa}_s \, d\vec{\kappa}_i \, \delta(\vec{\kappa}_s + \vec{\kappa}_i) \, g(\vec{\kappa}_s, \omega; \vec{\rho}_o, z_o) \, g(\vec{\kappa}_i, \omega; \vec{\rho}_2, z_2)$$

$$\propto e^{i\frac{\omega}{c}(s_0 + s_i)} \int d\vec{\kappa}_s \, d\vec{\kappa}_i \, \delta(\vec{\kappa}_s + \vec{\kappa}_i)$$

$$\times \int_{\text{lens}} d\vec{\rho}_l \int_{\text{source}} d\vec{\rho}_S \, e^{i\vec{\kappa}_s \cdot \vec{\rho}_S} G\left(|\vec{\rho}_S - \vec{\rho}_l|, \frac{\omega}{c d_1}\right)$$

$$\times \, G\left(|\vec{\rho}_l|, \frac{\omega}{c f}\right) G\left(|\vec{\rho}_l - \vec{\rho}_o|, \frac{\omega}{c s_o}\right)$$

$$\times \int_{\text{source}} d\vec{\rho}_S \, e^{i\vec{\kappa}_i \cdot \vec{\rho}_S} G\left(|\vec{\rho}_S - \vec{\rho}_2|, \frac{\omega}{c d_2}\right), \qquad (7.7.3)$$

where we have ignored all the proportional constants. Completing the double integral of $d\vec{\kappa}_s$ and $d\vec{\kappa}_s$

$$\int d\vec{\kappa}_s \, d\vec{\kappa}_i \, \delta(\vec{\kappa}_s + \vec{\kappa}_i) \, e^{i\vec{\kappa}_s \cdot \vec{\rho}_S} \, e^{i\vec{\kappa}_i \cdot \vec{\rho}_S} \sim \delta(\vec{\rho}_S - \vec{\rho}_S), \qquad (7.7.4)$$

Equation (7.7.3) becomes

$$\Psi(\vec{\rho}_o, \vec{\rho}_2) \propto \int_{\text{lens}} d\vec{\rho}_l \int_{\text{source}} d\vec{\rho}_S \, G\left(|\vec{\rho}_2 - \vec{\rho}_S|, \frac{\omega}{c d_2}\right) G\left(|\vec{\rho}_S - \vec{\rho}_l|, \frac{\omega}{c d_1}\right)$$

$$\times \, G(|\vec{\rho}_l|, \frac{\omega}{c f}) G\left(|\vec{\rho}_l - \vec{\rho}_o|, \frac{\omega}{c s_o}\right). \qquad (7.7.5)$$

We then apply the properties of the Gaussian functions and complete the integral on $d\vec{\rho}_S$ by assuming a large enough transverse size of source to be treated as infinity,

$$\Psi(\vec{\rho}_o, \vec{\rho}_2) \propto \int_{\text{lens}} d\vec{\rho}_l \, G\left(|\vec{\rho}_2 - \vec{\rho}_l|, \frac{\omega}{cs_i}\right) G\left(|\vec{\rho}_l|, \frac{\omega}{cf}\right) G\left(|\vec{\rho}_l - \vec{\rho}_o|, \frac{\omega}{cs_o}\right).$$

$$(7.7.6)$$

Although the signal and idler propagate to different directions along two optical arms, Interestingly, Green's function in Eq. (7.7.6) is equivalent to that of a classical imaging setup, if we imagine the fields start propagating from a point $\vec{\rho}_o$ on the object plane to the lens and then stop at point $\vec{\rho}_2$ on the imaging plane. The mathematics is consistent with our previous qualitative analysis of the experiment.

The integral on $d\vec{\rho}_l$ yields a point-to-point relationship between the object plane and the image plane that is defined by the Gaussian thin-lens equation:

$$\int_{\text{lens}} d\vec{\rho}_l \, G\left(|\vec{\rho}_l|, \frac{\omega}{c}\left[\frac{1}{s_o} + \frac{1}{s_i} - \frac{1}{f}\right]\right) e^{-i\frac{\omega}{c}\left(\frac{\vec{\rho}_o}{s_o} + \frac{\vec{\rho}_i}{s_i}\right) \cdot \vec{\rho}_l} \propto \delta\left(\vec{\rho}_o + \frac{\vec{\rho}_i}{m}\right),$$

$$(7.7.7)$$

where the integral is approximated to infinity and the Gaussian thin-lens equation of $1/s_o + 1/s_i = 1/f$ is applied. We have also defined $m = s_i/s_o$ as the magnification factor of the imaging system. The function $\delta(\vec{\rho}_o + \vec{\rho}_i/m)$ indicates that a point of $\vec{\rho}_o$ on the object plane corresponds to a unique point of $\vec{\rho}_i$ on the image plane. The two vectors pointed to opposite directions and the magnitudes of the two vectors hold a ratio of $m = |\vec{\rho}_i|/|\vec{\rho}_o|$.

If the finite size of the imaging lens has to be taken into account (finite diameter D), the integral yields a point-spread function of $somb(x)$:

$$\int_{\text{lens}} d\vec{\rho}_l \, e^{-i\frac{\omega}{c}\left(\frac{\vec{\rho}_o}{s_o} + \frac{\vec{\rho}_i}{s_i}\right) \cdot \vec{\rho}_l} \propto \text{somb}\left(\frac{R}{s_o} \frac{\omega}{c}\left[\vec{\rho}_o + \frac{\vec{\rho}_i}{m}\right]\right), \qquad (7.7.8)$$

where $\text{somb}(x) = 2J_1(x)/x$, $J_1(x)$ is the first-order Bessel function and R/s_o is named as the numerical aperture. The point-spread function turns the point-to-point correspondence between the object plane and the image plane into a point-to-"spot" relationship and thus limits the spatial resolution. This point has been discussed in detail in last section.

Therefore, by imposing the condition of the Gaussian thin-lens equation, the transverse biphoton effective wavefunction is approximated as a δ

function:

$$\Psi(\vec{\rho}_o, \vec{\rho}_i) \propto \delta\left(\vec{\rho}_o + \frac{\vec{\rho}_i}{m}\right), \tag{7.7.9}$$

where $\vec{\rho}_o$ and $\vec{\rho}_i$, again, are the transverse coordinates on the object plane and the image plane, respectively, defined by the Gaussian thin-lens equation. Thus, the second-order spatial correlation function $G^{(2)}(\vec{\rho}_o, \vec{\rho}_i)$ turns to be

$$G^{(2)}(\vec{\rho}_o, \vec{\rho}_i) = |\,\Psi(\vec{\rho}_o, \vec{\rho}_i)\,|^2 \propto \left|\delta\left(\vec{\rho}_o + \frac{\vec{\rho}_i}{m}\right)\right|^2. \tag{7.7.10}$$

Equation (7.7.10) indicates a point to point EPR correlation between the object plane and the image plane, i.e., if one observes the signal photon at a position of $\vec{\rho}_o$ on the object plane, the idler photon can only be found at a certain unique position of $\vec{\rho}_i$ on the image plane satisfying $\delta(\vec{\rho}_o + \vec{\rho}_i/m)$ with $m = s_i/s_o$.

We now include an object-aperture function, a collection lens and a photon counting detector D_1 into the optical transfer function of arm-1 as shown in Fig. 7.7.1.

First, we treat the collection-lens-D_1 package as a "bucket" detector. The "bucket" detector integrates all $\Psi(\vec{\rho}_o, \vec{\rho}_2)$ that pass the object aperture $A(\vec{\rho}_o)$ as a joint-photodetection event. This process is equivalent to the following convolution:

$$R_{1,2} \propto \left| \int_{\text{obj}} d\vec{\rho}_o\, A(\vec{\rho}_o)\, \Psi(\vec{\rho}_o, \vec{\rho}_i) \right|^2$$

$$\simeq \left| A(\vec{\rho}_o) \otimes \delta\left(\vec{\rho}_o + \frac{\vec{\rho}_2}{m}\right) \right|^2 = \left| A\left(\frac{-\vec{\rho}_2}{m}\right) \right|^2, \tag{7.7.11}$$

where $\otimes$ means convolution, again, D_2 is scanning in the image plane, $\vec{\rho}_2 = \vec{\rho}_i$. Equation (7.7.11) indicates a magnified (or demagnified) image of the object-aperture function by means of the joint-detection events between distant photodetectors D_1 and D_2. The "$-$" sign in $A(-\vec{\rho}_i/m)$ indicates opposite orientation of the image. The model of "bucket" detector is a good and realistic approximation.

Second, we calculate Green's function from the source to D_1 in detail by including the object-aperture function, the collection lens and the photon counting detector D_1 into arm-1. Green's function of Eq. (7.7.1) becomes

$$g(\vec{\kappa}_s, \omega_s; \vec{\rho}_1, z_1 = d_1 + s_o + f_{\text{coll}})$$

$$= e^{i\frac{\omega_s}{c} z_1} \int_{\text{obj}} d\vec{\rho}_o \int_{\text{lens}} d\vec{\rho}_l \int_{\text{source}} d\vec{\rho}_S \left\{ \frac{-i\omega_s}{2\pi c d_1} e^{i\vec{\kappa}_s \cdot \vec{\rho}_S} G\left(|\vec{\rho}_S - \vec{\rho}_l|, \frac{\omega_s}{c d_1} \right) \right\}$$

$$\times \; G\left(|\vec{\rho}_l|, \frac{\omega_s}{cf} \right) \left\{ \frac{-i\omega_s}{2\pi c s_o} G\left(|\vec{\rho}_l - \vec{\rho}_o|, \frac{\omega_s}{c s_o} \right) \right\} A(\vec{\rho}_o)$$

$$\times \; G\left(|\vec{\rho}_o|, \frac{\omega_s}{c f_{\text{coll}}} \right) \left\{ \frac{-i\omega_s}{2\pi c f_{\text{coll}}} G\left(|\vec{\rho}_o - \vec{\rho}_1|, \frac{\omega_s}{c f_{\text{coll}}} \right) \right\}, \qquad (7.7.12)$$

where f_{coll} is the focal length of the collection lens and D_1 is placed on the focal point of the collection lens. Repeating the previous calculation, we obtain the transverse biphoton effective wavefunction:

$$\Psi(\vec{\rho}_1, \vec{\rho}_2) \propto \int_{\text{obj}} d\vec{\rho}_o \, A(\vec{\rho}_o) \, \delta\left(\vec{\rho}_o + \frac{\vec{\rho}_2}{m} \right) = A(\vec{\rho}_o) \otimes \delta\left(\vec{\rho}_o + \frac{\vec{\rho}_2}{m} \right),$$

$$(7.7.13)$$

which is the same as Eq. (7.7.11). Note that in Eq. (7.7.13) we have ignored the phase factors which have no contribution to the formation of the image. The joint detection counting rate, $R_{1,2}$, between photon counting detectors D_1 and D_2 is thus

$$R_{1,2} \propto G^{(2)}(\vec{\rho}_1, \vec{\rho}_2) \propto \left| A(\vec{\rho}_o) \otimes \delta\left(\vec{\rho}_o + \frac{\vec{\rho}_2}{m} \right) \right|^2 = \left| A\left(\frac{-\vec{\rho}_2}{m} \right) \right|^2, \quad (7.7.14)$$

where, again, $\vec{\rho}_2 = \vec{\rho}_i$.

The physical process corresponding to the above convolution can be summarized as follows. Due to the unique point-to-point correlation between the object plane and the image plane, whenever the bucket detector receives a signal photon that is either transmitted, scattered or reflected from a unique point of the object, the scanning point photodetector D_2 or a CCD element that receives the idler photon identifies the coordinate of $\vec{\rho}_o$ and the value of the aperture function $A(\vec{\rho}_o)$ for that joint-photodetection event. For instance, at time t, the bucket detector receives a signal photon that is either transmitted, scattered, or reflected from a unique point $\vec{\rho}_o$ of the object plane within a coincidence time window. The joint detection of the idler photon by the scanning point detector D_2 or a CCD element, with known coordinate $\vec{\rho}_i$, identifies $\vec{\rho}_o$ immediately for that event. At time t', the bucket detector receives another signal photon that is either transmitted, scattered, or reflected from another unique point $\vec{\rho'}_o$ of the object plane within the coincidence window, and the joint detection of the idler photon by the scanning point detector D_2 or a CCD element, with known coordinate $\vec{\rho}'_i$, identifies $\vec{\rho}'_o$ immediately for that event. The probability of receiving a joint-photodetection event at $(\vec{\rho}_o, \vec{\rho}_i)$ and at $(\vec{\rho}'_o, \vec{\rho}'_i)$ is proportional to the value of the aperture function $A(\vec{\rho}_o)$ and $A(\vec{\rho}'_o)$, respectively. Accumulating a large number of joint-detection events at each transverse coordinate on the image plane, the aperture function $A(\vec{\rho}_o)$ is thus reproduced in the joint detection as a function of $\vec{\rho}_i$.

As we have discussed earlier, the point-to-point EPR correlation is the result of the coherent superposition of the biphoton probability amplitudes. In principle, one signal–idler pair contains all the necessary biphoton probability amplitudes that generate the ghost image. We name this kind of image as *two-photon coherent* image to distinguish the *two-photon incoherent* image of thermal light.

7.8　Turbulence-free Positioning and Clock Synchronization

It is well known that atmospheric turbulence is harmful for precise timing and positioning applications, such as Lidar and time transfer. Random variations in the composition or density of the atmosphere lead to variations in its index of refraction, known as optical turbulence, and thus vary the relative phases between different optical paths of the radiation. It is interesting, using entangled biphoton as the radiation source, any refraction index, length, or phase variations along the optical paths of the biphoton would not have any harmful effect on the two-photon temporal correlation and thus its positioning and timing applications. In addition, utilizing the δ-function like temporal correlation of entangled biphoton, millimeter and pico-second accuracy is potentially achievable for quantum Lidar, or for nonlocal clock synchronization.

Consider a simplified experimental setup in which the signal-idler photon pair is generated collinerlly and propagated in collimation. The signal and idler are received by two point-like photon counting detectors D_1 (signal) and D_2 (idler), respectively and jointly, for the measurements of the second-order temporal (longitudinal) coherence function $G^{(2)}(\tau_1 - \tau_2)$. The two-photon state of the signal-idler photon pair can be approximate as

$$|\Psi\rangle \simeq \Psi_0 \sum_{s,i} \delta(\omega_s + \omega_i - \omega_p)\, a_s^\dagger(\omega_s)\, a_i^\dagger(\omega_i)\, |0\rangle \qquad (7.8.1)$$

The two-photon effective wavefunction of the biphoton system $\Psi(\tau_1, \tau_2)$ is thus calculated as

$$\Psi(\tau_1, \tau_2) \simeq \Psi_0 \sum_{s,i} \delta\left(\omega_s + \omega_i - \omega_p\right) e^{-i\omega_s \tau_1} e^{-i\omega_i \tau_2}, \tag{7.8.2}$$

representing a superposition of a large number of two-photon amplitudes. It is not difficult to find from this superposition that if all two-photon amplitudes of the signal-idler experience the same turbulence along the same optical path with the same phase variations, the two-photon effective wavefunction, and thus the temporal second-order correlation, will be invariant. We consider this the unique condition for a turbulence-free correlation measurement. The experimental realization for turbulence-free positioning and timing, including nonlocal time transfer or clock synchronization, must achieve the above condition: to be sure all two-photon amplitudes in Eq. (7.8.2) experience the same turbulence along the same optical path.

Taking previous result in Chapter 6, the two-photon effective wavefunction can be written as

$$\Psi(\tau_1; \tau_2) \cong \Psi_0 \, e^{-i(\omega_s^0 \tau_1 + \omega_i^0 \tau_2)} \int d\nu \, f(\nu) \, e^{-i\nu(\tau_1 - \tau_2)}$$

$$= \left[\Psi_0 \, e^{-\frac{i}{2}(\omega_s^0 + \omega_i^0)(\tau_1 + \tau_2)}\right] \left[\mathcal{F}_{\tau_1 - \tau_2}\left\{f(\nu)\right\} e^{-\frac{i}{2}(\omega_s^0 - \omega_i^0)(\tau_1 - \tau_2)}\right] \tag{7.8.3}$$

where we have replaced the δ-function with a realistic phase matching function, see section 6.4; $\mathcal{F}_{\tau_1 - \tau_2}\left\{f(\nu)\right\}$ is the Fourier transform of the spectrum amplitude function $f(\nu)$. The temporal (longitudinal) correlation function $G^{(2)}(\tau_1 - \tau_2)$ is thus

$$G^{(2)}(\tau_1 - \tau_2) \propto \left|\mathcal{F}_{\tau_1 - \tau_2}\left\{f(\nu)\right\}\right|^2,$$

which is a δ-function-like function in the case of SPDC due to its super-wide spectrum.

Figure 7.8.1 shows two histograms measured in a nonlocal clock synchronization experiment. A histogram reports the number of joint-photodetection events, or coincidence counts, vs photo-detection time difference $t_1 - t_2$. Theoretically, the spectral bandwidth of the SPDC produces in a much narrower $G^{(2)}(t_1 - t_2)$; however, the slow response time and the jitter of the photodetectors cannot be avoid from the measurement. In Fig. 7.8.1, the solid line is a theoretical fitting curve of $G^{(2)}(t_1 - t_2)$ by taking into account the broadening effect of the photon counting detectors

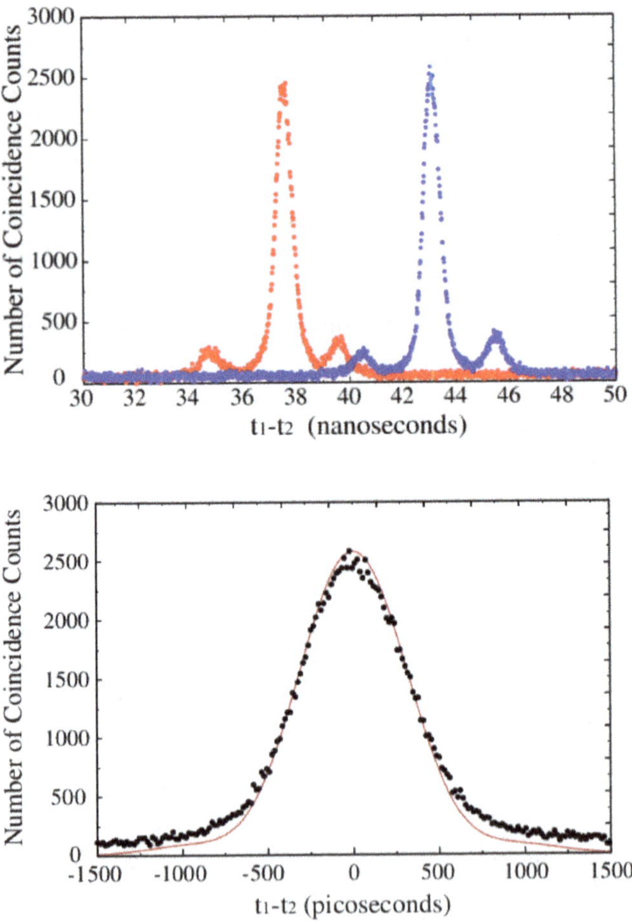

Fig. 7.8.1 Upper: Typical histograms, number of counts vs photo-detection time difference $t_1 - t_2$, for different optical delays. The histogram is in nanosecond scale. Lower: Central peak of the measured distribution function of $t_1 - t_2$. The solid line is a theoretical fitting curve of $G^{(2)}(t_1 - t_2)$ by taking into account the broadening contributions of the response time and the jitter of the photon counting detectors. The histogram is in picosecond scale.

D_1 and D_2. The statistical fitting curve concluded roughly ± 1 ps resolution, corresponding to ± 0.3 mm resolution of positioning at ~ 3 km distance.

Figure 7.8.2 shows two sets of histograms measured by a turbulence-free and interference-noise-free Quantum Lidar. This Lidar transmits and

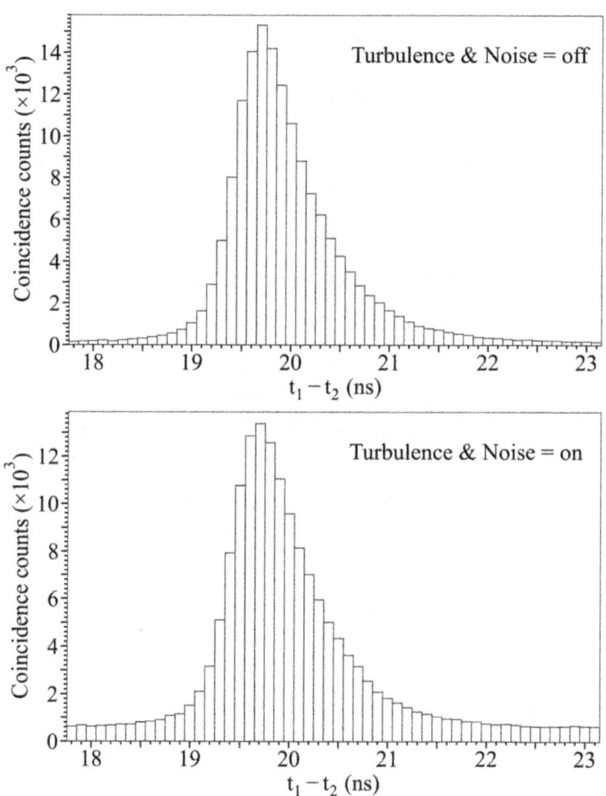

Fig. 7.8.2 Typical histograms, number of counts vs photo-detection time difference $t_1 - t_2$, for the positioning of a distant target. Upper: without turbulence and interference-noise. Lower: with significant turbulence and interference-noise. No temporal shifts were observed from the central peaks of the two histograms, except the background noise. Both measurements achieved centimeter accuracy for a "flat" target.

receives entangled photon pairs and measures their second-order temporal correlation function $G^{(2)}(t_1 - t_2)$. The entangled signal-idler photon pairs are generated from a continuous wave (CW) laser pumped collinear-degenerate SPDC, and both the signal photon and the idler photon are in the form of continuous wave. Nevertheless the temporal correlation measurement of the biphoton at different distances show a uncertainty in the order of a few hundred picoseconds, mainly determined by the photon counting detecters and the resolution of the coincidence measurement device, corresponding to a centimeter-scale precise positioning. More interestingly, the measured correlation function $G^{(2)}(t_1 - t_2)$ is "turbulence-free" and "interference-noise-free". In the measurements of Fig. 7.8.2, the introduced atmospheric turbulence was strong enough to blur out the interference of double-slit interferometer; the applied maximum interference-noise, limited by the damage threshold of the photon counting detectors, was about 20 times stronger then the Lidar signal.

7.9   Delayed Choice Quantum Eraser — Two-photon Interference of Entangled Photon Pair

Quantum eraser, proposed by Scully and Drühl in 1982, is an interesting "thought experiment" challenging the "basic mystery" of quantum mechanics: wave-particle duality. So far, several quantum erasers have been demonstrated with experimental results supporting the ideas of Scully and Drühl.

A double-slit type quantum eraser experiment, closing to the original Scully-Drühl thought experiment of 1982, is illustrated in Fig. 7.9.1. A pair of entangled photons, photon 1 and photon 2, is excited by a weak laser pulse either from atom A, which is located in slit A, or from atom B, which is located in slit B. Photon 1, propagates to the right, is registered by detector D_0, which can be scanned by a step motor along its x_0-axis for the examination of interference fringes. Photon 2, propagating to the left,

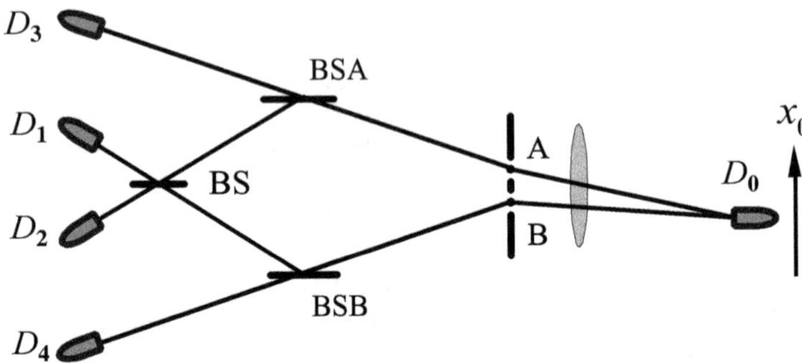

Fig. 7.9.1   Quantum erasure: a thought experiment of Scully-Drühl. A pair of entangled photons is emitted from either atom A or atom B by atomic cascade decay. The experimental condition guarantees no interference fringes is observable in the single detector counting rate of D_0. The "clicks" at D_1 or D_2 erase the which-path information, thus helping to restore the interference even after the "click" of D_0. On the other hand, the "clicks" at D_3 or D_4 record which-slit information. Thus, no observable interference is expected with the help of these "clicks".

is injected into a beamsplitter. If the pair is generated in atom A, photon 2 will follow the A path meeting BSA with 50% chance of being reflected or transmitted. If the pair is generated in atom B, photon 2 will follow the B path meeting BSB with 50% chance of being reflected or transmitted. In view of the 50% chance of being transmitted by either BSA or BSB, photon 2 is detected by either detector D_3 or D_4. The registration of D_3 or D_4 provides which-path information (path A or path B) on photon 2 and in turn provides which-path information for photon 1 because of the entanglement nature of the two-photon state generated by atomic cascade decay. Given a reflection at either BSA or BSB photon 2 will continue to follow its A or B path to meet another 50–50 beamsplitter BS and then be detected by either detectors D_1 or D_2.

The experimental condition was arranged in such a way that no interference is observable in the single counting rate of D_0., i.e., the distance between A and B is large enough to be "distinguishable" for D_0 to learn which-path information of photon 1. When the states of a photon associated with slit-A and slit-B are distinguishable, the photon carries which-path information. If the states of a photon associated with slit-A and slit-B are indistinguishable, the photon carries both-paths information.[5]

However, the "clicks" at D_1 or D_2 will erase the which-path information of photon 1 and help to restore the interference. On the other hand, the "clicks" at D_3 or D_4 record which-path information. Thus, no observable interference is expected with the help of these "clicks". It is interesting to note that both the "erasure" and "recording" of the which-path information can be made as a "delayed choice": the experiment is designed in such a way that L_0, the optical distance between atoms A, B and detector D_0, is much shorter than L_A (L_B), which is the optical distance between atoms A, B and the beamsplitter BSA (BSB) where the "which-path" or "both-paths"

---

[5] In the early times of quantum theory, Einstein and Bohr had serious debates based on the double-slit interference phenomenon. Now a days, "which-path" and/or "both-paths" information is still a topic of debate. What do we mean which-path and both-paths information? For physicists, at least for Einstein and Bohr, when a photon passes through slit-A and slit-B with distinguishable states, the photon carries which-path information. If its states associated with slit-A and slit-B are indistinguishable, the photon carries both-paths information. In quantum optics, when the separation between slit-A and slit-B is greater than the transverse coherence length of the field, the states of a photon associated with slit-A and slit-B are distinguishable; therefore, the photon carries which-path information. The idea of delayed choice quantum eraser is to erase this information after the annihilation of the photon itself by restoring the interference.

"choice" is made randomly by photon 2. Thus, after the annihilation of photon 1 at D_0, photon 2 is still on its way to BSA (BSB), i.e., "which-path" or "both-path" choice is "delayed" compare to the detection of photon 1. After the annihilation of photon 1, we look at these "delayed" detection events of D_1, D_2, D_3, and D_4 which have constant time delays, $\tau_i \simeq (L_i - L_0)/c$, $i = 1, 2, 3, 4$, relative to the triggering time of D_0. L_i is the optical distance between atoms A, B and the ith photodetector. According to Scully and Drühl, the "joint-detection" counting rate R_{01} (joint-detection rate between D_0 and D_1) and R_{02} will show an interference pattern as a function of the position of D_0 on its x-axis. This reflects the wave nature (both-path) of photon 1. However, no interference fringes will be observable in the joint detection counting events R_{03} and R_{04} when scanning detector D_0 along its x-axis. This is as would be expected because we have now inferred the particle (which-path) property of photon 1. It is important to emphasize that all four joint detection rates R_{01}, R_{02}, R_{03}, and R_{04} are recorded at the same time during one scanning of D_0. That is, in the present experiment, we "learn" both-path (interference) and which-path (particle-like) information from the same setup of measurement.

It should be mentioned that (1) the "choice" in this experiment is not actively switched by the experimentalist during the measurement. The "delayed choice" associated with either the wave or particle behavior of photon 1 is "randomly" made by photon 2. The experimentalist simply looks at which detector D_1, D_2, D_3 or D_4 is triggered by photon 2 to determine either wave or particle properties of photon 1 after the annihilation of photon 1; (2) the photodetction event of photon 1 at D_0 and the delayed choice event of photon 2 at BSA (BSB) are space-like separated events. The "coincidence" time window is chosen to be much shorter than the distance between D_0 and BSA (BSB). Within the joint-detection time window, it is impossible to have the two events "communicating".

Kim *et al.* realized the above random delayed choice quantum eraser in 2000. The schematic diagram of the experimental setup of Kim *et al.* is shown in Fig. 7.9.2. Instead of atomic cascade decay, SPDC is used to prepare the entangled two-photon state. In the experiment, a 351.1 nm Argon ion pump laser beam is divided by a double-slit and directed onto a type-II phase matching nonlinear crystal BBO at regions A and B. A pair of 702.2 nm orthogonally polarized signal-idler photon is generated either from region A or region B. The width of the region is about $a = 0.3$ mm and the distance between the center of A and B is about $d = 0.7$ mm. A Glen-Thompson prism is used to split the orthogonally polarized signal

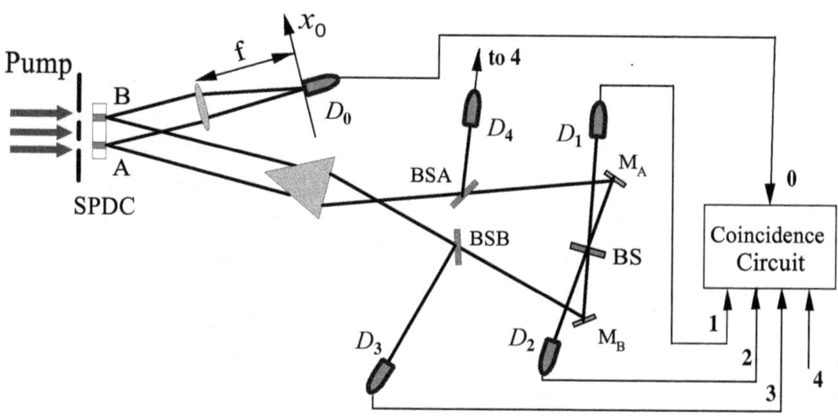

Fig. 7.9.2 Delayed choice quantum eraser: Schematic of an actual experimental setup of Kim *et al.* Pump laser beam is divided by a double-slit and makes two regions A and B inside the SPDC crystal. A pair of signal-idler photons is generated either from the A or B region. The "delayed choice" to observe either wave or particle behavior of the signal photon is made randomly by the idler photon about 7.7 ns after the detection of the signal photon.

and idler. The signal photon (photon 1, coming either from A or B) propagates through lens LS to detector D_0, which is placed on the Fourier transform plane of the lens. The use of lens LS is to achieve the "far field" condition, but still keep a short distance between the slit and the detector D_0. Detector D_0 can be scanned along its x-axis by a step motor for the observation of interference fringes. The idler photon (photon 2) is sent to an interferometer with equal-path optical arms. The interferometer includes a prism PS, two 50–50 beamsplitters BSA, BSB, two reflecting mirrors M_A, M_B, and a 50–50 beamsplitter BS. Detectors D_1 and D_2 are placed at the two output ports of the BS, respectively, for erasing the which-path information. The triggering of detectors D_3 and D_4 provides which-path information for the idler (photon 2) and, in turn, which-path information for the signal (photon 1). The detectors are fast avalanche photodiodes with less than 1 ns rise time and about 100 ps jitter. A constant fractional discriminator is used with each of the detectors to register a single photon whenever the leading edge of the detector output pulse is above the threshold. Coincidences between D_0 and D_j ($j = 1, 2, 3, 4$) are recorded, yielding the joint detection counting rates R_{01}, R_{02}, R_{03}, and R_{04}.

In the experiment, the optical delay $(L_{A,B} - L_0)$ is chosen to be $\simeq 2.3\,\mathrm{m}$, where L_0 is the optical distance between the output surface of *BBO* and detector D_0, and L_A (L_B) is the optical distance between the output surface of the BBO and the beamsplitter BSA (BSB). This means that any information (which-path or both-path) one can infer from photon 2 must be at least 7.7 ns later than the registration of photon 1. Compared to the 1 ns response time of the detectors, 2.3 m delay is thus enough for "delayed erasure". Although there is an arbitrariness about when a photon is detected, it is safe to say that the "choice" of photon 2 is delayed with respect to the detection of photon 1 at D_0 since the entangled photon pair is created simultaneously.

Figure 7.9.3, reports the joint detection rates R_{01} and R_{02}, indicating the regaining of standard Young's double-slit interference pattern. An expected π phase shift between the two interference patterns is clearly shown in the measurement. The single detector counting rates of D_0 and D_1 are recorded simultaneously. Although interference is observed in the joint detection counting rate, there is no significant modulation in any of the single detector counting rate during the scanning of D_0. R_0 is a

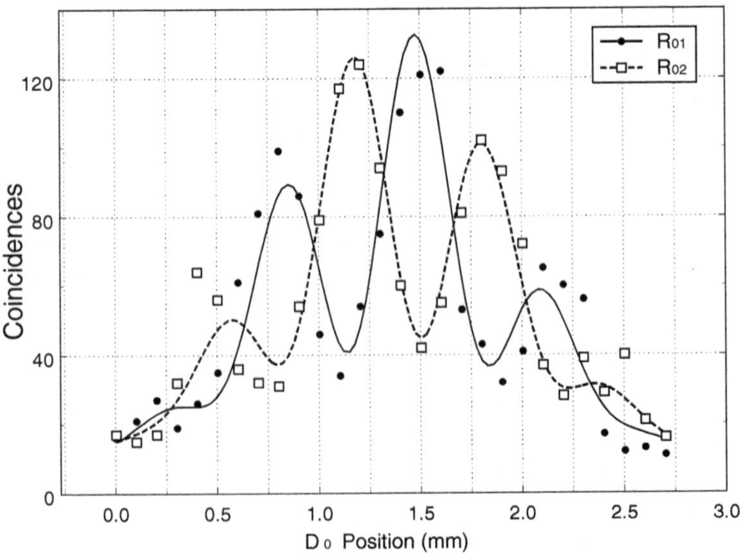

Fig. 7.9.3 Joint detection rates R_{01} and R_{02} against the x coordinates of detector D_0. Standard Young's double-slit interference patterns are observed. Note the π phase shift between R_{01} and R_{02}. The solid line and the dashed line are theoretical fits to the data.

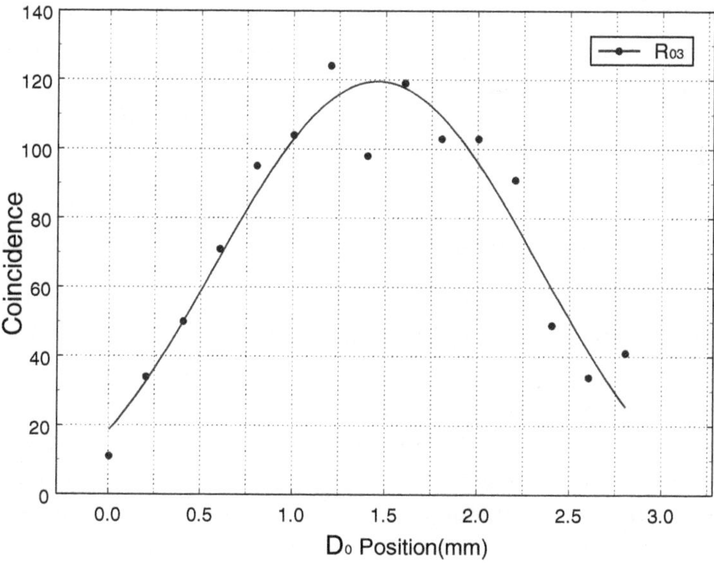

Fig. 7.9.4 Joint detection counting rate of R_{03}. Absence of interference is clearly demonstrated. The solid line is a sinc-function fit.

constant during the scanning of D_0. The absence of interference in the single detector counting rate of D_0 is simply because the separation between slits-A and slit-B is much greater than the coherence length of the single field.

Figure 7.9.4 reports a typical R_{03} (R_{04}), joint detection counting rate between D_0 and "which-path detector" D_3 (D_4). An absence of interference is clearly demonstrated. The fitting curve of the experimental data indicates a sinc-function like envelope of the standard Young's double slit interference-diffraction pattern. Two features should bring to our attention that (1) there is no observable interference modulation as expected, and (2) the curve is different from the constant single detector counting rate of D_0.

The experimental result is surprising from a classical point of view. The result, however, is easily explained in the contents of quantum theory. In this experiment, there are two kinds of very different interference phenomena: single-photon interference and two-photon interference. As we have discussed earlier, single-photon interference is the result of the superposition between single-photon amplitudes, and two-photon interference is the results of the superposition between two-photon amplitudes. Quantum mechanically, single-photon amplitude and two-photon amplitude represent very different measurements and, thus, very different physics.

In this regard, we analyze the experiment by answering the following questions:

(1)Why is there no observable interference in the single-detector counting rate of D_0?

This question belongs to single-photon interferometry. The absence of interference in single-detector counting rate of D_0 is very simple: the separation between slit-A and slit-B is much greater than the spatial coherence length of the the signal field. The single-photon states of the signal photon associated with slit-A and slit-B are distinguishable. Therefore, the signal photon carries "which-slit information. The idea of delayed choice quantum eraser is to erase this information after the annihilation of the signal photon it self.

(2) Why is there observable interference in the joint detection counting rate of R_{01} and R_{02}?

This question belongs to two-photon interferometry. Two-photon interference is very different from single-photon interference. Two-photon interference involves the addition of different yet indistinguishable two-photon amplitudes. The coincidence counting rate R_{01}, again, is proportional to the probability P_{01} of joint detecting the signal-idler pair by detectors D_0 and D_1,

$$R_{01} \propto P_{01} = \langle \Psi | E_0^{(-)} E_1^{(-)} E_1^{(+)} E_0^{(+)} | \Psi \rangle = \big| \langle 0 | E_1^{(+)} E_0^{(+)} | \Psi \rangle \big|^2. \quad (7.9.1)$$

To simplify the mathematics, we use the following "two-mode" expression for the state, bearing in mind that the transverse momentum δ-function will be taken into account.

$$| \Psi \rangle = \epsilon \, [a_s^\dagger a_i^\dagger e^{i\varphi_A} + b_s^\dagger b_i^\dagger \, e^{i\varphi_B}] \, | 0 \rangle$$

where ϵ is a normalization constant that is proportional to the pump field and the nonlinearity of the SPDC crystal, φ_A and φ_B are the phases of the pump field at A and B, and $a_j^\dagger$ $(b_j^\dagger)$, $j = s, i$, are the photon creation operators for the lower (upper) mode in Fig. 7.9.2.

In Eq. (7.9.1), the fields at the detectors D_0 and D_1 are given by:

$$E_0^{(+)} = a_s \, e^{ikr_{A0}} + b_s \, e^{ikr_{B0}}$$

$$E_1^{(+)} = a_i \, e^{ikr_{A1}} + b_i \, e^{ikr_{B1}} \tag{7.9.2}$$

where r_{Aj} (r_{Bj}), $j = 0, 1$, are the optical path lengths from region A (B) to the jth detector. Substituting the biphoton state and the field operators into Eq. (7.9.1),

$$R_{01} \propto \left| e^{i(kr_A + \varphi_A)} + e^{i(kr_B + \varphi_B)} \right|^2 = |\Psi_A + \Psi_B|^2$$

$$= 1 + \cos\left[k(r_A - r_B)\right] \simeq \cos^2\left(x_0 \pi d / \lambda z_0\right) \tag{7.9.3}$$

where, $r_A = r_{A0} + r_{A1}$, $r_B = r_{B0} + r_{B1}$; Ψ_A and Ψ_B are the two-photon effective wavefunctions of path A and path B, representing the two different yet indistinguishable probability amplitudes to produce a joint-photodetection event of D_0 and D_1, indicating a two-photon interference. In Dirac's language: a signal-idler photon pair interferes with the pair itself.

To calculate the diffraction effect of a single-slit, again, we need an integral of the effective two-photon wavefunction over the slit width (the superposition of infinite number of probability amplitudes results in a click-click joint detection event):

$$R_{01} \propto \left| \int_{-a/2}^{a/2} dx_{AB} \, e^{-ik\,r(x_0, x_{AB})} \right|^2 \cong \operatorname{sinc}^2(x_0 \pi a / \lambda z_0) \tag{7.9.4}$$

where $r(x_0, x_{AB})$ is the distance between points x_0 and x_{AB}, x_{AB} belongs to the slit's plane, and the far-field condition is applied.

Repeating the above calculations, the combined interference-diffraction joint detection counting rate for the double-slit case is given by:

$$R_{01} \propto \operatorname{sinc}^2(x_0 \pi a / \lambda z_0) \cos^2(x_0 \pi d / \lambda z_0). \tag{7.9.5}$$

If the finite size of the detectors are taken into account, the interference visibility will be reduced.

(3) Why is there no observable interference in the joint detection counting rate of R_{03} and R_{04}?

This question belongs to two-photon interferometry. From the view of two-photon physics, the absence of interference in the joint detection counting rate of R_{03} and R_{04} is obvious: only one two-photon amplitude contributes to the joint detection events.

In fact, the two-photon states of the signal-idler pair associated with slit-A and slit-B are "indistinguishable" and thus carries "both-paths" information. The "both-paths" information can be read by means of the joint photodetection events between D_0 and D_1 as well as between D_0 D_2. These joint-photodetection events are the results of the superposition between the indistinguishable two-photon amplitudes. The "both-paths" information can be erased by means of the joint-photodetection events between D_0 and D_3 as well as between D_0 D_4 which contains only one two-photon amplitude.

Bibliography

Alley C.O. and Y.H. Shih, *Foundations of Quantum Mechanics in the Light of New Technology*, in M. Namiki (Ed.), Physical Society of Japan, Tokyo, 47 (1986); Y.H. Shih and C.O. Alley, *Phys. Rev. Lett.* **61**, 2921 (1988).

Dirac P.A., *The Principle of Quantum Mechanics*, Oxford University Press, 1982.

Franson J.D., *Phys. Rev. Lett.* **62**, 2205 (1989).

Hong C.K., Z.Y. Ou and L. Mandel, *Phys. Rev. Lett.* **59**, 2044 (1987); Z.Y. Ou and L. Mandel, *Phys. Rev. Lett.* **62**, 50 (1988).

Kiess T.E., Y.H. Shih, A.V. Sergienko, and C.O. Alley, *Phys. Rev. Lett.* **71**, 3893 (1993).

Kwiat P.G., A.M. Steinberg, and R.Y. Chiao, *Phys. Rev. A* **45**, 7729 (1992).

Kwiat P.G., A.M. Steinberg, and R.Y. Chiao, *Phys. Rev. A* **47**, 2472 (1993).

Kwiat P.G. *et al. Phys. Rev. Lett.* **75**, 4337 (1995).

Magyar G. and L. Mandel, *Nature* **198**, 255 (1963); R.L. Pfleegor and L. Mandel, *Phys. Rev.* **159**, 1084 (1967).

Rubin M.H., D.N. Klyshko, Y.H. Shih, and A.V. Sergienko, *Phys. Rev. A* **50**, 5122 (1994).

Shih Y.H. and A.V. Sergienko, *Phys. Rev. A* **50**, 2564 (1994); A.V. Sergienko, Y.H. Shih, and M.H. Rubin, *JOSAB* **12**, 859 (1995).

Shih Y.H., Two-photon entanglement and quantum reality, in B. Bederson and H. Walther (Eds.), *Advances in Atomic, Molecular, and Optical Physics*, Academic Press, Cambridge, 1997.

Strekalov D.V., A.V. Sergienko, D.N. Klyshko, and Y.H. Shih, *Phys. Rev. Lett.* **74**, 3600 (1995).

Strekalov D.V. *et al. Phys. Rev. A* **54**, R1 (1996).

Strekalov D.V., T.B. Pittman, and Y.H. Shih, *Phys. Rev. A* **57**, 567 (1998); T.B. Pittman *et al. Phys. Rev. Lett.* **77**, 1917 (1996).

Tapster P.R., J.G. Rarity, and P.C.M. Owens, *Phys. Rev. Lett.* **73**, 1923 (1994).

Chapter 8

Two-Photon Interferometry II: Interference of Randomly Paired Photons

The study of entangled states greatly advanced our understanding about two-photon interferometry. Two-photon interference is not the interference between two photons, it is about a pair of photons interfering with the pair itself. In the language of quantum theory, two-photon interference is the result of coherent superposition between different yet indistinguishable two-photon probability amplitudes. Is the concept of two-photon amplitude applicable only to the entangled states? Is two-photon interference occurring only with entangled photon pair? The answer is negative. In this chapter we study two-photon interference of "thermal field", or "thermal light". Thermal light is usually created from a natural light source, such as the sun, containing a large number of randomly radiated spontaneous atomic transitions. Thermal light is traditionally defined as "classical" light.[1] It is not the philosophy of this book to classify light into quantum and classical. Our interests are at (1) generalize the quantum theory of two-photon interferometry to thermal radiation: a randomly created and randomly paired photons interfering with the pair itself; (2) distinguish the quantum mechanical concept of nonlocal two-photon interference from the classical concept of local statistical correlation of light caused in the preparation process; (3) recognize the relationship between two-photon interference and intensity fluctuation correlation of thermal field: two-photon interference of thermal light is observed from the intensity fluctuation correlation, however, this type of intensity fluctuation correlation is not preprepared at the

---

[1]There exist a number of definitions to classify "classical light" and "quantum light". One of the commonly used definitions considers thermal light classical because its positive P-function.

373

light source, instead this type of intensity fluctuation correlation is the result of two-photon interference. We may name it two-photon interference induced intensity fluctuation correlation. Thermal light may not be named as "correlated" radiation at all. There is no physical reason to assume the randomly created and randomly distributed photons in thermal state preprepared with "correlation", either "bunching" or "anti-bunching", at the radiation source.

Under certain conditions, the observed second-order interference can be factorized into a product of two individual classic first-order interferences. In this case, the interference is not only observable from the intensity fluctuation correlation measurement, or from the photon number fluctuation correlation measurement, of two photodetectors, but also observable in the counting rate of each individual photodetector. We consider this kind of two-photon interference as trivial second-order phenomenon. We avoid this type of second-order interference in this chapter by arranging the experimental condition in such a way under which no classic first-order interferences are observable.

It may not be uneasy to accept the quantum concept of two-photon interference for "classical" thermal radiation. However, this is not the first time in the history of physics we apply quantum mechanical concepts to "classical" light. We should not forget that it was Planck's theory of blackbody radiation originated the theory of quantum physics. Indeed, the radiation Planck dealt with was "classical" thermal light.

8.1 Two-Photon Interference between Spatially Separated Incoherent Thermal Fields

We start from a simple Young's double-slit, or double-pinhole, interferometer which is schematically illustrated in Fig. 8.1.1. This interferometer is the same as the classic Young's double-slit, or double-pinhole, interferometer, except it has two point-like photodetectors scannable on the observation plane. In this interferometer, we are not only measuring $\langle n(x_1) \rangle \propto \langle I(x_1) \rangle$ and $\langle n(x_2) \rangle \propto \langle I(x_2) \rangle$, but also $\langle \Delta n(x_1) \Delta n(x_2) \rangle \propto \langle \Delta I(x_1) \Delta I(x_2) \rangle$. In addition, we arrange an experimental condition to achieve $G^{(1)}(\mathbf{r}_A, t_A; \mathbf{r}_B, t_B) = 0$ by making the separation between the upper slit-A and the lower slit-B much greater than the spatial coherence length of the thermal field, $d \gg l_c$. In this case, (1) no first-order interferences are observable from $\langle n(x_1) \rangle \propto \langle I(x_1) \rangle$ and $\langle n(x_2) \rangle \propto \langle I(x_2) \rangle$;

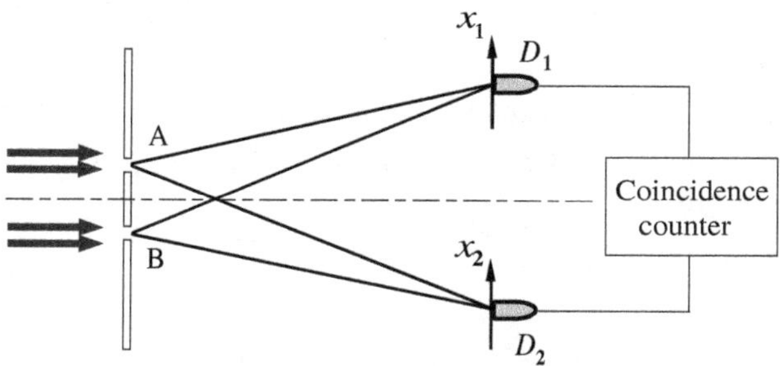

Fig. 8.1.1 Schematic of a simple Young's double-slit, or double-pinhole, interference experiment. The separation between the upper slit-A and the lower slit-B is much greater than the spatial coherence length of the thermal field, $d \gg l_c$, i.e., the thermal fields at pinhole A and at pinhole B are first-order incoherent $G^{(1)}(\mathbf{r}_A, t_A; \mathbf{r}_B, t_B) = 0$, and thus $\langle \Delta n_A \Delta n_B \rangle \propto \langle \Delta I_A \Delta I_B \rangle = 0$. Two point-like photodetectors, D_1 and D_2, are scannable along their x-axes in the far-field of the interferometer. The single-detector counting rates of D_1 and D_2 are monitored, respectively, and the coincidence counting rate of D_1 and D_2 is monitored jointly, during the scanning of D_1 and D_2. To simplify the mathematics, we assume line-like slits, or pint-like pinholes.

and (2) $\langle \Delta n_A \Delta n_B \rangle \propto \langle \Delta I_A \Delta I_B \rangle = 0$, i.e., the radiations do not have any intensity fluctuation correlation at the input port of the interferometer. Under the experimental conditions (1) and (2), do we expect observing interference in the photon number fluctuation correlation measurement $\langle \Delta n(x_1) \Delta n(x_2) \rangle \propto \langle \Delta I(x_1) \Delta I(x_2) \rangle$ when scanning D_1 and/or D_2 along their x-axis?

In the view of quantum theory of light, the single-photodetection counting rate of D_1 and D_2, respectively, monitors the first-order interference, which is the result of a superposition between single-photon amplitudes. The coincidence-photodetection counting rate of D_1 and D_2, jointly, observes the second-order interference, which is the result of superposition between two-photon amplitudes. In this experiment, fields A and B are first-order incoherent, i.e., the first-order mutual coherence function $G^{(1)}(\mathbf{r}_A, t_A; \mathbf{r}_B, t_B) = 0$ and thus no classic Young's double-slit interferences are observable from the measurements of $\langle n(x_1) \rangle \propto \langle I(x_1) \rangle$ and $\langle n(x_2) \rangle \propto \langle I(x_2) \rangle$, respectively. Although the counting rate of D_1 and D_2 are both constants during their scanning, two-photon interference will produce observable second-order interference modulations in the joint photodetection of D_1 and D_2 while $G^{(1)}(\mathbf{r}_A, t_A; \mathbf{r}_B, t_B) = 0$ and $\langle \Delta n_A \Delta n_B \rangle \propto \langle \Delta I_A \Delta I_B \rangle = 0$.

(I) **Two-photon Young's interference in Einstein's granularity picture of light:**

In the following, we calculate the two-photon Young's interference in the joint measurement of $\langle n_1 n_2 \rangle \propto \langle I_1 I_2 \rangle \propto G^{(2)}(x_1, t_1; x_2, t_2)$ in Einstein's granularity picture:

$$
G^{(2)}(x_1, t_1; x_2, t_2)
$$

$$
= \left\langle \sum_{m,n,p,q} E_m^*(x_1, t_1) E_n(x_1, t_1) E_p^*(x_2, t_2) E_q(x_2, t_2) \right\rangle
$$

$$
= \sum_{m=n} E_m^*(x_1, t_1) E_m(x_1, t_1) \sum_{m=n} E_n^*(x_2, t_2) E_n(x_2, t_2)
$$

$$
+ \sum_{m \neq n} E_m^*(x_1, t_1) E_n(x_1, t_1) E_n^*(x_2, t_2) E_m(x_2, t_2)
$$

$$
\simeq \sum_{m,n} \left| E_m(x_1, t_1) E_n(x_2, t_2) + E_n(x_1, t_1) E_m(x_2, t_2) \right|^2
$$

$$
= \sum_{m,n} \left| (E_{mA1} + E_{mB1})(E_{nA2} + E_{nB2}) \right.
$$

$$
\left. + (E_{nA1} + E_{nB1})(E_{mA2} + E_{mB2}) \right|^2
$$

$$
= \sum_{m,n} \left| E_{mA1}E_{nA2} + E_{mA1}E_{nB2} + E_{mB1}E_{nA2} + E_{mB1}E_{nB2} \right.
$$

$$
\left. + E_{nA1}E_{mA2} + E_{nA1}E_{mB2} + E_{nB1}E_{mA2} + E_{nB1}E_{mB2} \right|^2, \quad (8.1.1)
$$

where we have approximated the sum of $(m \neq n)$ to the sum of (m, n) by ignoring the $m = n$ terms, which is reasonable for a large number of subfields. We have also used short-hand notation E_{mAj} and E_{mBj}, $j = 1, 2$, to specify the subfields at D_j that coming from pinhole-A and pinhole-B, respectively. Completing the sum for each term in the $|E_{mB1}E_{nA2} + \cdots + E_{nA1}E_{mB2}|^2$, we find that a number of cross terms vanish due to the experimental condition $d > l_c$. We may group the surviving terms into the following four superpositions:

$$
G^{(2)}(x_1, t_1; x_2, t_2)
$$

$$
= \sum_{m,n} \left| E_{mA1}E_{nA2} + E_{nA1}E_{mA2} \right|^2 + \sum_{m,n} \left| E_{mB1}E_{nB2} + E_{nB1}E_{mB2} \right|^2
$$

$$
+ \sum_{m,n} \left| E_{mA1}E_{nB2} + E_{nB1}E_{mA2} \right|^2 + \sum_{m,n} \left| E_{mB1}E_{nA2} + E_{nA1}E_{mB2} \right|^2
$$

$$\equiv G_{AA}^{(2)}(x_1, t_1; x_2, t_2) + G_{BB}^{(2)}(x_1, t_1; x_2, t_2)$$

$$+ G_{AB}^{(2)}(x_1, t_1; x_2, t_2) + G_{BA}^{(2)}(x_1, t_1; x_2, t_2). \tag{8.1.2}$$

We may calculate $G_{A1,A2}^{(2)}$, $G_{B1,B2}^{(2)}$, $G_{A1,B2}^{(2)}$, and $G_{B1,A2}^{(2)}$, separately, as four groups of distinguishable two-photon interferences. Focusing on the second-order spatial correlation and simplifying the the mathematics, in the following calculations, we assume perfect second-order temporal coherence with idealized photodetectors and idealized correlation circuit:

$$G_{AA}^{(2)}(x_1; x_2) = \sum_{m,n} \left| E_m g_{mA}(x_1) E_n g_{nA}(x_2) + E_n g_{nA}(x_1) E_m g_{mA}(x_2) \right|^2$$

$$\propto \sum_{m} |E_m g_{mA}(x_1)|^2 \sum_{n} |E_n g_{nA}(x_2)|^2$$

$$+ \sum_{n} |E_n g_{nA}(x_1)|^2 \sum_{m} |E_m g_{mA}(x_2)|^2$$

$$+ \sum_{m,n} E_m^* g_{mA}^*(x_1) E_n g_{nA}(x_1) E_n^* g_{nA}^*(x_2) E_m g_{mA}(x_2)$$

$$+ \sum_{m,n} E_n^* g_{nA}^*(x_1) E_m g_{mA}(x_1) E_m^* g_{mA}^*(x_2) E_n g_{nA}(x_2)$$

$$\propto \langle I(x_1) \rangle \langle I(x_2) \rangle + \langle \Delta I(x_1) \Delta I(x_2) \rangle$$

$$= G_0^{\langle I_1 \rangle \langle I_2 \rangle} + G_0^{\langle \Delta I_1 \Delta I_2 \rangle}, \tag{8.1.3}$$

which is the result of a superposition between two-photon amplitudes (1) the mth subfield passes pinhole-A and then propagates to D_1 and the nth subfield passes pinhole-A and then propagates to D_2 and (2) the mth subfield passes pinhole-A and then propagates to D_2 and the nth subfield passes pinhole-A and then propagates to D_1. Note that we have especially labeled the two constants with upper indexes of $\langle n_1 \rangle \langle n_2 \rangle$ and $\langle n_1 n_2 \rangle$, although they have equal values:

$$G_{AB}^{(2)}(x_1; x_2) = \sum_{m,n} \left| E_m g_{mA}(x_1) E_n g_{nB}(x_2) + E_n g_{nB}(x_1) E_m g_{mA}(x_2) \right|^2$$

$$\propto \sum_{m} |E_m g_{mA}(x_1)|^2 \sum_{n} |E_n g_{nB}(x_2)|^2$$

$$+ \sum_{n} |E_n g_{nA}(x_1)|^2 \sum_{m} |E_m g_{mB}(x_2)|^2$$

$$+ \sum_{m,n} E_m^* g_{mA}^*(x_1) E_n g_{nB}(x_1) E_n^* g_{nB}^*(x_2) E_m g_{mA}(x_2)$$

$$+ \sum_{m,n} E_n^* g_{nA}^*(x_1) E_m g_{mB}(x_1) E_m^* g_{mB}^*(x_2) E_n g_{nA}(x_2)$$

$$= \langle I(x_1) \rangle \langle I(x_2) \rangle + \langle \Delta I(x_1) \Delta I(x_2) \rangle$$

$$= G_0^{\langle I_1 \rangle \langle I_2 \rangle} + G_0^{\langle \Delta I_1 \Delta I_2 \rangle} \cos \frac{2\pi d}{\lambda z}(x_1 - x_2), \tag{8.1.4}$$

which is the result of a superposition between two-photon amplitudes; (3) the mth subfield passes pinhole-A and then propagates to D_1 and the nth subfield passes pinhole-B and then propagates to D_2; (4) the mth subfield passes pinhole-A and then propagates to D_2 and the nth subfield passes pinhole-B and then propagates to D_1.

In Eqs. (8.1.3) and (8.1.4), $g_m(x_j)$ is Green's function that propagates the mth subfield from the source to D_j at (x_j), $j = 1, 2$:

$$g_{mA}(x_j) = g_m(\mathbf{r}_A) g_A(x_j),$$

$$g_{nB}(x_j) = g_n(\mathbf{r}_B) g_B(x_j),$$

with $g_m(\mathbf{r}_A)$ $[g_n(\mathbf{r}_B)]$ representing Green's function propagating the mth (nth) subfield from the mth (nth) sub-source to pinhole-A (pinhole-B), $g_A(x_j)$ and $g_B(x_j)$ Green's functions propagating the subfields from pinhole-A and pinhole-B to D_j at (x_j), respectively, where A labels the upper path (passing pinhole-A), B labels the lower path (passing pinhole-B). Note that in the above calculations, we have assumed line-like slits, or point-like pinholes, i.e., the size of the slits or pinholes can be approximated to be infinitely small.

$G_{BB}^{(2)}(x_1; x_2)$ and $G_{BA}^{(2)}(x_1; x_2)$ can be calculated in a similar way:

$$G_{BB}^{(2)}(x_1; x_2) = \sum_{m,n} |E_m g_{mB}(x_1) E_n g_{nB}(x_2) + E_n g_{nB}(x_1) E_m g_{mB}(x_2)|^2$$

$$= G_0^{\langle I_1 \rangle \langle I_2 \rangle} + G_0^{\langle \Delta I_1 \Delta I_2 \rangle}, \tag{8.1.5}$$

which is the result of a superposition between two-photon amplitudes (5) the mth subfield passes pinhole-B and then propagates to D_1 and the nth subfield passes pinhole-B and then propagates to D_2; (6) the mth subfield passes pinhole-B and then propagates to D_2 and the nth subfield passes pinhole-B and then propagates to D_1:

$$G_{BA}^{(2)}(x_1;x_2) = \sum_{m,n} |E_m g_{mB}(x_1) E_n g_{nA}(x_2) + E_n g_{nA}(x_1) E_m g_{mB}(x_2)|^2$$

$$= G_0^{\langle I_1 \rangle \langle I_2 \rangle} + G_0^{\langle \Delta I_1 \Delta I_2 \rangle} \cos \frac{2\pi d}{\lambda z}(x_1 - x_2), \qquad (8.1.6)$$

which is the result of a superposition between two-photon amplitudes (7) the mth subfield passes pinhole-B and then propagates to D_1 and the nth subfield passes pinhole-A and then propagates to D_2; (8) the mth subfield passes pinhole-B and then propagates to D_2 and the nth subfield passes pinhole-A and then propagates to D_1.

When scanning D_1 and D_2 along their x-axes, taking into account all possible joint photodetection events contributed from $G_{A1,A2}^{(2)}$, $G_{B1,B2}^{(2)}$, $G_{A1,B2}^{(2)}$, and $G_{B1,A2}^{(2)}$, the jointly measured photon number correlation, which is proportional to the intensity correlation, of D_1 and D_2 is thus

$$\langle n_1(x_1) n_2(x_2) \rangle \propto 1 + \frac{1}{2} \cos \frac{2\pi d}{\lambda z}(x_1 - x_2). \qquad (8.1.7)$$

A sinusoidal two-photon interference pattern with 50% contrast is observable from the joint measurement $\langle n_1 n_2 \rangle$, despite the experimental condition of $d > l_c$.

Adding only the photon number fluctuation correlations, or the intensity fluctuation correlations, calculated in $G_{A1,A2}^{(2)}$, $G_{B1,B2}^{(2)}$, $G_{A1,B2}^{(2)}$, and $G_{B1,A2}^{(2)}$, the jointly measured photon number fluctuation correlation, which is proportional to the intensity fluctuation correlation, of D_1 and D_2 is therefore

$$\langle \Delta n_1(x_1) \Delta n_2(x_2) \rangle \propto 1 + \cos \frac{2\pi d}{\lambda z}(x_1 - x_2). \qquad (8.1.8)$$

A sinusoidal two-photon interference pattern with 100% visibility is observable from the joint measurement of $\langle \Delta n_1 \Delta n_2 \rangle$, despite the experimental condition of $d > l_c$.

Differ from classic Young's double-pinhole interference, it is interesting to find that Green's functions $g_m(\mathbf{r}_A, t_A)$ and $g_m(\mathbf{r}_B, t_B)$, which propagate the subfields from the source plane to the double-pinhole plane, have no effects on the two-photon interferences. What is the reason for $g_m(\mathbf{r}_A, t_A)$ and $g_m(\mathbf{r}_B, t_B)$ contribute to the first-order interference but not to the second-order interference? Let us first exam the classic Young's double-pinhole interferences which are, respectively, measured by D_1 and D_2:

$$\langle I(x_j)\rangle = \sum_m \left|E_m \, g_m(x_j)]\right|^2$$

$$= \sum_m \left|E_m \, \frac{1}{\sqrt{2}}\left[g_{mA}(x_j) + g_{mB}(x_j)\right]\right|^2$$

$$= \sum_m |E_m|^2 \left|\frac{1}{\sqrt{2}}\left[g_m(\mathbf{r}_A)g_A(x_j) + g_m(\mathbf{r}_B)g_B(x_j)\right]\right|^2$$

$$\simeq I_0 \left[1 + \mathrm{sinc}\frac{\pi d \Delta\theta_s}{\lambda} \cos\frac{2\pi d}{\lambda z}x_j\right], \tag{8.1.9}$$

where $g_m(x_j)$ is Green's function that propagates the mth subfield from the source to D_j at (x_j), $j = 1,2$. The mth subfield, or photon, has two different yet indistinguishable alternatives to produce a photodetection event of D_j: (1) the mth subfield, or photon, passes upper-pinhole A and then is annihilated by D_j and (2) the mth subfield, or photon, passes lower-pinhole B and then is annihilated by D_j:

$$E(x_j) = \sum_m E_m(x_j) = \sum_m \frac{E_m}{\sqrt{2}}\left[g_{mA}(x_j) + g_{mB}(x_j)\right]]. \tag{8.1.10}$$

In Eq. (8.1.9), the sinc-function is defined as $\mathrm{sinc}(x) \equiv \sin(x)/x$, which is the result of $\sum_m g_m^*(\mathbf{r}_A)g_m(\mathbf{r}_B)$ in the cross term of the above first-order superposition:

$$\sum_m g_m^*(\mathbf{r}_A)g_m(\mathbf{r}_B) \propto \int_{-\frac{\theta_s}{2}}^{\frac{\theta_s}{2}} d\theta_m \, e^{-i\frac{2\pi d}{\lambda}\theta_m} \simeq \mathrm{sinc}\frac{\pi d \Delta\theta_s}{\lambda}, \tag{8.1.11}$$

where θ_m is the 1-D angular coordinate of the mth subfield and $\Delta\theta_s$ is the angular diameter of the light source relative to the double-pinhole interferometer. The sinc-function determines the visibility, or the contrast, of the classic Young's double-pinhole interference. The interference reaches its maximum visibility when the argument of the sinc-function is assigned with a value closer to zero. This can be achieved by applying a point-like source, $\Delta\theta_s \sim 0$, or other mechanisms to obtain $(d\Delta\theta_s)/\lambda \sim 0$. The interference vanishes when setting the argument of the sinc-function with a value of π or greater, for example achieving $d \gg \lambda/\Delta\theta_s$ as we have arranged in this experiment. Similar calculation of the above sinc-function has been given in Chapter 6.

On the other hand, in the joint measurement of $\langle n(x_1)n(x_2)\rangle$, it is the cross terms of two-photon superposition in $G^{(2)}_{A1,B2}$ and $G^{(2)}_{B1,A2}$ contribute sinusoidal modulations. For example the cross term of $G^{(2)}_{AB}(x_1,x_2)$:

$$\sum_{m,n} E_m^* g_m^*(x_A)g_A^*(x_1)E_n g_n(x_B)g_B(x_1)$$

$$\times E_n^* g_n^*(x_B)g_B^*(x_2)E_m g_m(x_A)g_A(x_2)$$

$$\propto \sum_{m,n}\left[g_m^*(x_A)g_m(x_A)\,g_n(x_B)g_n^*(x_B)\right]$$

$$\times g_A^*(x_1)g_B(x_1)g_A(x_2)g_B^*(x_2)$$

$$\propto e^{-i\frac{2\pi d}{\lambda z}(x_1-x_2)}, \tag{8.1.12}$$

where $g_m^*(x_A)g_m(x_A)$ and $g_n(x_B)g_n^*(x_B)$ are normalized to one as usual, which means the visibility of the two-photon interference does not depend on the angular size of the light source and thus the spatial coherence length l_c.

Young's interference calculated above from Einstein's picture is consistent with that calculated from quantum mechanics. For comparison, two simple quantum mechanical analyses are given as follows:

(II) **Two-photon Young's interference in single-photon state representation**:

In the following, we calculate $\langle n(x_1,t_1)\,n(x_2,t_2)\rangle$ from the single-photon state representation:

$$|\Psi\rangle \simeq \prod_m \left(|0\rangle + \epsilon\,\hat{a}_m^\dagger|0\rangle\right)$$

$$\simeq |0\rangle + \epsilon\sum_m \hat{a}_m^\dagger|0\rangle + \epsilon^2\sum_{m<n}\hat{a}_m^\dagger\hat{a}_n^\dagger|0\rangle + \cdots.$$

Under weak light condition, we may keep the necessary lowest-order terms for calculating the counting rate of D_1 and D_2, respectively, and the coincidence counting rate of D_1 and D_2, jointly. The joint photodetection counting rate is proportional to $G^{(2)}(x_1, t_1; x_2, t_2)$,

$$G^{(2)}(x_1, t_1; x_2, t_2)$$

$$= \langle\langle \hat{E}^{(-)}(x_1,t_1)\hat{E}^{(-)}(x_2,t_2)\hat{E}^{(+)}(x_2,t_2)\hat{E}^{(+)}(x_1,t_1)\rangle_{\text{QM}}\rangle_{\text{En}}$$

$$\simeq \left\langle \left| \langle 0| \sum_p [\hat{E}^{(+)}_{pA2} + \hat{E}^{(+)}_{pB2}] \sum_q [\hat{E}^{(+)}_{qA1} + \hat{E}^{(+)}_{qB1}] \sum_{m,n} \hat{a}^\dagger_m \hat{a}^\dagger_n |0\rangle \right|^2 \right\rangle_{\text{En}}$$

$$= \sum_{m,n} \big| \psi_{mA1}\psi_{nA2} + \psi_{mA1}\psi_{nB2} + \psi_{mB1}\psi_{nA2} + \psi_{mB1}\psi_{nB2}$$

$$+ \psi_{nA1}\psi_{mA2} + \psi_{nA1}\psi_{mB2} + \psi_{nB1}\psi_{mA2} + \psi_{nB1}\psi_{mB2} \big|^2 \qquad (8.1.13)$$

with

$$\psi_{mAj} = \langle 0|\hat{a}_m \hat{a}^\dagger_m|0\rangle e^{ik(r_{mA}+r_{Aj})}, \quad \psi_{nBj} = \langle 0|\hat{a}_n \hat{a}^\dagger_n|0\rangle e^{ik(r_{nB}+r_{Bj})},$$

the effective wavefunctions of the mth photon and the nth photon measured by D_j, where the subindex A and B label slit-A and slit-B, respectively, r_{mA} (r_{nB}) is the optical path between the mth (nth) sub-source and slit-A (slit-B), r_{Aj} (r_{Bj}) is the optical path connecting slit-A (slit-B) and D_j. Again, to simplify the mathematics, perfect second-order temporal correlation, or monochromatic thermal field, is applied to the above second-order spatial coherence calculation, as usual.

Similar to the calculation in Einstein's granularity picture, completing the sum for each term in the $|\psi_{mB1}\psi_{nA2} + \cdots + \psi_{nA1}\psi_{mB2}|^2$, we find that a number of cross terms vanish due to the experimental condition $d > l_c$. We may group the surviving terms into the following four superpositions:

$$G^{(2)}(x_1, t_1; x_2, t_2)$$

$$= \sum_{m,n} \big| \psi_{mA1}\psi_{nA2} + \psi_{nA1}\psi_{mA2} \big|^2 + \sum_{m,n} \big| \psi_{mB1}\psi_{nB2} + \psi_{nB1}\psi_{mB2} \big|^2$$

$$+ \sum_{m,n} \big| \psi_{mA1}\psi_{nB2} + \psi_{nB1}\psi_{mA2} \big|^2 + \sum_{m,n} \big| \psi_{mB1}\psi_{nA2} + \psi_{nA1}\psi_{mB2} \big|^2$$

$$\equiv G^{(2)}_{AA}(x_1, t_1; x_2, t_2) + G^{(2)}_{BB}(x_1, t_1; x_2, t_2)$$

$$+ G^{(2)}_{AB}(x_1, t_1; x_2, t_2) + G^{(2)}_{B}(x_1, t_1; x_2, t_2). \qquad (8.1.14)$$

Under the assumption of perfect second-order temporal coherence, it is not difficult to see that the two-photon interferences in $G^{(2)}_{AA}(x_1, x_2)$ and

$G_{BB}^{(2)}(x_1, x_2)$ contribute two constants to the joint photodetection of D_1 and D_2, for example,

$$G_{AA}^{(2)}(x_1, x_2) = \sum_{m,n} \left| \psi_{mA1}\psi_{nA2} + \psi_{nA1}\psi_{mA2} \right|^2$$

$$\propto \sum_m \left| \psi_{mA}(x_1) \right|^2 \sum_n \left| \psi_{nA}(x_2) \right|^2$$

$$+ \sum_{m,n} \psi_{mA}^*(x_1)\psi_{nA}(x_1)\psi_{nA}^*(x_2)\psi_{mA}(x_2)$$

$$+ \sum_n \left| \psi_{nA}(x_1) \right|^2 \sum_m \left| \psi_{mA}(x_2) \right|^2$$

$$+ \sum_{m,n} \psi_{mA}(x_1)\psi_{nA}^*(x_1)\psi_{nA}(x_2)\psi_{mA}^*(x_2)$$

$$= G_0^{\langle n_1 \rangle \langle n_2 \rangle} + G_0^{\langle \Delta n_1 \Delta n_2 \rangle}, \tag{8.1.15}$$

where, again, we have especially labeled the two constants with upper indexes of $\langle n_1 \rangle \langle n_2 \rangle$ and $\langle n_1 n_2 \rangle$, although they have equal values.

It is $G_{AB}^{(2)}(x_1, x_2)$ and $G_{BA}^{(2)}(x_1, x_2)$ contribute a two-photon interference to the joint photodetection of D_1 and D_2, for example,

$$G_{AB}^{(2)}(x_1, x_2) = \sum_{m,n} \left| \psi_{mA1}\psi_{nB2} + \psi_{nB1}\psi_{mA2} \right|^2$$

$$= \sum_m \left| \psi_{mA}(x_1) \right|^2 \sum_n \left| \psi_{nB}(x_2) \right|^2$$

$$+ \sum_{m,n} \psi_{mA}^*(x_1)\psi_{nB}(x_1)\psi_{nB}^*(x_2)\psi_{mA}(x_2)$$

$$+ \sum_n \left| \psi_{nB}(x_1) \right|^2 \sum_m \left| \psi_{mA}(x_2) \right|^2$$

$$+ \sum_{m,n} \psi_{mA}(x_1)\psi_{nB}^*(x_1)\psi_{nB}(x_2)\psi_{mA}^*(x_2)$$

$$= G_0^{\langle n_1 \rangle \langle n_2 \rangle} + G_0^{\langle \Delta n_1 \Delta n_2 \rangle} \cos \frac{2\pi d}{\lambda z}(x_1 - x_2). \tag{8.1.16}$$

The photon number correlation which is proportional to the second-order spatial coherence function is therefore a two-photon interference pattern with 50% contrast (33% visibility):

$$\langle n_1(x_1) n_2(x_2) \rangle \propto 1 + \frac{1}{2} \cos \frac{2\pi d}{\lambda z}(x_1 - x_2). \tag{8.1.17}$$

If the correlation measurement is designed for the photon number fluctuations only, the four cross terms of the superposition in $G_{AA}^{(2)}(x_1, x_2)$, $G_{BB}^{(2)}(x_1, x_2)$, $G_{AB}^{(2)}(x_1, x_2)$ and $G_{BA}^{(2)}(x_1, x_2)$ contribute a two-photon interference pattern with 100% visibility (100% contrast):

$$\langle \Delta n_1(x_1) \, \Delta n_2(x_2) \rangle \propto 1 + \cos \frac{2\pi d}{\lambda z}(x_1 - x_2). \tag{8.1.18}$$

(III) Two-photon Young's interference in coherent state representation:

Assuming a large number of randomly created photons, such as a natural radiation; or a large number of randomly created groups of identical photons, such as a pseudo-thermal field scattered from laser beam by rotating ground glass, in thermal state:

$$|\widetilde{\Psi}\rangle = \prod_m |\{\alpha_m\}\rangle \tag{8.1.19}$$

which is written in the quantum coherent state representation. Coherent state representation simplifies the calculation of second-order coherence function significantly, and also made it possible to calculate the interference of a pair of two groups of identical photons interfering with the pair itself. The coherent state may represent a group of identical photons with $\bar{n} = |\alpha| \gg 1$. To not bring additional confusion to the simple physics we intended to discuss here, we restrict our discussion at single-photon level, $\bar{n} = |\alpha| \ll 1$. Under the condition of $\bar{n} = |\alpha| \ll 1$, we may consider the state a single-photon state in which m labels the mth quantized radiation or photon created from the mth atomic transition.

Following Glauber's theory, the probability to produce a joint photodetection event at D_1 and D_2 shown in Fig. 8.1.1 is proportional to the second-order coherence function $G^{(2)}(x_1, t_1; x_2, t_2)$:

$$G^{(2)}(x_1, t_1; x_2, t_2)$$

$$= \left\langle \langle \hat{E}^{(-)}(x_1, t_1) \hat{E}^{(-)}(x_2, t_2) \hat{E}^{(+)}(x_2, t_2) \hat{E}^{(+)}(x_1, t_1) \rangle_{\mathrm{QM}} \right\rangle_{\mathrm{En}}$$

$$= \left\langle \left| \langle \widetilde{\Psi} | \hat{E}^{(+)}(x_2, t_2) \hat{E}^{(+)}(x_1, t_1) | \widetilde{\Psi} \rangle \right|^2 \right\rangle_{\mathrm{En}}$$

$$= \left\langle \left| \prod_m \langle\{\alpha_m\}| \sum_p [\hat{E}_{pA}^{(+)}(x_2, t_2) + \hat{E}_{pB}^{(+)}(x_2, t_2)] \right.\right.$$

$$\left.\left. \times \sum_q [\hat{E}_{qA}^{(+)}(x_1, t_1) + \hat{E}_{qB}^{(+)}(x_1, t_1)] \prod_n |\{\alpha_n\}\rangle \right|^2 \right\rangle_{En} \qquad (8.1.20)$$

$$= \sum_{m,n} \left| \psi_{mA1}\psi_{nA2} + \psi_{mA1}\psi_{nB2} + \psi_{mB1}\psi_{nA2} + \psi_{mB1}\psi_{nB2} \right.$$

$$\left. + \psi_{nA1}\psi_{mA2} + \psi_{nA1}\psi_{mB2} + \psi_{nB1}\psi_{mA2} + \psi_{nB1}\psi_{mB2} \right|^2$$

with

$$\psi_{mAj} = \alpha_m e^{ik(r_{mA}+r_{Aj})}, \quad \psi_{nBj} = \alpha_n e^{ik(r_{nB}+r_{Bj})},$$

the effective wavefunctions of the mth photon and the nth photon measured by D_j, where the subindex A and B label slit-A and slit-B, respectively, r_{mA} (r_{nB}) is the optical path between the mth (nth) sub-source and slit-A (slit-B), r_{Aj} (r_{Bj}) is the optical path connecting slit-A (slit-B) and D_j. To simplify the mathematics, perfect second-order temporal correlation, or monochromatic thermal field is applied to the above second-order spatial coherence calculation, as usual.

Similar to the calculation in Einstein's granularity picture, completing the sum for each term in the $|\psi_{mB1}\psi_{nA2} + \cdots + \psi_{nA1}\psi_{mB2}|^2$, we find that a number of cross terms vanish due to the experimental condition $d > l_c$. We may group the surviving terms into the following four superpositions:

$$G^{(2)}(x_1, t_1; x_2, t_2)$$

$$= \sum_{m,n} \left| \psi_{mA1}\psi_{nA2} + \psi_{nA1}\psi_{mA2} \right|^2 + \sum_{m,n} \left| \psi_{mB1}\psi_{nB2} + \psi_{nB1}\psi_{mB2} \right|^2$$

$$+ \sum_{m,n} \left| \psi_{mA1}\psi_{nB2} + \psi_{nB1}\psi_{mA2} \right|^2 + \sum_{m,n} \left| \psi_{mB1}\psi_{nA2} + \psi_{nA1}\psi_{mB2} \right|^2$$

$$\equiv G_{AA}^{(2)}(x_1, t_1; x_2, t_2) + G_{BB}^{(2)}(x_1, t_1; x_2, t_2)$$

$$+ G_{AB}^{(2)}(x_1, t_1; x_2, t_2) + G_B^{(2)}(x_1, t_1; x_2, t_2). \qquad (8.1.21)$$

The above nonlocal superposition represents a quantum phenomenon, namely, "two-photon interference": two randomly created and randomly paired photons interfering with the pair itself. Different from Einstein's picture which is still under the framework of electromagnetic theory of light, effective wavefunction represents a concept of quantum mechanics: probability amplitude.

It is unnecessary to repeat the rest of calculations that we have completed in Einstein's granularity picture, replacing the quantized subfield with the effective wavefunction of photon, and assuming perfect second-order temporal coherence, we obtain the same results:

$$\langle n_1(x_1)n_2(x_2)\rangle \propto 1 + \frac{1}{2}\cos\frac{2\pi d}{\lambda z}(x_1 - x_2),$$

$$\langle \Delta n_1(x_1)\Delta n_2(x_2)\rangle \propto 1 + \cos\frac{2\pi d}{\lambda z}(x_1 - x_2). \tag{8.1.22}$$

(IV) Experimental demonstration of two-photon Young's interferometer

Smith and Shih demonstrated the above two-photon Young's double-slit interference phenomenon recently. Their experimental setup is schematically illustrated in Fig. 8.1.2.

A typical measured two-photon interference pattern of Smith and Shih is reported in Fig. 8.1.3, which is in good agreement with the above predictions that are calculated either from Einstein's granularity picture of light, or

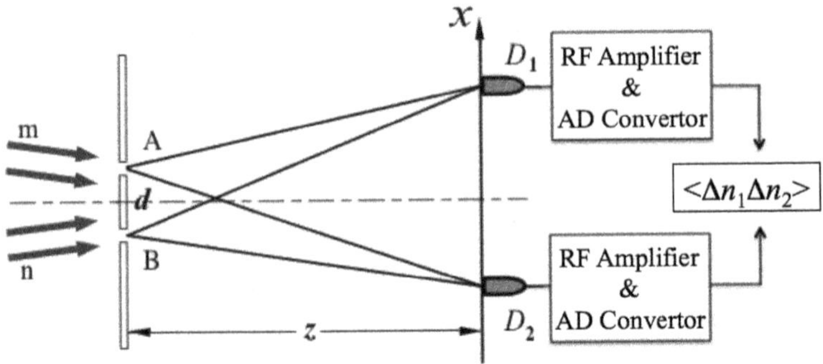

Fig. 8.1.2 Two-photon Young's double-slit interference experiment. The interferometer is a standard Young's double-slit interferometer, except (1) the measurements are not only $\langle n(x_1)\rangle \propto \langle I(x_1)\rangle$ and $\langle n(x_2)\rangle \propto \langle I(x_2)\rangle$ but also $\langle \Delta n(x_1)\Delta n(x_2)\rangle \propto \langle \Delta I(x_1)\Delta I(x_2)\rangle$. (2) The separation between the upper slit-A and the lower slit-B is much greater than the spatial coherence length of the thermal field, $d \gg l_c$. Consequently, no first-order interferences are observable from $\langle n(x_1)\rangle \propto \langle I(x_1)\rangle$ and $\langle n(x_2)\rangle \propto \langle I(x_2)\rangle$. Do we expect observing interference in the photon number fluctuation correlation measurement $\langle \Delta n(x_1, t_1)\Delta n(x_2, t_2)\rangle \propto \langle \Delta I(x_1, t_1)\Delta I(x_2, t_2)\rangle$ when scanning D_1 and/or D_2 along the x-axis?

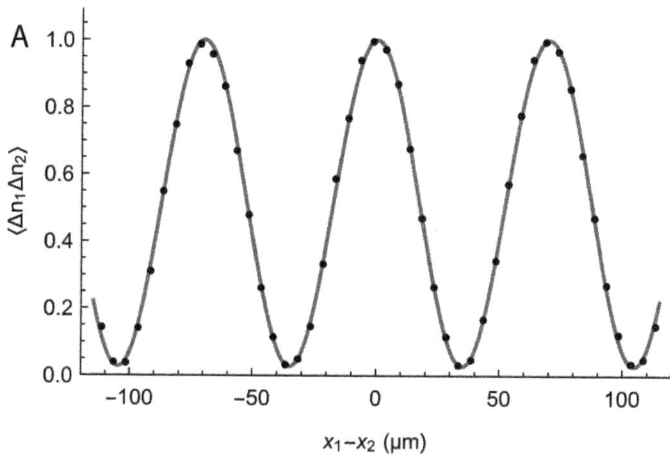

Fig. 8.1.3 A typically two-photon Young's interference pattern is observed from $\langle \Delta n(x_1) \Delta n(x_2) \rangle$ under the experimental condition of $d \gg l_c$. During the scanning of D_1 and D_2, their single-detector counts are also monitored $\langle n(x_1) \rangle \sim$ constant and $\langle n(x_2) \rangle \sim$ constant. The absence of the first-order interferences guarantees $G_{AB}^{(1)} = 0$ and $\langle \Delta n_A \Delta n_B \rangle \propto \langle \Delta I_A \Delta I_B \rangle = 0$.

from quantum theory of light in coherent state representation and in single-photon state representation.

In fact, the first two-photon Young's double-slit interference phenomenon with spatially incoherent thermal fields was experimental demonstration in 2004 by Scarcelli *et al.* Their experimental setup is shown in Fig. 8.1.4. In that experiment, the measured pseudo-thermal fields, E_A and E_B, can be treated as independent incoherent thermal radiations coming from two independent thermal sources A and B. The two independent incoherent thermal sources are simulated by a He–Ne laser beam, a double-slit, and a fast rotating ground glass: a He–Ne laser beam is first impinging on slit-A and slit-B, and a converging lens is followed to image slit-A and slit-B, respectively, onto the fast rotating ground glass. The two images of slit-A and slit-B can be treated as two independent incoherent thermal light sources with completely independent random fluctuations. It is easy to see that their experimental setup has achieved $G_{AB}^{(1)} = 0$ and $\langle \Delta I_A \Delta I_B \rangle = 0$. Two transversely scannable optical fibers, each coupled with a photon counting detector, are facing the simulated two weak thermal sources A and B in the far-field. The single-detector counting rate and the joint-detection counting rate of D_1 and D_2 are monitored, respectively, during

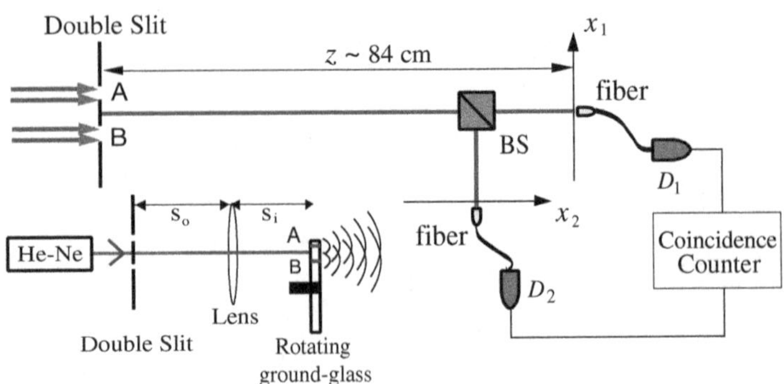

Fig. 8.1.4 Schematic setup of a nontrivial two-photon interference experiment of Scarcelli *et al.* (2004). The experiment is prepared in such a way that slit-A and slit-B can be treated as two independent incoherent thermal light sources with completely independent random fluctuations $\langle \Delta I_A \Delta I_B \rangle = 0$ while fields from A and B are mutually incoherent with $G_{AB}^{(1)} = 0$. No first-order interferences are observable from single-detector counting rates of D_1 and D_2, respectively. Do we expect observing interference in the coincidence counting rate of D_1 and D_2 when scanning D_1 and/or D_2 along their x-axis?

the scanning of the two fiber tips. In their measurement, x_1 and x_2 were scanned with equal value but moved to opposite directions, i.e., $x_1 = -x_2$.

The measured two-photon interference–diffraction pattern is shown in the upper plot of Fig. 8.1.5. It is interesting to see that the interference–diffraction pattern is twice as narrow as the standard interference–diffraction pattern of He–Ne light and with interference modulation twice faster as that of the standard pattern, as if it was produced by a source of light with half the wavelength of the He–Ne laser. For comparison, a first-order interference–diffraction pattern of a He–Ne laser beam is also shown in the lower plot of Fig. 8.1.5. In Fig. 8.1.5 the solid line represents a fitting curve of standard interference–diffraction. The visibility of the pattern is about $\sim(28 \pm 1)\%$ (contrast $\sim(36 \pm 1)\%$). During the measurement, the single-detector counting rates of D_1 and D_2 were both observed as constants over the entire scanning range, which demonstrates the absence of any first order mutual-coherence between fields A and B.

It is worth mentioning that in the experiment of Scarcelli *et al.*, their slit-A and slit-B cannot be treated as line-like slits. Taking into account of the finite width of the sources A and B, their 1-D interference–diffraction pattern is expected to be

$$g^{(2)}(x_1, x_2) = 1 + \text{sinc}^2\left[\frac{\pi b(x_1 - x_2)}{\lambda(z - z_0)}\right]\cos^2\left[\frac{\pi d}{\lambda z}(x_1 - x_2)\right], \qquad (8.1.23)$$

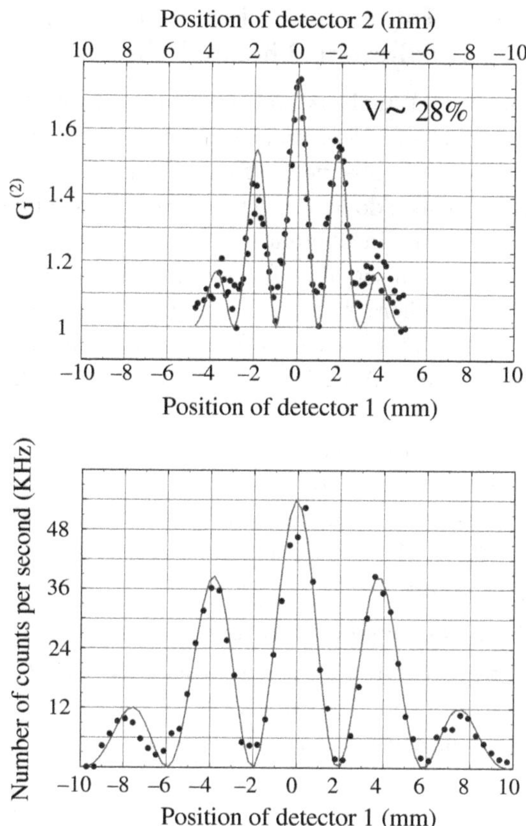

Fig. 8.1.5   Upper: Normalized second-order interference–diffraction pattern vs position of the detectors when $G^{(1)}_{AB} = 0$. In this measurement, x_1 and x_2 were chosen with equal value but opposite direction. Lower: Equivalent first-order interference–diffraction pattern of a He–Ne laser beam satisfying $G^{(1)}_{AB} \simeq 1$.

where b is the 1-D width of the light sources A and B, d is the separation distance between the incoherent light sources A and B. The observed two-photon interference agrees with the theoretical expectation within their experimental error.

If Scarcelli's experiment was measuring photon number fluctuation correlation $\langle \Delta n(x_1)\Delta n(x_2)\rangle$, which is proportional to the intensity fluctuation correlation $\langle \Delta I(x_1)\Delta I(x_2)\rangle$, the two-photon interference visibility may achieve $\sim 100\%$ while no first-order interferences are observable in the measurements of D_1 and D_2, respectively.

8.2 Two-Photon Interference between Temporally Separated Incoherent Thermal Fields

Based on the theoretical analysis and experimental demonstration of the above Young's double-slit interference phenomenon, we conclude that two-photon interference is observable from spatially separated incoherent thermal fields. Now we ask: do we expect to observe two-photon interference from temporally separated incoherent thermal fields? For instance, no first-order interferences are observable from a Mach–Zehnder interferometer when $(L - S)/c > \tau_c$, where τ_c is the coherence time of the thermal field. Is it possible to observe interference modulation from the joint measurement of $\langle n(z_1, t_1)n(z_2, t_2)\rangle$ or $\langle \Delta n(z_1, t_1)\Delta n(z_2, t_2)\rangle$ under the experimental condition of $(L - S)/c > \tau_c$?

In the view of quantum theory of light, we do expect to observe two-photon interference from temporally separated incoherent thermal fields in the measurement of photon number fluctuation correlation $\langle \Delta n(z_1, t_1)\Delta n(z_2, t_2)\rangle$, which is proportional to intensity fluctuation correlation $\langle \Delta I(z_1, t_1)\Delta I(z_2, t_2)\rangle$.

(I) Two-photon interference in Einstein's granularity picture:

In the following, we calculate the second-order coherence $G^{(2)}(z_1, t_1; z_2, t_2)$, which is proportional to $\langle n(z_1, t_1)n(z_2, t_2)\rangle$, for the Mach–Zehnder interferometer of Fig. 8.2.1 in Einstein's granularity picture:

$$G^{(2)}(z_1, t_1; z_2, t_2)$$

$$= \left\langle \sum_{m,n,p,q} E_m^*(z_1, t_1)E_n(z_1, t_1)E_p^*(z_2, t_2)E_q(z_2, t_2) \right\rangle$$

$$= \sum_{m=n} E_m^*(z_1, t_1)E_m(z_1, t_1) \sum_{m=n} E_n^*(z_2, t_2)E_n(z_2, t_2)$$

$$+ \sum_{m \neq n} E_m^*(z_1, t_1)E_n(z_1, t_1)E_n^*(z_2, t_2)E_m(z_2, t_2)$$

$$\simeq \sum_{m,n} \left| E_m(z_1, t_1)E_n(z_2, t_2) + E_n(z_1, t_1)E_m(z_2, t_2) \right|^2$$

$$= \sum_{m,n} \left| (E_{mL1} + E_{mS1})(E_{nL2} + E_{nS2}) + (E_{nL1} + E_{nS1})(E_{mL2} + E_{mS2}) \right|^2$$

$$= \sum_{m,n} \left| E_{mL1}E_{nL2} + E_{mL1}E_{nS2} + E_{mS1}E_{nL2} + E_{mS1}E_{nS2} \right.$$

$$\left. + E_{nL1}E_{mL2} + E_{nL1}E_{mS2} + E_{nS1}E_{mL2} + E_{nS1}E_{mS2} \right|^2, \tag{8.2.1}$$

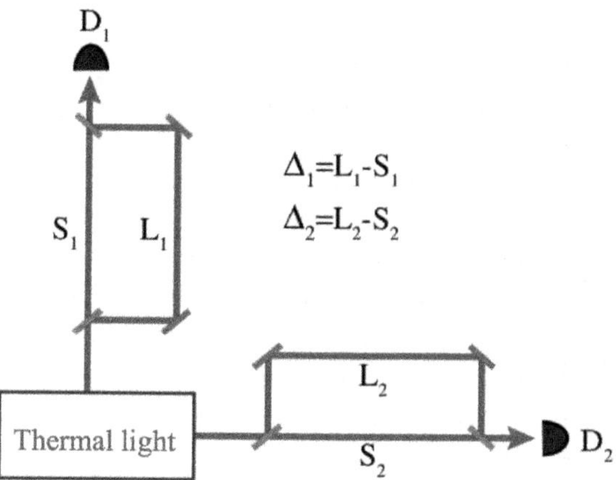

Fig. 8.2.1   Schematic setup of a two-photon Mach–Zehnder interferometer. Under the experimental condition $(L - S)/c > \tau_c$, no first-order interferences are observable from the measurements of $\langle n_1 \rangle$ and $\langle n_2 \rangle$, respectively. Is it possible to observe interference from the measurement of $\langle n_1 n_2 \rangle$ or $\langle \Delta n_1 \Delta n_2 \rangle$?

where we have approximated the sum of $(m \neq n)$ to the sum of (m, n) by ignoring the $m = n$ term, which is reasonable for a large number of subfields; we have also used short-hand notation E_{mLj} and E_{mSj}, $j = 1, 2$, to specify the subfields at D_j that coming from the long path and the short path of the interferometer, respectively. Completing the sum for each term in the $|E_{mL1}E_{nL2} + \cdots + E_{nS1}E_{mS2}|^2$, we find that a number of cross terms vanish due to the experimental condition of $(L - S)/c > \tau_c$. We may group the surviving terms into the following four superpositions,

$$G^{(2)}(z_1, t_1; z_2, t_2)$$

$$= \sum_{m,n} \left| E_{mL1}E_{nL2} + E_{nL1}E_{mL2} \right|^2 + \sum_{m,n} \left| E_{mS1}E_{nS2} + E_{nS1}E_{mS2} \right|^2$$

$$+ \sum_{m,n} \left| E_{mL1}E_{nS2} + E_{nS1}E_{mL2} \right|^2 + \sum_{m,n} \left| E_{mS1}E_{nL2} + E_{nL1}E_{mS2} \right|^2$$

$$\equiv G^{(2)}_{LL}(z_1, t_1; z_2, t_2) + G^{(2)}_{SS}(z_1, t_1; z_2, t_2)$$

$$+ G^{(2)}_{LS}(z_1, t_1; z_2, t_2) + G^{(2)}_{SL}(z_1, t_1; z_2, t_2). \tag{8.2.2}$$

$G^{(2)}_{LL}(z_1, t_1; z_2, t_2)$ is the result of a two-photon superposition between amplitudes: (1) The mth wavepacket passes through the long path of the interferometer and excites a photoelectron at D_1, while the nth wavepacket passes through the long path of the interferometer and excites a

 Quantum Entanglement and Interferometry

photoelectron at D_2; (2) the nth wavepacket passes through the long path of the interferometer and excites a photoelectron at D_1, while the mth wavepacket passes through the long path of the interferometer and excites a photoelectron at D_2.

$G_{SS}^{(2)}(z_1, t_1; z_2, t_2)$ is the result of a two-photon superposition between amplitudes: (3) The mth wavepacket passes through the short path of the interferometer and excites a photoelectron at D_1, while the nth wavepacket passes through the short path of the interferometer and excites a photoelectron at D_2; (4) the nth wavepacket passes through the short path of the interferometer and excites a photoelectron at D_1, while the mth wavepacket passes through the short path of the interferometer and excites a photoelectron at D_2.

$G_{LS}^{(2)}(z_1, t_1; z_2, t_2)$ is the result of a two-photon superposition between amplitudes: (5) The mth wavepacket passes through the long path of the interferometer and excites a photoelectron at D_1, while the nth subfield passes through the short path of the interferometer and excites a photoelectron at D_2; (6) the nth wavepacket passes through the short path of the interferometer and excites a photoelectron at D_1, while the mth wavepacket passes through the long path of the interferometer and excites a photoelectron at D_2.

$G_{SL}^{(2)}(z_1, t_1; z_2, t_2)$ is the result of a two-photon superposition between amplitudes: (7) The mth wavepacket passes through the short path of the interferometer and excites a photoelectron at D_1, while the nth wavepacket passes through the long path of the interferometer and excites a photoelectron at D_2; (8) the nth wavepacket passes through the long path of the interferometer and excites a photoelectron at D_1, while the mth wavepacket passes through the short path of the interferometer and excites a photoelectron at D_2.

It is interesting to find that $G_{LL}^{(2)}$ and $G_{SS}^{(2)}$, respectively, contribute two constants to measurement of $G^{(2)}(z_1, t_1; z_2, t_2)$; one to $\langle I_1 \rangle \langle I_2 \rangle$ and one to $\langle \Delta I_1 \Delta I_2 \rangle$ if the measurement satisfies certain experimental condition:

$$G_{LL}^{(2)}(z_1, t_1; z_2, t_2)$$

$$= \sum_{m,n} \left| \mathcal{F}_{\tau_{L1} - t_m}\{a_m(\nu)\} e^{-i\omega_0(\tau_{L1} - t_m)} \mathcal{F}_{\tau_{L2} - t_n}\{a_n(\nu)\} e^{-i\omega_0(\tau_{L2} - t_n)} \right.$$

$$\left. + \mathcal{F}_{\tau_{L1} - t_n}\{a_n(\nu)\} e^{-i\omega_0(\tau_{L1} - t_n)} \mathcal{F}_{\tau_{L2} - t_m}\{a_m(\nu)\} e^{-i\omega_0(\tau_{L2} - t_m)} \right|^2$$

$$= \sum_{m} \left| \mathcal{F}_{\tau_{L1} - t_m}\{a_m(\nu)\} \right|^2 \sum_{n} \left| \mathcal{F}_{\tau_{L2} - t_n}\{a_n(\nu)\} \right|^2$$

$$+ \sum_n \left| \mathcal{F}_{\tau_{L1}-t_n}\{a_n(\nu)\} \right|^2 \sum_m \left| \mathcal{F}_{\tau_{L2}-t_m}\{a_m(\nu)\} \right|^2$$

$$+ \sum_m \mathcal{F}^*_{\tau_{L1}-t_m}\{a_m(\nu)\} \mathcal{F}_{\tau_{L2}-t_m}\{a_m(\nu)\}$$

$$\times \sum_n \mathcal{F}^*_{\tau_{L2}-t_n}\{a_n(\nu)\} \mathcal{F}_{\tau_{L1}-t_n}\{a_n(\nu)\}$$

$$+ \sum_m \mathcal{F}_{\tau_{L1}-t_m}\{a_m(\nu)\} \mathcal{F}^*_{\tau_{L2}-t_m}\{a_m(\nu)\}$$

$$\times \sum_n \mathcal{F}_{\tau_{L2}-t_n}\{a_n(\nu)\} \mathcal{F}^*_{\tau_{L1}-t_n}\{a_n(\nu)\}$$

$$= \langle I_1 \rangle \langle I_2 \rangle + \langle \Delta I_1 \Delta I_2 \rangle, \tag{8.2.3}$$

where

$$\langle \Delta I_1 \Delta I_2 \rangle \simeq \int dt_m\, \mathcal{F}^*_{\tau_{L1}-t_m}\{a_m(\nu)\} \mathcal{F}_{\tau_{L2}-t_m}\{a_m(\nu)\}$$

$$\times \int dt_n\, \mathcal{F}^*_{\tau_{L2}-t_n}\{a_n(\nu)\} \mathcal{F}_{\tau_{L1}-t_n}\{a_n(\nu)\}$$

$$+ \int dt_m\, \mathcal{F}_{\tau_{L1}-t_m}\{a_m(\nu)\} \mathcal{F}^*_{\tau_{L2}-t_m}\{a_m(\nu)\}$$

$$\times \int dt_n\, \mathcal{F}_{\tau_{L2}-t_n}\{a_n(\nu)\} \mathcal{F}^*_{\tau_{L1}-t_n}\{a_n(\nu)\}$$

$$\simeq 2 \left| \mathcal{F}_{[(t_1-t_2)-(L_1-L_2)/c]}\{a^2(\nu)\} \right|^2. \tag{8.2.4}$$

Obviously, $\langle \Delta I_1 \Delta I_2 \rangle$ is observable only when $\Delta(t_1 - t_2) < 2\pi/\Delta\omega$, where $\Delta\omega$ is the spectral bandwidth of the input thermal field. $\Delta(t_1 - t_2)$ is determined by the response times of D_1 and D_2, therefore, a relatively narrowband spectrum is necessary for "slow" photodetectors.

$G^{(2)}_{LS}$ and $G^{(2)}_{SL}$ may contribute constants or sinusoidal interferences to the measurement of $G^{(2)}(z_1, t_1; z_2, t_2)$, depending on the experimental setups and conditions:

$$G^{(2)}_{LS}(z_1, t_1; z_2, t_2)$$

$$= \sum_{m,n} \Big| \mathcal{F}_{\tau_{L1}-t_m}\{a_m(\nu)\} e^{-i\omega_0(\tau_{L1}-t_m)} \mathcal{F}_{\tau_{S2}-t_n}\{a_n(\nu)\} e^{-i\omega_0(\tau_{S2}-t_n)}$$

$$- \mathcal{F}_{\tau_{S1}-t_n}\{a_n(\nu)\} e^{-i\omega_0(\tau_{S1}-t_n)} \mathcal{F}_{\tau_{L2}-t_m}\{a_m(\nu)\} e^{-i\omega_0(\tau_{L2}-t_m)} \Big|^2$$

$$= \sum_m \left| \mathcal{F}_{\tau_{L1}-t_m} \{a_m(\nu)\} \right|^2 \sum_n \left| \mathcal{F}_{\tau_{S2}-t_n} \{a_n(\nu)\} \right|^2$$

$$+ \sum_n \left| \mathcal{F}_{\tau_{S1}-t_n} \{a_n(\nu)\} \right|^2 \sum_m \left| \mathcal{F}_{\tau_{L2}-t_m} \{a_m(\nu)\} \right|^2$$

$$+ \sum_m \mathcal{F}^*_{\tau_{L1}-t_m} \{a_m(\nu)\} \mathcal{F}_{\tau_{L2}-t_m} \{a_m(\nu)\}$$

$$\times \sum_n \mathcal{F}^*_{\tau_{S2}-t_n} \{a_n(\nu)\} \mathcal{F}_{\tau_{S1}-t_n} \{a_n(\nu)\} e^{i\omega_0[(\tau_{L1}-\tau_{L2})-(\tau_{S1}-\tau_{S2})]}$$

$$+ \sum_m \mathcal{F}_{\tau_{L1}-t_m} \{a_m(\nu)\} \mathcal{F}^*_{\tau_{L2}-t_m} \{a_m(\nu)\}$$

$$\times \sum_n \mathcal{F}_{\tau_{S2}-t_n} \{a_n(\nu)\} \mathcal{F}^*_{\tau_{S1}-t_n} \{a_n(\nu)\} e^{-i\omega_0[(\tau_{L1}-\tau_{L2})-(\tau_{S1}-\tau_{S2})]}$$

$$= \langle I_1 \rangle \langle I_2 \rangle + \langle \Delta I_1 \Delta I_2 \rangle, \tag{8.2.5}$$

where

$$\langle \Delta I_1 \Delta I_2 \rangle$$

$$\simeq \int dt_m \, \mathcal{F}^*_{\tau_{L1}-t_m} \{a_m(\nu)\} \mathcal{F}_{\tau_{L2}-t_m} \{a_m(\nu)\}$$

$$\times \int dt_n \, \mathcal{F}^*_{\tau_{S2}-t_n} \{a_n(\nu)\} \mathcal{F}_{\tau_{S1}-t_n} \{a_n(\nu)\} e^{i\omega_0[(\tau_{L1}-\tau_{L2})-(\tau_{S1}-\tau_{S2})]}$$

$$+ \int dt_m \, \mathcal{F}_{\tau_{L1}-t_m} \{a_m(\nu)\} \mathcal{F}^*_{\tau_{L2}-t_m} \{a_m(\nu)\}$$

$$\times \int dt_n \, \mathcal{F}_{\tau_{S2}-t_n} \{a_n(\nu)\} \mathcal{F}^*_{\tau_{S1}-t_n} \{a_n(\nu)\} e^{-i\omega_0[(\tau_{L1}-\tau_{L2})-(\tau_{S1}-\tau_{S2})]}$$

$$= \mathcal{F}_{[(t_1-t_2)-(L_1-L_2)/c]} \{a^2(\nu)\} \, \mathcal{F}^*_{[(t_1-t_2)-(S_1-S_2)/c]} \{a^2(\nu)\}$$

$$\times e^{i\omega_0[(\tau_{L1}-\tau_{L2})-(\tau_{S1}-\tau_{S2})]}$$

$$+ \mathcal{F}^*_{[(t_1-t_2)-(L_1-L_2)/c]} \{a^2(\nu)\} \, \mathcal{F}_{[(t_1-t_2)-(S_1-S_2)/c]} \{a^2(\nu)\}$$

$$\times e^{-i\omega_0[(\tau_{L1}-\tau_{L2})-(\tau_{S1}-\tau_{S2})]}$$

$$= 2\,\mathrm{Re}\left[\mathcal{F}_{[(t_1-t_2)-(L_1-L_2)/c]} \{a^2(\nu)\} \, \mathcal{F}^*_{[(t_1-t_2)-(S_1-S_2)/c]} \{a^2(\nu)\} \right]$$

$$\times \cos\{\omega_0[(L_2-S_2)-(L_1-S_1)]/c\}. \tag{8.2.6}$$

Again, $\langle \Delta I_1 \Delta I_2 \rangle$ is observable only when $\Delta(t_1-t_2) < 2\pi/\Delta\omega$, where $\Delta\omega$ is the spectral bandwidth of the input thermal field. $\Delta(t_1-t_2)$ is determined by the response times of D_1 and D_2, therefore, a relatively narrowband spectrum is necessary for "slow" photodetectors.

Assuming the experimental condition for observing $\langle \Delta I_1 \Delta I_2 \rangle$ is satisfied, scanning the two-photon Mach–Zehnder interferometer of Fig. 8.2.1 around $[(L_2 - S_2) - (L_1 - S_1)] \sim 0$, taking into account all possible joint photodetection events contributed from $G_{LL}^{(2)}$, $G_{SS}^{(2)}$, $G_{LS}^{(2)}$, and $G_{SL}^{(2)}$, the jointly measured photon number correlation of D_1 and D_2 is thus

$$\langle n_1 n_2 \rangle \propto 1 + \frac{1}{2} \cos \left\{ \omega_0 [(L_2 - S_2) - (L_1 - S_1)]/c \right\}. \tag{8.2.7}$$

A sinusoidal two-photon interference pattern with 50% contrast is observable from the measurement of photon number correlation $\langle \Delta I_1 \Delta I_2 \rangle$ of temporally separated incoherent thermal fields when scan the Mach–Zehnder interferometer of Fig. 8.2.1 around $[(L_2 - S_2) - (L_1 - S_1)] \sim 0$.

Adding only the photon number fluctuation correlations we have calculated from $G_{LL}^{(2)}$, $G_{SS}^{(2)}$, $G_{LS}^{(2)}$, and $G_{SL}^{(2)}$, the jointly measured photon number fluctuation correlation of D_1 and D_2 is therefore

$$\langle \Delta n_1 \Delta n_2 \rangle \propto 1 + \cos \left\{ \omega_0 [(L_2 - S_2) - (L_1 - S_1)]/c \right\}. \tag{8.2.8}$$

A sinusoidal two-photon interference pattern with 100% visibility is observable from the measurement of photon number fluctuation correlation $\langle \Delta I_1 \Delta I_2 \rangle$ of temporally separated incoherent thermal fields when scan the two-photon Mach–Zehnder interferometer of Fig. 8.2.1 around $[(L_2 - S_2) - (L_1 - S_1)] \sim 0$.

(II) Two-photon interference in single-photon state representation:

In the following, we calculate the interference for the same two-photon Mach–Zehnder interferometer of Fig. 8.2.1 in single-photon state representation, under the same experimental condition of $(L - S)/c > \tau_c$. The quantum state of the fields prepared at the source can be approximated as

$$|\Psi\rangle \simeq \prod_m \left(|0\rangle + \epsilon \int d\omega \, \hat{a}_m^\dagger(\omega)|0\rangle \right)$$

$$\simeq |0\rangle + \epsilon \sum_m \int d\omega \, \hat{a}_m^\dagger(\omega)|0\rangle + \epsilon^2 \sum_{m<n} \int d\omega \, d\omega' \, \hat{a}_m^\dagger(\omega)\hat{a}_n^\dagger(\omega')|0\rangle + \cdots.$$

$$\tag{8.2.9}$$

Under weak light condition, we may keep the necessary lowest-order terms of the state for calculating the counting rate of D_1 and D_2, respectively, and the coincidence counting rate of D_1 and D_2, jointly. The joint photodetection counting rate is proportional to the second-order coherence function $G^{(2)}(z_1, t_1; z_2, t_2)$:

$$G^{(2)}(z_1, t_1; z_2, t_2)$$

$$= \left\langle \langle \hat{E}_1^{(-)} \hat{E}_2^{(-)} \hat{E}_2^{(+)} \hat{E}_1^{(+)} \rangle_{\text{QM}} \right\rangle_{\text{En}}$$

$$\simeq \left\langle \left| \langle 0| \sum_p [\hat{E}_p^{(+)}(\tau_{L2}) + \hat{E}_p^{(+)}(\tau_{S2})] \sum_q [\hat{E}_q^{(+)}(\tau_{L1}) + \hat{E}_q^{(+)}(\tau_{S1})] \right. \right.$$

$$\left. \left. \times \sum_{m,n} \int d\omega\, d\omega'\, \hat{a}_m^\dagger(\omega)\hat{a}_n^\dagger(\omega')|0\rangle \right|^2 \right\rangle_{\text{En}}$$

$$= \sum_{m,n} \left| \psi_m(\tau_{L1})\psi_n(\tau_{L2}) + \psi_m(\tau_{L1})\psi_n(\tau_{S2}) \right.$$

$$+ \psi_m(\tau_{S1})\psi_n(\tau_{L2}) + \psi_m(\tau_{S1})\psi_n(\tau_{S2})$$

$$+ \psi_n(\tau_{L1})\psi_m(\tau_{L2}) + \psi_n(\tau_{L1})\psi_m(\tau_{S2})$$

$$\left. + \psi_n(\tau_{S1})\psi_m(\tau_{L2}) + \psi_n(\tau_{S1})\psi_m(\tau_{S2}), \right|^2, \tag{8.2.10}$$

with

$$\psi_m(\tau_{Lj}) = \langle 0|\hat{E}_m^{(+)}(\tau_{Lj}) \int d\omega\, \hat{a}_m^\dagger(\omega)|0\rangle,$$

$$\psi_n(\tau_{Sj}) = \langle 0|\hat{E}_n^{(+)}(\tau_{Sj}) \int d\omega'\, \hat{a}_n^\dagger(\omega')|0\rangle, \tag{8.2.11}$$

the effective wavefunctions of the mth photon and the nth photon measured by D_j, where the L and S label long path and short path of the interferometer, respectively, and $\hat{E}_m^{(+)}(\tau_{Lj})$ and $\hat{E}_n^{(+)}(\tau_{Sj})$, $j = 1, 2$, are the field operators associated with the long path and the short path of the Mach–Zehnder interferometer:

$$\hat{E}_m^{(+)}(\tau_{Lj}) = \int d\omega\, \hat{a}_m(\omega)e^{i\omega(\tau_{Lj} - t_m)},$$

$$\hat{E}_n^{(+)}(\tau_{Sj}) = \int d\omega'\, \hat{a}_n(\omega')e^{i\omega'(\tau_{Sj} - t_n)}. \tag{8.2.12}$$

Note, in Eq. (8.2.10), we have approximated $\sum_{m<n} \cdots$ to $\sum_{m,n}$ by ignoring the $m = n$ terms and by renormalizing the state.

Similar to the calculation in Einstein's granularity picture, completing the sum for each term in the $|\psi_m(\tau_{L1})\psi_n(\tau_{L2}) + \cdots + \psi_n(\tau_{S1})\psi_m(\tau_{S2})|^2$, we find that a number of cross terms vanish due to the experimental condition

$(L - S)/c > \tau_c$. We may group the surviving terms into the following four superpositions:

$$G^{(2)}(z_1, t_1; z_2, t_2) = \sum_{m,n} \left| \psi_m(\tau_{L1})\psi_n(\tau_{L2}) + \psi_n(\tau_{L1})\psi_m(\tau_{L2}) \right|^2$$

$$+ \sum_{m,n} \left| \psi_m(\tau_{S1})\psi_n(\tau_{S2}) + \psi_n(\tau_{S1})\psi_m(\tau_{S2}) \right|^2$$

$$+ \sum_{m,n} \left| \psi_m(\tau_{L1})\psi_n(\tau_{S2}) + \psi_n(\tau_{S1})\psi_m(\tau_{L2}) \right|^2$$

$$+ \sum_{m,n} \left| \psi_m(\tau_{S1})\psi_n(\tau_{L2}) + \psi_n(\tau_{L1})\psi_m(\tau_{S2}) \right|^2$$

$$\equiv G^{(2)}_{LL}(z_1, t_1; z_2, t_2) + G^{(2)}_{SS}(z_1, t_1; z_2, t_2)$$

$$+ G^{(2)}_{LS}(z_1, t_1; z_2, t_2) + G^{(2)}_{SL}(z_1, t_1; z_2, t_2), \qquad (8.2.13)$$

where

$$G^{(2)}_{LL}(z_1, t_1; z_2, t_2) = \sum_{m,n} \left| \psi_m(\tau_{L1})\psi_n(\tau_{L2}) + \psi_n(\tau_{L1})\psi_m(\tau_{L2}) \right|^2, \quad (8.2.14)$$

$$G^{(2)}_{SS}(z_1, t_1; z_2, t_2) = \sum_{m,n} \left| \psi_m(\tau_{S1})\psi_n(\tau_{S2}) + \psi_n(\tau_{S1})\psi_m(\tau_{S2}) \right|^2, \quad (8.2.15)$$

$$G^{(2)}_{LS}(z_1, t_1; z_2, t_2) = \sum_{m,n} \left| \psi_m(\tau_{L1})\psi_n(\tau_{S2}) + \psi_n(\tau_{S1})\psi_m(\tau_{L2}) \right|^2, \quad (8.2.16)$$

$$G^{(2)}_{SL}(z_1, t_1; z_2, t_2) = \sum_{m,n} \left| \psi_m(\tau_{S1})\psi_n(\tau_{L2}) + \psi_n(\tau_{L1})\psi_m(\tau_{S2}) \right|^2. \quad (8.2.17)$$

Repeating the calculations in Einstein's granularity picture by replacing Einstein's subfields with the quantum effective wavefunctions, assuming idealized experimental condition, taking into account all possible joint photodetection events contributed from $G^{(2)}_{LL}$, $G^{(2)}_{SS}$, $G^{(2)}_{LS}$, and $G^{(2)}_{SL}$, the jointly measured photon number correlation of D_1 and D_2 is calculated to be

$$\langle n_1 n_2 \rangle \propto 1 + \frac{1}{2} \cos \frac{\omega_0}{c} [(L_2 - S_2) - (L_1 - S_1)],$$

which is the same as Eq. (8.2.7). A sinusoidal two-photon interference pattern with 50% contrast is observable from the measurement of photon number correlation of temporally separated incoherent thermal fields when scan the two-photon Mach–Zehnder interferometer of Fig. 8.2.1 around $[(L_2 - S_2) - (L_1 - S_1)] \sim 0$.

Adding only the photon number fluctuation correlations we have calculated from $G_{LL}^{(2)}$, $G_{SS}^{(2)}$, $G_{LS}^{(2)}$, and $G_{SL}^{(2)}$, assuming idealized experimental condition, the jointly measured photon number fluctuation correlation of D_1 and D_2 is therefore

$$\langle \Delta n_1 \Delta n_2 \rangle \propto 1 + \cos \frac{\omega_0}{c}[(L_2 - S_2) - (L_1 - S_1)],$$

which is the same as Eq. (8.2.8). A sinusoidal two-photon interference pattern with 100% visibility is observable from the measurement of photon number fluctuation correlation of temporally separated incoherent thermal fields when scan the two-photon Mach–Zehnder interferometer of Fig. 8.2.1 around $[(L_2 - S_2) - (L_1 - S_1)] \sim 0$.

Ihn *et al.* demonstrated such an unbalanced two-photon Mach–Zehnder interferometer recently, which is schematically shown in Fig. 8.2.2. In their experiment, a standard pseudo-thermal field is generated by focusing a laser beam onto a rotating ground glass disk. The laser is an external cavity diode laser operating at 780 nm, and is frequency locked to the $5S_{1/2}(F = 3) - 5P_{3/2}(F' = 4)$ transition of the 85Rb atomic energy levels. The coherence length of the pseudo-thermal field is measured approximately 120 m ($\tau_c \sim 573$ ns) in an optical fiber. To satisfy condition (*b*), the coincidence time window in their experiment is set 15 ns, which is much shorter than $\tau_c \sim 573$ ns. In their experiment, the horizontally polarized thermal light beam is first split by a fiber beam splitter (FBS). Each beam

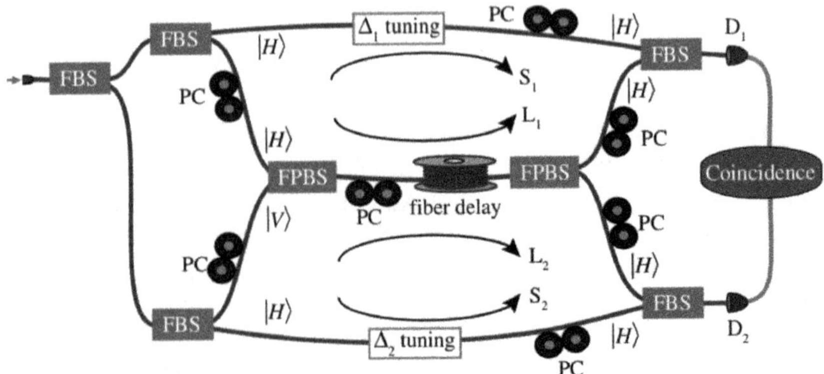

Fig. 8.2.2 Schematic of an unbalanced two-photon Mach–Zehnder interferometer demonstrated by Ihn *et al.* recently. A standard pseudo-thermal radiation of coherence time $\tau_c \sim 572$ ns ($l_c \sim 120$ m) is generated by focusing a laser beam onto a rotating ground glass disk. The coincidence time window is set 15 ns.

is then sent through an unbalanced Mach–Zehnder interferometer (UMZI) with a long and a short optical fiber path. The UMZI consists of FBSs, fiber polarizing beam splitters (FPBSs), fiber polarization controllers (PCs), and optical fibers. The short paths, S1 and S2, each contain a 1-m-long optical fiber and a free-space delay line, labeled as Δ_1 tuning or Δ_2 tuning, controlled by a piezoactuator for phase modulation. The long paths, L_1 and L_2, each include a long fiber spool of length 200 m, 400 m, 600 m, or 800 m. The long paths L_1 and L_2 of the two UMZIs physically share the same fiber spool. The L_1 and L_2 paths, instead, are defined by the polarization states $|H\rangle$ and $|V\rangle$, respectively, by using PCs and FPBSs. Finally, the Δ_1 and Δ_2 delays are scanned by applying voltages to the piezoactuators while observing the single and coincidence counting rates of the two detectors D_1 and D_2.

Figure 8.2.3 reports three sets of their observed two-photon interferences from the measurement of $\langle n_1 n_2 \rangle \propto \langle I_1 I_2 \rangle$ with optical fiber delays

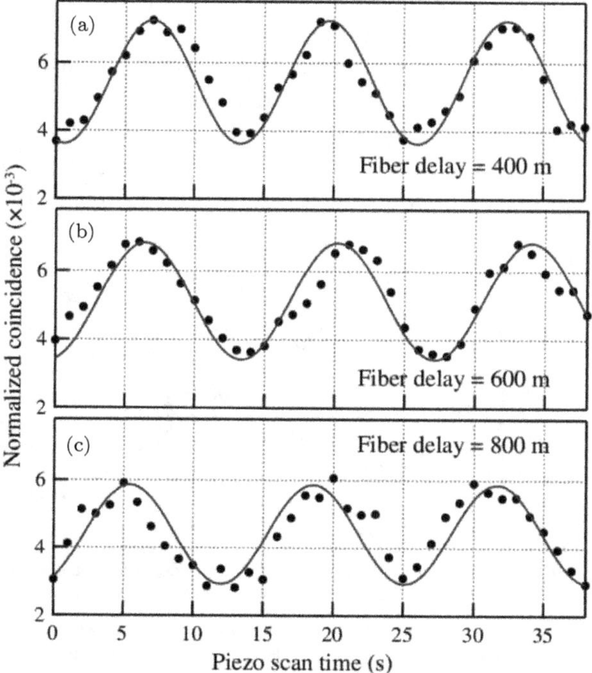

Fig. 8.2.3   Observed two-photon interferences from the measurement of $\langle n_1 n_2 \rangle \propto \langle I_1 I_2 \rangle$ with optical fiber delays $L-S = 400$ m, 600 m, and 800 m. The visibility becomes $\sim 100\%$ in the measurement of $\langle \Delta n_1 \Delta n_2 \rangle \propto \langle \Delta I_1 \Delta I_2 \rangle$. Note that a 120 m optical fiber delay is sufficient to completely remove the classic first-order Mach–Zehnder interferences.

$L - S = 400$ m, 600 m, and 800 m. The visibility becomes $\sim100\%$ in the measurement of $\langle\Delta n_1 \Delta n_2\rangle \propto \langle\Delta I_1 \Delta I_2\rangle$. Note that a 120 m optical fiber delay is sufficient to completely remove the classic first-order Mach–Zehnder interferences.

8.3   Two-Photon Anti-Correlation of Incoherent Thermal Fields

In this section, we discuss a slightly different two-photon interference phenomenon: anti-correlation of incoherent thermal fields. We start from the modified double-slit-Mach–Zehnder interferometer of Fig. 8.3.1. In this modified double-slit interferometer, a large angular sized thermal light source is used to illuminate slit-A and slit-B. We keep the experimental condition of $d \gg l_c$, achieving $G_{AB}^{(1)} = 0$ and thus $\langle\Delta n_A \Delta n_B\rangle \propto \langle\Delta I_A \Delta I_B\rangle = 0$. The incoherent thermal fields from slit-A and slit-B are injected into the two input ports of a beamsplitter. Two photon counting detectors D_1 and D_2 are placed at the two output ports of the beamsplitter for joint measurement of either photon number (intensity) correlation, $\langle n_1 n_2\rangle$, or photon number fluctuation (intensity fluctuation) correlation, $\langle\Delta n_1 \Delta n_2\rangle$. If E_A and E_B are first-order coherent, i.e., $G_{AB}^{(1)} \sim 1$, the setup of Fig. 8.3.1 is equivalent a Mach–Zehnder interferometer. Classic interference patterns are expected from the single-detector counting rates of D_1 and D_2, respectively, when scanning the beamsplitter around its

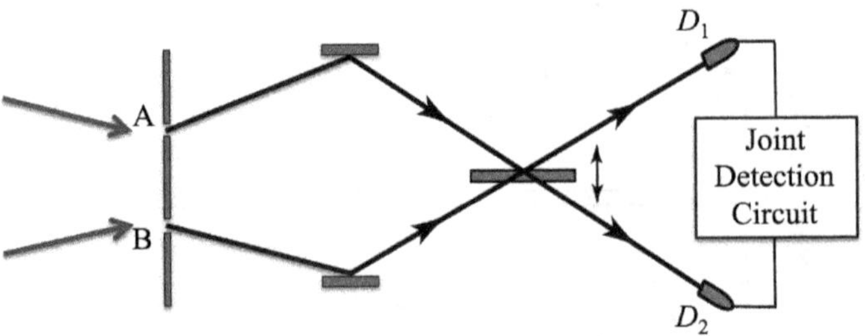

Fig. 8.3.1   Schematic of a modified double-slit-Mach–Zehnder interferometer. A large angular sized thermal light source is applied. Under the experimental condition of $d \gg l_c$, E_A and E_B achieve first-order incoherent $G_{AB}^{(1)} = 0$, and thus $\langle\Delta n_A \Delta n_B\rangle \propto \langle\Delta I_A \Delta I_B\rangle = 0$. The incoherent thermal fields from slit-A and slit-B are injected from the two input ports of the beamsplitter. Two photon counting detectors D_1 and D_2 are placed at the two output ports of the beamsplitter for joint measurement of either photon number (intensity) correlation, $\langle n_1 n_2\rangle$, or photon number fluctuation (intensity fluctuation) correlation, $\langle\Delta n_1 \Delta n_2\rangle$.

balanced position. However, the experimental condition of $G_{AB}^{(1)} = 0$ forces the single-detector counting rates constants, $\langle n_j \rangle \propto \langle I_j \rangle \sim$ constant. Under the experimental condition of $G_{AB}^{(1)} = 0$ and $\langle \Delta n_A \Delta n_B \rangle \propto \langle \Delta I_A \Delta I_B \rangle = 0$, we ask the following three questions:

(1) Do we expect observing a sinusoidal two-photon interference from the joint measurement of D_1 and D_2 when scanning the beamsplitter around its balanced position?

In the view of quantum theory of light, the probability of observing a joint photodetection event of D_1 and D_2 is proportional to the second-order coherence function

$$G^{(2)}(z_1, t_1; z, t_2) = \left\langle \langle \hat{E}^{(-)}(z_1, t_1) \hat{E}^{(-)}(z_2, t_2) \hat{E}^{(+)}(z_2, t_2) \hat{E}^{(+)}(z_1, t_1) \rangle_{\mathrm{QM}} \right\rangle_{\mathrm{En}}$$

$$= \left\langle \left| \langle \Psi | \hat{E}^{(+)}(z_2, t_2) \hat{E}^{(+)}(z_1, t_1) | \Psi \rangle \right|^2 \right\rangle_{\mathrm{En}}$$

$$= \left\langle \left| \langle \Psi | [\hat{E}_{A1}^{(+)} + \hat{E}_{B1}^{(+)}][\hat{E}_{A2}^{(+)} - \hat{E}_{B2}^{(+)}] | \Psi \rangle \right|^2 \right\rangle_{\mathrm{En}}. \tag{8.3.1}$$

The "$-$" sign comes from the beamsplitter. Substituting the thermal state of the randomly created and randomly paired two photons, either in the single-photon state representation or in the coherent state representation, and completing the ensemble average, we have

$$G^{(2)}(z_1, t_1; z_2, t_2)$$

$$= \sum_{m,n} \left| \psi_{mA1}\psi_{nA2} + \psi_{nA1}\psi_{mA2} \right|^2 + \sum_{m,n} \left| \psi_{mB1}\psi_{nB2} + \psi_{nB1}\psi_{mB2} \right|^2$$

$$+ \sum_{m,n} \left| \psi_{mA1}\psi_{nB2} - \psi_{nB1}\psi_{mA2} \right|^2 + \sum_{m,n} \left| \psi_{mB1}\psi_{nA2} - \psi_{nA1}\psi_{mB2} \right|^2$$

$$= G_{AA}^{(2)}(z_1, t_1; z_2, t_2) + G_{BB}^{(2)}(z_1, t_1; z_2, t_2)$$

$$+ G_{AB}^{(2)}(z_1, t_1; z_2, t_2) + G_{BA}^{(2)}(z_1, t_1; z_2, t_2). \tag{8.3.2}$$

We may calculate $G_{AA}^{(2)}$, $G_{BB}^{(2)}$, $G_{AB}^{(2)}$, and $G_{BA}^{(2)}$ separately, as four groups of distinguishable two-photon interferences. Following early sections and chapters, it is not difficult to find out that the two-photon interferences in $G_{AA}^{(2)}$ and $G_{BB}^{(2)}$ contribute two constants to the joint photodetection of D_1 and D_2. $G_{AB}^{(2)}$ and $G_{BA}^{(2)}$, however, may contribute a sinusoidal two-photon interference:

$$G_{AB}^{(2)}(z_1, t_1; z_2, t_2)$$

$$= \sum_{m,n} \left| \psi_{mA1}\psi_{nB2} - \psi_{nB1}\psi_{mA2} \right|^2$$

$$= \sum_{m,n} \left| \mathcal{F}_{\tau_{A1}-t_m}\{a_m(\nu)\} e^{-i\omega_0(\tau_{A1}-t_m)} \mathcal{F}_{\tau_{B2}-t_n}\{a_n(\nu)\} e^{-i\omega_0(\tau_{B2}-t_n)} \right.$$

$$\left. - \mathcal{F}_{\tau_{B1}-t_n}\{a_n(\nu)\} e^{-i\omega_0(\tau_{B1}-t_n)} \mathcal{F}_{\tau_{A2}-t_m}\{a_m(\nu)\} e^{-i\omega_0(\tau_{A2}-t_m)} \right|^2 .$$

$$(8.3.3)$$

Examine the cross-interference term of $G_{AB}^{(2)}(z_1, t_1; z_2, t_2)$:

$$\sum_{m,n} \mathcal{F}^*_{\tau_{A1}-t_m}\{a_m(\nu)\} e^{i\omega_0(\tau_{A1}-t_m)} \mathcal{F}^*_{\tau_{B2}-t_n}\{a_n(\nu)\} e^{i\omega_0(\tau_{B2}-t_n)}$$

$$\times \mathcal{F}_{\tau_{B1}-t_n}\{a_n(\nu)\} e^{-i\omega_0(\tau_{B1}-t_n)} \mathcal{F}_{\tau_{A2}-t_m}\{a_m(\nu)\} e^{-i\omega_0(\tau_{A2}-t_m)}$$

$$= \sum_{m,n} \mathcal{F}^*_{\tau_{A1}-t_m}\{a_m(\nu)\} \mathcal{F}^*_{\tau_{B2}-t_n}\{a_n(\nu)\} \, \mathcal{F}_{\tau_{B1}-t_n}\{a_n(\nu)\} \mathcal{F}_{\tau_{A2}-t_m}\{a_m(\nu)\}$$

$$\times e^{i\omega_0[(\tau_{A1}-\tau_{A2})-(\tau_{B1}-\tau_{B2})]}$$

$$= \sum_{m,n} \mathcal{F}^*_{\tau_{A1}-t_m}\{a_m(\nu)\} \mathcal{F}^*_{\tau_{B2}-t_n}\{a_n(\nu)\} \, \mathcal{F}_{\tau_{B1}-t_n}\{a_n(\nu)\} \mathcal{F}_{\tau_{A2}-t_m}\{a_m(\nu)\}$$

$$\times 1, \qquad\qquad\qquad (8.3.4)$$

we conclude immediately that no sinusoidal two-photon interference is observable from the joint measurement of D_1 and D_2 when scanning the beamsplitter around its balanced position.

(2) Do we expect observing two-photon correlation or anti-correlation from the joint measurement of D_1 and D_2 when scanning the beamsplitter around its balanced position?

Due to the "$-$" sign in Eq. (8.3.3), we conclude that if there exists any two-photon correlation, it must be anti-correlation. Observing anti-correlation, obviously, the cross-interference term of the superposition in Eq. (8.3.3) must be able to achieve a nonzero value. Assuming an incoherent CW thermal field is applied to the two-photon double-slit-Mach–Zehnder interferometer in Fig. 8.3.1, we may model t_m and t_n, the radiation time of the mth and the nth photon, randomly but continuously, and approximate the sum into an integral as usual:

$$\sum_{m,n} \mathcal{F}^*_{\tau_{A1}-t_m}\{a_m(\nu)\}\mathcal{F}^*_{\tau_{B2}-t_n}\{a_n(\nu)\}\,\mathcal{F}_{\tau_{B1}-t_n}\{a_n(\nu)\}\mathcal{F}_{\tau_{A2}-t_m}\{a_m(\nu)\}$$

$$\simeq \int dt_m\,\mathcal{F}^*_{(t_1-t_m)-\frac{z_{A1}}{c}}\{a_m(\nu)\}\mathcal{F}_{(t_2-t_m)-\frac{z_{A2}}{c}}\{a_m(\nu)\}$$

$$\times \int dt_n\,\mathcal{F}^*_{(t_2-t_n)-\frac{z_{B2}}{c}}\{a_n(\nu)\}\mathcal{F}_{(t_1-t_n)-\frac{z_{B1}}{c}}\{a_n(\nu)\}$$

$$\simeq \mathcal{F}_{(t_1-t_2)-\frac{z_{A1}-z_{A2}}{c}}\{a^2(\nu)\}\mathcal{F}^*_{(t_1-t_2)-\frac{z_{B1}-z_{B2}}{c}}\{a^2(\nu)\}. \tag{8.3.5}$$

Interestingly, the above two Fourier transforms always "overlap" resulting in nonzero values unless $\Delta(t_1 - t_2) > 2\pi/\Delta\omega$. This means the cross-interference term of $G^{(2)}_{AB}(z_1,t_1;z_2,t_2)$ and $G^{(2)}_{BA}(z_1,t_1;z_2,t_2)$ always contribute a negative value ~ 1, i.e., anti-correlation, to the correlation measurement of a CW thermal radiation with relatively narrow spectrum. However, no anti-correlation "dip" is observable when scanning the beamsplitter around its balanced position. Although no anti-correlation "dip" is observable, this two-photon interference induced anti-correlation is still useful, for example, to demonstrate Bell-type polarization correlation. In 2015, Peng and Shih experimentally observed Bell-type polarization correlation from the measurement of intensity-fluctuation correlation of CW pseudo-thermal field in an Alley–Shih type two-photon interferometer. In their experiment, no anti-correlation "dip" was necessary.

(3) How to observe correlation "peak" or anti-correlation "dip" by scanning the beamsplitter from its balanced position?

In fact, an anti-correlation "dip" is observable from the double-slit-Mach–Zehnder interferometer of Fig. 8.3.1, if the input radiation is a short pulsed thermal field. A short puled thermal field forces the mth and the nth wavepackets to be created simultaneously, $t_m \sim t_n$. Examine the cross-interference term of $G^{(2)}_{AB}(z_1,t_1;z_2,t_2)$ by assuming $t_m \sim t_n = t_0$:

$$\sum_{m,n} \mathcal{F}^*_{\tau_{A1}-t_0}\{a(\nu)\}e^{i\omega_0(\tau_{A1}-t_0)}\mathcal{F}^*_{\tau_{B2}-t_0}\{a(\nu)\}e^{i\omega_0(\tau_{B2}-t_0)}$$

$$\times \mathcal{F}_{\tau_{B1}-t_0}\{a(\nu)\}e^{-i\omega_0(\tau_{B1}-t_0)}\mathcal{F}_{\tau_{A2}-t_0}\{a(\nu)\}e^{-i\omega_0(\tau_{A2}-t_0)}$$

$$\simeq \sum_{t_0} \mathcal{F}^*_{\tau_{A1}-t_0}\{a(\nu)\}\mathcal{F}_{\tau_{B1}-t_0}\{a(\nu)\} \sum_{t_0} \mathcal{F}^*_{\tau_{B2}-t_0}\{a(\nu)\}\mathcal{F}_{\tau_{A2}-t_0}\{a(\nu)\}$$

$$\simeq \left|\mathcal{F}_\delta\{a^2(\nu)\}\right|^2, \tag{8.3.6}$$

where δ is the optical delay between path A and path B, $\delta = (z_{Aj} - z_{Bj})/c$, $j = 1, 2$.

In a set of recent experiments, Chen *et al.* observed the anti-correlation "dip" from the measurement of incoherent pulsed thermal field. The experimental setup of Chen *et al.* is schematically shown in Fig. 8.3.2. The light source is a short pulsed pseudo-thermal light source which consists of a CW mode-locked laser beam with ~200 fs pulses at a 78 MHz repetition rate and a fast rotating diffusing ground glass. The rotating diffusing ground glass scatters the laser beam into a large number of transversely randomly distributed point-like sub-sources. A large number of subfields scattered from these point-like sub-sources are than collected by two optical fiber tips A and B and coupled into a 50/50 optical fiber beamsplitter. This experiment measures $\langle n(z_1, t_1) n(z_2, t_2) \rangle$ or $\langle \Delta n(z_1, t_1) \Delta n(z_2, t_2) \rangle$ at the output ports of the 50/50 optical fiber beamsplitter BS2. It is BS2 in Fig. 8.3.2 introduces a "$-$" sign between its two input fields and produces a destructive two-photon superposition that is shown in Eqs. (8.3.2) and (8.3.3). If the two input fiber tips A and B are placed within the longitudinal temporal coherence length and the transverse coherence area of the thermal field at the output ports of the first beamsplitter BS1, i.e., $G^{(1)}(\mathbf{r}_A, t_A; \mathbf{r}_B, t_B) \neq 0$, this setup is equivalent to a Mach–Zehnder interferometer. D_1 and D_2, respectively, will each observe first-order

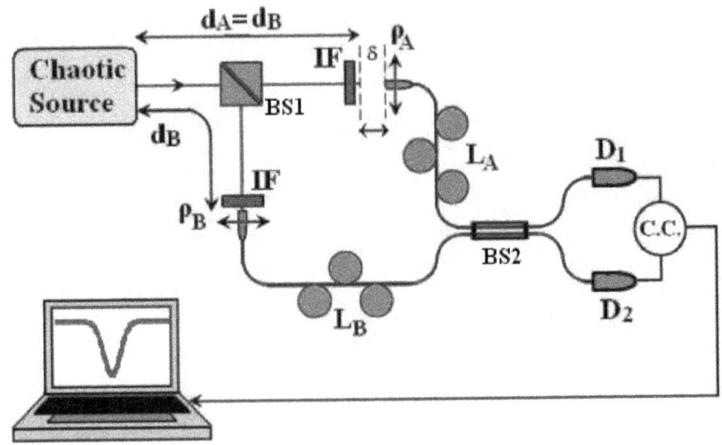

Fig. 8.3.2   Schematic setup of an anti-correlation measurement from pulsed incoherent thermal fields by Chen *et al.* (2009).

interference as a function of the optical delay δ when scanning the fiber tip A along its longitudinal axis. Consequently, the joint photodetection of D_1 and D_2 produces an interference that is factorizable into two first-order interferences. However, in this experiment we decided to move the fiber tip A outside the transverse coherence area to force $G^{(1)}(\mathbf{r}_A, t_A; \mathbf{r}_B, t_B) = 0$. Under this condition, the input radiation at the fiber tips A and B are incoherent, there would be no first-order interference observable from D_1 and D_2, respectively, even in the case of "single-exposure" or instantaneous measurement.

Although the experimental condition achieved $G^{(1)}_{AB} = 0$ and $\langle \Delta n_A \Delta n_B \rangle \propto \langle \Delta I_A \Delta I_B \rangle = 0$, an anti-correlation function was observed from pulsed incoherent thermal fields E_A and E_B in the joint photodetection counting rate of D_1 and D_2: the measurement of $\langle n_1 n_2 \rangle \propto \langle \Delta I_1 \Delta I_2 \rangle$ yielded a "dip" as function of the optical delay δ and the bandwidth of the spectral filters IF, while the counting rate of D_1 and D_2 both kept constants. Figure 8.3.3 shows two typical measured anti-correlation functions of the experiment of Chen *et al.*, corresponding to two different spectrum bandwidths of the pulsed thermal field, in photon number correlation.

The following experimental details of Chen *et al.* may be helpful for understanding the physics behind anti-correlation of thermal field:

(1) **The source**: The light source is a standard pseudo-thermal source that was developed since the 1960s and used widely in HBT correlation measurements. The source consists of a mode-locked laser beam with $\sim$200 fs pulses at a 78 MHz repetition rate and a fast rotating diffusing ground glass. The linearly polarized laser beam is enlarged transversely onto the ground glass with a diameter of 4.5 mm. The enlarged laser beam is scattered by the fast rotating diffusing ground into a large number of transversely randomly distributed point-like sub-sources to simulate a near-field, pseudo-thermal radiation source: a large number of randomly distributed incoherent wavepackets with random relative phases are then collected by the optical fiber tips A and B and coupled into the optical fiber beamsplitter.

(2) **The interferometer**: A 50/50 beam splitter (BS1) is used to split the pulsed thermal field into transmitted and reflected radiations which are then coupled into two identical polarization-controlled single-mode fibers A and B respectively. The fiber tips are located $\sim$200 mm from the ground glass, i.e. $d_A = d_B \sim 200$ mm. At this distance, the angular size of the source $\Delta\theta$ is $\sim$22.5 mili-radian (1.29°) with respect to each input fiber tip, which satisfies the Fresnel near-field condition. Two identical narrowband spectral

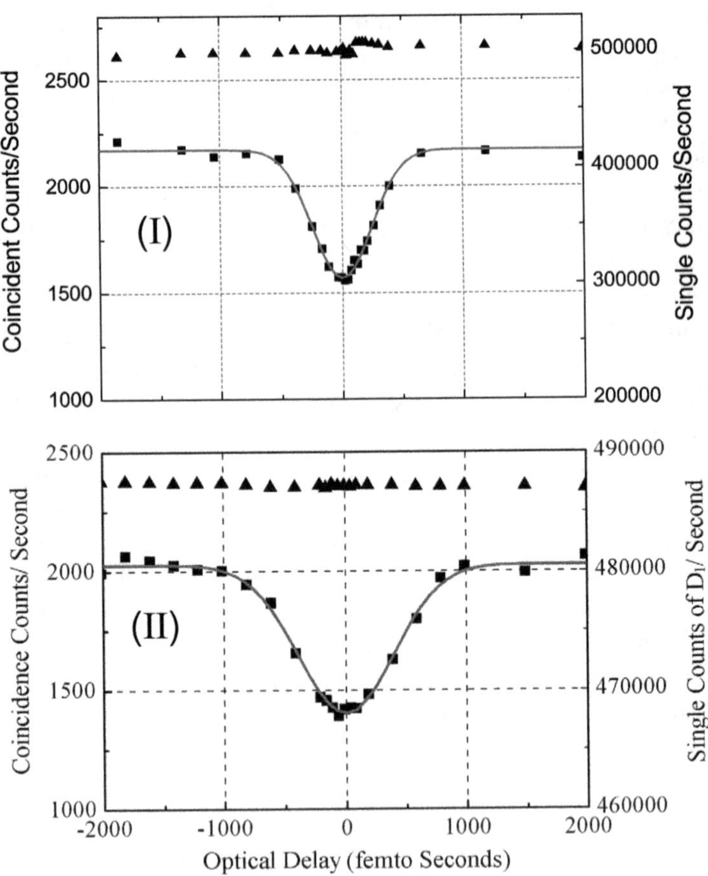

Fig. 8.3.3 Typical observed anti-correlation "dips" in *photon number* correlation measurements with different bandwidth of spectrum: $\tau_c \sim 345$ fs for (I), $\tau_c \sim 541$ fs for (II). The temporal width of the "dip" is determined by the bandwidth of the spectral filters (IF). The counting rate of D_1 and D_2 both kept constants during the scanning of δ. The observed contrast of "dip" is $\sim 100\%$ in photon number fluctuation correlation measurement.

filters (IF) are placed in front of the two fiber tips A and B. The transverse and longitudinal coordinates of the input fiber tips are both scannable by step-motors. The output ends of the two fibers can be directly coupled into the photon counting detectors D_1 and D_2, respectively, for near-field HBT correlation measurements, or coupled into the two input ports of a 50/50 single-mode optical fiber beamsplitter (BS2) for the anti-correlation measurement.

(3) **The measurement**: Two steps of measurements were made. The purpose of step-one is to confirm the pseudo-thermal light source produces thermal field by means of temporal and spatial HBT correlations. In step-one measurement, the output ends of the two fibers A and B are directly coupled into the photon counting detectors D_1 and D_2 for photon number correlation and photon number fluctuation correlation measurement, while scanning the input fiber tips longitudinally and transversely. Thermal radiation can easily be distinguished from a laser beam by examining its second-order coherence function $G^{(2)}(\mathbf{r}_A, t_A; \mathbf{r}_B, t_B)$, which is characterized experimentally by the coincidence counting rate that counts the joint photodetection events at space–time points $(\mathbf{r}_A, t_A)$ and $(\mathbf{r}_B, t_B)$. Figure 8.3.4 reports two measured second-order correlations at zero and at 1000 revolutions per minute (rpm) of the rotating ground glass. This measurement guarantees a typical HBT correlation of thermal field at rotation speeds of 1000 rpm of the ground glass, indicating the thermal nature of the light source. In this measurement, we have also located experimentally the longitudinal and transverse coordinates of the fiber tips A and B for achieving the maximum coincidence counting rate, corresponding to the maximum second-order correlation.

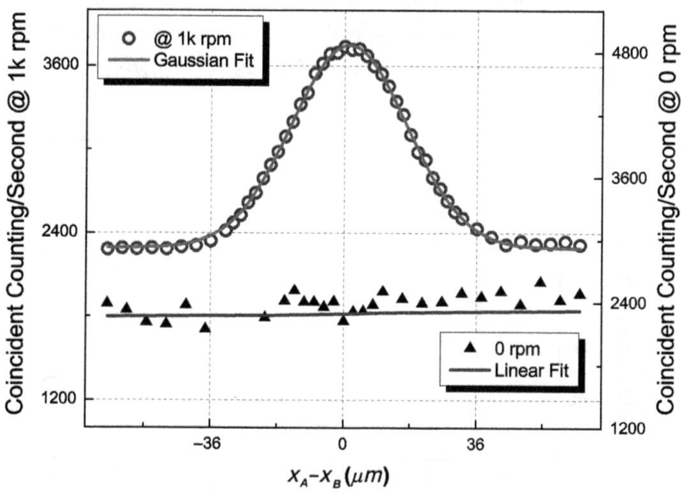

Fig. 8.3.4   Measurement of HBT correlation of $G^{(2)}(x_A - x_B)$ on the planes of pinhole-A and pinhole-B for pseudo-thermal light (ground glass rotated at 1000 rpm) and for laser beam (ground glass stopped at 0 rpm). Here, x_A and x_B are the x-component of $\vec{\rho}_A$ and $\vec{\rho}_B$.

In step-two, we couple the 50/50 fiber beamsplitter (BS2) into the setup as shown in Fig. 8.3.2. This measurement was done in two steps. We first measured the first-order interference at $\vec{\rho}_A = \vec{\rho}_B$ by scanning the input fiber tip A longitudinally in the neighborhood of $d_A \sim 200$ mm. There is no surprise to have first-order interference in the counting rates of D_1 and D_2 respectively. When choosing $\vec{\rho}_A = \vec{\rho}_B$, the two input fiber tips are coupled within the spatial coherence area of the radiation field; we have effectively built a Mach–Zehnder interferometer. We then move the input fiber tip A transversely from $\vec{\rho}_A = \vec{\rho}_B$ to $|\vec{\rho}_A - \vec{\rho}_B| \gg l_c$, where l_c is the transverse coherence length of the thermal field. (In most of the measurements, $|\vec{\rho}_A - \vec{\rho}_B|$ was chosen to be $|\vec{\rho}_A - \vec{\rho}_B| \geq 40 l_c$.) Then we scan the input fiber tip A again longitudinally in the neighborhood of $d_A \sim 200$ mm. The optical delay between the plane $z = d_A \sim 200$ mm and the scanning input fiber tip A is labeled as δ in Fig. 8.3.2. Under the experimental condition of $G^{(1)}(\mathbf{r}_A, t_A; \mathbf{r}_B, t_B) = 0$ and $G^{(2)}(\mathbf{r}_A, t_A; \mathbf{r}_B, t_B) = \text{constant}$, there is no surprise we lose any first-order interference. However, it is indeed a surprise that an anti-correlation is observed in terms of the joint photodetection of D_1 and D_2 as a function of the optical delay δ that is reported in Fig. 8.3.3. In these measurements, $|\vec{\rho}_A - \vec{\rho}_B| \sim 40 l_c$. The observed contrast of the "dips" are $\sim 50\%$ in photon number correlation measurements and are $\sim 100\%$ in photon number fluctuation correlation measurements. Figure 8.3.5 shows another observed photon number fluctuation anti-correlation "dip". The visibility of the "dip" is recognized as $(93.2 \pm 5.1)\%$.

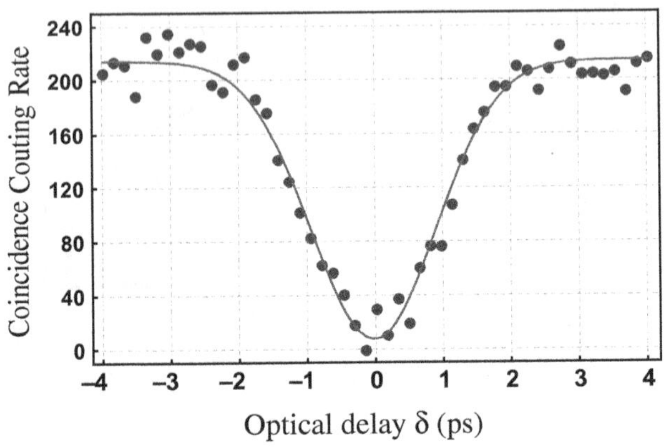

Fig. 8.3.5 Typical observed anti-correlation function of "dip" in *photon number fluctuation* correlation measurement. The visibility of the "dip" is calculated to be $(93.2 \pm 5.1)\%$. The fitting curve agrees with the theoretical prediction well within the experimental error.

8.4 Two-Photon Interference with Incoherent Orthogonal Polarized Thermal Fields

In this section, we discuss two-photon interference with orthogonal polarized incoherent thermal fields. We start from the analysis of a simple experiment, which is schematically shown in Fig. 8.4.1. The experimental setup is similar to that of the historical two-photon polarization interferometer of Alley–Shih, except the use of orthogonal polarized incoherent thermal light, instead of entangled photon pair. If the input radiation at points A and B are first-order coherent, this setup is equivalent to a classic polarization Mach–Zehnder interferometer. D_1 and D_2 will each observe first-order interference, and consequently, the joint photodetection of D_1 and D_2 outputs an interference that is the product of the two first-order interferences. To avoid this trivial effect to happen, E_A and E_B are managed mutually incoherent, $G^{(1)}(\mathbf{r}_A, t_A; \mathbf{r}_B, t_B) = 0$. The incoherent thermal radiations A and B are prepared in the same way as that in the previous experiment of Chen *et al.* by moving the fiber tip A transversely outside of the coherence area of the thermal field, which has been described in detail in previous sections. Similarly, achieving an experimental condition of $|\bar{\rho}_A - \bar{\rho}_B| \geq 40 l_c$, there is no question that we will lose any first-order interferences measured by D_1 and D_2, respectively. In addition, D_1 and D_2, respectively, should observe randomly polarized

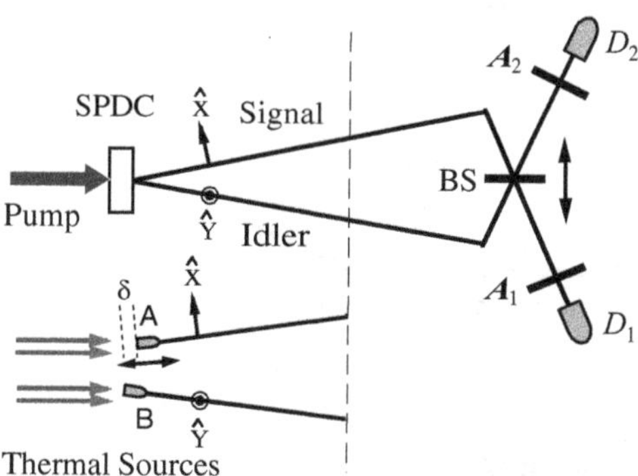

Fig. 8.4.1 Schematic of a typical two-photon interferometer with orthogonal polarized *incoherent thermal light*. The setup is the same as the historical two-photon polarization interferometer of Alley–Shih, except the use of randomly created and randomly photon pair in thermal state, instead of entangled photon pair. The use of *incoherent radiations* E_A and E_B guarantee $G^{(1)}(\mathbf{r}_A, t_A; \mathbf{r}_B, t_B) = 0$ and $G^{(2)}(\mathbf{r}_A, t_A; \mathbf{r}_B, t_B) = $ constant. No classic Mach–Zehnder interferences are observable by D_1 and D_2, respectively. Can we expect to observe either space–time correlation or polarization correlation from the joint measurement of D_1 and D_2?

light too. The incoherent $\hat{x}$ and $\hat{y}$ polarized radiations will pass A_1 and A_2 randomly according to Malus' law. The total intensity measured by D_1 and D_2, respectively, should be $\sin^2 \theta_j + \cos^2 \theta_j = 1$, $j = 1, 2$, in any chosen orientation of the polarizer. Again, the question is in the second-order correlation measurement. Under the experimental condition of $G^{(2)}(\mathbf{r}_A, t_A; \mathbf{r}_B, t_B) = \text{constant}$, i.e., $\langle \Delta I_A \Delta I_B \rangle = 0$, there should be no observable HBT type correlations. What can we expect from the measurement of $G^{(2)}(\mathbf{r}_1, t_1; \mathbf{r}_2, t_2)$? Can we expect to observe either space–time correlation or polarization correlation from the joint measurement of D_1 and D_2?

(I) **Experimental realization of Chen *et al.*:**

Chen *et al.* provided a positive answer to the above question in another recent experiment: two-photon interference with incoherent orthogonal polarized thermal fields produce both space–time correlation and polarization correlation from the joint measurement of D_1 and D_2. Their schematic experimental setup is illustrated in Fig. 8.4.2. Some of the experimental details are emphasized as follows:

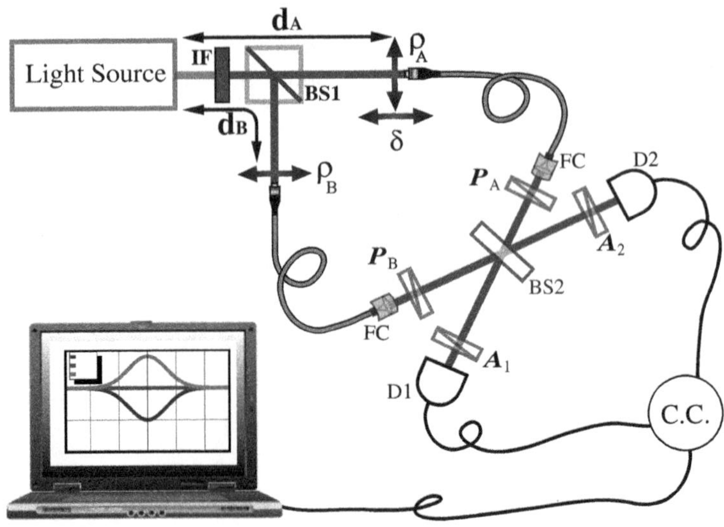

Fig. 8.4.2 Schematic experimental setup of Chen *et al.* in which two surprises were observed from first-order and second-order incoherent orthogonal polarized thermal fields: (1) anti-correlation "dip" and correlation "peak" and (2) Bell-type polarization correlation. The experimental condition of *first-order and second-order incoherence* between radiations A and B is achieved by moving fiber tip A transversely outside of the spatial coherence area of the field. In this experiment, most of the measurements were performed at $|\bar{\rho}_A - \bar{\rho}_B| \geq 40 l_c$.

(1) **The source**: The incoherent pseudo-thermal source is the same as that in the anti-correlation experiment of Chen *et al.*, which consists of a mode-locked laser beam with $\sim$200 fs pulses at a 78 MHz repetition rate and a fast rotating diffusing ground glass. The laser beam is enlarged transversely onto the ground glass with a diameter of 4.5 mm. The enlarged laser beam is scattered by the fast rotating diffusing ground into a large number of transversely randomly distributed point-like sub-sources to simulate a near-field, pseudo-thermal radiation source. A large number of randomly distributed incoherent wavepackets with random relative phases are then collected by the optical fiber tips A and B and coupled into the optical fiber beamsplitter.

(2) **The polarization state**: A set of polarization bases of $\hat{x}$ and $\hat{y}$ is defined by two Glen–Thompson polarizers which are coupled with fiber A and fiber B, respectively. The $\hat{x}$ polarized radiation A and the $\hat{y}$ polarized radiation B are injected onto a 50–50% beamsplitter from its opposite input ports in a near-normal incidence configuration.[2] The beamsplitter is measured with 50% reflection and 50% transmission for both $\hat{x}$ and $\hat{y}$ polarization.

(3) **The polarization analyzers**: Two Glen–Thompson polarizers are coupled into the two output ports of the 50–50% beamsplitter, each followed by a photon counting detector, D_1 and D_2, for the polarization correlation measurement. The polarization analyzer can be set at any angle θ relative to the $\hat{x}$. The $\hat{x}$ direction is defined as $0°$ in the above analysis.

The measurement of Chen *et al.* produces two interesting effects in one experimental setup of Fig. 8.4.2:

(1) "Unexpected" anti-correlation "dip" and correlation "peak" are observed in the joint photodetection counting rate of D_1 and D_2, $\langle n_1 n_2 \rangle$, as a function of the optical delay δ, which is defined in Fig. 8.4.1. The anti-correction "dip" is observed when P_1 and P_2 were oriented at $\theta_1 = 45°, \theta_2 = 45°$. The correlation "peak" is obtained when P_1 and P_2 are oriented at $\theta_1 = 45°, \theta_2 = 135°$.

---

[2] To achieve 50–50% reflection-transmission for both $\hat{x}$ and $\hat{y}$ polarization, near-normal incidence configuration is a better choice. This configuration is the same as that of the historical experiment of Alley and Shih which observed Bell-type polarization correlation from an orthogonal polarized photon pair of SPDC. The first a few historical "dip" experiments, such as the Hong–Ou–Mandel "dip" measurement, adopted this near-normal incidence configuration from the two-photon interferometer of Alley and Shih. In fact, it is unnecessary to choose near-normal incidence beamsplitter for a biphoton interferometer which uses only one polarization (type-I SPDC) for "dip" measurement.

(2) Bell-type polarization correlation are observed as a function of the relative angle $(\theta_1 - \theta_2)$ of the two polarization analyzers under the condition of $\delta = 0$,

$$R_c(\theta_1, \theta_2) \propto \langle n_1(\theta_1)n_2(\theta_2)\rangle \propto \left[1 + V\sin^2(\theta_1 - \theta_2)\right], \qquad (8.4.1)$$

where V is the contrast of the sinusoidal modulation, $R_c(\theta_1, \theta_2)$ is the coincidence counting rate of D_1 and D_2, which is obtained by manipulating the relative orientation of A_1 and A_2. Figure 8.4.4 is a typical observed polarization correlation reported by Chen *et al.* Furthermore, if the measurement is on photon number fluctuation correlation $\langle \Delta n_1(\theta_1)\Delta n_2(\theta_2)\rangle \propto \Delta R_c(\theta_1, \theta_2)$,

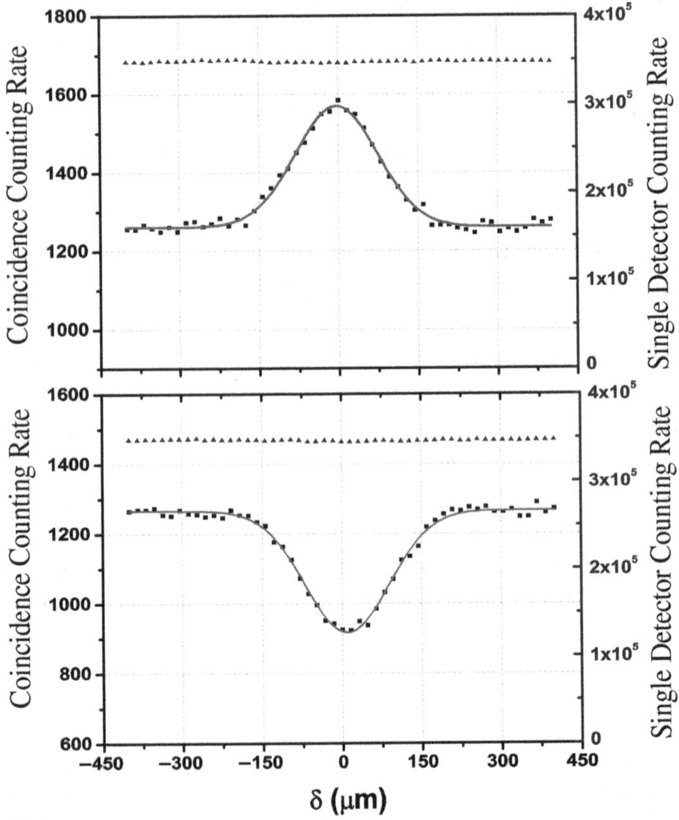

Fig. 8.4.3   Typical observed correlation function of "peak" and anti-correlation function of "dip" as functions of the optical delay δ. The correlation "peak" is obtained when P_1 and P_2 were oriented at $\theta_1 = 45°, \theta_2 = 135°$. The anti-corrrection "dip" is observed when P_1 and P_2 were oriented at $\theta_1 = 45°, \theta_2 = 45°$.

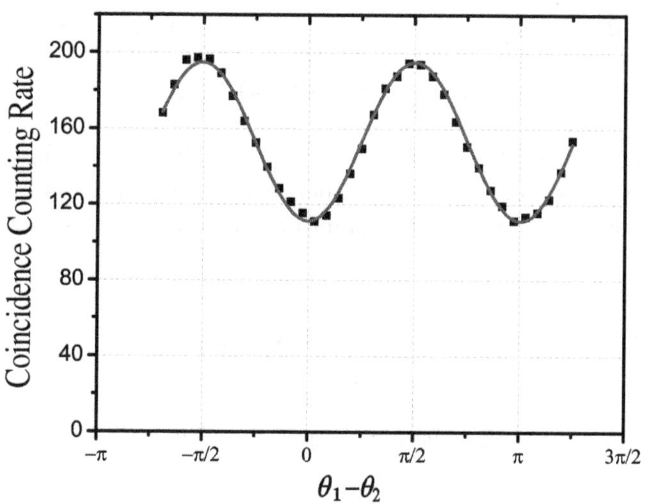

Fig. 8.4.4 Typical observed polarization correlation $\langle n_1(\theta_1)n_2(\theta_2)\rangle$ as a function of $\theta_1 - \theta_2$. In this measurement, θ_2 was fixed at $45°$ relative to $\hat{x}$ and $\hat{y}$ and θ_1 was rotated at each chosen angle θ_1. If the measurement is for photon number fluctuation correlation $\langle \Delta n_1(\theta_1)\Delta n_2(\theta_2)\rangle$, the visibility of the sinusoidal function of $\theta_1 - \theta_2$ becomes $\sim 100\%$.

the visibility of the sinusoidal function of $\theta_1 - \theta_2$ becomes $\sim 100\%$,

$$\Delta R_c(\theta_1, \theta_2) \propto \langle \Delta n_1(\theta_1)\Delta n_2(\theta_2)\rangle \propto \sin^2(\theta_1 - \theta_2). \tag{8.4.2}$$

Quantum theory gives a reasonable interpretation to the observation: The observation is a two-photon interference phenomenon, the result of a superposition between indistinguishable two-photon amplitudes. Taking the result of early sections of this chapter, we start from $G^{(2)}(\delta; \theta_1, \theta_2)$ with the polarization analyzers oriented at θ_1 and θ_2, respectively,

$$G^{(2)}(\theta_1, \theta_2; \delta)$$

$$= \sum_{m,n} \Big| [(\hat{\theta}_1 \cdot \hat{x})\psi_{mA1} - (\hat{\theta}_1 \cdot \hat{y})\psi_{mB1}][(\hat{\theta}_2 \cdot \hat{x})\psi_{nA2} + (\hat{\theta}_1 \cdot \hat{y})\psi_{nB2}]$$

$$+ [(\hat{\theta}_1 \cdot \hat{x})\psi_{nA1} - (\hat{\theta}_1 \cdot \hat{y})\psi_{nB1}][(\hat{\theta}_2 \cdot \hat{x})\psi_{mA2} + (\hat{\theta}_2 \cdot \hat{y})\psi_{mB2}] \Big|^2$$

$$= \sum_{m,n} \Big| (\hat{\theta}_1 \cdot \hat{x})(\hat{\theta}_2 \cdot \hat{x})\psi_{mA1}\psi_{nA2} + (\hat{\theta}_1 \cdot \hat{x})(\hat{\theta}_2 \cdot \hat{y})\psi_{mA1}\psi_{nB2}$$

$$- (\hat{\theta}_1 \cdot \hat{y})(\hat{\theta}_2 \cdot \hat{x})\psi_{mB1}\psi_{nA2} - (\hat{\theta}_1 \cdot \hat{y})(\hat{\theta}_2 \cdot \hat{y})\psi_{mB1}\psi_{nB2}$$

$$+ (\hat{\theta}_1 \cdot \hat{x})(\hat{\theta}_2 \cdot \hat{x})\psi_{nA1}\psi_{mA2} + (\hat{\theta}_1 \cdot \hat{x})(\hat{\theta}_2 \cdot \hat{y})\psi_{nA1}\psi_{mB2}$$

$$- (\hat{\theta}_1 \cdot \hat{y})(\hat{\theta}_2 \cdot \hat{x})\psi_{nB1}\psi_{mA2} - (\hat{\theta}_1 \cdot \hat{y})(\hat{\theta}_2 \cdot \hat{y})\psi_{nB1}\psi_{mB2} \Big|^2. \qquad (8.4.3)$$

Completing the sum for each term in the $|\ \dots\ |^2$, we find that a number of cross terms vanish due to the experimental condition $|\bar{\rho}_A - \bar{\rho}_B| \gg l_c$. We may group the surviving terms into the following four superpositions:

$$G^{(2)}(\theta_1, \theta_2; \delta)$$

$$= \sum_{m,n} \Big|(\hat{\theta}_1 \cdot \hat{x})(\hat{\theta}_2 \cdot \hat{x})\psi_{mA1}\psi_{nA2} + (\hat{\theta}_1 \cdot \hat{x})(\hat{\theta}_2 \cdot \hat{x})\psi_{nA1}\psi_{mA2}\Big|^2$$

$$+ \sum_{m,n} \Big|(\hat{\theta}_1 \cdot \hat{y})(\hat{\theta}_2 \cdot \hat{y})\psi_{mB1}\psi_{nB2} + (\hat{\theta}_1 \cdot \hat{y})(\hat{\theta}_2 \cdot \hat{y})\psi_{nB1}\psi_{mB2}\Big|^2$$

$$+ \sum_{m,n} \Big|(\hat{\theta}_1 \cdot \hat{x})(\hat{\theta}_2 \cdot \hat{y})\psi_{mA1}\psi_{nB2} - (\hat{\theta}_1 \cdot \hat{y})(\hat{\theta}_2 \cdot \hat{x})\psi_{nB1}\psi_{mA2}\Big|^2$$

$$+ \sum_{m,n} \Big|(\hat{\theta}_1 \cdot \hat{y})(\hat{\theta}_2 \cdot \hat{x})\psi_{mB1}\psi_{nA2} - (\hat{\theta}_1 \cdot \hat{x})(\hat{\theta}_2 \cdot \hat{y})\psi_{nA1}\psi_{mB2}\Big|^2$$

$$\equiv G^{(2)}_{AA}(\theta_1, \theta_2; \delta) + G^{(2)}_{BB}(\theta_1, \theta_2; \delta) + G^{(2)}_{AB}(\theta_1, \theta_2; \delta) + G^{(2)}_{BA}(\theta_1, \theta_2; \delta).$$
$$(8.4.4)$$

The four groups of probabilities $G^{(2)}_{AB}(\theta_1, \theta_2; \delta)$, $G^{(2)}_{BA}(\theta_1, \theta_2; \delta)$, $G^{(2)}_{AA}(\theta_1, \theta_2; \delta)$, and $G^{(2)}_{BB}(\theta_1, \theta_2; \delta)$ are schematically pictured in Fig. 8.4.5: (a) and (b) represent $G^{(2)}_{AB}(\theta_1, \theta_2; \delta)$ and $G^{(2)}_{BA}(\theta_1, \theta_2; \delta)$ — the probability of triggering a joint photodetection event by two wavepackets, one coming from A and the other coming from B. (c) and (d) represent $G^{(2)}_{AA}(\theta_1, \theta_2; \delta)$ and $G^{(2)}_{BB}(\theta_1, \theta_2; \delta)$ — the probability of triggering a joint photodetection event of D_1 and D_2 by two wavepackets both coming from A or both coming from B. In (a) and (b), the two wavepackets are both transmitted or both reflected at the beamsplitter. In (c) and (d), one wavepacket is transmitted, while another wavepacket is reflected at the beamsplitter. When $\delta = 0$, the two-photon amplitudes in (a) and (b) completely "overlap" and thus completely "indistinguishable" for a joint photodetection event of D_1 and D_2. When $\delta > 0$ but $< l_c$, the two-photon amplitudes in (a) and (b) partially "overlap" partially and thus partially "indistinguishable", until $\delta \geq l_c$.

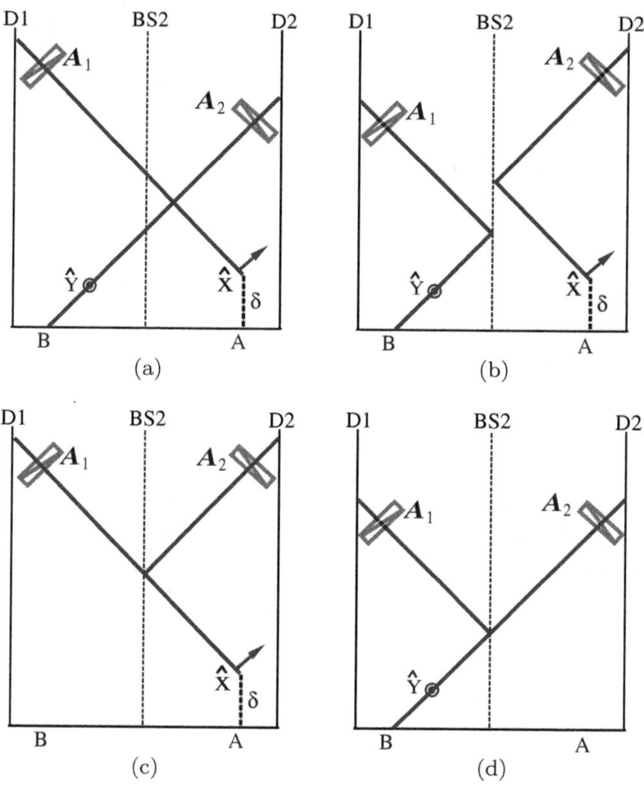

Fig. 8.4.5   Feynman diagram of the measurement. Note that when $\delta = 0$, the two-photon amplitudes in (a) and (b) "overlap" and thus indistinguishable for a joint photodetection event of D_1 and D_2.

Examining $G_{AB}^{(2)}(\theta_1, \theta_2)$ and $G_{BA}^{(2)}(\theta_1, \theta_2)$,

$$G_{AB}^{(2)}(\theta_1, \theta_2; \delta)$$

$$\propto \sum_{m,n} \left| (\hat{\theta}_1 \cdot \hat{y})(\hat{\theta}_2 \cdot \hat{x})\, \psi_{mA1}^T(\delta)\psi_{nB2}^T - (\hat{\theta}_1 \cdot \hat{x})(\hat{\theta}_2 \cdot \hat{y})\, \psi_{nB1}^R \psi_{mA2}^R(\delta) \right|^2$$

$$(8.4.5)$$

indicating a superposition between two-photon amplitudes: (a) The mth $\hat{x}$ polarized wavepacket from A is delayed with δ and then is transmitted onto D_1, while the nth $\hat{y}$ polarized wavepacket from B is transmitted onto D_2, and (b) the mth $\hat{x}$ polarized wavepacket from A is delayed by δ and then reflected onto D_2, while the nth $\hat{y}$ polarized wavepacket from B is reflected onto D_1:

$$G^{(2)}_{BA}(\theta_1, \theta_2; \delta)$$
$$\propto \sum_{m,n} \left| (\hat{\theta}_1 \cdot \hat{y})(\hat{\theta}_2 \cdot \hat{x}) \psi^R_{mB1} \psi^R_{nA2}(\delta) - (\hat{\theta}_1 \cdot \hat{x})(\hat{\theta}_2 \cdot \hat{y}) \psi^T_{nA1}(\delta) \psi^T_{mB2} \right|^2$$

$$(8.4.6)$$

indicating a superposition between two-photon amplitudes: (a) The mth $\hat{y}$ polarized wavepacket from B is delayed with δ and then is reflected onto D_1, while the nth $\hat{x}$ polarized wavepacket from A is reflected onto D_2, and (b) the mth $\hat{y}$ polarized wavepacket from A is delayed by δ and then transmitted onto D_2, while the nth $\hat{x}$ polarized wavepacket from A is transmitted onto D_1.

According to Eqs. (8.4.5) and (8.4.6), in the two-photon interferometer of Fig. 8.4.2 with orthogonal polarized incoherent thermal fields, we are able to observe two types of two-photon interference effects:

(1) Anti-correlation "dip" and correlation "peak"

In the measurement of anti-correlation "dip", see Fig. 8.4.6, we choose $\theta_1 = 45°$ and $\theta_2 = 45°$, Eq. (8.4.5) becomes

$$G^{(2)}_{AB}(45°, 45°; \delta) \propto \sum_{m<n} \left| \psi^T_{mA1}(\delta) \psi^T_{nB2} - \psi^R_{nB1} \psi^R_{mA2}(\delta) \right|^2. \qquad (8.4.7)$$

The relative delay between wavepackets ψ^*_{mA1} $[\psi_{nB2}]$ and ψ_{nB1} $[\psi_{mA2}]$ produces an anti-correlation "dip" as a function of the optical delay δ is thus expected in the coincidence counting rate R_{AB}.

In the measurement of correlation "peak", we choose $\theta_1 = 45°$ and $\theta_2 = 135°$, Eq. (8.4.5) becomes

$$G^{(2)}_{AB}(45°, 135°; \delta) \propto \sum_{m,n} \left| \psi^T_{mA1}(\delta) \psi^T_{nB2} + \psi^R_{nB1} \psi^R_{mA2}(\delta) \right|^2, \qquad (8.4.8)$$

which is the same as Eq. (8.4.7), except with a "+" sign.

It is easy to see that $G^{(2)}_{BA}(45°, 45°; \delta)$ and $G^{(2)}_{BA}(45°, 135°; \delta)$ have the same contribution to the anti-correlation "dip" and the correlation "peak". The calculation of the anti-correlation "dip" and the correlation "peak" are similar to that of the anti-correlation "dip" in previous section. The experimentally observed anti-correlation "dip" and correlation "peak" of Chen *et al.* have been shown in Fig. 8.4.3.

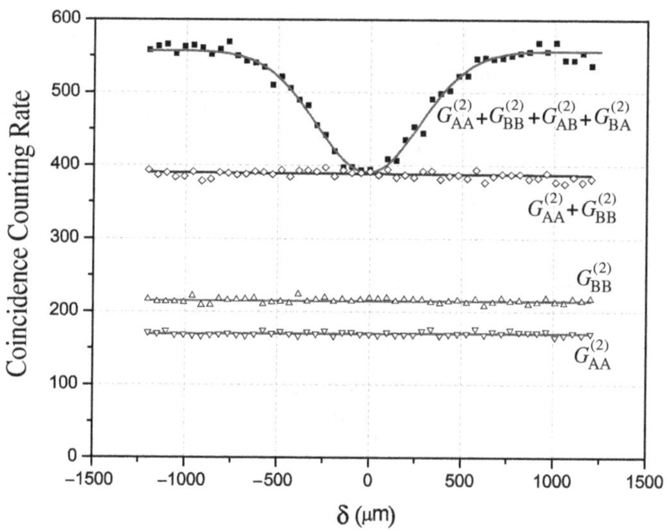

Fig. 8.4.6   The observed $G^{(2)}_{AA} + G^{(2)}_{BB} + G^{(2)}_{AB} + G^{(2)}_{BA}$, and $G^{(2)}_{AA}$, $G^{(2)}_{BB}$. $G^{(2)}_{AA}$ and $G^{(2)}_{BB}$ are measured by blocking the other source. Subtracting the contributions of $G^{(2)}_{AA}$ and $G^{(2)}_{BB}$ the contrast of the observed anti-correlation function becomes ~100%. The anti-correction "dip" is observed when A_1 and A_2 were oriented at $\theta_1 = \theta_2 = 45°$.

(2) Polarization correlation

In this measurement, we make $\delta = 0$ to achieve complete overlapping between $\psi^T_{mA1}\psi^T_{nB2}$ and $\psi^R_{nB1}\psi^R_{mA2}$. Equation (8.4.5) becomes

$$G^{(2)}_{AB}(\theta_1, \theta_2; 0) \propto \left|(\hat{\theta}_1 \cdot \hat{y})(\hat{\theta}_2 \cdot \hat{x}) - (\hat{\theta}_1 \cdot \hat{x})(\hat{\theta}_2 \cdot \hat{y})\right|^2$$

$$\propto \sin^2(\theta_1 - \theta_2), \tag{8.4.9}$$

corresponding to the measurement of the Bell state

$$|\Psi^{(-)}\rangle = \frac{1}{\sqrt{2}}\left[|X_1\rangle|Y_2\rangle - |Y_1\rangle|X_2\rangle\right].$$

When $\delta = 0$, we find $\psi^R_{mB1}\psi^R_{nA2}(\delta)$ and $\psi^T_{nA1}(\delta)\psi^T_{mB2}$ are also overlap completely, $G^{(2)}_{BA}(\theta_1, \theta_2; 0)$ has the same contribution to the polarization correlation measurement as that of $G^{(2)}_{AB}(\theta_1, \theta_2; 0)$.

Considering the trivial contributions from the other two terms $G^{(2)}_{AA}$ and $G^{(2)}_{BB}$, the contrast of the Bell-type correlation is reduced from 100% to 50%:

$$G^{(2)}_{AB}(\theta_1, \theta_2) + G^{(2)}_{BA}(\theta_1, \theta_2) \propto \left[1 + \sin^2(\theta_1 - \theta_2)\right]. \tag{8.4.10}$$

Since the contribution of $G_{AA}^{(2)}$ and $G_{BB}^{(2)}$ can be measured experimentally by blocking the other source, in principle, the contribution of $G_{AB}^{(2)} + G_{BA}^{(2)}$ can be isolated from $G_{AA}^{(2)}$ and $G_{BB}^{(2)}$. It is interesting to find that after subtracting the contributions of R_{AA} and R_{BB} from the total joint photodetection counting rate R_c, the visibilities of the "dip" and the Bell-type polarization correlation resume to $\sim 100\%$ in the experiment of Chen *et al.*

It is interesting, we also find that the visibility of the anti-correlation "dip" and the Bell-type polarization correlation become 100% in the measurement of photo number fluctuation correlation $\langle \Delta n_1 \Delta n_2 \rangle$. This is easily seen by adding the cross-interference terms of the superpositions in $G_{AB}^{(2)}(\theta_1, \theta_2; \delta)$, $G_{BA}^{(2)}(\theta_1, \theta_2; \delta)$, $G_{AA}^{(2)}(\theta_1, \theta_2)$ and $G_{BB}^{(2)}(\theta_1, \theta_2)$,

$$\langle \Delta n_1(\theta_1, \delta) \Delta n_2(\theta_2, \delta) \rangle$$

$$\propto [(\hat{\theta}_1 \cdot \hat{x})(\hat{\theta}_2 \cdot \hat{x})]^2 \psi_{mA1}^* \psi_{nA1}(\delta) \psi_{nA2}^*(\delta) \psi_{mA2}(\delta)$$

$$+ [(\hat{\theta}_1 \cdot \hat{y})(\hat{\theta}_2 \cdot \hat{y})]^2 \psi_{mB1}^* \psi_{nB1} \psi_{nB2}^* \psi_{mB2}$$

$$- (\hat{\theta}_1 \cdot \hat{x})(\hat{\theta}_2 \cdot \hat{y})(\hat{\theta}_1 \cdot \hat{y})(\hat{\theta}_2 \cdot \hat{x}) \psi_{mA1}^*(\delta) \psi_{nB1} \psi_{nB2}^* \psi_{mA2}(\delta)$$

$$- (\hat{\theta}_1 \cdot \hat{y})(\hat{\theta}_2 \cdot \hat{x})(\hat{\theta}_1 \cdot \hat{x})(\hat{\theta}_2 \cdot \hat{y}) \psi_{mB1}^* \psi_{nA1}(\delta) \psi_{nA2}^*(\delta) \psi_{mB2}. \quad (8.4.11)$$

(1) 100% contrast Anti-correlation "dip" observed from the measurement of $\langle \Delta n_1 \Delta n_2 \rangle$, see Fig. 8.4.7,

Choosing $\theta_1 = \theta_2 = 45°$, Eq. (8.4.11) turns to be

$$\langle \Delta n_1(45°, \delta) \Delta n_2(45°, \delta) \rangle$$

$$\propto \psi_{mA1}^*(\delta) \psi_{nA1}(\delta) \psi_{nA2}^*(\delta) \psi_{mA2}(\delta) + \psi_{mB1}^* \psi_{nB1} \psi_{nB2}^* \psi_{mB2}$$

$$- \psi_{mA1}^*(\delta) \psi_{nB1} \psi_{nB2}^* \psi_{mA2}(\delta) - \psi_{mB1}^* \psi_{nA1}(\delta) \psi_{nA2}^*(\delta) \psi_{mB2}.$$

$$(8.4.12)$$

The photon number fluctuation correlation $\langle \Delta n_1(45°, \delta) \Delta n_2(45°, 0) \rangle$ yields a zero value when $\delta = 0$, under which all four two-photon effective wavefunctions overlap completely in space–time. However, when δ increases, the values of the first two terms of Eq. (8.4.12) have no change, while the values of the second two terms decrease because the pulsed conjugated wavepackets $\psi_{mA1}^*(\delta) \, [\psi_{nB2}^*]$ and $\psi_{nB1} \, [\psi_{mA2}(\delta)]$ start shift away from each other. When the conjugated wavepackets are separated completely at D_1 and D_2, i.e., one wavepacket is delayed beyond the coherence time τ_c of the

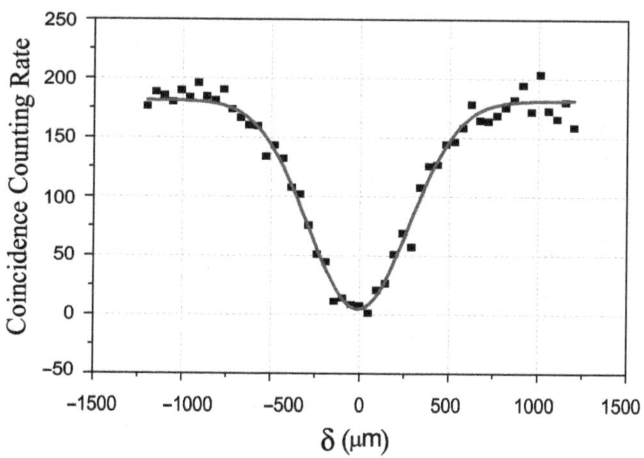

Fig. 8.4.7   In the measurement of photo number fluctuation correlation $\langle \Delta n_1 \Delta n_2 \rangle$, the contrast of the anti-correlation is close to 100%.

thermal field, the two cross-interference terms vanish. An anti-correlation "dip" is thus observed as a function of the optical delay δ.

(2) 100% contrast Bell-type polarization correlation observed from the measurement of $\langle \Delta n_1 \Delta n_2 \rangle$, see Fig. 8.4.8,

Choosing $\delta = 0$, under which all four two-photon effective wavefunctions overlap completely in space–time, Eq. (8.4.11) turns to be

$$\langle \Delta n_1(\theta_1, \delta) \Delta n_2(\theta_2) \rangle$$

$$\propto [(\hat{\theta}_1 \cdot \hat{x})(\hat{\theta}_2 \cdot \hat{x})]^2 + [(\hat{\theta}_1 \cdot \hat{y})(\hat{\theta}_2 \cdot \hat{y})]^2$$

$$- (\hat{\theta}_1 \cdot \hat{x})(\hat{\theta}_2 \cdot \hat{y})(\hat{\theta}_1 \cdot \hat{y})(\hat{\theta}_2 \cdot \hat{x}) - (\hat{\theta}_1 \cdot \hat{y})(\hat{\theta}_2 \cdot \hat{x})(\hat{\theta}_1 \cdot \hat{x})(\hat{\theta}_2 \cdot \hat{y})$$

$$= (\cos\theta_1 \cos\theta_2)^2 + (\sin\theta_1 \sin\theta_2)^2$$

$$- (\sin\theta_1 \cos\theta_2 \cos\theta_1 \sin\theta_2) - (\sin\theta_1 \cos\theta_2 \cos\theta_1 \sin\theta_2)$$

$$= \sin^2(\theta_1 - \theta_2). \tag{8.4.13}$$

(II) **Experimental realization of Peng and Shih:**

Chen *et al.* demonstrated the two-photon interference induced anti-correlation "dip" and correlation "peak" as well as Bell-type polarization correlation by using a short pulsed pseudo-thermal light source. In fact, if the experimental purpose is only to observe Bell-type polarization correlations, short pulsed thermal radiation is unnecessary. We concluded

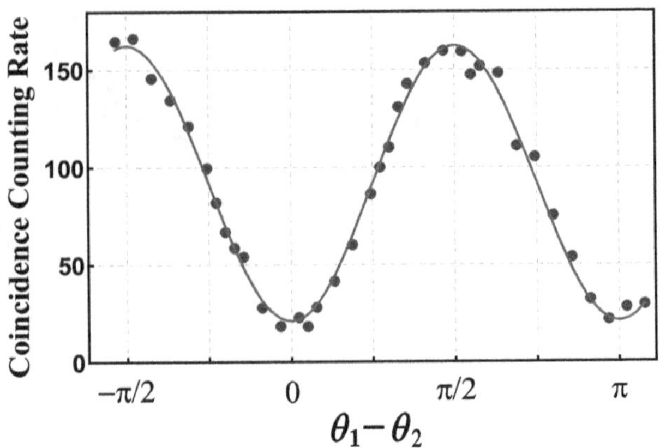

Fig. 8.4.8   In the measurement of photon number fluctuation correlation $\langle \Delta n_1 \Delta n_2 \rangle$, the visibility of the observed polarization correlation as a function of $\theta_1 - \theta_2$ is greater then 71%, indicating the behavior of a two-photon Bell state $\Psi^{(-)}$. 71% has been considered as the border between "quantum" and "classical" in some theoretical concerns. In this measurement θ_2 was fixed at 45° relative to $\hat{x}$ and $\hat{y}$, θ_1 was rotated at each chosen angle θ_1. Either <71% or >71%, its nonlocal two-photon interference nature never change.

in the previous section that two-photon interference of CW thermal field is capable of producing anti-correlation or correlation, and thus Bell states, from photon number fluctuation correlation measurements, although no anti-correlation "dip" or correlation "peak" are observable. In 2015, Peng and Shih demonstrated the Bell-type polarization correlation in an Alley–Shih interferometer from a CW pseudo-thermal field. A schematic setup of the experiment is shown in Fig. 8.4.9. Figure 8.4.10 reports the observed Bell correlation.

A large number of circular polarized wavepackets at the single-photon level, such as the mth and the nth, come from a standard pseudo-thermal light source consisting of a circularly polarized 633 nm CW laser beam and a fast rotating ground glass (GG). The diameter of the laser beam is about 2 mm. The randomly distributed wave packets pass two pinholes P_H and P_V with two linear polarizers oriented at a horizontal polarization and vertical polarization, respectively. The circular polarized wave packets have 50% chance to pass the upper pinhole P_H with horizontal polarization and 50% chance to pass the lower pinhole P_V with vertical polarization. The separation between the two pinholes is much greater than the coherence length of the pseudo-thermal field. Therefore, (1) the H-polarization and V-polarization are first-order incoherent and the mixture of the two

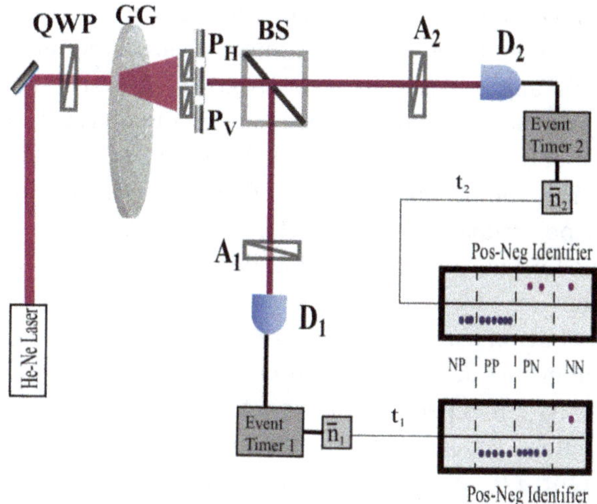

Fig. 8.4.9   Schematic setup of the 2015 experiment of Peng and Shih: Bell correlation of thermal fields in photon-number fluctuations.

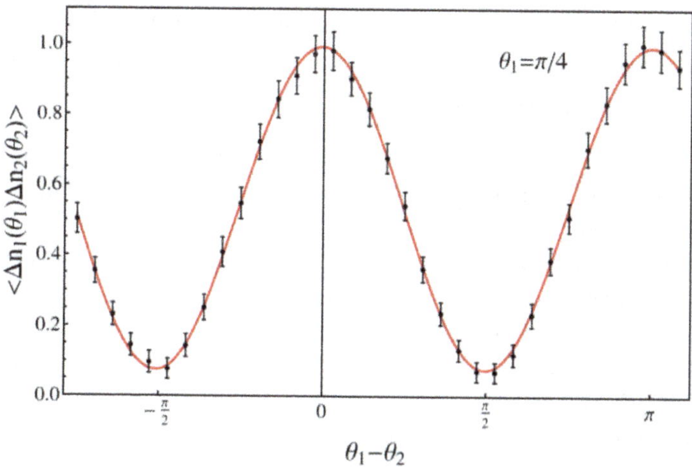

Fig. 8.4.10   Experimental observation of a Bell correlation with 92.5% contrast, indicating the behavior of a two-photon Bell state. The black dots are experimental data and the red sinusoidal curve is a theoretical fitting. The horizontal axis labels $\theta_1 - \theta_2$ while θ_1 was fixed at 45°, the vertical axis reports the normalized number fluctuation correlation $\langle \Delta n_1(\theta_1)\Delta n_2(\theta_2)\rangle$.

polarizations results in a unpolarized field and (2) the fluctuations of the H-polarization and the V-polarization are completely independent and random without any correlation. The H-polarized and V-polarized fields are then aligned spatially overlapped as an unpolarized thermal field. A 50–50% nonpolarizing beamsplitter (BS) is used to divide the unpolarized thermal field, i.e., the 50–50% mixture of the two polarizations, into arms 1 and 2. Two polarization analyzers A_1, oriented at θ_1, and A_2, oriented at θ_2, followed by two photon counting detectors D_1 and D_2 centered at the optical axis, are placed into arms 1 and 2 for the measurement of the Bell-type polarization correlation. The registration time and the number of photodetection events of D_1 and D_2 at each jth time window are recorded, respectively, by two independent but synchronized event timers. The width of the time window, Δt_j, can be adjusted from nanoseconds to milliseconds. For each detector, D_1 and D_2, at each chosen polarization angle, θ_1 and θ_2, the mean photon number, $\bar{n}_1$ and $\bar{n}_2$, is calculated from $\bar{n} = (\sum_{j=1}^{N} n_j)/N$, where N is the total number of time windows recorded for each data point in which θ_1 and θ_2 are set at certain chosen angles. In addition, the counting rates of D_1 and D_2 are monitored to be constants, independent of the chosen angles of θ_1 and θ_2. The number fluctuation correlation is calculated from

$$\langle \Delta n_1(\theta_1) \Delta n_2(\theta_2) \rangle = \frac{1}{N} \left[\sum_{j=1}^{j=N} \Delta n_{1j}(\theta_1) \Delta n_{2j}(\theta_2) \right]. \qquad (8.4.14)$$

The above two experiments seem to show that two randomly created and randomly paired photons in thermal state can simulate the behavior of an entangled pair of photons by means of two-photon constructive and destructive interferences and by means of Bell-type polarization correlation. In fact, it is not too difficult to construct the complete set of Bell states by modifying the experimental setup of Chen *et al.* as well as Peng and Shih, slightly. Furthermore, if the goal of the experiment is aimed for space–time correlation or Bell-type polarization correlation only but not aimed at observing anti-correlation "dip" or correlation "peak", pulsed pseudo-thermal source may not be necessary. Anti-correlation does not have to be associated with a "dip" anyway. CW thermal light sources are able to produce anti-correlation, correlation, Bell-type polarization correlation in their photon number fluctuations under certain experimental conditions.

How could randomly created and randomly paired photons in thermal state produce the same correlation as that of an entangled state? The above analysis of the anti-correlation experiment and the Bell-type polarization correlation measurement may be helpful for us to find an answer. In fact, we have concluded the same physics in terms of two-photon interference: a pair of photons interfering with the pair itself. The superposition of two-photon amplitudes of a photon pair, either in a thermal state or in an entangled state, physically, all corresponding to different yet indistinguishable alternative ways for a pair of photons to produce a joint photodetection event.

8.5   Turbulence-Free Two-Photon Interferometer

It is well known that optical turbulence is harmful in optical observations. Random variations in the composition or density of the medium lead to changes in the index of refraction, known as optical turbulence, and thus vary the relative phases between different optical paths of an interferometer. These variations may "blur" the interference pattern partially or completely, thus reducing the sensitivity and effectiveness of an interferometer. Optical turbulence is particularly detrimental for extremely sensitive interferometers, interferometric spectrometers, and other interferometric sensors. For instance those used in gravitational-wave detection like the Laser Interferometer Gravitational-Wave Observatory (LIGO) which must be placed in a high-cost vacuum. In this section, we discuss the physics of turbulence-free interferometer.

We start from a recent experiment of Smith and Shih which demonstrated a turbulence-free double-slit interferometer. The experimental setup is schematically depicted in Fig. 8.5.1. This interferometer looks like a classic Young's double-slit interferometer except that it has two point-like scannable photon counting detectors, D_1 and D_2, rather than one. Together they measure the photon number fluctuation correlation $\langle \Delta n(x_1) \Delta n(x_2) \rangle$ which is proportional to the intensity fluctuation correlation $\langle \Delta I(x_1) \Delta I(x_2) \rangle$. In fact, this interferometer is able to produce three outputs corresponding to two types of measurements: (1) $\langle n(x_1) \rangle \propto \langle I(x_1) \rangle$ and $\langle n(x_2) \rangle \propto \langle I(x_2) \rangle$, corresponding to the measurement of mean intensities at D_1 and D_2, respectively, and (2) $\langle \Delta n(x_1) \Delta n(x_2) \rangle \propto \langle \Delta I(x_1) \Delta I(x_2) \rangle$, corresponding to the measurement of intensity fluctuation correlation at D_1 and D_2, jointly. In this interferometer, we managed to have the spatial coherence length, l_c, of the thermal field much smaller than the separation, d, between the upper slit (slit-A) and the lower slit (slit-B), $l_c \ll d$, where $l_c = \lambda/\Delta\theta_s$ with λ the wavelength of the monochromatic thermal radiation, and $\Delta\theta_s$ the angular diameter of the

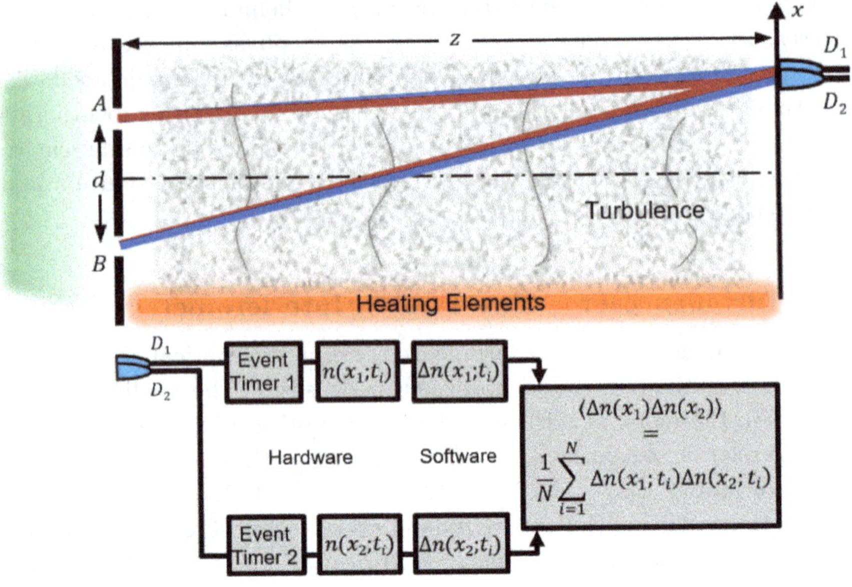

Fig. 8.5.1   A turbulence-free double-slit interferometer. This interferometer has two types of output (1) $\langle n(x_1)\rangle \propto \langle I(x_1)\rangle$ and $\langle n(x_2)\rangle \propto \langle I(x_2)\rangle$, and (2) $\langle \Delta n(x_1)\Delta n(x_2)\rangle \propto \langle \Delta I(x_1)\Delta I(x_2)\rangle$. Due to the experimental condition of $d \gg l_c$, no interferences are observable from $\langle I(x_1)\rangle$ and $\langle I(x_2)\rangle$. However, a turbulence-free interference with 100% visibility is observed from the measurement of $\langle \Delta n(x_1)\Delta n(x_2)\rangle$. The observed interference is a two-photon phenomenon: a random pair of photons interfering with the pair itself. In the figure, the superposed two different yet indistinguishable two-photon amplitudes are indicated by red and blue colors. When the detectors are scanning in the neighborhood of $x_1 \approx x_2$, the red amplitude and the blue amplitude "overlap" which means the pair experience the same phase variations, and thus turbulence-free.

light source. Consequently, no first-order interferences are observable from $\langle I(x_1)\rangle$ and $\langle I(x_2)\rangle$. However, an interference pattern with 100% visibility is observed from the measurement of $\langle \Delta n(x_1)\Delta n(x_2)\rangle \propto \langle \Delta I(x_1)\Delta I(x_2)\rangle$ and the interference is insensitive to any index-phase variations in the optical path of the interferometer and also atmospheric vibration induced phase variations, namely, turbulence-free when scanning D_1 and D_2 in the neighborhood of $x_1 \sim x_2$.

To simplify the photon number fluctuation correlation circuit, we have employed a standard monochromatic pseudo-thermal light source consisting of a rotating ground glass and a single-frequency laser beam of wavelength $\lambda = 532$ nm. Millions of tiny diffusers within the rotating ground glass scatter the laser beam into many independent wave packets, or subfields, at the single-photon level with random relative phases, artificially simulating a natural thermal light source such as the sun. Directly following the ground glass is an adjustable pinhole used to control the transverse size of the light source, allowing us to alter the spatial coherence length of the thermal field. A double-slit with $d = 2.5$ mm and line-like slits is then placed 1.6 m after the pinhole. Using this, we simulated a thermal light source with an angular diameter of $\Delta\theta_s \approx 0.00156$ and thus obtained a spatial coherence length of $l_c = \lambda/\Delta\theta_s \approx 0.34$ mm, satisfying $d \gg l_c$. In this case, no interference is observable from $\langle n(x_1)\rangle \propto \langle I(x_1)\rangle$ and $\langle n(x_2)\rangle \propto \langle I(x_2)\rangle$. The atmospheric turbulences are introduced by heating elements of a toaster oven. These heating elements produce enough heat to rapidly vary the air density above them, thereby causing variations in the index of refraction. The resulting optical path variations "blur out" completely the classic interference pattern observed in the measurement of $\langle n(x_j)\rangle$ when $l_c > d$, where $j = 1, 2$ labels the measurements by D_1 and D_2. A Photon Number Fluctuation Correlation (PNFC) circuit is used to measure the photon number fluctuation correlation for each chosen value of $(x_1 - x_2)$. The PNFC circuit has two synchronized event timers to record the registration times of each photodetection event of D_1 and D_2. The time axes of the event timers can be divided into a sequence of time windows Δt, each labeled by time t_i, for $i = 1, 2, \ldots, N$. The software first calculates the mean photon number for each detector, $\overline{n}_1$ and $\overline{n}_2$, and then calculates the photon number fluctuations for each ith time window, $\Delta n_1(t_i) = n_1(t_i) - \overline{n}_1$ and $\Delta n_2(t_i) = n_2(t_i) - \overline{n}_2$, which can either be positive or negative. The photon number fluctuation correlation is then calculated from

$$\langle \Delta n(x_1)\Delta n(x_2)\rangle = \frac{1}{N}\sum_{i=1}^{N}\Delta n_1(t_i)\Delta n_2(t_i), \qquad (8.5.1)$$

where N is the total number of time windows for a data point of a chosen x_1 and x_2.

Figure 8.5.2 reports a set of typical experimental results observed from photon number fluctuation correlation. Figure 8.5.2(a) is a measurement of $\langle \Delta n(x_1)\Delta n(x_2)\rangle$ when the heating elements were powered off, i.e., no turbulence present, and Fig. 8.5.2(b) is the same measurement of $\langle \Delta n(x_1)\Delta n(x_2)\rangle$ when the heating elements were powered on, i.e., strong turbulence present. When D_1 and D_2 were scanned in the neighborhood of $x_1 \sim x_2$, the visibility of the interference pattern with turbulence present was 94.3±0.2%, which is consistent with the visibility without turbulence present, 94.6±0.2%.

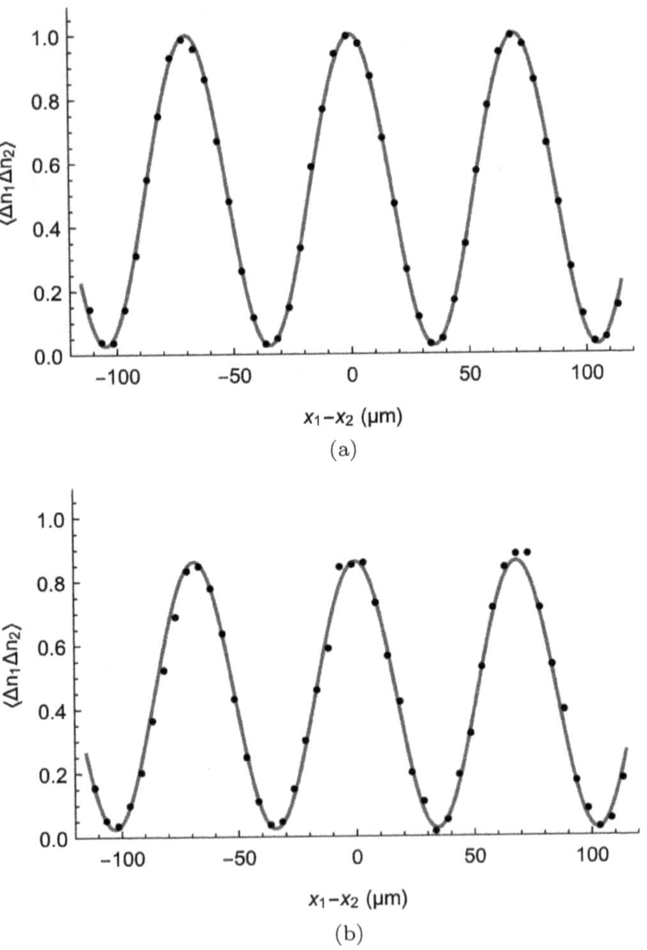

Fig. 8.5.2   Typical measurement of turbulence-free interference: (a) Without turbulence, the measurement of photon number fluctuation correlation produces an interference pattern with 94.6±0.2% visibility, when $l_c \ll d$. (b) When turbulence was introduced, the interference pattern remained with 94.3±0.2% visibility.

To demonstrate that the turbulence is strong enough to blur the classic interference present in, we removed the rotating ground glass and directed the unaltered laser beam in the TEM_{00} mode directly onto the double-slit. The spatial coherence length of a TEM_{00} mode laser beam is as large as the transverse size of the beam itself, equivalent to having a thermal source of $\Delta\theta_s \rightarrow 0$ or $l_c \rightarrow \infty$, satisfying the condition of $l_c > d$. Figure 8.5.3(a) reports a typical measured result of $\langle n(x_j) \rangle$ when the heating elements

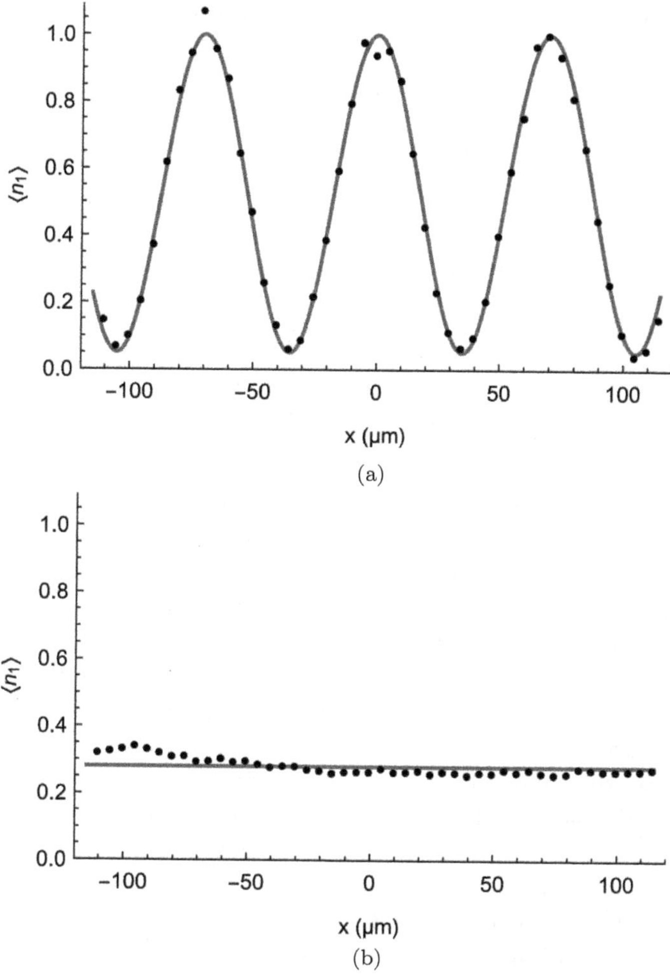

Fig. 8.5.3 Typical measurement to confirm strong enough turbulence is present: (a) Without turbulence, the classic Young's double-slit interferometer produces an interference pattern when $l_c > d$. (b) When turbulence is present, it "blurs" the interference pattern completely.

were powered off, i.e., without turbulence. Figure 8.5.3(b) reports the same measurement of $\langle n(x_j) \rangle$ but now with the heating elements powered on. It is clear that the interference pattern is completely blurred out by the turbulence. This result guarantees the turbulence introduced by our heating elements is strong enough to demonstrate the turbulence-free nature of our new type of interferometer.

To see why the measurement of $\langle \Delta n(\mathbf{r}_1, t_1) \Delta n(\mathbf{r}_2, t_2) \rangle$ is turbulence-free, following previous section, we examine the second-order coherence function $G^{(2)}(x_1, t_1; x_2, t_2)$:

$$G^{(2)}(x_1, t_1; x_2, t_2) = G^{(2)}_{AA}(x_1, t_1; x_2, t_2) + G^{(2)}_{BB}(x_1, t_1; x_2, t_2)$$
$$+ G^{(2)}_{AB}(x_1, t_1; x_2, t_2) + G^{(2)}_{BA}(x_1, t_1; x_2, t_2). \quad (8.5.2)$$

We first calculate $G^{(2)}_{AB}(x_1, t_1; x_2, t_2)$ in Einstein's picture,

$$G^{(2)}_{AB}(x_1, t_1; x_2, t_2) = \sum_{m,n} |E_m g_m(x_A, t_A) g_A(x_1, t_1) E_n g_n(x_B, t_B) g_B(x_2, t_2)$$

$$+ E_m g_m(x_A, t_A) g_A(x_2, t_2) E_n g_n(x_B, t_B) g_B(x_1, t_1)|^2,$$
$$(8.5.3)$$

indicating a superposition between two different yet indistinguishable two-photon amplitudes: (1) the mth wavepacket passing through slit-A then propagates to D_1 while the nth wavepacket passing through slit-B then propagates to D_2 and (2) the mth wavepacket passing through slit-A then propagates to D_2 while the nth wavepacket passing through slit-B then propagates to D_1. We may consider amplitude (1) the blue path and amplitude (2) the red path in Fig. 8.5.1, respectively. It is easy to find that the blue path and the red path overlap in the neighborhood of $x_1 \sim x_2$. In this case, the randomly created and randomly paired two wavepackets experience the same turbulence and thus the same phase variations. The phase variations associated with the two-photon amplitudes (1) and (2) cancel each other results in a turbulence-free two-photon interference.

The cross-interference term of $G^{(2)}_{AB}(x_1, t_1; x_2, t_2)$ yields a turbulence-free sinusoidal modulation of $(x_1 - x_2)$ in the measurement of $\langle \Delta I(x_1) \Delta I(x_2) \rangle$:

$$\langle \Delta n_{AB}(x_1) \Delta n_{AB}(x_2) \rangle \propto \langle \Delta I_{AB}(x_1) \Delta I_{AB}(x_2) \rangle$$

$$= \sum_{m,n} E_m^* g_m^*(x_A) g_A^*(x_1) E_n^* g_n^*(x_B) g_B^*(x_2)$$

$$\times E_m g_m(x_A) g_A(x_2) E_n g_n(x_B) g_B(x_1)$$

$$= \sum_{m \neq n} |E_m|^2 |E_n|^2 |g_m(x_A)|^2 |g_n(x_B)|^2$$

$$\times g_A^*(x_1) g_B(x_1) g_A(x_2) g_B^*(x_2)$$

$$= I_0^2 \, \cos \frac{2\pi d}{\lambda z} (x_1 - x_2), \quad (8.5.4)$$

where we have ignored the temporal variables by assuming perfect second-order temporal coherence as usual. Similar to the case of Section 10.1, unlike the classic Young's double-slit interference, we find Green's functions that propagate the fields from the source to the double-slit do not have any contributions to the two-photon interference because of $|g_m(x_A, t_A)|^2 = |g_n(x_B, t_B)|^2 = 1$.

In addition to the above alternatives for the mth and the nth subfields to produce a joint photodetection event of D_1 and D_2, the randomly created and randomly paired mth and nth subfields, or photons, may have a second superposition, $G_{BA}^{(2)}(x_1, t_1; x_2, t_2)$, to produce a joint photodetection event at (x_1, t_1) and (x_2, t_2), corresponding to the following two-photon amplitudes: (3) The nth subfield passing through slit-A and then propagates to D_1 and the mth subfield passes slit-B and then propagates to D_2; (4) the nth subfield passes slit-A and then propagates to D_2 and the mth subfield passes slit-B and then propagates to D_1. This superposition has the same mathematical expression as Eq. (8.5.3), except switching A and B. The cross terms of this superposition produces the same turbulence-free sinusoidal modulation in the measurement of $\langle \Delta I(x_1)\Delta I(x_2)\rangle$.

The randomly created and randomly paired mth and nth subfields, or photons, may have a third superposition, $G_{AA}^{(2)}(x_1, t_1; x_2, t_2)$, to produce a joint photodetection event at (x_1, t_1) and (x_2, t_2): (5) The mth subfield passes slit-A then propagates to D_1 and the nth subfield passes slit-A and then propagates to D_2; (6) the mth subfield passes slit-A and then propagates to D_2 and the nth subfield passes slit-A and then propagates to D_1:

$$G_{AA}^{(2)}(x_1; x_2) = \sum_{m,n} |E_m g_m(x_A) g_A(x_1) E_n g_n(x_A) g_A(x_2)$$

$$+ E_n g_n(x_A) g_A(x_1) E_m g_m(x_A) g_A(x_2)|^2. \tag{8.5.5}$$

The cross-interference term of $G_{AA}^{(2)}(x_1; x_2)$ yields a turbulence-free constant to the measurement of $\langle \Delta n(x_1)\Delta n(x_2)\rangle$,

$$\langle \Delta n_{AA}(x_1)\Delta n_{AA}(x_2)\rangle \propto \langle \Delta I_{AA}(x_1)\Delta I_{AA}(x_2)\rangle$$

$$= \sum_{m,n} |E_m^* g_m^*(x_A) g_A^*(x_1) E_n^* g_n^*(x_A) g_A^*(x_2)$$

$$\times E_n g_n(x_A) g_A(x_1) E_m g_m(x_A) g_A(x_2)|^2$$

$$= \sum_{m \neq n} |E_m|^2 |E_n|^2 |g_m(x_A)|^2 |g_n(x_A)|^2 |g_A(x_1)|^2 ||g_A(x_2)|^2$$

$$= I_0^2. \tag{8.5.6}$$

The randomly created and randomly paired mth and nth subfields, or photons, may have a fourth superposition, $G_{BB}^{(2)}(x_1, t_1; x_2, t_2)$, to produce a joint photodetection event at (x_1, t_1) and (x_2, t_2): (7) The mth subfield passes slit-B and then propagates to D_1 and the nth subfield passes slit-B and then propagates to D_2; (8) the mth subfield passes slit-B and then propagates to D_2 and the nth subfield passes slit-B and then propagates to D_1. This superposition has the same mathematical expression as Eq. (8.5.5), except switching A and B. The cross terms of this superposition also contribute a turbulence-free constant to the measurement of $\langle \Delta I(x_1) \Delta I(x_2) \rangle$.

Adding the above four turbulence-free contributions to the intensity fluctuation correlation measurement $\langle \Delta I(x_1) \Delta I(x_2) \rangle$, we have an observable turbulence-free interference in the neighborhood of $x_1 \sim x_2$ with 100% visibility:

$$\langle \Delta n(x_1) \Delta n(x_2) = \langle \Delta n_{AB}(x_1) \Delta n_{AB}(x_2) \rangle + \langle \Delta n_{BA}(x_1) \Delta n_{BA}(x_2)$$

$$+ \langle \Delta n_{AA}(x_1) \Delta n_{AA}(x_2) \rangle + \langle \Delta n_{BB}(x_1) \Delta n_{BB}(x_2) \rangle$$

$$= I_0^2 \left[1 + \cos \frac{2\pi d}{\lambda z} (x_1 - x_2) \right]. \tag{8.5.7}$$

According to the quantum theory of light, there is no suppress to achieve turbulence-free in a two-photon interferometer: What we need is to make the superposed indistinguishable two-photon amplitudes experience the same turbulence so that any refraction index, length, or phase variations along the optical paths of the interferometer do not have any effect on the two-photon interference. Examining the two-photon double-slit interferometer in Fig. 8.5.1 and the superpositions of Eqs. (8.5.3) and (8.5.5), we find that all four groups of superposed two-photon amplitudes are "overlapped", respectively, when scanning D_1 and D_2 in the neighborhood of $x_1 \sim x_2$. In principle, this mechanism of achieving turbulence-free two-photon interference is able to apply to other types of interferometers and interferometric sensors, making them turbulence-free as well. To avoid atmospheric turbulence and vibrations, many interferometers and interferometric sensors, such as the gravitational wave detector, have to be maintained within complicated, high cost vacuum systems. With a turbulence-free interferometer, these complicated and expensive systems would no longer be required.

8.6 Turbulence Induced Turbulence-Free Two-Photon Interference of Laser Beam

In Section 11.5, we analyzed a turbulence-free double-slit interferometer in which incoherent thermal field was able to produce turbulence-free two-photon interference pattern from the second-order measurement of photon number fluctuation correlation $\langle \Delta n_1 \Delta n_2 \rangle$, or intensity fluctuation correlation $\langle \Delta I_1 \Delta I_2 \rangle$, while no classic interference was observable from the first-order measurement of mean intensities $\langle I_1 \rangle$ and $\langle I_2 \rangle$. In that experiment, the input light of the interferometer was in thermal state. Can we observe turbulence-free interference from an interferometer that employs coherent laser beam as the light source? The second-order coherence of laser beam has been studied since the invention of the laser. Different from thermal field, the photon number fluctuation correlation, or intensity fluctuation correlation of a laser beam is always zero, $\langle \Delta n_1 \Delta n_2 \rangle \propto \langle \Delta I_1 \Delta I_2 \rangle = 0$. Thermal field was traditionally considered as Gaussian field, and the nontrivial second-order correlation of thermal field was explained as an intrinsic property of Gaussian field. Unfortunately, the laser field is nonGaussian and can be approximated as a coherent state.

When a coherent laser beam is incident on a double-slit, without turbulence, classic Young's double-slit interference can be easily observed from the measurement of mean photon number $\langle n \rangle$, or mean intensity $\langle I \rangle$. When optical turbulence is introduced into the interferometer, it may blur the interference pattern completely. The turbulence introduces random phase shifts following slit-A and slit-B that vary rapidly, randomly, and independently. This turns a single coherent state, representing a group of identical photons, into a mixture of two separate, distinguishable groups of identical photons in coherent states A and B with varying random relative phases from the turbulence. The incoherent superposition of coherent state A and coherent state B is unable to produce any classic interference pattern. Is it possible to observe turbulence-free second-order interference from a laser-based interferometer? Perhaps, no one would expect the same turbulence-free two-photon interference mechanism of thermal light to be applicable to a laser-based interferometer. Perhaps, no one would even expect observing nontrivial second-order correlation from a laser beam since laser field is non-Gaussian.

Surprisingly, in a recent experiment, Smith and Shih observed turbulence-free two-photon interference from the second-order correlation measurement of photon number fluctuations, or intensity fluctuations, of an Young's double-slit interferometer which not only employed a laser beam as the light source but also was under the influence of strong turbulence. How could a measurement of photon number fluctuation correlation, or intensity fluctuation correlation, on a laser beam produce nontrivial sinusoidal function? Why is this interference pattern seemingly turbulence-free but also only present due to the turbulence itself? We address these questions in this section after describing the experimental observations of Smith and Shih.

The experimental setup of Smith and Shih is depicted in Fig. 8.6.1. An Nd:YVO$_4$ laser was used to produce a continuous wave (CW) beam in the TEM$_{00}$ spatial mode at $\lambda = 532$ nm. A beam expander with a well designed spatial filter was used to increase the diameter of the TEM$_{00}$ laser beam. The expanded beam was incident on a standard Young's double-slit interferometer with slit separation of $d = 2.5$ mm. The slits are narrow enough to be treated as lines. In this experiment, optical turbulence was introduced by a set of kilowatt heating elements beneath the optical paths of the interferometer. The heating elements heat the air and introduce random airflow and random optical index variations, i.e., optical turbulence, between the double-slit and the observation plane. The turbulence is strong enough to blur the classic interference pattern but not strong enough to thermalize the laser beam into Gaussian field. To detect the radiation at more precise spatial locations, point-like tips of single-mode optical

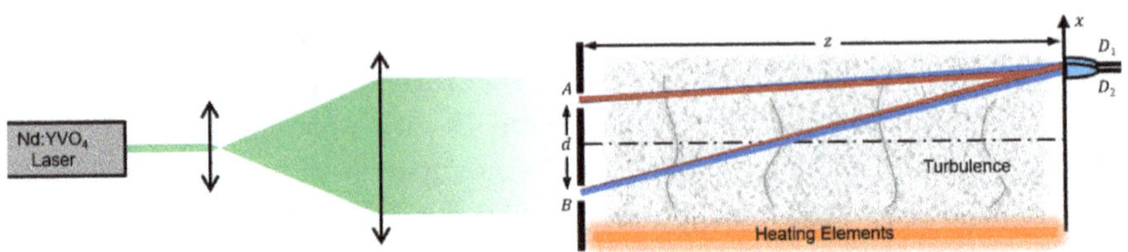

Fig. 8.6.1 Experimental Setup. Light emitted from a CW Yttrium Vanadate (Nd:YVO$_4$) laser in the TEM$_{00}$ spatial mode is incident on a double-slit with slit separation d, a beam expander is used to enlarge the laser beam to diameter D achieving $D \gg d$. Two scannable single-photon detectors, D_1 and D_2, are placed on the far-field observation plane of the double-slit interferometer. The electronics interfaced with D_1 and D_2 can simultaneously obtain mean photon number, $\langle n(x_j) \rangle$, photon number correlation, $\langle n(x_1)n(x_2) \rangle$, and photon number fluctuation correlation, $\langle \Delta n(x_1)\Delta n(x_2) \rangle$. A lab-made atmospheric turbulence is introduced between the double-slit and the photodetector. The turbulence is strong enough to blur the classic interference pattern but not strong enough to thermalize the laser beam into Gaussian field.

fiber were used to interface the coherent light with or without turbulence into the single-photon counting detectors. A Photon Number Fluctuation Correlation (PNFC) circuit uses a series of measurements ($\sim$300,000) to determine the mean photon number and photon number fluctuations for each detector, while simultaneously calculating the photon number correlation, $\langle n(x_1)n(x_2)\rangle$, and photon number fluctuation correlation, $\langle \Delta n(x_1)\Delta n(x_2)\rangle$.

No surprises were observed from the measurement of first-order classic interference. As expected, when the heating elements were powered off, the observed visibility of the classic interference from D_1 and D_2, respectively, was $\sim$100% and when the heating elements were powered on, the interference pattern was blurred out by the turbulence. This confirms that the optical paths from each slit were experiencing random disturbances from the turbulence and thus introduce random phases into the two optical paths. The second-order measurements of the photon number correlation $\langle n(x_1)n(x_2)\rangle$ and the photon number fluctuation correlation $\langle \Delta n(x_1)\Delta n(x_2)\rangle$ were more interesting. When the heating elements were powered off, we observed $\sim$100% visibility interference in the measurement of $\langle n(x_1)n(x_2)\rangle$ (Fig. 8.6.2(a)) while the measurement of $\langle \Delta n(x_1)\Delta n(x_2)\rangle$ yielded a constant of $\sim$0, as seen in Fig. 8.6.3(a). When the heating elements were powered on, interference in the measurement of $\langle n(x_1)n(x_2)\rangle$ was still present; however, the visibility was reduced (Fig. 8.6.2(b)), and simultaneously, an interference pattern appeared in the measurement of $\langle \Delta n(x_1)\Delta n(x_2)\rangle$, as seen in Fig. 8.6.3(b). The turbulence seems to have produced interference in the measurement of photon number fluctuation correlation $\langle \Delta n(x_1)\Delta n(x_2)\rangle$. More interestingly, the interference pattern contains not only "correlation" of $\langle \Delta n_1 \Delta n_2\rangle > 0$, but also "anti-correlation" of $\langle \Delta n_1 \Delta n_2\rangle < 0$ within one period scanning of the interference pattern, as seen in Fig. 8.6.3(b).

The observation of classic interferences from coherent laser beam without turbulence are easily understood. In the following, we analyze the measurement processes of mean photon number $\langle n(x_j)\rangle \propto \langle I(x_j)\rangle$ and photon number fluctuation correlation $\langle \Delta n(x_1)\Delta n(x_2)\rangle \propto \langle \Delta I(x_1)\Delta I(x_2)\rangle$ from the turbulence disturbed double-slit interferometer of Fig. 8.6.1.

(I) **Einstein's picture of light**:

A laser beam contains a group of large number identical photons, usually approximated in coherent state. When turbulence is introduced into an Young's double-slit interferometer behind slit-A and slit-B, one group of

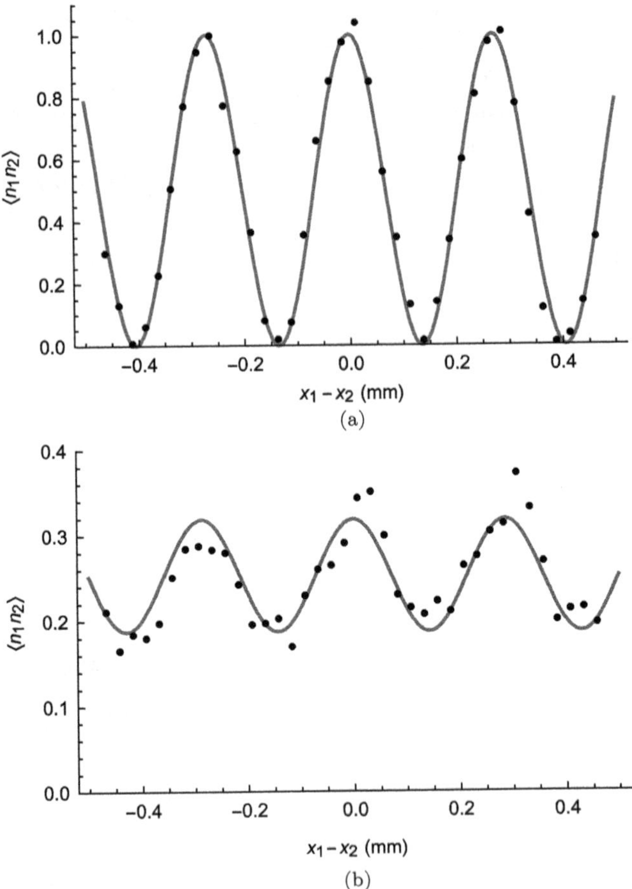

Fig. 8.6.2 Typical measurement of photon number correlation. Each data point is estimated from ∼300,000 measurements. (a) When the heating elements were powered off: ∼100% visibility interference was observed. (b) When the heating elements were powered on: interference visibility was significantly reduced.

identical photons is divided into two distinguishable groups of identical photons with random relative phases. In Einstein's picture, we may consider $E_A(x_j, t_j)$ and $E_B(x_j, t_j)$, $j = 1, 2$, distinguishable subfields:

$$E_A(x_j, t_j) = \int d\omega \, E_A(\omega) \, g_A(\omega; x_j, t_j) \, e^{-i\varphi_{Aj}(t_j)},$$

$$E_B(x_j, t_j) = \int d\omega \, E_B(\omega) \, g_B(\omega; x_j, t_j) \, e^{-i\varphi_{Bj}(t_j)}, \qquad (8.6.1)$$

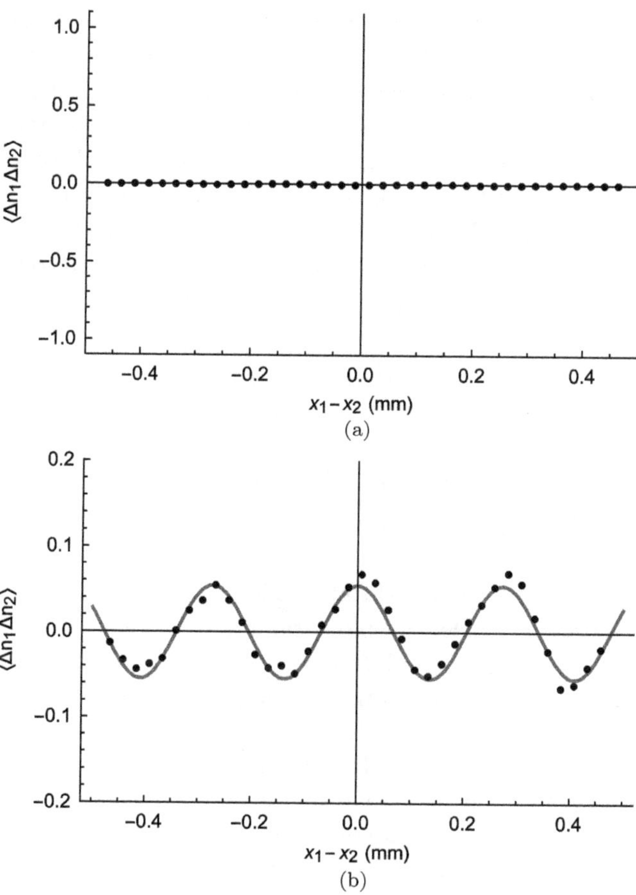

Fig. 8.6.3 Typical measurement of photon number fluctuation correlation. The mean photon number and photon number fluctuation for each data point is estimated from ∼300,000 measurements. (a) When the heating elements were powered off: No interference and correlation was observed from $\langle \Delta n(x_1)\Delta n(x_2)\rangle$. (b) When the heating elements were powered on: An interference pattern appeared in the measurement of $\langle \Delta n(x_1)\Delta n(x_2)\rangle$ with "correlation", corresponding to constructive interference, and "anti-correlation", corresponding to destructive interference, of photon number fluctuations.

where $g_k(\omega; x_j, t_j)$, $k = A, B$, $j = 1, 2$, is Green's function propagating the ω mode of the kth subfield $E_k(\omega)$ from slit-k to point-like photodetector D_j, $\varphi_{kj}(t_j)$ is the turbulence induced random phase shift along path-kj

$$\varphi_{kj}(t_j) \simeq \frac{\omega}{c} \int dr_{kj}\, \delta n(r_{kj}, t_j). \tag{8.6.2}$$

The integral is along the path of k to j, the refraction index of air is approximated $n_0 \simeq 1$, $\delta n(r_{kj}, t_j)$ represents the turbulence induced variation of the refraction index along path-kj, which takes a random value from time to time and from measurement to measurement. In the experiment of Smith and Shih, each data point is estimated from $\sim 300{,}000$ measurements, the experimentally executed time average can be treated as an ensemble average. The ensemble average of the first-order incoherent superposition of the two turbulence affected subfields gives a constant mean value for the measurement of $\langle n(x_j) \rangle$:

$$
\begin{aligned}
\langle n(x_j) \rangle \propto \langle I(x_j) \rangle &\propto \left\langle \left| E_A(x_j, t_j) + E_B(x_j, t_j) \right|^2 \right\rangle \\
&= \left\langle \left| E_A(x_j, t_j) \right|^2 \right\rangle + \left\langle \left| E_B(x_j, t_j) \right|^2 \right\rangle \\
&\quad + \left\langle E_A^*(x_j, t_j) E_B(x_j, t_j) \right\rangle + \left\langle E_A(x_1, t_j) E_B^*(x_1, t_j) \right\rangle \\
&\propto n_0.
\end{aligned}
\tag{8.6.3}
$$

Although the fields passing through slit-A and slit-B are initially in phase, the turbulence affects each one randomly. While $[\varphi_{Aj}(t_j) - \varphi_{Bj}(t_j)]$ varies from time to time and from measurement to measurement, the time average or ensemble average of the cross terms yields $\langle E_A^*(x_j, t_j) E_B(x_j, t_j) \rangle = \langle E_A(x_1, t_j) E_B^*(x_1, t_j) \rangle = 0$ when taking into account all possible random relative phases. In reality, even for $\sim 300{,}00$ measurements, $[\varphi_{Aj}(t_j) - \varphi_{Bj}(t_j)]$ may not take all possible random values and may not vanish completely. The remaining cross terms $\langle E_A^*(x_j, t_j) E_B(x_j, t_j) \rangle + \langle E_A(x_1, t_j) E_B^*(x_1, t_j) \rangle \neq 0$ are usually treated as "noise" or "photon number fluctuations".

Now, we calculate the correlation of $\langle n(x_1) n(x_2) \rangle$ under turbulence:

$$
\begin{aligned}
\langle n(x_1) n(x_2) \rangle &\propto \langle I(x_1) I(x_2) \rangle \\
&= \left\langle E^*(x_1, t_1) E(x_1, t_1) E^*(x_2, t_2) E(x_2, t_2) \right\rangle \\
&= \left\langle [E_A^*(x_1, t_1) + E_B^*(x_1, t_1)][E_A(x_1, t_1) + E_B(x_1, t_1)] \right. \\
&\quad \left. \times [E_A^*(x_2, t_2) + E_B^*(x_2, t_2)]^*[E_A(x_2, t_2) + E_B(x_2, t_2)] \right\rangle
\end{aligned}
\tag{8.6.4}
$$

resulting in 16 terms of expectation values to calculate. For brevity, we drop terms that have no contribution to the measurement of $\langle n(x_1) n(x_2) \rangle$ or cannot survive the ensemble average by taking into account all possible turbulence introduced random phases along the A-path and the B-path:

$$\langle n(x_1)n(x_2)\rangle \propto \langle I(x_1)I(x_2)\rangle$$

$$\propto \langle |E_A(x_1,t_1)|^2|E_A(x_2,t_2)|^2\rangle + \langle |E_B(x_1,t_1)|^2|E_B(x_2,t_2)|^2\rangle$$

$$+ \langle |E_A(x_1,t_1)|^2|E_B(x_2,t_2)|^2\rangle + \langle |E_A(x_2,t_2)|^2|E_B(x_1,t_1)|^2\rangle$$

$$+ \langle E_A^*(x_1,t_1)E_B(x_1,t_1)E_B^*(x_2,t_2)E_A(x_2,t_2)\rangle$$

$$+ \langle E_A(x_1,t_1)E_B^*(x_1,t_1)E_B(x_2,t_2)E_A^*(x_2,t_2)\rangle$$

$$= \langle n(x_1)\rangle\langle n(x_2)\rangle + \langle \Delta n(x_1)\Delta n(x_2)\rangle, \tag{8.6.5}$$

where

$$\langle \Delta n(x_1)\Delta n(x_2)\rangle \propto \langle \Delta I(x_1)\Delta I(x_2)\rangle$$

$$= \langle E_A^*(x_1,t_1)E_B(x_1,t_1)E_B^*(x_2,t_2)E_A(x_2,t_2)\rangle$$

$$+ \langle E_A(x_1,t_1)E_B^*(x_1,t_1)E_B(x_2,t_2)E_A^*(x_2,t_2)\rangle$$

$$\propto \cos\frac{2\pi d}{\lambda z}(x_1 - x_2), \tag{8.6.6}$$

when scanning D_1 in the neighborhood of D_2. It is easy to find that the above sinusoidal modulation in the measurement of $\langle \Delta n(x_1)\Delta n(x_2)\rangle$ comes from the cross-interference terms of the following superposition:

$$\langle |E_A(x_1,t_1)E_B(x_2,t_2) + E_A(x_2,t_2)E_B(x_1,t_1)|^2\rangle \tag{8.6.7}$$

corresponding to two different yet indistinguishable alternatives for the two distinguishable groups of identical photons, represented by subfields $E_A(x_j,t_j)$ and $E_B(x_j,t_j)$ in Einstein's picture, to produce a join photodetection event of D_1 and D_2: (1) Subfield-A is detected at D_1 and subfield-B is detected at D_2 and (2) subfield-A is detected at D_2 and subfield-B is detected at D_1; namely, a pair of distinguishable coherent subfields interfering with the pair itself. Note that, at first glance, the cross term of this superposition would average to zero due to the random phases present. It has been discussed in previous section that by scanning D_1 in the neighborhood of D_2 such that $x_1 \sim x_2$, the optical paths of the two alternatives overlap in space–time and experience the same turbulence. The turbulence induced random phases in the intensity fluctuation correlation would have no contribution to this superposition.

It is interesting to find from Eq. (8.6.6), the interference results in a positive value of $\langle \Delta n(x_1)\Delta n(x_2)\rangle > 0$, indicating a "correlation", when the above two alternatives superpose constructively, and results in a negative value of $\langle \Delta n(x_1) \Delta n(x_2) \rangle < 0$, indicating an "anti-correlation", when the above two alternatives superpose destructively. The constructive superposition forces the measured photon number fluctuate to the same, positive–positive or negative–negative, direction, while the destructive interference forces the measured photon number fluctuate to opposite, positive–negative or negative–positive, directions: If one fluctuates positively, the other one must fluctuates negatively and vice versa.

(II) **Quantum coherent state representation:**

A laser beam of TEM$_{00}$ mode contains a group of large number of identical photons. We approximate the state of the group of identical photons a multi-mode coherent states

$$|\Psi\rangle = \prod_\omega |\alpha(\omega)\rangle. \tag{8.6.8}$$

The field operator at coordinate (x_j, t_j) of the jth photodetector

$$\hat{E}^{(+)}(x_j, t_j) = \int d\omega \, \hat{a}(\omega) \, g_A(\omega; x_j, t_j) \, e^{-i\varphi_{Aj}(t_j)}$$

$$+ \int d\omega \, \hat{a}(\omega) \, g_B(\omega; x_j, t_j) \, e^{-i\varphi_{Bj}(t_j)}$$

$$= \hat{E}_A^{(+)}(x_j, t_j) + \hat{E}_B^{(+)}(x_j, t_j), \tag{8.6.9}$$

where $g_A(\omega; x_j, t_j)$ and $g_B(\omega; x_j, t_j)$ are Green's functions (or propagators) which propagate the ω mode of the state from slit-A and slit-B to the photodetector D_j at space–time coordinate (x_j, t_j). With the help of the quantum state and the field operators, the second-order coherence function $G^{(2)}(x_1, t_1; x_1, t_2)$ is calculated as follows:

$$G^{(2)}(x_1, t_1; x_1, t_2)$$

$$= \langle\langle \Psi| \, \hat{E}^{(-)}(x_1, t_1) \hat{E}^{(-)}(x_2, t_2) \hat{E}^{(+)}(x_2, t_2) \hat{E}^{(+)}(x_1, t_1) \, |\Psi\rangle\rangle_{\text{En}}$$

$$= \langle\langle \Psi| [\hat{E}_A^{(-)}(x_1, t_1) + \hat{E}_B^{(-)}(x_1, t_1)][\hat{E}_A^{(-)}(x_2, t_2) + \hat{E}_B^{(-)}(x_2, t_2)]$$

$$\times [\hat{E}_A^{(+)}(x_2, t_2) + \hat{E}_B^{(+)}(x_2, t_2)][\hat{E}_A^{(+)}(x_1, t_1) + \hat{E}_B^{(+)}(x_1, t_1)]|\Psi\rangle\rangle_{\text{En}}. \tag{8.6.10}$$

Equation (8.6.10) results in sixteen expectations to evaluate. For brevity, we drop terms that have no contribution to the measurement of $\langle n(x_1) n(x_2) \rangle$

or cannot survive the ensemble average by taking into account all possible turbulence introduced random phases along the A-path and the B-path:

$$G^{(2)}(\mathbf{r}_1, t_1; \mathbf{r}_1, t_2)$$

$$= \left\langle |\Psi_A(x_1, t_1)|^2 |\Psi_A(x_2, t_2)|^2 \right\rangle_{\text{En}} + \left\langle |\Psi_B(x_2, t_2)|^2 |\Psi_B(x_1, t_1)|^2 \right\rangle_{\text{En}}$$

$$+ \left\langle |\Psi_A(\mathbf{r}_1, t_1)|^2 |\Psi_B(\mathbf{r}_2, t_2)|^2 \right\rangle_{\text{En}} + \left\langle |\Psi_A(\mathbf{r}_2, t_2)|^2 |\Psi_B(\mathbf{r}_1, t_1)|^2 \right\rangle_{\text{En}}$$

$$+ \left\langle \Psi_A^*(\mathbf{r}_1, t_1) \, \Psi_B^*(\mathbf{r}_2, t_2) \, \Psi_A(\mathbf{r}_2, t_2) \, \Psi_B(\mathbf{r}_1, t_1) \right\rangle_{\text{En}}$$

$$+ \left\langle \Psi_A(\mathbf{r}_1, t_1) \, \Psi_B(\mathbf{r}_2, t_2) \, \Psi_A^*(\mathbf{r}_2, t_2) \, \Psi_B^*(\mathbf{r}_1, t_1) \right\rangle_{\text{En}}$$

$$\propto \langle n(\mathbf{r}_1, t_1) \rangle \langle n(\mathbf{r}_2, t_2) \rangle + \langle \Delta n(\mathbf{r}_1, t_1) \Delta n(\mathbf{r}_2, t_2) \rangle, \qquad (8.6.11)$$

where $\Psi_k(x_j, t_j)$ is the effective wavefunction of the group-k identical photons, and

$$\langle \Delta n(x_1) \Delta n(x_2) \rangle$$

$$= \left\langle \Psi_A^*(\mathbf{r}_1, t_1) \, \Psi_B^*(\mathbf{r}_2, t_2) \, \Psi_A(\mathbf{r}_2, t_2) \, \Psi_B(\mathbf{r}_1, t_1) \right\rangle_{\text{En}}$$

$$+ \left\langle \Psi_A(\mathbf{r}_1, t_1) \, \Psi_B(\mathbf{r}_2, t_2) \, \Psi_A^*(\mathbf{r}_2, t_2) \, \Psi_B^*(\mathbf{r}_1, t_1) \right\rangle_{\text{En}} \qquad (8.6.12)$$

is the cross terms of the following superposition of "two-photon" effective wavefunctions:

$$\left\langle \left| \Psi_A(\mathbf{r}_1, t_1) \, \Psi_B(\mathbf{r}_2, t_2) + \Psi_A(\mathbf{r}_2, t_2) \, \Psi_B(\mathbf{r}_1, t_1) \right|^2 \right\rangle \qquad (8.6.13)$$

corresponding to two different yet indistinguishable alternatives for the two distinguishable groups of identical photons to produce a join photodetection event of D_1 and D_2: (1) group A of identical photons propagate to D_1 and group B of identical photons propagate to D_2, and (2) group A of identical photons propagate to D_2 and group B of identical photons propagate to D_1, indicating the interference of two distinguishable groups of identical photons.

When scanning D_1 and D_2 in the neighborhood of $x_1 \sim x_2$, we find the measurements $\langle n(x_1) n(x_2) \rangle$ and $\langle \Delta n(x_1) \Delta n(x_2) \rangle$ yield the same results as those calculated from Einstein's picture and agreeing well with the experimental observations of Smith and Shih. It is interesting to find that the interference results in a positive value of $\langle \Delta n(x_1) \Delta n(x_2) \rangle > 0$, indicating a "correlation", when the group-pair interferences with the pair itself constructively, and results in a negative value of $\langle \Delta n(x_1) \Delta n(x_2) \rangle < 0$, indicating an "anti-correlation", when the group-pair interferences with the pair itself destructively. The constructive superposition forces the measured photon number fluctuate to the same, positive–positive or negative–negative, direction, while the destructive interference forces the measured photon number fluctuate to opposite, positive–negative or negative–positive, directions: If one fluctuates positively the other one must fluctuates negatively and vice versa.

As discussed in early sections, photon number fluctuation correlation, or intensity fluctuation correlation, is typically observable from thermal field by means of two-photon interference: A pair of randomly created and randomly distributed photons, or a pair of distinguishable groups of identical photons, interfering with the pair itself. When the light detected is of a group of indistinguishable identical photons in coherent state, there are no "distinguishable groups" causing fluctuations and fluctuation correlation. However, under strong turbulence, variance of the turbulence introduce random phases to coherent radiations that passing through slit-A and slit-B and thus causing two distinguishable groups of identical photons in state A and state B. The observations of the experiment of Smith and Shih is the result of a pair of distinguishable groups of identical photons interfering with the group-pair itself. It should be emphasized that the turbulence-induced interference is different from the two-photon interference of thermal field. When a thermal light source is used, a randomly paired photons can pass through the same slit and contribute a constant term to the measurement of photon number fluctuation correlation of D_1 and D_2. However, when a coherent source is used and the turbulence is not strong enough to thermalize the coherent field, all of the photons passing through a single slit are in the same coherent state, preventing any photon number fluctuation correlation between D_1 and D_2 resulting from identical photons passing through a single slit. In rare cases, there may exist a condition with turbulence strong enough to introduce different random phases to different sub-groups of identical photons following a single slit, but that was not the condition in Smith and Shih's experiment and was not taken into account in the above calculation.

8.7 Delayed Choice Quantum Eraser — Two-Photon Interference of Randomly Created Photons in Thermal State

We have discussed a delayed choice quantum eraser in Section 7.5. In that experiment, entangled photon pairs was employed to erase the which-path information. Now we ask, what would happen if we replace the entangled photons with randomly created and randomly paired photons, or wavepackets, in thermal state? Can a randomly paired photons in thermal state erase the which path information? The answer is positive. A random delayed choice quantum eraser with thermal light has been demonstrated by Peng *et al.* recently.

The schematic of the experiment of Peng *et al.* is illustrated in Fig. 8.7.1. The experimental setup is almost the asme as that of the experiment of Kim *et al.* of 2000, except (1) the entangled signal-idler photon pair is replaced by two randomly created and randomly paired photons in a thermal state; (2) the joint photodetection measures the photon number fluctuation correlation, or the intensity fluctuation correlation, resulting from two-photon interferences.

The experimental setup in Fig. 8.7.1 can be divided into four parts: a thermal light source, a Young's double-slit interferometer, a Mach–Zehnder-like interferometer, and a photon number fluctuation correlation (PNFC) measurement circuit. (1) The light source is a standard pseudo-thermal source which consists of a He–Ne laser beam ($\sim$2 mm diameter) and a rotating ground glass (GG). Within the $\sim$2 mm diameter spot, the ground glass contains millions of tiny diffusers. A large number of randomly distributed sub-fields, or wavepacket, are scattered from millions of randomly distributed tiny diffusers with random phases. The pseudo-thermal field then passes a double-slit which is about 25cm away from the GG. (2) The double-slit has a slit-width 150 μm, and a slit-separation $d = 0.7$ mm (distance between the center of two slits). The spatial coherence length of the pseudo-thermal field on the double slit plane is $l_c = \lambda/\Delta\theta \sim 160\,\mu \ll d$, which guarantees the two fields E_A and E_B are spatially incoherent. Under this experimental condition, (1) no first-order classic Young's interference is observable; (2) the photon number fluctuates correlatively only within slit-A or slit-B. Therefore, the states of the photons associated with slit-A and slit-B are distinguishable and carries "which-slit" information. We may learn the which-slit information from either a photon number measurement or from a photon number-fluctuation correlation measurement. A lens, f, is placed following the double-slit. On the focal-plane of the lens a scannable point-like photodetector D_0 is used to learn the which-slit information or to observe the Young's double-slit interference pattern. (3) The Mach–Zehnder-like interferometer and the photodetectors D_1, D_2, are used to "erase" the which-slit information. Simultaneously, the joint-detection between D_0 and D_3 or D_4 are used to "read" the which-slit information. All five photodetectors are photon-counting detectors working

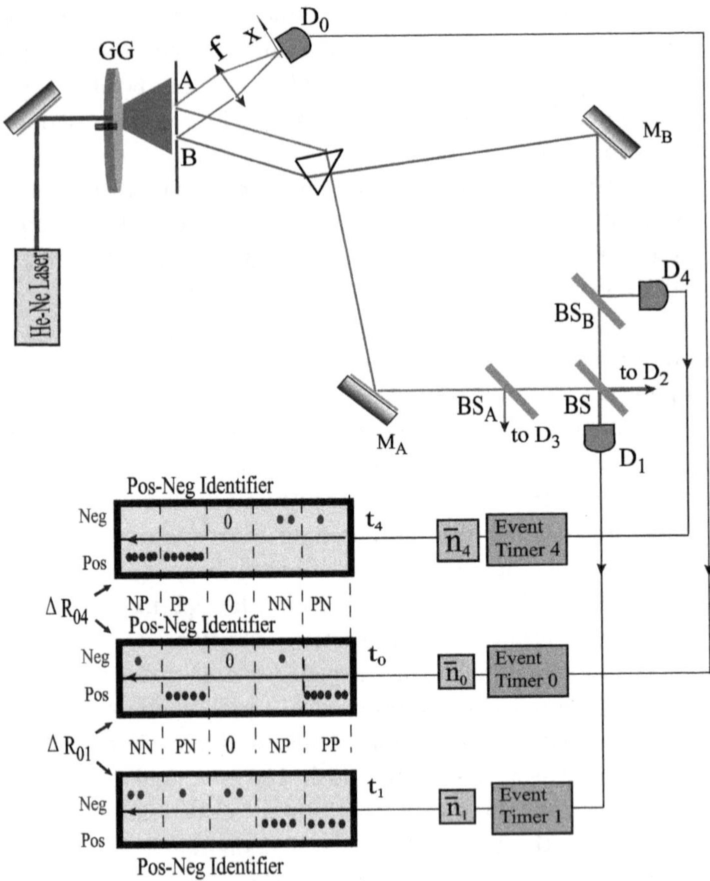

Fig. 8.7.1 Schematic of a random delayed choice quantum eraser. The He–Ne laser beam spot on the rotating ground glass has a diameter of ~ 2 mm. A double-slit, with slit-width 150 μm and slit-separation $d = 0.7$ mm, is placed ~ 25 cm away from the GG. The spatial coherence length of the pseudo-thermal field on the double-slit plane is $l_c = \lambda/\Delta\theta \sim 160 \ \mu \ll d$, which guarantees the two fields E_A and E_B are spatially incoherent. The states of the photons associated with slit-A and slit-B are distinguishable and thus carries "which-slit" information. All beamsplitters are 50/50 nonpolarizing beamsplitters. The two fields from the two slits may propagate to detector D_0 which is transversely scanned on the focal plan of lens f for observing the interference patten of the double-slit interferometer; and may also pass along a Mach–Zehnder-like interferometer and finally reach at D_1 or D_4 (D_2 or D_3). A photon number fluctuation correlation (PNFC) measurement circuit is followed to evaluate the photon number fluctuation correlations measured by D_0-D_1 and D_0-D_4 (or D_0-D_2 and D_0-D_3).

at single-photon level. The Mach–Zehnder-like interferometer has three beamsplitters, BS, BS_A and BS_B, all of them are 50/50 nonpolarizing beamsplitters. Moreover, the detectors are fast avalanche photodiodes with rise time less than 1 ns, and the path delay between BS_A or BS_B, and D_0 is ≈ 1.5 m which ensure that, at each joint-detection measurement, when a photon chooses to be reflected (read which-way) or transmitted (erase which-way) at BS_A or BS_B, it is already 5 ns later than the annihilation of its partner at D_0. Comparing the 1 ns rise time, we are sure this is a "delayed choice" made by that photon. (4) The PNFC circuit consists of five synchronized "event-timers" which record the registration times of D_0, D_1, D_2, D_3 and D_4. A positive–negative fluctuation identifier follows each event-timer to distinguish "positive-fluctuation" Δn^+, from "negative-fluctuation" Δn^-, for each photodetector within each coincidence time window. The photon number fluctuation correlations of D_0-D_1: $\Delta R_{01} = \langle \Delta n_0 \Delta n_1 \rangle$ and D_0-D_4: $\Delta R_{04} = \langle \Delta n_0 \Delta n_4 \rangle$ are calculated, accordingly and respectively, based on their measured positive–negative fluctuations. The detailed description of the PNFC circuit can be found in the references.

The experimental observation of ΔR_{04} is reported in Fig. 8.7.2. The data excludes any possible existing interferences. This measurement means the coincidences that contributed to ΔR_{04} must have passed through slit-B. Figure 8.7.3 reports a typical experimental result of ΔR_{01}: a typical

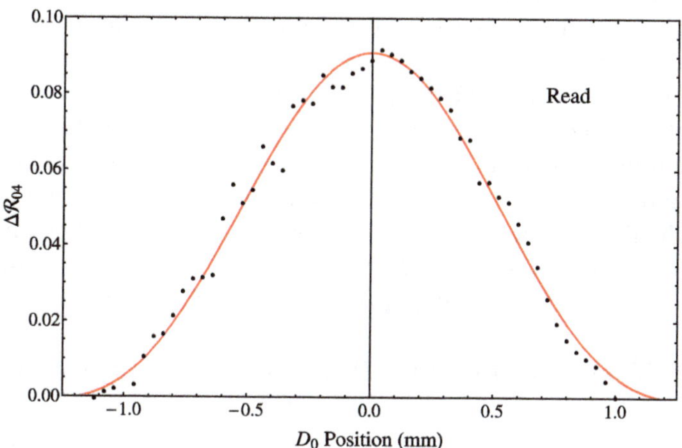

Fig. 8.7.2   The measured ΔR_{04} by scanning D_0 on the observation plane of the Young's double-slit interferometer. The black dots are experimental data, the red line is the theoretical fitting with Eq. (8.7.8).

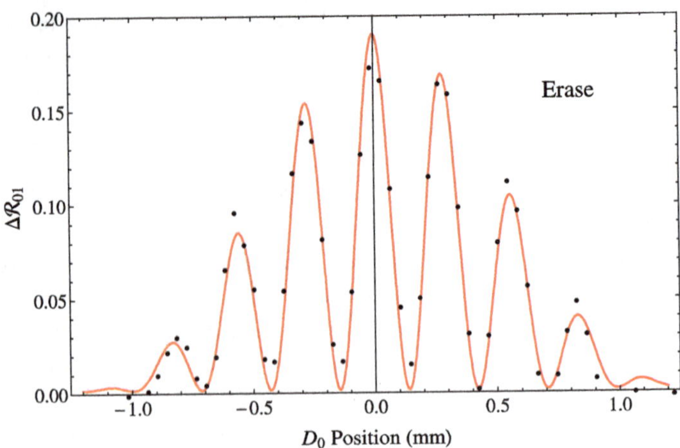

Fig. 8.7.3   The measured ΔR_{01} as a function of the transverse coordinate of D_0. The black dots are experimental data, the red line is the theoretical fitting with Eq. (8.7.9).

double-slit interference–diffraction pattern. The 100% visibility of the sinusoidal modulation indicates complete erasure of the which-slit information.

Assuming a random pair of sub-fields at single-photon level, such as the mth and nth wavepackets, is scattered from the mth and the nth sub-sources located at transverse coordinates $\vec{\rho}_{0m}$ and $\vec{\rho}_{0n}$ of the ground glass and fall into the coincidence time windows of D_0–D_1 and D_0–D_4, the mth wavepacket may propagate to the double-slit interferometer and the nth wavepacket may pass through the Mach–Zehnder, or vice versa. Under the experimental condition of spatial incoherence between E_A and E_B, the which-slit information is learned from the photon number fluctuation correlation measurements $\Delta R_{04} = \langle \Delta n_0 \Delta n_4 \rangle = \langle \Delta n_{B0} \Delta n_{B4} \rangle$ of D_0–D_4, and no interference is observable by scanning D_0. It is interesting that the which-slit information are erasable in the photon number fluctuation correlation measurements of $\Delta R_{01} = \langle \Delta n_0 \Delta n_1 \rangle$ of $D_0 - D_1$, resulting in a reappeared interference pattern as a function of the scanning coordinate of D_0.

The field operator at detector D_0 can be written in the following form in terms of the subfields:

$$\hat{E}^{(+)}(\mathbf{r}_0, t_0) = \hat{E}_A^{(+)}(\mathbf{r}_0, t_0) + \hat{E}_B^{(+)}(\mathbf{r}_0, t_0)$$

$$= \sum_m \left[\hat{E}_{mA}^{(+)}(\mathbf{r}_0, t_0) + \hat{E}_{mB}^{(+)}(\mathbf{r}_0, t_0) \right]$$

$$= \sum_m \int d\mathbf{k} \, \hat{a}_m(\mathbf{k}) \big[g_m(\mathbf{k}; \mathbf{r}_A, t_A) g_A(\mathbf{k}; \mathbf{r}_0, t_0)$$

$$+ g_m(\mathbf{k}; \mathbf{r}_B, t_B) g_B(\mathbf{k}; \mathbf{r}_0, t_0) \big], \tag{8.7.1}$$

where $g_m(\mathbf{k}; \mathbf{r}_s, t_s)$ is a Green's function which propagates the mth subfield from the mth sub-source to the sth slit ($s = A, B$). $g_s(\mathbf{k}; \mathbf{r}_0, t_0)$ is another Green's function that propagates the field from the sth slit to detector D_0. It is easy to notice that, although there are two ways a photon can be detected at D_0, due to the first order incoherence of E_A and E_B, there should be no interference at the detection plane.

D_4 (D_3) in the experiment can only receive photons from slit-B (slit-A), so the field operator is then:

$$\hat{E}^{(+)}(\mathbf{r}_4, t_4) = \sum_m \hat{E}_{mB}^{(+)}(\mathbf{r}_4, t_4)$$

$$= \sum_m \int d\mathbf{k} \, \hat{a}_m(\mathbf{k}) g_m(\mathbf{k}; \mathbf{r}_B, t_B) g_B(\mathbf{k}; \mathbf{r}_4, t_4). \tag{8.7.2}$$

The detector D_1 (D_3), however, can receive photons from both slit-A and slit-B through the Mach–Zehnder-like interferometer, so the field operator has two terms:

$$\hat{E}^{(+)}(\mathbf{r}_1, t_1) = \sum_m \big[\hat{E}_{mA}^{(+)}(\mathbf{r}_1, t_1) + \hat{E}_{mB}^{(+)}(\mathbf{r}_1, t_1) \big]$$

$$= \sum_m \int d\mathbf{k} \, \hat{a}_m(\mathbf{k}) \big[g_m(\mathbf{k}; \mathbf{r}_A, t_A) g_A(\mathbf{k}; \mathbf{r}_1, t_1)$$

$$+ g_m(\mathbf{k}; \mathbf{r}_B, t_B) g_B(\mathbf{k}; \mathbf{r}_1, t_1) \big]. \tag{8.7.3}$$

Based on the state, either in the single-photon state representation or in the coherent state representation, and the field operators of Eqs. (8.7.1)– (8.7.3), we apply the Glauber–Scully theory to calculate the photon number fluctuation correlation or the second-order coherence function $G^{(2)}(\mathbf{r}_0, t_0; \mathbf{r}_\alpha, t_\alpha)$ from the coincidence measurement of D_0 and D_α,($\alpha = 1, 2, 3, 4$):

$$G^{(2)}(\mathbf{r}_0, t_0; \mathbf{r}_\alpha, t_\alpha)$$

$$= \Big\langle \langle \Psi | E^{(-)}(\mathbf{r}_0, t_0) E^{(-)}(\mathbf{r}_\alpha, t_\alpha) E^{(+)}(\mathbf{r}_\alpha, t_\alpha) E^{(+)}(\mathbf{r}_0, t_0) | \Psi \rangle \Big\rangle_{\mathrm{Es}}$$

$$= \left\langle \langle \Psi | \sum_m E_m^{(-)}(\mathbf{r}_0, t_0) \sum_n E_n^{(-)}(\mathbf{r}_\alpha, t_\alpha) \right.$$

$$\left. \times \sum_q E_q^{(+)}(\mathbf{r}_\alpha, t_\alpha) \sum_p E_p^{(+)}(\mathbf{r}_0, t_0) | \Psi \rangle \right\rangle_{\mathrm{Es}}$$

$$= \sum_m \psi_m^*(\mathbf{r}_0, t_0)\psi_m(\mathbf{r}_0, t_0) \sum_n \psi_n^*(\mathbf{r}_\alpha, t_\alpha)\psi_n(\mathbf{r}_\alpha, t_\alpha)$$

$$+ \sum_{m,n} \psi_m^*(\mathbf{r}_0, t_0)\psi_n(\mathbf{r}_0, t_0)\psi_n^*(\mathbf{r}_\alpha, t_\alpha)\psi_m(\mathbf{r}_\alpha, t_\alpha)$$

$$= \langle n_0 \rangle \langle n_\alpha \rangle + \langle \Delta n_0 \Delta n_\alpha \rangle. \tag{8.7.4}$$

Here, $\psi_m(\mathbf{r}_\alpha, t_\alpha)$ is the effective wavefunction of the mth subfield at $(\mathbf{r}_\alpha, t_\alpha)$. In the case of $\alpha = 1, 2$

$$\psi_m(\mathbf{r}_\alpha, t_\alpha) = \psi_{mA\alpha} + \psi_{mB\alpha}$$

$$= \int d\mathbf{k}\alpha_m(\mathbf{k}) \big[g_m(\mathbf{k}; \mathbf{r}_A, t_A) g_A(\mathbf{k}; \mathbf{r}_\alpha, t_\alpha)$$

$$+ g_m(\mathbf{k}; \mathbf{r}_B, t_B) g_B(\mathbf{k}; \mathbf{r}_\alpha, t_\alpha) \big]. \tag{8.7.5}$$

This shows that the measured effective wavefunction $\psi_m(\mathbf{r}_\alpha, t_\alpha)$ is the result of a superposition between two alternative amplitudes in terms of path-A and path-B, $\psi_{m\alpha} = \psi_{mA\alpha} + \psi_{mB\alpha}$. When $\alpha = 4$ (or $\alpha = 3$), the effective wavefunction has only one amplitude

$$\psi_m(\mathbf{r}_4, t_4) = \psi_{mB4} = \int d\mathbf{k}\alpha_m(\mathbf{k}) g_m(\mathbf{k}; \mathbf{r}_B, t_B) g_B(\mathbf{k}; \mathbf{r}_4, t_4). \tag{8.7.6}$$

From Eq. (8.7.4) and the measurement circuit in Fig. 8.7.1, it is easy to find that what we measure in this experiment is the photon number fluctuation correlation:

$$\langle \Delta n_0 \Delta n_\alpha \rangle = \sum_{m,n} \psi_m^*(\mathbf{r}_0, t_0)\psi_n(\mathbf{r}_0, t_0)\psi_n^*(\mathbf{r}_\alpha, t_\alpha)\psi_m(\mathbf{r}_\alpha, t_\alpha). \tag{8.7.7}$$

We thus obtain

$$\Delta R_{04} \propto \langle \Delta n_0 \Delta n_4 \rangle = \sum_{n \neq m} \psi_{mB0}^* \psi_{nB0} \psi_{nB4}^* \psi_{mB4} \propto \mathrm{sinc}^2(x\pi a/\lambda f),$$

$$\tag{8.7.8}$$

indicating a diffraction pattern which agrees with the experimental observation of Fig. 8.7.2.

In the case of $\alpha = 1, 2$, we obtain

$$\Delta R_{01} \propto \langle \Delta n_0 \Delta n_1 \rangle$$

$$\propto \sum_{n \neq m} \left[\psi^*_{mA0} \psi_{nA0} \psi^*_{nA1} \psi_{mA1} + \psi^*_{mB0} \psi_{nB0} \psi^*_{nB1} \psi_{mB1} \right.$$

$$\left. + \psi^*_{mA0} \psi_{nB0} \psi^*_{nB1} \psi_{mA1} + \psi^*_{mB0} \psi_{nA0} \psi^*_{nA1} \psi_{mB1} \right]$$

$$\propto \mathrm{sinc}^2(x\pi a/\lambda f) \cos^2(x\pi d/\lambda f), \tag{8.7.9}$$

which agrees with the experimental observation in Fig. 8.7.3.

8.8 Ghost Frequency Comb — Randomly Paired Indistinguishable Photon Groups Interferes with the Pair Itself

Since the development of frequency combs led by John Hall and Theodor Hänsch, researchers have developed a wide array of practical applications for these light sources. The precise structure of a frequency-time comb enables precision spectroscopy that makes it possible to measure time and frequency of light more accurately than ever before. A different type of frequency-time comb, which we label as a Ghost Frequency Comb (GFC), was demonstrated recently by Joshi et al. Their experiment employed a CW fiber laser with half a million longitudinal cavity-modes. The laser is in continuous wave operation and there is no pulse structure in the laser beam. Surprisingly, a comb-like function was observed from the nonlocal intensity correlation measurement between two point-like, independent photodetectors.

Figure 8.8.1 is a schematic setup of the measurement of Joshi at al. The laser used in the experiment is a homemade CW fiber ring laser with a cavity length of 11.7 kilometer, which produces half a million evenly-spaced longitudinal cavity-modes at central wavelength 1550 nm. Without mode-locking, the relative phases between cavity-modes are all random. Unlike a traditional frequency comb, the laser beam does not consist of any pulse train but instead it is a continuous wave. The laser beam is divided into two by a fiber beamsplitter, then passed through kilometers long fiber delays, respectively, and fed into two point-like photodetectors, D_1 and D_2 for joint nonlocal correlation measurement. As expected, D_1 and D_2, individually and respectively, measures a constant signal. The joint nonlocal correlation measurement between D_1 and D_2, however, observes a set of well-defined, comb-like, ultra-narrow peaks as a function of the temporal

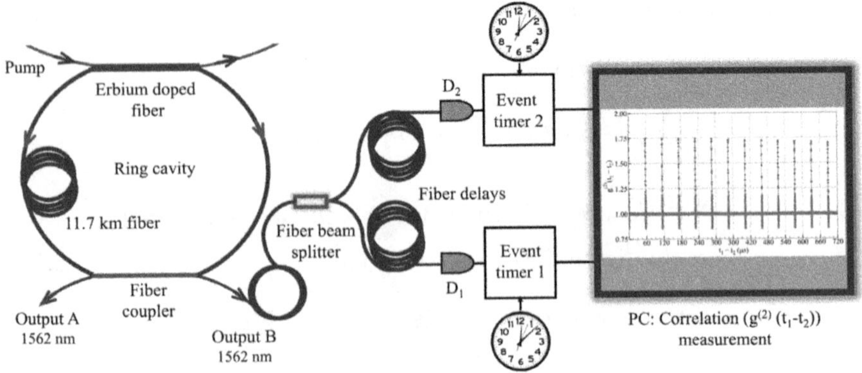

Fig. 8.8.1   Schematic setup for experimental observation of ghost frequency comb. Light from a fiber ring laser is directed into a fiber beam splitter. The outputs of the beam splitter pass through fiber optic delays and are fed into two photodetectors D_1 and D_2. The event timer for each detector, timed by two independent clocks, produces a time series of the photodetection events of that detector, which can be further analyzed by a computer (PC). In this experiment, D_1 and D_2 are both analog detectors, the analog signals are digitized by two 50GHz A-to-D convertors.

delay between the two photodetections. This comb-like correlation has been labeled as Ghost Frequency Comb (GFC). The experimentally measured GFC is illustrated in Fig. 8.8.2.

The second-order temporal coherence function of multi-cavity-mode laser beam has been calculated in Chapter 5. Here, we give a brief review and simply treat the second-order coherence function as nonlocal intensity correlation measured at space–time coordinates (r_1, t_1) and (r_2, t_2)

$$\langle I(\tau_1)I(\tau_2)\rangle = \langle I(\tau_1)\rangle\langle I(\tau_2)\rangle + \langle \Delta I(\tau_1)\Delta I(\tau_2)\rangle, \qquad (8.8.1)$$

where, again, $\tau_j = t_j - r_j/c$, $j = 1, 2$.

First, we show that the measured intensity by the jth detector is a constant. Consider the measured electromagnetic field at the jth detector is a superposition of a large number of longitudinal cavity-modes:

$$E(\tau_j) = \sum_m^N E_m(\tau_j) = \sum_m^N E_0(\omega_m)e^{-i\omega_m \tau_j}e^{i\varphi_m}, \qquad (8.8.2)$$

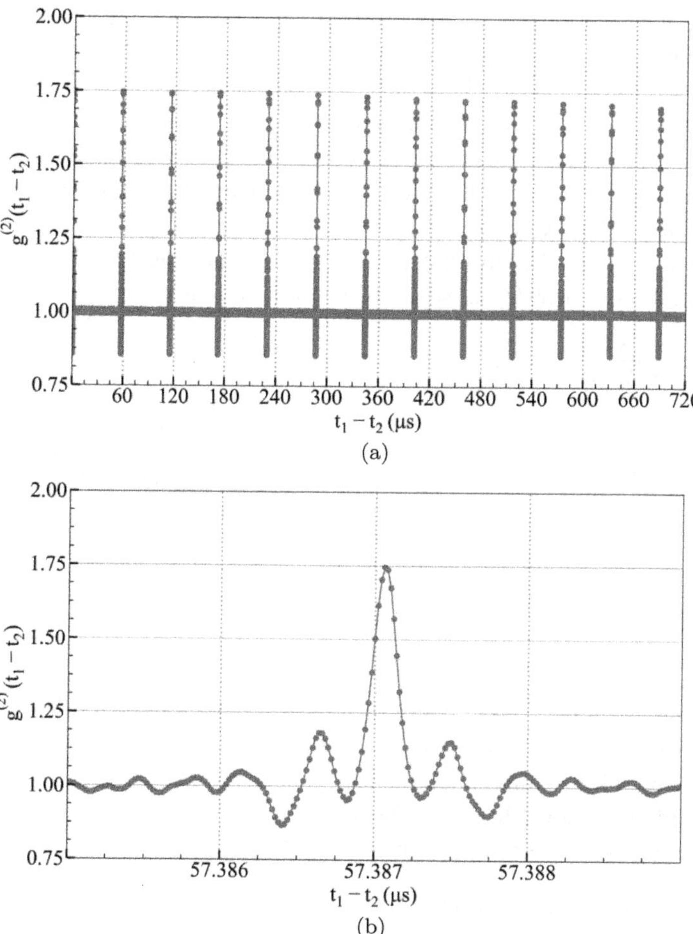

Fig. 8.8.2 (a) Typical ghost frequency comb observed from the nonlocal normalized temporal correlation measurements for the setup in Fig. 8.8.1. In this measurement, the two fiber delay lines have equal length. The zero-order ($n = 0$) GFC peak was observed at $\tau_1 - \tau_2 = 0$, or $t_1 - t_2 \simeq 0$ when $r_1 \simeq r_2$. A least squares fitting concluded the GFC has a period of $57,386.064 \pm 0.001$ ns, equivalent to a beat frequency of $17,425.8336 \pm 0.0003$ Hz. This value is close to the theoretical estimate of $17,464.26 \pm 62.86$ Hz. In both plots, the standard deviation at each value of the correlation is too small to be displayed as an error bar. (b) A closer look at the second GFC peak. Due to the relatively slow response time of the photodetector and electronics, the measured correlation temporal width is wider than expected.

where m denotes individual longitudinal cavity-modes, φ_m represents the random phase of the mth mode. The case of random phases of the cavity modes is distinctly different from that of mode-locked lasers. Mode-locked lasers produce traditional frequency combs due to the mode-locking mechanism $\varphi_m = \varphi_0$. The coherent superposition of the longitudinal cavity-modes results in a clear pulse train in the laser beam. When the phases φ_m are all independent and vary randomly, as is the case for the CW laser used in this experiment, the superposition is incoherent. With this, the intensity measured at detector D_j is expected to be a constant:

$$
\begin{aligned}
I(\tau_j) &= \sum_{m,n}^{N} E_m^*(\tau_j) E_n(\tau_j) \\
&= \left\langle \left[\sum_{m=0}^{N-1} E_0^*(\omega_m) e^{im\omega_b \tau_j} e^{-i\varphi_m} \right] \left[\sum_{n=0}^{N-1} E_0(\omega_n) e^{-in\omega_b \tau_j} e^{i\varphi_n} \right] \right\rangle \\
&= \sum_{m=n}^{N-1} |E_0(\omega_m)|^2 \\
&= I_0.
\end{aligned}
\tag{8.8.3}
$$

Due to the random relative phases between the cavity-modes, only the terms with $m = n$ survive from the ensemble average result in a constant expectation value of intensity.

Second, we show the intensity correlation measured by D_1 and D_2, jointly, is a comb-like function of $\tau_1 - \tau_2$. We start from

$$
\begin{aligned}
\langle I(\tau_1) I(\tau_2) \rangle &= \langle E^*(\tau_1) E(\tau_1) E^*(\tau_2) E(\tau_2) \rangle \\
&= \left\langle \sum_m^N E_m^*(\tau_1) \sum_n^N E_n(\tau_1) \sum_p^N E_q^*(\tau_2) \sum_q^N E_p(\tau_2) \right\rangle \\
&= \sum_m^N E_m^*(\tau_1) E_m(\tau_1) \sum_n^N E_n^*(\tau_2) E_n(\tau_2) \\
&\quad + \sum_{m \neq n}^N E_m^*(\tau_1) E_n(\tau_1) E_n^*(\tau_2) E_m(\tau_2) \\
&= \langle I(\tau_1) \rangle \langle I(\tau_2) \rangle + \langle \Delta I(\tau_1) \Delta I(\tau_2) \rangle.
\end{aligned}
\tag{8.8.4}
$$

Due to the random relative phases between the cavity-modes, the only surviving terms are the terms where $m = n$ and $p = q$, as well as the terms where $m = q$ and $n = p$. We recognized immediately that the leading term in Eq. (8.8.4) is simply the product of the mean intensities measured by D_1 and D_2, respectively; and the end term in Eq. (8.8.4) are what are known as the intensity fluctuation correlation:

$$\langle \Delta I(\tau_1) \Delta I(\tau_2) \rangle = I_0^2 \sum_{m \neq n}^{N} e^{i\omega_m \tau_1} e^{-i\omega_n \tau_1} e^{i\omega_n \tau_2} e^{-i\omega_m \tau_2} = I_0^2 \left| \sum_{m}^{N} e^{i\omega_m \tau} \right|^2 ,$$

$$(8.8.5)$$

where $\tau \equiv \tau_1 - \tau_2 = (t_1 - t_2) - (r_1 - r_2)/c$, and all constant quantities in the expression have been absorbed into I_0^2. Using a known exponential sum formula, the result can be calculated as

$$\langle \Delta I(\tau_1) \Delta I(\tau_2) \rangle = I_0^2 \left| \sum_{m=0}^{N-1} e^{im\omega_b \tau} \right|^2 = I_0^2 \frac{\sin^2(N\omega_b \tau/2)}{\sin^2(\omega_b \tau/2)}. \qquad (8.8.6)$$

The normalized second-order coherence function is therefore

$$g^{(2)}(\tau) = 1 + \frac{\sin^2(N\omega_b \tau/2)}{N^2 \sin^2(\omega_b \tau/2)}. \qquad (8.8.7)$$

Examining $g^{(2)}(\tau)$, it appears that the intensity fluctuations of the laser beam are correlated with $g^{(2)} > 1$ only within these periodic, precise, and narrow time windows. The intensity fluctuations of the laser beam become uncorrelated with $g^{(2)} = 1$, as long as the relative delay of the two photodetections, $\tau = \tau_1 - \tau_2$, falls into the region between the periodic sharp correlation peaks. Does any CW laser beam have such peculiar statistical behavior with respect to intensity fluctuations? Of course not. In fact, in a CW laser beam, the intensity fluctuations are much smaller than the intensity, $|\Delta n| \ll \bar{n}$, and can be ignored. A question naturally arises: Can the observed GFC be considered as correlation of intensity fluctuations of the CW laser beam?

If the GFC is not caused by the statistical intensity fluctuations of the laser beam, then what is the cause of the observed GFC? From the perspective of quantum coherence, the observed GFC is the result of nonlocal interference. To illustrate this, we rewrite Eq. (8.8.4) as follows:

$$\langle I(\tau_1)I(\tau_2)\rangle = \langle E^*(\tau_1)E(\tau_1)E^*(\tau_2)E(\tau_2)\rangle$$

$$= \left\langle \sum_m^N E_m^*(\tau_1) \sum_n^N E_n(\tau_1) \sum_p^N E_q^*(\tau_2) \sum_q^N E_p(\tau_2) \right\rangle$$

$$= \sum_m^N E_m^*(\tau_1)E_m(\tau_1) \sum_n^N E_n^*(\tau_2)E_n(\tau_2)$$

$$+ \sum_{m\neq n}^N E_m^*(\tau_1)E_n(\tau_1)E_n^*(\tau_2)E_m(\tau_2)$$

$$= \sum_{m,n} \left| \frac{1}{\sqrt{2}}\left[E_m(\tau_1)E_n(\tau_2) + E_n(\tau_1)E_m(\tau_2)\right] \right|^2. \tag{8.8.8}$$

The observed GFC does correspond to the cross-interference term of Eq. (8.8.8). Equation (8.8.8) specifies two different yet indistinguishable alternatives for a randomly paired cavity modes to produce a joint-detection of D_1 and D_2: (1) one or more identical photons from the mth cavity mode triggers D_1 at (r_1, t_1) while one or more identical photons from the nth cavity mode triggers D_2 at (r_2, t_2); and (2) one or more identical photons from the mth cavity mode triggers D_2 at (r_2, t_2) while one or more identical photons from the nth cavity mode triggers D_1 at (r_1, t_1). The second-order coherence function, or correlation function, of a multi-mode CW laser beam is indeed the result of a nonlocal interference: a randomly created and randomly paired identical photon groups interferes with the pair itself. Apparently, the observation of GFC is strong evidence of nonlocal two-photon interference.

Besides its fundamental interests, the bright GFC has also been recognized as an important contribution to the fields of precision spectroscopy and applications based on nonlocal quantum correlations. Superior to entangled photon pairs, measurements of entangled laser beams do not rely on photon counting and can be performed over greater distance in shorter time with higher resolution and accuracy. In this section, we discuss an application of non-local time transfer that has been experimentally demonstrated recently.

Assume two clocks are carried by Space Station 1 and Space Station 2. A CW laser beam in the ground laboratory is divided into two paths and directed to two photodetectors, D_1 in Space Station 1 and D_2 in Space Station 2, through two individual telescopes. The individual time history records can be brought together to the ground laboratory through a classical communication channel for comparison and subsequent correlation calculations. When the two clocks are synchronized, the joint photodetection between D_1 and D_2 will show an expected set of GFC peaks. If the clocks lose their synchronization, one can rematch the records by adjusting one of the clocks until the expected set of GFC peaks is achieved. The clocks can be adjusted and kept synchronized accordingly. A linear least squares fitting of the comb-like sharp correlation peaks is able to help to achieve higher resolution and higher accuracy in the nonlocal clock synchronization.

A proof-of-concept experimental demonstration of nonlocal clock synchronization was recently reported by Joshi *et al.* Their experimental setup is the same as in Fig. 8.8.1, except that different optical delay lines are used. The experimental result is reported in Fig. 8.8.3. In this measurement, 1 km and 5 km fiber delay lines are used for D_1 and D_2, respectively, to simulate the nonlocal condition. The optical distance between D_1 and D_2 is therefore 4 km (with some degree of uncertainty in the lengths of the delay fibers). The sub-figure (b) is a least-squared fitting of the experimental data. Notice that the comb-function has multipole periodic correlation peaks at $\omega_b \tau/2 = n\pi$, for $n = 0, \pm 1, \pm 2, \ldots, \pm(N-1)$. This can be written as

$$(t_1 - t_2)_n = \frac{1}{\nu_b} n + (r_1 - r_2)/c, \qquad (8.8.9)$$

where we have used $\omega_b = 2\pi\nu_b$ and defined $(t_1 - t_2)_n$ as the measured value of $t_1 - t_2$ at the nth comb peak. The linear fitting matches Eq. (8.8.9) with high accuracy. The GFC has a period of $57,386.066 \pm 0.001$ ns, equivalent to a beat frequency of $17,425.8392 \pm 0.0003$ Hz. It is also noticed that the intersection of the fitting line on the $t_1 - t_2$ axis is $19,569.025 \pm 0.004$ ns, corresponding to $3,999.1195 \pm 0.0008$ m optical distance with the given index of refraction. The accuracy of the measurements, characterized by the error in the least squares fitting, is in the order of picoseconds to sub-picoseconds with a data acquisition time of milliseconds.

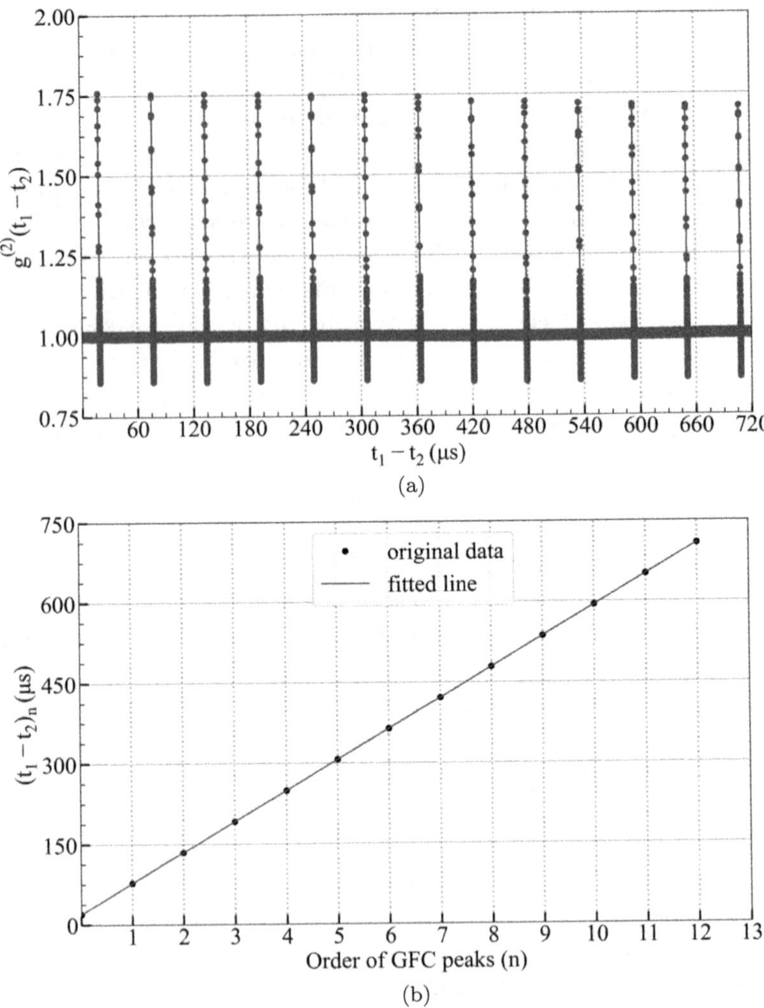

Fig. 8.8.3 (a) Measured GFC for principle demonstration of nonlocal clock synchronization. In this measurement, 1 km and 5 km fiber delay lines are used for D_1 and D_2 to simulate the nonlocal condition. The optical distance between D_1 and D_2 is 4 km. The standard deviation of the measurement at each value of the correlation is too small to be displayed as an error bar. (b) A least squares fitting of the experimental data (an accuracy fitting of Eq. (8.8.9)). The least squares fitting concluded that the GFC has a period of $57,386.066 \pm 0.001$ ns, equivalent to a beat frequency of $17,425.8392 \pm 0.0003$ Hz. Note that the intersection of the fitting line on the $t_1 - t_2$ axis is $19,569.025 \pm 0.004$ ns, corresponding to $3,999.1195 \pm 0.0008$ m optical distance between D_1 and D_2. The accuracy of the measurements, characterized by the error in the least squares fitting, is in the order of picoseconds to sub-picoseconds with a data acquisition time of milliseconds.

8.9   A New Type of Quantum Eraser

The observation of GFC can be considered the experimental realization of a new type of quantum eraser. Examining Fig. 8.9.1, we ask two simple questions: (1) Do we expect to observe optical beats from the individual measurement of D_1 and D_2, respectively? (2) Do we expect to observed optical beats from the correlation measurement of D_1 and D_2, jointly? The first question is easy to answer: although the CW laser beam contains a large number of periodic cavity modes, their random phases prevent the observation of the optical beats between these cavity modes. The random phases of the cavity modes, each represents a group of indistinguishable photons, is marked in the phase space as the "which-path" information. Therefore, no optical beats are observable from the CW laser beam. The second question may not be easy to answer. However, the observation of GFC has given us a solid answer: the optical beats between cavity modes, or between groups of indistinguishable photons, is observable from the correlation measurement of the CW laser beam as a Ghost Frequency Comb. The "two-photon" interference, i.e., a pair of indistinguishable photon groups interfering with the pair itself, is able to erase the which-path information in the phase space.

In fact, Fig. 8.9.1 is the schematic setup of the new type of quantum eraser. The light source used in this experiment is a laboratory-assembled multi-cavity-mode CW fiber laser. The laser assembly consists of, among other things, an erbium doped fiber amplifier and a single-mode fiber ring cavity with adjustable length. The spectrum bandwidth of the TEM_{00} laser beam is measured to be $\Delta f \simeq 8.6$ GHz, with the spectrum peak being in the telecom C band (in the range of $1562 - 1563$ nm). In this experiment, the ring cavity length is chosen to be 11.7 km. The expected value of longitudinal-mode separation for this cavity is $\omega_b/2\pi = 17464.26$ Hz, with a small margin of error resulting from the uncertainties associated with the fiber used in the cavity. For 8.6 GHz spectrum bandwidth, this means $N = 492554$ cavity modes, or approximately $N \simeq 500000$ modes. The laser beam with these half-a-million cavity modes is passed through a fiber beamsplitter and fed into two identical point-like analog photodetectors D_1 and D_2, each of which has a response spectrum bandwidth of 5 GHz. A fiber launch cable of approximately 10 km length is used with D_2 as a delay line to ensure that the observation time of D_2 is delayed after the registration of the indistinguishable photon group at D_1. The output photocurrents of the photodiodes, $i_1(t_1) \propto I_1(t_1)$ and $i_2(t_2) \propto I_2(t_2)$, together with their

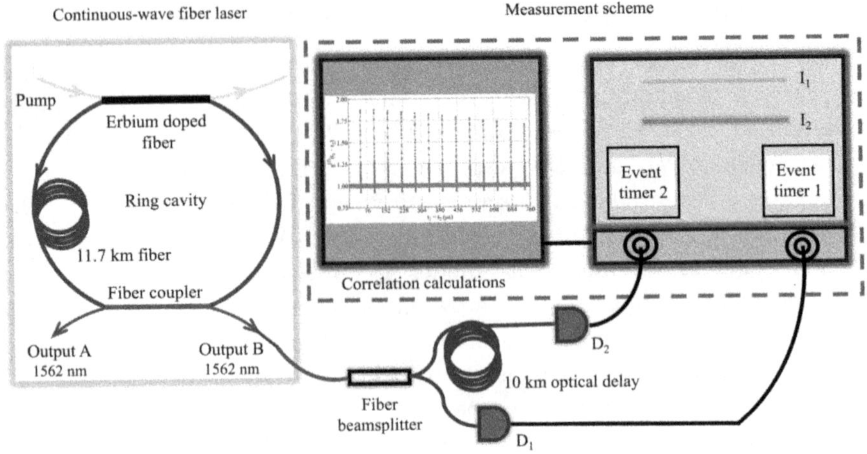

Fig. 8.9.1 Schematic setup of the new type of quantum eraser. No optical beats are observable from the multi-cavity-mode CW laser beam due to the path markers, in phase space, of each group of indistinguishable photons. However, the which-path markers can be erased in the intensity correlation measurement of the CW laser beam.

registration times t_1 and t_2, are processed by the measurement scheme which involves high speed event timers for registering the photodetection events; and a computer for the post-processing of the data to get the intensity correlation as a function of $t_1 - t_2$.

It is interesting to find from the measurements that, although both $I_1(t_1)$ and $I_2(t_2)$ are constants, a set of well-defined comb-like ultra-narrow peaks, which has been introduced as a *ghost frequency comb*, is obtained from the measured intensity correlation $\langle I_1(t_1)I_2(t_2)\rangle$, as shown in Fig. 8.9.2. The observed sharp peaks repeat at a frequency 17425.7673 ± 0.0001 Hz, which agrees with the estimated beat frequency between the cavity modes within the margin of error. Apparently, the which-path information of each group of identical photons is erased from the phase space by a correlation measurement.

In the following, we present a brief theory that first shows how the path marker of each group of indistinguishable photons prevents an observer from observing the optical beats. We then demonstrate how correlation measurements can erase the which-path information and recover the optical beats.

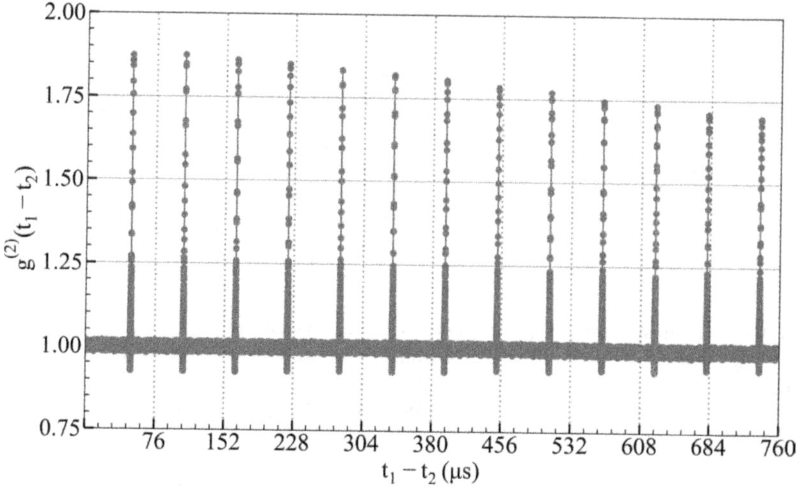

Fig. 8.9.2   Measurement of normalized intensity correlation, which represents the erased optical beats between cavity modes. In this measurement, a 10 km fiber delay line is used for D_2 to ensure that the observation time of D_2 is delayed after the registration of the indistinguishable photon group at D_1. The standard error of the measurement at each value of the correlation is significantly small.

Consider the state of a CW laser beam with multi-cavity-modes as a "pseudovector" $|\tilde{\Psi}\rangle$ in the coherent state space

$$|\tilde{\Psi}\rangle = \prod_m |\Psi_m\rangle = \prod_m |\alpha_m(\omega_m)\rangle, \qquad (8.9.1)$$

where m labels the mth cavity mode, or the mth group of identical photons with frequency ω, and $|\alpha_m(\omega)\rangle$ is an eigenstate of the annihilation operator. The eigenstate satisfies $\hat{a}_m(\omega)|\alpha_m(\omega)\rangle = \alpha_m(\omega)|\alpha_m(\omega)\rangle$ with a complex eigenvalue $\alpha_m(\omega) = a_m(\omega)e^{i\varphi_m}$, where $a_m(\omega)$ is the amplitude and φ_m is the random phase attributed to the mth mode. Fig 8.9.3 illustrates the marking of the paths of the mth and nth photon groups in phase space.

The field operators for the measurements involved in this experiment are

$$\hat{E}^{(+)}(r_j, t_j) = \sum_m \hat{a}_m(\omega)\, g_m(\omega; r_j, t_j)$$

$$\hat{E}^{(-)}(r_j, t_j) = \sum_m \hat{a}_m^\dagger(\omega)\, g_m^*(\omega; r_j, t_j), \qquad (8.9.2)$$

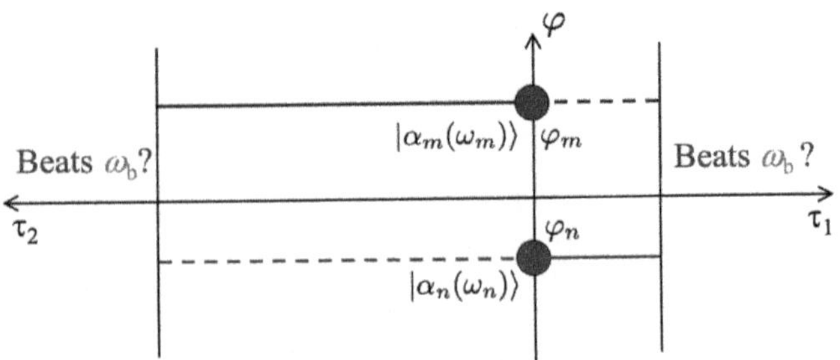

Fig. 8.9.3   In phase space, for the photo-detection events of D_1 and D_2, respectively, φ_m and φ_n mark the paths of the mth and nth photon groups. However, for the joint-detection event of D_1 and D_2, the two "two-photon" paths, indicated by the joint solid lines and the joint dashed lines, are indistinguishable because the pair come from points α_m and α_n jointly.

where $g_m(\omega; r_j, t_j)$ is the Green's function that propagates the mth mode from the source to space-time coordinate (r_j, t_j) of the jth detector D_j.

The first-order coherence function, which is proportional to the measured intensity at the jth detector, is calculated as follows

$$G^{(1)}(r_j, t_j; r_j, t_j) = \langle \hat{E}^{(-)}(r_j, t_j) \hat{E}^{(+)}(r_j, t_j) \rangle$$

$$= \left\langle \sum_m e^{-i\varphi_m} \psi_m^*(r_j, t_j) \sum_n e^{i\varphi_n} \psi_n(r_j, t_j) \right\rangle. \quad (8.9.3)$$

Here, we have deliberately indicated the random phase of the mth (nth) group indistinguishable photons by $e^{i\varphi_m}$ ($e^{i\varphi_n}$); and introduced the "effective wavefunction" of the mth group of indistinguishable photons that is measured by D_j as

$$\psi_m(r_j, t_j) = a_m(\omega)\, g_m(\omega; r_j, t_j). \quad (8.9.4)$$

The first-order coherence function of the laser beam is therefore

$$G^{(1)}(r_j, t_j; r_j, t_j)$$

$$= \left\langle \sum_m e^{-i\phi_m} \psi_m^*(r_j, t_j) \sum_n e^{i\phi_n} \psi_n(r_j, t_j) \right\rangle$$

$$= \left\langle \sum_{m=n} |\psi_m(r_j, t_j)|^2 \right\rangle + \left\langle \sum_{m \neq n} e^{i(\phi_n - \phi_m)} \psi_m^*(r_j, t_j) \psi_n(r_j, t_j) \right\rangle$$

$$= \sum_{m=n} |\psi_m(r_j, t_j)|^2 \quad (8.9.5)$$

which is a constant that sums all sub-intensities of the half-a-million modes. The second term in Eq. (8.9.5) vanishes when taking into account all possible random phases of the modes. The optical beats are unobservable due to the which-path markers of each group of indistinguishable photons in phase space.

Next, we calculate the second order coherence function, or correlation function,

$$
\begin{aligned}
G^{(2)}&(r_1, t_1; r_2, t_2) \\
&= \langle \hat{E}^{(-)}(r_1, t_1)\hat{E}^{(-)}(r_2, t_2)\hat{E}^{(+)}(r_2, t_2)\hat{E}^{(+)}(r_1, t_1) \rangle \\
&= \Big\langle \sum_m e^{-i\phi_m} \psi_m^*(r_1, t_1) \sum_n e^{-i\phi_n} \psi_n^*(r_2, t_2) \\
&\qquad \times \sum_q e^{i\phi_q} \psi_q(r_2, t_2) \sum_p e^{i\phi_p} \psi_p(r_1, t_1) \Big\rangle \\
&= \sum_m \psi_m^*(r_1, t_1)\psi_m(r_1, t_1) \sum_n \psi_n^*(r_2, t_2)\psi_n(r_2, t_2) \\
&\quad + \sum_{m \neq n} \psi_m^*(r_1, t_1)\psi_n(r_1, t_1)\psi_n^*(r_2, t_2)\psi_m(r_2, t_2) \\
&= \sum_{m,n} \left| \frac{1}{\sqrt{2}} [\psi_m(r_1, t_1)\psi_n(r_2, t_2) + \psi_n(r_1, t_1)\psi_m(r_2, t_2)] \right|^2 .
\end{aligned}
$$

(8.9.6)

It is interesting to find (1) which-path markers no longer contribute to the calculation; (2) $G^{(2)}(r_1, t_1; r_2, t_2)$ is the result of a superposition between two different yet indistinguishable quantum probability amplitudes: (1) one or more indistinguishable photons from the mth cavity mode triggers D_1 at (r_1, t_1) while one or more indistinguishable photons from the nth cavity mode triggers D_2 at (r_2, t_2); and (2) one or more indistinguishable photons from the mth cavity mode triggers D_2 at (r_2, t_2) while one or more identical photons from the nth cavity mode triggers D_1 at (r_1, t_1). The second-order coherence function, or correlation function, of a multi-mode CW laser beam is thus the result of a nonlocal interference: a randomly created and randomly paired groups of indistinguishable photons interferes with the pair itself.

The cross-interference term of Eq. (8.9.6) is the "nontrivial" contribution to the second-order coherence function $G^{(2)}(r_1, t_1; r_2, t_2)$, corresponding to the intensity fluctuation correlation of the CW laser beam measured jointly by D_1 and D_2,

$$\langle \Delta I(r_1,t_1)\Delta I(r_2,t_2)\rangle = \sum_{m\neq n} \psi_m^*(r_1,t_1)\,\psi_n(r_1,t_1)\,\psi_n^*(r_2,t_2)\,\psi_m(r_2,t_2)$$

$$\propto \frac{\sin^2(N\omega_b\tau/2)}{\sin^2(\omega_b\tau/2)}. \tag{8.9.7}$$

The detailed calculation has been given in early discussions. Likewise, the leading term of Eq. (8.9.6) simply represents the product of the mean intensity at each detector, which is a constant:

$$\langle I(r_1,t_1)\rangle\langle I(r_2,t_2)\rangle = G^{(1)}(r_1,t_1;r_1,t_1)G^{(1)}(r_2,t_2;r_2,t_2)$$

$$= \sum_m |\psi_m(r_1,t_1)|^2 \sum_n |\psi_n(r_2,t_2)|^2. \tag{8.9.8}$$

The *normalized* second-order coherence function is thus:

$$g^{(2)}(\tau) \propto 1 + \frac{\sin^2(N\omega_b\tau/2)}{N\,\sin^2(\omega_b\tau/2)}, \tag{8.9.9}$$

where ω_b is the *beat* frequency between neighboring cavity-modes, N is the total number of cavity-modes, and $\tau \equiv \tau_1-\tau_2 = (t_1-t_2)-(r_1-r_2)/c$, with c as the propagating speed of light in the medium. Eq. (8.9.9) agrees well with the experimentally measured $g^{(2)}(\tau)$ shown in Fig. 8.9.2. Apparently, "two-photon" interference concealed in the intensity correlation measurement is able to erase the which-path information, in phase space, of a group of indistinguishable photons even after the registration of the indistinguishable photons.

How does "two-photon" interference erase the which-path information of a group of indistinguishable photons? In addition to the above mathematical based analysis, Fig. 8.9.3 may provide an intuitive explanation. In phase space, for the photo-detection events of D_1 and D_2, respectively, φ_m and φ_n mark the paths of the mth and nth identical photon groups. However, for the joint-detection event of D_1 and D_2, the two "two-photon" paths, indicated by the joint solid lines and the joint dashed lines, are indistinguishable because the pair come from points α_m and α_n jointly.

Bibliography

Alley C.O. and Y.H. Shih, *Foundations of Quantum Mechanics in the Light of New Technology*, in M. Namiki (Ed.), Physical Society of Japan, Tokyo, 47 (1986); Y.H. Shih and C.O. Alley, *Phys. Rev. Lett.*, **61**, 2921, (1988).

Chen H., T. Peng, S. Karmakar, Z.D. Xie, and Y.H. Shih, "Observation of Anti-Correlation in Incoherent Thermal Light Fields", *Phys. Rev. A*, **84**, 033835 (2011).

Chen H., T. Peng, S. Karmaker, and Y.H. Shih, "Simulation of Bell States with Incoherent Thermal Light", *New J. Phys.*, **13**, 083018 (2011).

Ihn Y.S., Y. Kim, V. Tamma, and Y.H. Kim, "Second-Order Temporal Interference with Thermal Light: Interference Beyond the Coherence Time", *Phys. Rev. Lett.*, **119**, 263603 (2017).

Joshi B., T.A. Smith, and Y.H. Shih, *Phys. Rev. A* **110**, L031702 (2024).

Joshi B., T.A. Smith, and Y.H. Shih, *Appl. Phys. Lett.*, **125**, 241105 (2025).

Peng T., H. Chen, Y.H. Shih, and M.O. Scully, *Phys. Rev. Lett.*, **112**, 180401 (2014).

Peng T. and Y.H. Shih, "Bell Correlation of Thermal Fields in Photon-Number Fluctuations", *Europhys. Lett.*, **112**, 60006 (2015).

Scarcelli G., A. Valencia, and Y.H. Shih, "Two-photon Interference with Thermal Light", *Europhys. Lett.*, **68**, 618 (2004).

Smith T. and Y.H. Shih, "Turbulence-Free Double-slit Interferometer", *Phys. Rev. Lett.*, **120**, 063606 (2018).

Smith T. and Y.H. Shih, "Turbulence Induced Two-photon Interference", *APL Photonics*, **5**, 121302 (2020).

Index